C000183036

The Avro Vulcan

The Avro Vulcan

Tim McLelland

Crécy Publishing Ltd

Published by Crécy Publishing Ltd 2019

First published 2007

A CIP record for this book is available from the British Library

ISBN 978 191080 9273

Printed in Bulgaria by Multiprint

Crécy Publishing Limited

1a Ringway Trading Estate, Shadowmoss Road, Manchester M22 5LH

www.crecy.co.uk

Front cover:
B.1A XH497 No.9sqn *RAF via Terry Panopalis*

Rear cover from top:
A suitably posed RAF PR photograph showing a factory-fresh 83 Squadron Vulcan.

A post-Operation 'Corporate' publicity photograph showing a 'Black Buck' Vulcan at Waddington complete with a representative selection of weapons that the Vulcan could carry.

XL448 of 35 Sqn at Binbrook in July 1975. *Fred Martin*

Front flap top: A Waddington Wing 44 Sqn B.2, XM575, banks towards the camera on a misty June morning in 1981 to present a near-perfect plan view of its upper wing surfaces. *Fred Martin*

Front flap lower:
The co-pilot's position in the Vulcan B.2. This excellent ultra-wide-angle view gives a slightly false impression of spaciousness – the Vulcan's cockpit is, in fact, very cramped. *Paul Tomlin*

Back flap:
The distinctive smoky trail of a Vulcan made visual detection rather too easy, and became a familiar sight to air show audiences.

Frontispiece:
XH558 was the RAF's very last Vulcan and until 2015 was the sole remaining airworthy example. *Tim McLelland*

Table of Contents

Introduction

FEW military aircraft can be described as being charismatic. Warplanes are functional machines, and most are recognised purely for their technical merits rather than any aesthetic appeal. There are, of course, exceptions, such as the immortal Spitfire and perhaps even the legendary Lighting interceptor. But for generations of post-war aircraft enthusiasts and countless RAF air and ground crews, it is the Vulcan that will for ever be remembered as the one aircraft that was guaranteed to make an impression. A combination of elegant design and sheer brute power made the Vulcan a 'show-stopper' no matter where the aircraft appeared, be it at air shows across the United Kingdom, exercises and goodwill visits to the furthest corners of the globe, or ceremonial flypasts above Buckingham Palace. Undoubtedly, the Vulcan is a true icon, with a fan base that many aspiring pop stars can only envy. And yet the Vulcan is a killing machine, designed to guarantee the slaughter of millions of humans on an almost unimaginable scale. It is this dichotomy that makes the Vulcan a fascinating subject, and one that has inspired numerous books and countless magazine and newspaper articles throughout its sixty-year history.

This book is my third written account of the Vulcan's history. Naturally it is not possible to re-write historical facts, and having told the Vulcan's story before, I'm obliged to repeat much of my previous work in order to re-trace the development and service history of Avro's magnificent creation. Likewise, many other books have traced the same story, some quite successfully, while others have simply served to repeat factual inaccuracies or even create new ones. I have endeavoured to clarify as many details as space permits, but of course it would be impossible to dwell on every technical aspect of the Vulcan's activities or systems without the luxury of an infinite number of pages, and while the more intricate and esoteric details of the aircraft's career might well be fascinating to some readers, they would inevitably be rather less interesting to many others. However, I have placed the Vulcan's story in a proper context, explaining how and why the aircraft was created and the political and scientific background from which it emerged. In order to give the story

An atmospheric shot of a Vulcan climbing away from a slow flyby at Scampton. *Joe L'Estrange*

A Vulcan B.1A being prepared for flight at Scampton during the mid-1960s.

a little more human interest I have also included first-hand accounts from people who worked on the Vulcan, and crews who flew in the aircraft during its service life. Additionally, this book also includes a scattering of technical data together with extracts from the RAF's Vulcan Manual, the 'owner's manual', which gives a more specific appraisal of how some of the Vulcan's systems were operated. Hopefully, this combination of material will provide a varied and interesting overview of a truly outstanding aircraft.

In order to complete this book I have relied upon the assistance and support of countless individuals who have freely given their time and advice, while others have kindly released photographs from their collections for publication. Space precludes a complete list of all those who have helped to create this publication, but to all those who offered advice and time, and written and pictorial material, I am of course extremely grateful. However, I must thank some specific individuals who went far beyond the 'call of duty' to make this book happen, and these are:

> Artist and archivist Richard Caruana; Vulcan enthusiast Craig Jackson; aviation author Glenn Sands; former aircrew John Reeve, Bruce Woodruff, Les Aylott and Joe L'Estrange; David Fildes at Avro Heritage; Dave Griffiths and Richard Clarkson at the Vulcan Restoration Trust at Southend; Andy Leitch, creator of the outstanding 'Vulcans in Camera' website; former V-Force crew Ron Evans, Roy Brocklebank and Ray Evans; photographers Shaun Connor and Terry Senior; and contributors Bill Pearsey, Denis O'Brien, Robin Walker, Paul Tomlin, Michel Haines, Mike Freer and Ken Elliott.

A dramatic view from Scampton's main entrance as a Vulcan B.2 sweeps over the Lancaster gate guard, adjacent to the guard room.

Tim McLelland

XM575 over 'Bomber Country' making a flypast over Lincoln Cathedral.

Four Vulcans from Coningsby pictured during a dispersal exercise at Wittering. Although Wittering was both a Valiant and Victor base, it was Vulcan aircraft that regularly deployed to the ORP area at the eastern end of the single runway, close to the A1 trunk road.

Right: Operation 'Black Buck' Vulcan, XM607, is pictured performing in the hands of Joe L'Estrange at Greenham Common. *Michael J. Freer*

Below: The unforgettable sight (and sound) of a four-Vulcan scramble as demonstrated by Waddington Wing Vulcans at Finningley's 1982 Battle of Britain 'At Home' Day. *Fred Martin*

CHAPTER ONE

Birth of a Bomb

IN order to properly describe the true origins of the Vulcan, one has to go way back to the dark days of 1939 and the fascinating (if not confusing) world of particle physics; Otto Frisch, an Austrian scientist who was studying at the Institute of Theoretical Physics in Copenhagen (having already been forced to leave Nazi Germany), was invited to relocate to England, in order to continue his molecular physics studies at the University of Birmingham. Having studied in Denmark under the leadership of physicist Niels Bohr, he had established the basic principles of nuclear fission whilst spending Christmas with his aunt Lise Meitner, herself a respected scientist who had been conducting experiments with fellow physicists Otto Hahn and Fritz Strassman. Together they had produced a report describing the creation of barium from the collision of uranium nuclei and neutrons. After conducting further experiments and calculations based on the report, Frisch and Meitner concluded that in order to have created barium, the impact of the neutron upon the uranium atom must have actually elongated the shape of the nucleus. Nuclei contain protons that (thanks to their electrical charge) try to repel each other, but strong surface tension normally holds the protons together. Frisch and Meitner realised that the elongation of the nucleus would allow the electrical forces to overcome the surface tension, enabling the nucleus to split into two similarly sized portions. Frisch described this process as 'nuclear fission'. Significantly, the mass of the two halves created by the fission process is slightly less than that of the original nucleus, and this shortfall of mass (according to Einstein's famous calculations) represents an output of energy produced by the creation of the two fragments. Frisch and Meitner calculated that the actual amount of energy involved in this fission process would (proportionally speaking) be surprisingly large.

The impressive sight of Britain's first atomic explosion during Operation 'Hurricane'.

As this discovery was being explored, Niels Bohr's experiments and studies had led him to conclude that the uranium fission process was entirely due to the presence of uranium 235, a particularly rare isotope found in relatively small quantities within the more familiar uranium 238 metal. It was established that, during the fission process, secondary neutrons were created, and if these same neutrons were able to initiate further fusions, then a 'chain reaction' could occur, creating more and more fission and a huge release of energy. However, because the process relied on uranium 235, the chain reaction process could not occur naturally because of the isotope's relative scarcity when compared to the more common uranium 238 isotope (occurring at a rate of something like just one in 140 parts). Frisch accepted this conclusion but speculated as to just how much uranium 235 would actually be needed to start the chain reaction process. Working with Rudolf Peierls, Professor of Physics at Birmingham University, Frisch discovered that the necessary amount was astonishingly small – less than half a kilogram, in fact ('About a pound,' he later recalled). Clearly, this discovery had the potential to form the basis of an incredibly efficient source of power and (at least in theory) a hugely powerful weapon of war.

Frisch and Peierls reported their findings to Henry Tizard, who at that time was Chairman of the Committee on the Scientific Survey of Air Defence. It was somewhat ironic that Tizard had recently championed the development and introduction of radar, a subject that both Frisch and Peierls had also been keen to explore, but had been forbidden to study at University of Birmingham, because of their nationalities. Part of their report submitted to Tizard (made in March 1940) read as follows:

> 'The attached detailed report concerns the possibility of constructing a 'super bomb' which utilises the energy stored in atomic nuclei as a source of energy. The energy liberated in the explosion of such a super-bomb is about the same as that produced by the explosion of 1,000 tons of dynamite. This energy is liberated in a small volume in which it will, for an instant, produce a temperature comparable to that in the interior of the sun. The blast from such an explosion would destroy life in a wide area. The size of this area is difficult to estimate, but it will probably cover the centre of a big city. In addition, some part of the energy set free by the bomb goes to produce radioactive substances, and these will emit very powerful and dangerous radiations. The effect of these radiations is greatest immediately after the explosion, but it decays only gradually and even for days after the explosion any person entering the affected area will be killed. Some of this radioactivity will be carried along with the wind and will spread the contamination; several miles downwind this may kill people.'

Not surprisingly, Tizard was immediately impressed by the findings and made arrangements to set up a committee to explore the matter in more detail. Meeting for the first time in April 1940, the 'Maud Committee' (the peculiar code name came from Maud Rey, the former governess of Niels Bohr's children) tried to establish the potential of the new discovery, and its potential for providing Britain with a new weapon of war. Their 1941 report included the following:

> 'Work to investigate the possibilities of utilizing the atomic energy of uranium for military purposes has been in progress since 1939, and a stage has now been reached when it seems desirable to report progress. We should like to emphasize at the beginning of this report that we entered the project with more skepticism than belief, though we felt it was a matter which had to be investigated. As we proceeded we became more and more convinced that release of atomic energy on a large scale is possible and that conditions can be chosen which would make it a very powerful weapon of war. We have now reached the conclusion that it will be possible to make an effective uranium bomb which, containing some 25lb of active material, would be equivalent as regards destructive effect to 1,800 tons of TNT and would also release large quantities of radioactive substance, which would make places near to where the bomb exploded dangerous to human life for a long period. The bomb would be composed of an active constituent (referred to in what follows as U) present to the extent of about a part in 140 in ordinary Uranium. Owing to the very small difference in properties (other than explosive) between this substance and the rest of the Uranium, its extraction is a matter of great difficulty and a plant to produce 2-4 lb per day (or 3 bombs per month) is estimated to cost approximately £95,000,000, of which sum a considerable proportion would be spent on engineering, requiring labour of the same highly skilled character as is needed for making turbines.
>
> In spite of this very large expenditure we consider that the destructive effect, both material and moral, is so great that every effort should be made to produce bombs of this kind. As regards the time required we estimate that the material for the first bomb could be ready by the end of 1943. This of course assumes that no major difficulty of an entirely unforeseen character arises. Even if the war should end before the bombs are ready the effort would not be wasted, except in the unlikely event of complete disarmament, since no nation would care to risk being caught without a weapon of such decisive possibilities.
>
> This type of bomb is possible because of the enormous store of energy resident in atoms and because of the special properties of the active constituent of uranium. The explosion is very different in its mechanism from the ordinary chemical explosion, for it can occur only if the quantity of U is greater than a certain critical amount. Quantities of the material less than the critical amount are quite stable. Such quantities are therefore perfectly safe and this is a point which we wish to emphasize. On the other hand, if the amount of material exceeds the critical value it is unstable and a reaction will develop and multiply itself with enormous

rapidity, resulting in an explosion of unprecedented violence. Thus all that is necessary to detonate the bomb is to bring together two pieces of the active material each less than the critical size but which when in contact form a mass exceeding it.

In order to achieve the greatest efficiency in an explosion of this type, it is necessary to bring the two halves together at high velocity and it is proposed to do this by firing them together with charges of ordinary explosive in a form of double gun. The weight of this gun will of course greatly exceed the weight of the bomb itself, but should not be more than one ton, and it would certainly be within the carrying capacity of a modern bomber. It is suggested that the bomb (contained in the gun) should be dropped by parachute and the gun should be fired by means of a percussion device when it hits the ground. The time of drop can be made long enough to allow the aeroplane to escape from the danger zone, and as this is very large, great accuracy of aim is not required. The cost per lb of this explosive is so great it compares very favourably with ordinary explosives when reckoned in terms of energy released and damage done. It is, in fact, considerably cheaper, but the points which we regard as of overwhelming importance are the concentrated destruction which it would produce, the large moral effect, and the saving in air effort the use of this substance would allow, as compared with bombing with ordinary explosives.

One outstanding difficulty of the scheme is that the main principle cannot be tested on a small scale. Even to produce a bomb of the minimum critical size would involve a great expenditure of time and money. We are however convinced that the principle is correct, and whilst there is still some uncertainty as to the critical size it is most unlikely that the best estimate we can make is so far in error as to invalidate the general conclusions. We feel that the present evidence is sufficient to justify the scheme being strongly pressed. It will be seen from the foregoing that a stage in the work has now been reached at which it is important that a decision should be made as to whether the work is to be continued on the increasing scale which would be necessary if we are to hope for it as an effective weapon for this war. Any considerable delay now would retard by an equivalent amount the date by which the weapon could come into effect. We are informed that while the Americans are working on the uranium problem the bulk of their effort has been directed to the production of energy, as discussed in our report on uranium as a source of power, rather than to the production of a bomb. We are in fact co-operating with the United States to the extent of exchanging information, and they have undertaken one or two pieces of laboratory work for us. We feel that it is important and desirable that development work should proceed on both sides of the Atlantic irrespective of where it may be finally decided to locate the plant for separating the U, and for this purpose it seems desirable that certain members of the committee should visit the United States. We are informed that such a visit would be welcomed by the members of the United States committees which are dealing with this matter.'

It's interesting to note that at this point in history, the British Government was indeed openly co-operating with American scientists on all aspects of the new discovery (in fact, Niels Bohr was already sailing to New York while making his calculations to confirm the Frisch-Meitner findings early in 1939). Likewise, the committee regarded the discovery as a means of creating a new weapon for potential use against Nazi Germany, rather than being directed at any other unrecognised threat far in the future. Frisch later commented that:

> 'I have often been asked why I didn't abandon the project there and then, saying nothing to anybody. Why start on a project which, if it was successful, would end with the production of a weapon of unparalleled violence, a weapon of mass destruction such as the world had never seen? The answer is very simple. We were at war, and the idea was reasonably obvious; very probably some German scientists had had the same idea and were working on it.'

There was evidence to suggest that Germany was indeed showing some significant interest in the subject; a group of Paris-based scientists, who were also working on similar fission studies, had concluded that in order to slow the rapid progress of a chain reaction, a moderator would be required and that 'heavy water' (deuterium oxide) would be the ideal medium. The only source of heavy water was a hydroelectric station at Vemork in Norway, and it transpired that Germany had already offered to purchase the entire Norwegian stock of heavy water. It did not take much imagination to guess what might be happening inside Hitler's laboratories.

The findings of the Frisch-Peierls memorandum were handed to America as part of Henry Tizard's infamous mission to the US, which he made in September 1940. Although Britain was in the grip of a war with Nazi Germany, the United States was maintaining a neutral position towards the conflict, and, as Britain's future began to look increasingly bleak, governmental eyes were looking across the Atlantic to America's vast industrial resources. Tizard (with Churchill's grudging approval) instigated the concept of what was essentially a 'trade mission', which would give America access to a range of British technological developments in exchange for access to America's industrial know-how and resources. The concept was not new, and the mutual benefits of such exchanges had always been difficult to establish, but Tizard was convinced that the best approach would be to simply give Britain's technological information to the US, and essentially hope that co-operation was given in return. A wide range of information was handed over, which included hugely significant developments such as Whittle's jet engine, the cavity magnetron (vital for the development of radar), and much more. The Frisch-Peierls memorandum was only part of the mission's range of topics, and although it seems clear in retrospect that a huge variety of commercially valuable secrets were effectively handed to America (and Tizard's enthusiasm for the deal certainly was not shared by everyone in the UK – he found himself without a job when he returned home), the mission marked the beginnings of the 'Special Relationship' between the US and the UK, which remains solid even to this day. Although attention was therefore being paid to relations between America and Britain, and also potential developments inside Germany, little or no consideration was being given to the possibility that information might also make its way eastwards. In fact, most the Maud Committee's findings had already made their way across Europe a whole month previously, courtesy of Soviet agents.

The Tizard delegation also visited Columbia University to discuss the Frisch-Peierls memorandum with Enrico Fermi, a respected Italian physicist who was also investigating the newly discovered fission process. Fermi recalls:

> 'I remember very vividly the first month, January 1939, that I started working at the Pupin Laboratories because things began happening very fast. In that period, Niels Bohr was on a lecture engagement at the Princeton University and I remember one afternoon Willis Lamb came back very excited and said that Bohr had leaked out great news. The great news that had leaked out was the discovery of fission and at least the outline of its interpretation. Then, somewhat later that same month, there was a meeting in Washington where the possible importance of the newly discovered phenomenon of fission was first discussed in semi-jocular earnest as a possible source of nuclear power.'

It is interesting to note that by the time Tizard's delegation met with Fermi, he was still clearly looking at fission as a source of commercial power rather than as a potential weapon, even though other physicists had already reached much darker conclusions. In October 1939 a group of scientists had delivered a letter (signed by Albert Einstein) to President Roosevelt warning that:

> 'In the course of the last four months it has been made probable, through the work of Joliot in France as well as Fermi and Szilard in America, that it may become possible to set up a nuclear chain reaction in a large mass of uranium, by which vast amounts of power and large quantities of new radium-like elements would be generated. Now it appears almost certain that this could be achieved in the immediate future. This new phenomenon would also lead to the construction of bombs, and it is conceivable, though much less certain, that extremely powerful bombs of a new type may thus be constructed. A single bomb of this type, carried by boat and exploded in a port, might very well destroy the whole port together with some of the surrounding territory. However, such bombs might very well prove to be too heavy for transportation by air.'

Fermi remained primarily interested in the peaceful potential of nuclear power (through the creation of turbine steam), remaining sceptical of the Frisch-Peierls memorandum, and developments stagnated once Tizard's delegation returned to the UK. No further progress was made until Mark Oliphant, Professor of Physics at the University of Birmingham, made another trip to the United States in August 1941 to try and establish why there was still so little interest in the concept of creating an atomic bomb. A government committee had been set up to investigate fission research but it appeared to have simply ignored the findings of the Maud Committee. Oliphant recalled that:

> 'The minutes and reports had been sent to Lyman Briggs, who was the Director of the Uranium Committee, and we were puzzled to receive virtually no comment. I called on Briggs in Washington, only to find out that this inarticulate and unimpressive man had put the reports in his safe and had not shown them to members of his committee. I was amazed and distressed.'

Oliphant then met directly with the government's Uranium Committee, as committee member Samuel Allison recalls:

> 'Oliphant came to a meeting, and said "bomb" in no uncertain terms. He told us we must concentrate every effort on the bomb and said we had no right to work on power plants or anything but the bomb. The bomb would cost 25 million dollars, he said, and Britain did not have the money or the manpower, so it was up to us.'

Consequently, Columbia University was awarded $6,000 (although the money was not released until the spring of 1940 because of governmental fears surrounding the concept of 'foreigners' conducting secret research) and Fermi went on to construct the very first atomic pile at Stagg Field in Chicago, which finally went critical (ie the uranium chain reaction was allowed to begin) on 2 December 1942.

Even as this development was taking place, there was no obvious necessity for the US Government to enthusiastically pursue the concept of an atomic bomb. America was not at war even if Europe was. Further pressure from scientists continued to circulate until Vannevar Bush, Director of the Office of Scientific Research and Development, finally convinced Roosevelt (in October 1941) to conduct a fully funded effort to build an atomic weapon. The combination of scientific pressure, the Maud report and the clear possibility of entering into a war was finally enough to push America into action. A new committee was set up to oversee the project and to report to the President, and it was with more than a little irony that the committee met for the first time just one day before Japan attacked Pearl Harbor, and the United States finally entered the Second World War.

Roosevelt wrote to Churchill suggesting that efforts to develop an atomic bomb should be 'co-ordinated or even jointly conducted'. Given the amount of time that Britain had devoted to virtually pleading with America for such a move, it may seem remarkable that Churchill was less than enthusiastic. After months of consideration, the British Government's response was little more than a general willingness to co-operate with America, presumably on the assumption that America's research might still be largely geared towards atomic power developments rather than a bomb. However, the true cause of Churchill's lukewarm response was probably the British Government's miscalculation of the difficulty and cost of producing atomic weapons. A distinct 'go it alone' attitude prevailed, although when British scientists visited the US early in 1942 and were afforded full access to information that was then available, they were astounded by the rapid progress that was being made, especially in comparison to developments back in Britain.

Britain's position became increasingly difficult. As the complexities and cost of atomic research became clearer (the British programme being code-named as the Tube Alloys project), it was obvious to the government that full co-operation with America would be the only sensible way to proceed, but by the time Britain had reached this conclusion America's research and development had moved far ahead of Britain's, so there was no longer any obvious advantage for America in maintaining co-operation. Worse still, the US Army had taken over control of virtually all aspects of the US programme – the Manhattan Project – so all information to Britain slowly dried up. The situation was finally resolved through a series of diplomatic meetings, which were assisted by Roosevelt's personal inclination towards a more co-operative stance. He rightly accepted that, while American concerns over release of information were largely based on potential post-war commercial use of atomic power, Britain's interest was chiefly in the creation of an atomic bomb, and when Churchill assured Roosevelt that this was indeed the case (and that Britain didn't have any direct interest in obtaining America's data on nuclear power development), the situation was soon resolved, and a joint agreement between the two countries was signed by Roosevelt and Churchill on 19 August 1943 – the Quebec Agreement – which included the following:

> 'Articles of Agreement Governing Collaboration Between The Authorities of the U.S.A. and the U.K. in the Matter of Tube Alloys Whereas it is vital to our common safety in the present War to bring the Tube Alloys project to fruition at the earliest moments; and Whereas this maybe more speedily be achieved if all available British and American brains and resources are pooled; and Whereas owing to war conditions it would be an improvident use of war resources to duplicate plants on a large scale on both sides of the Atlantic and therefore a far greater expense has fallen upon the United States; It is agreed between us First, that we will never use this agency against each other. Secondly, that we will not use it against third parties without each other's consent. Thirdly, that we will not either of us communicate any information about Tube Alloys to third parties except by mutual consent. Fourthly, that in view of the heavy burden of production falling upon the United States as the result of a wise division of war effort, the British Government recognize that any post-war advantages of an industrial or commercial character shall be dealt with as between the United States and Great Britain on terms to be specified by the President of the United States to the Prime Minister of Great Britain. The Prime Minister expressly disclaims any interest in these industrial and commercial aspects beyond what may be considered by the President of the United States to be fair and just and in harmony with the economic welfare of the world. And Fifthly, that the following arrangements shall be made to ensure full and effective collaboration between the two countries in bringing the project to fruition.'

Following the signing of the Quebec Agreement, British scientists were soon absorbed into the infamous Manhattan Project in substantial numbers, working in close co-operation with their American counterparts. Information was still only exchanged between scientists on a strictly 'need to know' basis, but this compartmentalised approach was applied as a general security measure, rather than being any attempt to restrict British knowledge of the wider aspects of the project. In fact, Anglo-American co-operation was now better than ever, and during another meeting in 1944 Roosevelt and Churchill signed another agreement, which developed themes from the earlier Quebec Agreement. In essence, the 'Hyde Park Aide-Memoire' specified that Britain and America should continue joint military and industrial atomic energy development even after the war had ended, and also dismissed the idea of releasing information on Britain's nuclear programme (the Tube Alloys project) so that an international treaty of arms control could be set up. Inexplicably, Roosevelt did not even show the document to his advisors and it quietly disappeared into his private files for several years.

A typical weapons-sized sphere of plutonium encased in a graphite safety block.

The historic Trinity test explosion just a quarter of a second after detonation.

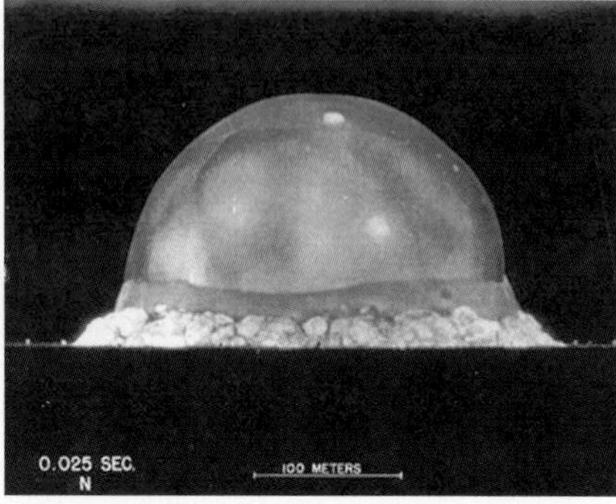

An aerial photograph of the Trinity test site following the detonation of the first nuclear device.

After meeting with the President a few days after the memoire was signed, Vannevar Bush suspected that Roosevelt was privately contemplating a post-war Anglo-American agreement to maintain complete secrecy on the atom bomb's development, thereby (at least in theory) controlling the stability of the rest of the world. Bush believed that this was a flawed policy and advised the Secretary of War, Henry L. Stimpson, that it would inevitably encourage the Soviets to develop their own bomb and ultimately lead to a catastrophic conflict. Further analysis only led to yet more confusion, and neither Bush nor Stimpson ever forwarded any views to the President. Subsequently, Stimpson commented that 'the atomic bomb might be a Frankenstein which would eat us up, or it might be the means by which the peace of the world would be helped in becoming secure'. But having accepted that the United States was effectively at a proverbial crossroads in terms of policy decision, it was far from clear which direction to take, and when Roosevelt died in April 1945 the situation remained as uncertain as ever.

Despite the political uncertainty, Britain's involvement in the Manhattan Project was now secured and rapid progress was made, culminating in the detonation of the world's first atomic device at 05:29 on 16 July 1945. This device (far bigger and cumbersome than anything that could be described as a 'bomb') was assembled deep in the heart of what is now the White Sands Missile Range near Alamogordo, in the New Mexico desert. This 'Trinity test' device employed an implosion system that utilised plutonium (a by-product of uranium created within an atomic 'pile') as its key fissile component. Plutonium had by now been determined to be a more efficient fissionable material (which could be created more easily than uranium 235), but the reactor-produced plutonium proved to be rather less efficient than expected, which meant that a simple gun implosion detonation simply would not work. The presence of additional neutrons meant that the plutonium would begin to pre-detonate, resulting in a bomb with a disappointingly low explosive yield.

The answer was to produce an implosion device that required the core of plutonium to be instantly compressed by conventional explosives. By carefully shaping a mix of fast and slow explosives into a series of 'lenses', an explosive shock wave could be accurately directed on to the plutonium core, ensuring that it was compressed simultaneously and equally on all sides. The result was a sufficiently efficient chain reaction fission process that required only a 10cm-diameter ball of reactor-grade plutonium. Another distinct advantage of the plutonium/implosion device was that it was also much safer to handle, avoiding all the obvious risks of the crude uranium gun device, which could easily be detonated by accident. However, because the creation of explosive lenses required a great deal of research and precision manufacturing (the explosive lenses had to be manufactured within millimetre-sized tolerances), the scientific director of the Manhattan Project, J. Robert Oppenheimer, decided that, although the uranium bomb was virtually guaranteed to function correctly, it would be unwise to proceed directly to the production and actual use of an implosion bomb, hence the Trinity test.

Informal bets were placed on the predicted outcome of the detonation, ranging from a complete failure through to a yield of 18kt (equivalent TNT). Hauled by a crane on to a 100-foot-high tower, the device was held aloft in order to maximise the destructive effect on the target area below, and to reduce the creation of radioactive fallout that would be generated by sweeping up ground debris. A huge and cumbersome steel canister nicknamed 'Jumbo' was manufactured to encase the device, so that the conventional explosion could be contained, should the chain reaction fail. However, confidence in the successful detonation of the 'gadget' eventually convinced the scientists to dispense with 'Jumbo', and ultimately the 240-ton case was simply positioned on a tower some 800 yards from the device, to monitor the effect of the explosion on it. The scheduled 4.00am detonation was delayed by thunderstorms, but by 5.00am conditions were good and the assembled scientists and military observers retired to various viewing locations, mostly around 10 miles from the device.

As predicted, the detonation was an unqualified success. The surrounding mountain ranges were briefly illuminated even more brightly than by the usual desert sunshine (some understandably confused local residents reported seeing the morning sun make a brief appearance before setting again), and 40 seconds after the detonation a powerful shock wave passed over the observers, eventually reaching out far into the desert, rattling windows some 200 miles away. One military observer later reported that:

> '...the lighting effects beggared description. The whole country was lighted by a searing light with the intensity many times that of the midday sun. It was golden, purple, violet, gray, and blue. It lighted every peak, crevasse and ridge of the nearby mountain range with a clarity and beauty that cannot be described but must be seen to be imagined.'

The desert sand had been melted into green-coloured glass (subsequently nicknamed 'Trinitite'), and although 'Jumbo' still remained intact, her supporting tower had been vaporised. The explosion had yielded 19kt. In order to disguise the true nature of the historic explosion, Alamogordo Air Base issued a press release reporting 'an explosion of a remotely located ammunitions dump, in which no one had been killed or injured'. Oppenheimer later recalled that as he watched the fireball rise into the early morning sky, he remembered a line from Hindu scriptures: 'I am become death, the destroyer of worlds'. Test director Kenneth Bainbridge's comments were rather more succinct. He simply said, 'Now we are all sons of bitches.'

When President Truman came into power, he knew nothing of the development of the atomic bomb, but once he was fully informed of the Manhattan Project's progress, he set up a committee to explore the possible ways in which the atomic bomb could be used against Japan, in order to expedite the end of the war, and possibly avoid the predicted (and unacceptable) casualties that would inevitably be created by an invasion of the Japanese mainland. Modern accounts of this period often suggest that the use of atomic bombs on Japan was little more than an experiment, designed to record the effects of atomic bombs on both people and targets, but in reality it is clear that Truman did consider the situation with great care before reaching any decisions as to whether atomic weapons should actually be used. Oppenheimer predicted that just one bomb might well kill 20,000 people, and felt that a military installation should be chosen as a target rather than simply dropping a bomb on a city. Others felt that an isolated part of the Japanese countryside should be chosen as a pre-publicised 'showcase' in order to demonstrate the bomb's power (and hopefully persuade Japan to capitulate without the need for further attacks), but the risk of publicising the bomber's arrival would have inevitably encouraged Japan to simply shoot it down. Of course there was also some risk that the bomb might simply fail to detonate (which would have been a huge embarrassment to the Allies if it had been publicised in advance), although the weapon was a relatively crude uranium device that relied on the simple gun-detonation method. Confidence in the success of this bomb's design was so high that no pre-drop tests were thought necessary, and the Manhattan team were convinced that the weapon would function as predicted. Doubts began to grow that Japan would ever consider a complete surrender unless a devastating and direct blow was delivered on Japanese cities, and when the Manhattan team reported that a suitable bomb could be made available by the beginning of August, there seemed to be no reason to avoid using it at the earliest opportunity.

On 6 August 1945 a Boeing B-29 from the specially formed 590th Composite Group took off from Tinian Island (the unit's forward base) and headed for the Japanese mainland some 1,500 miles away. Secured in the aircraft's bomb bay was 'Little Boy', a 9,700lb uranium bomb. Hiroshima had been chosen as the primary target, chiefly because of its significance as a military and communications centre but also because the area had been untouched by conventional bombing, which would enable observers to examine the destructive effects of the bomb. Accompanied by escorting observation aircraft, the 'Enola Gay' (named in honour of the aircraft Captain's mother) approached the target at 31,000 feet and released the weapon at 08:15 local time. Fitted with barometric and ground radar fusing, the bomb successfully detonated some 43 seconds later at a height of 1,900 feet above an Army parade field. A huge burst of light, heat and blast engulfed the city, instantly killing 70,000 people. As the 'Enola Gay' headed away from the impact point, Captain Paul Tibbets recalled that 'the city was hidden by that awful cloud ... boiling up, mushrooming, terrible and incredibly tall'. Ultimately, this single bomb was responsible for the deaths of more than 200,000 people and the complete devastation of 5 square miles of land. Truman immediately issued a statement on the bomb's use and served notice on Japan that if they did not surrender with immediate effect (as required by the Potsdam Declaration made on 26 July) then more Japanese cities would be attacked with similarly devastating results.

A second weapon was prepared for use, and on 9 August another B-29 (named 'Bockscar') departed from Tinian, this time carrying an implosion device (utilising plutonium) nicknamed 'Fat Man'. The primary target at Kokura was obscured by poor weather and the secondary target at Nagasaki was substituted, with the bomb released at 11:01 local time. The target point was missed by almost 2 miles, which spared a major portion of the city from the bomb's blast thanks to an intervening range of hills. Significantly more powerful than the uranium bomb (which yielded approximately 13kt), the plutonium device delivered an even more impressive 21kt, but because of the topography of the local area and the inaccurate delivery the resulting destruction was much the same (or even less) than that created by 'Little Boy'. Yet more bombing missions were planned, with another bomb being anticipated for use by the end of August, followed by three more in September and another three in October. But on the same day that Nagasaki was destroyed, Emperor Hirohito had already realised that surrender was his only option, stating that:

> '...the enemy has begun to employ a new and most cruel bomb, the power of which to do damage is, indeed, incalculable, taking the toll of many innocent lives. Should we continue to fight, not only would it result in an ultimate collapse and obliteration of the Japanese nation, but also it would lead to the total extinction of human civilization. Such being the case, how are we to save the millions of our subjects, or to atone ourselves before the hallowed spirits of our imperial ancestors? This is the reason why we have ordered the acceptance of the provisions of the Joint Declaration of the Powers.'

The Second World War was over.

Britain entered the immediate post-war period with a new socialist government. With a new committee set up to consider future nuclear policy, the country's new Prime Minister Attlee circulated these comments to the group during August 1945:

> 'A decision on major policy with regard to the atomic bomb is imperative. Until this is taken, civil and military departments are unable to plan. It must be recognized that the emergence of this weapon has rendered most of our post-war planning out of date.'

Clearly illustrating his belief in the potential value of atomic weapons, Attlee continued:

> 'We recognized, or some of us did before this war, that bombing would only be answered by counter bombing. We were right. Berlin and Magdeburg were the answer to London and Coventry. The answer to an atomic bomb on London is an atomic bomb on another great city. Scientists in other countries are certain in time to hit upon the secret. The most we may have is a few years' start. The question is, what use are we to make of that few years' start?'

While Attlee was trying to establish the government's policy on atomic weaponry, the military Chiefs of Staff were also carefully considering the impact of the new weapon's capabilities with respect to future offensive and defensive policy. As a direct result of their deliberations, they advised their Technical Warfare Committee to revise the Tizard

Committee report on future developments in methods of warfare, to take into account the development of the atomic bomb. The Tizard report was first created in response to a request made by the Chiefs of Staff in November 1944 to investigate potential future developments in weaponry design. Although Tizard had (as described previously) been involved with the development of atomic research, his committee had not been afforded any access to current developments vis-à-vis the ongoing Manhattan Project, and the completed report was therefore almost obsolete before it was completed, even though it did describe the potential for atomic bomb development together with high-speed and high-altitude bombers capable of delivering them. Most importantly, the original report touched upon the concept of nuclear deterrence:

'...the only answer that we can see to the atomic bomb is to be prepared to use it ourselves in retaliation. The knowledge that we were prepared, in the last resort, to do this might well deter an aggressive nation.'

The revised Tizard Report enabled the Chiefs of Staff to reach some fundamental conclusions, including key points such as these:

'Given sufficient accumulation in peace and adequate means of delivery, atomic and biological weapons might achieve decisive results with relatively small effort against the civil population of a nation without a clash between the major military forces or the exercise of sea power. Some five or ten atomic bombs landed on the target, with the prospect of more to follow, might well cause the evacuation of cities to an extent sufficiently seriously to sap the power of waging war by conventional means of any country physically or psychologically unorganized to meet such action. Without the moral backing of adequate military power in being, with which to limit or repel invasion, or to launch an effective counter-offensive, such attack might well lead to collapse. On the other hand, some hundreds of atomic weapons might fail to cause the collapse of a country suitably organized physically and psychologically, and morally reinforced by adequate military power in being.

There is no firm basis on which to assess the quantities of atomic and biological weapons required by any nation to bring about the collapse of another, and many of the factors involved are imponderable. Nevertheless, our estimate, based on such information as is at present available, leads us to believe that some 30-120 atomic bombs accurately delivered by the USSR might cause the collapse of the United Kingdom without invasion, whereas several hundred bombs might be required by the United States or the United Kingdom to bring about the collapse of the USSR. The number of bombs required to cause a similar collapse in the United States would probably be somewhat greater than for this country, but the problem of landing them accurately in the United States at the ranges involved is much greater.'

It is interesting to note that even by the time of the first Tizard report, attention had clearly shifted towards the USSR as a possible future aggressor. Although the origins of the British and American atomic programme were undoubtedly based on the fear that similar developments were undoubtedly beginning to take shape in Nazi Germany, it is important to note that, as the end of the Second World War, approached, the fear of Nazi atomic know-how was almost seamlessly replaced by an even greater fear of Soviet capabilities. Churchill had already expressed his fears in a telegram to President Truman sent during May 1945:

'I am profoundly concerned about the European situation. I learn that half the American Air Force has already begun to move to the Pacific theatre. The newspapers are full of great movements of the American armies out of Europe. Our armies also are, under previous arrangements, likely to undergo a marked reduction. The Canadian Army will certainly leave. The French are weak. Anyone can see that in a very short space of time our armed power on the Continent will have vanished, expect for moderate forces to hold down Germany. Meanwhile, what is to happen about Russia? I feel deep anxiety because of their misinterpretation of the Yalta decisions, their attitude towards Poland, their overwhelming influence in the Balkans, excepting Greece, the difficulties they make about Vienna, the combination of Russian power and the territories under their control or occupied, coupled with the Communist technique in so many other countries, and above all their power to maintain very large armies in the field for a long time. What will be the position in a year or two when the British and American armies have melted and the French have not yet been formed on any major scale, and when Russia may choose to keep two or three hundred divisions on active service? An iron curtain is drawn down upon their front. We do not know what is going on behind.'

There was some confidence both in Britain and America that the development of nuclear power and weaponry could be placed under international control, but the Chiefs of Staff clearly feared that international agreements could not put the atomic genie back into a proverbial bottle. In a minute to the Prime Minister submitted during October 1945, the Chiefs of Staff stated that:

'We must aim for international control – it is probably the only alternative to mutual destruction. Any international agreement into which we enter should include the most unequivocal and comprehensive rights of inspection. The whole concept of international control stands or falls on the efficacy of the arrangements for such an inspection. Russia is a country which appears to have both the natural resources and the remote areas for the secret development of atomic weapons. There is the obvious danger that we and the Americans might be led to agree not to produce atomic weapons while the Russians secretly carried out their research and production in the remote areas of the Soviet Union. The right of inspection will provide no security unless it is completely comprehensive. How this is to be achieved under the present Soviet system is the crux of the problem.'

Towards the end of the same minute, the Chiefs of Staff moved on to outline what would become the cornerstone of Britain's offensive and defensive strategy:

'It is clear that in the event of failure to secure an international agreement, possession of atomic weapons of our own would be vital to our security. The best method of defence against the new weapons is likely to be the deterrent effect that the possession of the means of retaliation would have on a potential aggressor. The Chiefs of Staff therefore consider that we should press ahead in the field of research and that it is essential that British production of atomic weapons should start as soon as possible. To delay production pending the outcome of negotiations regarding international control might well prove fatal to the security of the British Commonwealth.'

With the obvious need to develop atomic weapons now very clear, and the political will to embark upon such a programme gradually emerging, attention was drawn towards the complex, ambitious and expensive task at hand. Although British scientists had been an important part of the Manhattan Project (General Groves, who masterminded the project, later commented that there would probably have never been a bomb to drop on Hiroshima or Nagasaki had it not been for 'active and continuing British interest'), the post-war environment changed Britain's relationship with America yet again, and although involvement in the Manhattan Project had undoubtedly given Britain a great deal of technical knowledge, the compartmentalised nature of the activities at Los Alamos meant that Britain still did not have all the necessary knowledge to actually build her own atomic bomb. However, as Lord Chadwick, Chairman of the weaponry committee, which reported to the Ministry of Supply, rightly pointed out, the British scientists could not be 'expected to take amnesia tablets before returning home'.

Although the emerging Special Relationship between Britain and America had survived many ups and downs during the war, and a firm (if often undisclosed) military relationship continued to exist, the wider political climate was again

A well-known picture of the huge 20-kiloton 'Blue Danube' bomb, around which the Vulcan was effectively designed. The short-lived 'interim' 'Violet Club' bomb used the same Blue Danube carcass but would have yielded a rather more impressive 400 kilotons.

This model of the Blue Danube illustrates the weapon's internal layout and the size of the warhead package, which dictated the overall size of the bomb (and therefore the size of the aircraft designed to carry it).

pushing America back towards isolationism. An increasingly cynical American political opinion of Britain's new socialist government was only compounded when it was revealed in April 1946 that a British nuclear physicist had been secretly passing US nuclear research information directly to Soviet agents. Britain's hopes of nuclear co-operation with America effectively ended when the McMahon Bill was passed by the US Senate. Completely ignoring the earlier Hyde Park and Quebec agreements, the new Atomic Energy Act was an all-embracing piece of legislation that specifically prohibited the release of any scientific information relating to atomic power to a foreign nation. Senator McMahon subsequently stated that:

> 'The British contributed heavily to our own wartime atomic project, but due to a series of unfortunate circumstances, the nature of these agreements which made this contribution possible was not disclosed to me or my colleagues on the Senate Special Atomic Energy Committee at the time we framed the law in 1946.'

Whether this is an honest account of what really happened remains unclear, but it may well be that (as had happened on previous occasions) Britain's input had simply been overlooked within the machinations of America's huge political bureaucracy.

Prime Minister Attlee was both angry and astonished when the Atomic Energy Act was introduced, and immediately sent a telegram to Truman stating:

> 'Our continuing co-operation over raw material shall be balanced by an exchange of information which will give us, with all proper precautions with regard to security, that full information to which we believe we are entitled, both by the documents and by the history of our common efforts in the past.'

Truman, for reasons which still remain unclear, never even acknowledged the communication. This was probably the final sign that Britain was clearly going to have to 'go it alone' regardless of any continuing desire to work in co-operation with America. Attlee later commented that

> '...we had to hold up our position vis-à-vis the Americans. We couldn't allow ourselves to be totally in their hands and the position wasn't awfully clear always. There was the possibility of their withdrawing and becoming isolationist again. The manufacture of a British bomb was therefore at this stage essential to our defence.'

Foreign Secretary Ernest Bevin expressed Britain's position more succinctly: 'We've got to have it and it's got to have a bloody Union Jack on it.' Britain's most immediate action was to secure her own supply of raw uranium from the joint Anglo-American stocks that had first been established during the Second World War.

Dr William Penney, 'father' of the British nuclear bomb programme.

The US Government rightly believed that Britain would be able to secure good supplies of uranium thanks to favourable relations with Portugal, South Africa, Belgium and other Commonwealth countries, which possessed uranium mines in various locations, but in reality the access to such supplies would not be able to compete with the purchasing power of the American dollar. However, the very fact that Britain displayed an interest in securing a reliable supply or uranium emphasised to America that she was certainly prepared to embark upon her own nuclear programme regardless of America's indirect efforts to make the task more difficult and expensive than it needed to be. It was inevitable that, after the McMahon Act fiasco, Britain's relationship with America would improve again (it certainly could not have become any worse) once the US realised that Britain was clearly going to develop her own programme, regardless of external assistance or influence. Likewise there was a growing realisation in the US that commitments and agreements with Britain had not been honoured, combined with a growing shortage of uranium supplies from which Britain did not appear to be suffering. The rocky relationship was patched up during talks in January 1948 so that the US could gain access to British stocks of uranium in exchange for technical co-operation. But by the following year the spirit of co-operation had inexplicably broken down yet again to such an extent that Truman stated, 'We have got to protect our information and we must certainly try and see that the British do not have the information to build atomic weapons because they might be captured.' Of course, Truman's concerns came too late, and Britain had already decided to build her own bomb.

Cabinet-level approval of plans for new weapons finally led to the issue of an Air Staff Requirement OR1001, which emerged on 9 August 1946, calling for the development of a 'bomb employing the principle of nuclear fission'. Although the precise size and weight of the device was obviously unclear at this stage, experience with the Manhattan Project enabled a fair estimate to be established, and from this more precise dimensions of the bomb could be worked out. Original copies of the OR1001 no longer exist, but when the OR was re-issued a couple of years later it stated that the bomb should not exceed 290 inches in length, with a diameter of 60 inches. Weight limit was set at 10,000lb and it was specified that the weapon should be capable of release from heights of between 20,000 and 50,000 feet, and speeds between 150 and 500 knots. It was envisaged that flip-out fins would be used so that the overall dimensions of the bomb bay of the carrier aircraft would not need to be even larger than dictated by the specified bomb size. However, back in 1946 it was still impossible to confidently fix the dimensions of the device, the bomb, the weapons bay required to accommodate it, and the dimensions of the bomber aircraft into which it would be incorporated.

Despite the issue of OR1001, official governmental approval for the development of atomic weapons did not materialise until early in 1947. Lord Portal, Controller of Production of Atomic Energy, finally set the project into motion through a memo presented at 10 Downing Street on 8 January, which stated:

> 'I submit that a decision is required about the development of atomic weapons in this country. The Service Departments are beginning to move in the matter and certain sections of the press are showing interest in it. My organisation is charged solely with the production of fissile material, ie the filling that would go into any bomb that it was decided to develop. Apart from producing the filling, development of the bomb mechanism is a complex problem of nuclear physics and precision engineering on which some years of research and development will be necessary. I suggest that there are broadly three courses of action to choose from: a) Not to develop the atomic weapon at all, b) To develop the weapon by means of ordinary agencies in the Ministry of Supply and Service Departments, or c) To develop the weapon under special arrangements conducive to the utmost secrecy.
>
> I imagine that course a) above would not be favoured by HM Government in the absence of an international agreement on the subject. If course b) is adopted it will be impossible to conceal for long the fact that this development is taking place. Many interests are involved and the need for constant consultation with my organisation (which is the sole depository of atomic energy and atomic weapons derived from our wartime collaboration with the United States) would result in very many people, including scientists, knowing what was going on. Moreover, it would certainly not be long before the American authorities heard that we were developing the weapon "through the normal channels" and this might well seem to them another reason for reticence over technical matters, not only in the field of military uses of atomic energy, but also in the general "know-how" of the production of fissile material.'

After making further comments on possible courses of action, Portal continued:

> 'I therefore ask for direction on two points; first, whether research and development is to be undertaken and, if so, whether the arrangements outlined are to be adopted.'

With the support of government ministers, it was agreed that research and development work should indeed be undertaken, and at long last Britain's bomb project was formally initiated. Coincidentally, just one day earlier a rather distressed AVM Boothman's minute to Air Marshal Dickson illustrated that even though the development of an atomic bomb was now comfortably under way, the Air Staff still knew very little about it:

> 'I am very worried about the lack of information which exists or is indeed given to us about our own progress in the atomic field. As you know, we stated a requirement to the Ministry of Supply last September for an atomic bomb. CS(A) immediately got in touch with the Department of Atomic Energy and discussed the whole matter with them. It now transpires that there is no organisation in Great Britain to develop the military side of atomic energy and in the opinion of individuals in the Dept of Atomic Energy, there is not likely to be such an organisation for some time to come. The Air side of MoS have also been told that they will be ill advised to finalise the dimensions of the bomb bays of our future bombers until they obtain officially the probable dimensions of the bomb. In view of the fact that there is no organisation to do development on the bomb, things have now reached a complete impasse.'

Boothman continued:

> 'I also understand that Professor Penney, who is the only technical authority in this country on the design of atomic bombs (which he gleaned during his work in America) is in honour bound not to give this information

The hapless HMS *Plym* pictured shortly before the detonation of the first British atomic device, which was loaded inside the vessel.

away to anybody. We have therefore arrived at the Gilbertian situation in which we have asked for long-range bombers and atomic bombs to go inside them, but the one individual who is able to satisfy the major part of our demand is unable to start things going because there is no government organisation which can produce the necessary items, and also because of some wartime promise. In view of the fact that all our appreciations on future strategy hinge on the atomic bomb and on the dates when they will be available in quantity to ourselves and other powers, I would therefore ask your guidance as to the next step to take. The information which I have given above is all hearsay and I am at a loss to know what steps to take in order to get things moving.'

Of course a great deal had finally been happening, even though the Air Staff had not been made aware of any of it. An Official Committee on Atomic Energy had now been set up, and Dr Penney had already sent Lord Portal his proposals for atomic weapons development. A relatively unknown scientist during the war, William Penney had already completed a great deal of explosives research before he joined the Manhattan Project in America. As one of the most important contributors to the British aspects of the project (with extensive knowledge of blast and shockwave effects), the Americans made numerous attempts to retain Penney after the war but he declined their offers in favour of a return to Britain where he effectively assumed control of the embryonic atomic programme. Much of the preliminary development work was conducted under Penney's leadership, with a key RAF team being set up at Fort Halstead in Kent. Other establishments also became heavily involved in the project, most notably the MoS establishment at Aldermaston (where the bombs would ultimately be assembled for the RAF), Springfield (from where uranium would be prepared) and Farnborough (where the design of the bomb carcass would be developed). Of course production of plutonium was a fundamental part of the programme, and it was this requirement that led to the construction of the first British atomic pile at Windscale in Cumbria (a site now better known as Sellafield).

The time-scale required to set up Windscale and to begin production of plutonium was effectively the key element that set the time-scale for the development of the British bomb. The international political situation began to look increasingly troublesome, with the very beginnings of the Cold War taking place in 1948. The communist take-over of Czechoslovakia, deployment of USAF B-29s to the UK (with atomic bomb capability) and the restriction of Western access to Berlin all conspired to create a period of great tension, and when the USSR detonated its first atomic bomb in August 1949 there was certainly no longer any doubt in British minds that the decision to proceed with an atomic bomb programme had been a very sensible move. A third atomic pile was assembled and governmental eagerness to expedite the completion of the bomb programme was very evident, as illustrated by comments made by the Prime Minister:

'I attach to this expanded programme the same high degree of importance and urgency as I attached to the original. I hope nothing will be allowed to interfere with its realisation.'

However, it was not until early 1952 that the first plutonium became available, enabling the scientists and military staff to examine the issue of when and where the device would be tested. Despite the continually changing relationship with America, it was still envisaged that US test facilities in Nevada would be used for the first British test, and the US invited Britain to submit a formal request on this basis, but delays in responding, combined with many restrictions on the parameters of the test, effectively meant that the Americans would have been conducting the bomb test on Britain's behalf, rather than handling the project as a truly collaborative effort. One (un-named) State Department official commented that if he was British he 'would turn them down flat, would proceed to carry out the British test in Australia and that this would be the best possible thing for Britain, the British Empire and indirectly for the USA.'

Indeed, an area close to Australia was ultimately selected as the test site, although consideration was first given to a Canadian site in Manitoba where Penney could test the weapon in shallow water. His team believed that this would provide useful data on the possible effects of an enemy attack being conducted through the covert transportation of a weapon inside a sea vessel, and, given Britain's reliance on various ports and sea power, it was an idea with some merit, especially as the Americans had not made any similar tests. Unfortunately the Canadian site was too shallow to enable the project ships to reach the shore, and although a potential test area much closer to home (in north-east Scotland near Wick) was also briefly considered, it was the Monte Bello Islands, 50 miles from the coast of Australia, that were eventually selected. With a channel some 6 fathoms deep just off the shoreline, and with only seasonal human inhabitation (from pearl fishermen), the islands were a near-perfect solution. It certainly represented a much better option that a deal with the US, as the British Joint Services Mission in Washington expressed:

'When the British team arrived over here they would be subjected to so many petty restrictions and there would be so much red tape that in effect the Americans would explode our weapon for us and let us have only those results which they felt they could safely divulge.'

It was during 1950 that a date in October 1952 was fixed for the first test detonation, code-named Operation 'Hurricane'. HMS *Plym*, a retired Navy vessel, was selected as host for the plutonium device, and the various assemblies were loaded onto the ship at Shoeburyness in June 1952 before setting off for Australia. The all-important plutonium was finally delivered from Windscale to Aldermaston, where it was fabricated and transported to RAF Lyneham inside (rather comically) a furniture van. The plutonium core (in two sections) was then flown in a Hastings to RAF Seletar in Singapore, from where an RAF Sunderland completed the delivery to the Monte Bello Islands. The various components were then assembled on board HMS *Plym* while scientists set up a variety of test and recording equipment around the site. In addition to cameras, there were suitably positioned pieces of Lancaster and Spitfire fuselages and wings, some 200 petrol cans and toothpaste tubes, paint samples, clothing and thermometers, all designed to record every aspect of blast and heat effects. The Radiation Hazard Division constructed a variety of rather crude (but nonetheless effective) air sample filters, many fashioned from household vacuum cleaners.

Finally, at 09:30 on 3 October 1952, the device was detonated by cable from the nearly island of Trimouille. Ten seconds after detonation the ground observers emerged to take a look at the results of their efforts. Expecting to see the familiar mushroom cloud that had already become a familiar sight through newspaper images, they were somewhat surprised to see a rather odd cauliflower-shaped cloud, which (thanks to its positioning at sea level inside the ship) had sucked up huge amounts of mud and ground debris, which was billowing upwards and sideways. Many uninformed observers assumed that this was a result of some unspecified design modification made by Penny; indeed, more than a few members of the assembled team were surprised that the device had exploded at all, having become increasingly pessimistic during two months of frantic preparation.

HMS *Plym* had, of course, been vaporised, and after just four minutes the cloud of debris had reached 10,000 feet. One observer recalled that

> '...there was a blinding electric-blue light of such intensity I had not seen before or since. I pressed my hands hard to my eyes, then realised my hands were covering my eyes and this terrific light was actually passing through the tarpaulin, through the towel and through my head and body.'

Winston Churchill (who had by now returned to power as Prime Minister) sent a telegram to Penney that began 'Well done Sir William', and a simultaneous announcement was made by No 10 Downing Street that Penney had indeed been appointed Knight Commander of the Order of the British Empire. Inside the House of Commons, Churchill announced that

> '...the weapon was exploded in the morning of the third October. Thousands of tons of water and mud and rock from the sea bottom were thrown many thousands of feet into the air and a high tidal wave was caused.'

Britain had finally entered the atomic age.

CHAPTER TWO

Birth of a Bomber

WHILE the development of Britain's atomic bomb had been a long and sometimes uncertain business, often shrouded in uncertainty and confusion, it is also probably fair to say that the design and development of the aircraft destined to carry the bombs was equally complicated. Despite the fact that there were well-established procedures for identifying and selecting aircraft development programmes (whereas the bomb programme effectively had to be set up from scratch), the creation of a fleet of high-performance nuclear bombers was a brave step into the unknown.

As development of the atomic bomb progressed, attention was naturally directed towards the means by which the weapon could be delivered to its target. Although in retrospect it seems obvious that such devices would have to be carried by jet bombers, there was certainly no such conclusion at the time. Sir Arthur Harris speculated than 'an atomic exploder' could be brought into an enemy country piece by piece, to be assembled and detonated in almost any location where suitable cover could be found. Other commentators believed that the use of ships would be the ideal means of getting the bomb to its target, and it was this thinking that had influenced the first test-firing of the British atomic device as part of Operation 'Hurricane'. Perhaps only slightly more practical was the notion that a suitable missile could be built into which the atomic bomb could be incorporated, and the wartime Nazi development of the V-2 missile had undoubtedly been leading Germany's scientists towards the same idea. Germany had launched more than 4,000 V-2 rockets during the latter stages of the war and, when combined with more than 8,000 successful V-1 launches, it was clear that Germany had built up some considerable expertise in the use of liquid-fuelled guided missiles, whereas Britain had only the most basic grasp of small, solid-fuelled designs. Britain did conduct a fairly low-key investigation of captured German V-2s after the war, but American interest in the programme (and particularly the fates of the German rocket designers) meant that British interest eventually faded once the available sources of information had been explored.

Pictured during preparations for transportation to Hendon for preservation, the removal of XL318's nose section reveals the separation point for the forward cabin that would have been jettisonable, had the original Air Staff plan for a crew escape capsule been pursued. *Bruce Woodruff*

It was evident that the basic concept of delivering an atomic bomb by means of a ballistic missile was certainly sound, but on the basis of the V-2's design it was clear that there was no prospect of rocket technology being able to provide a short-term solution to Britain's requirements. The V-2's maximum payload was a mere 2,150lb, which was a long way from the predicted 10,000lb weapon being envisaged by the British scientists. Likewise, the V-2's range was woefully short, and there was little prospect of any rocket being designed with sufficient range to reach the heart of the Soviet Union, even if launched from British bases in Germany. Even the V-2's designer, Werner Von Braun, accepted that the V-2 was merely a high-altitude research vehicle that could only act as a stepping stone towards more ambitious designs. Von Braun was certainly right, and of course the V-2 eventually led to the creation of the mighty Saturn V rocket, which ultimately took NASA to the Moon.

However, British thinking was still very much centred on conventional approaches, which had been fostered through the years of the Second World War. Manned bomber aircraft were still regarded as the most practical means of delivering bombs to their targets, although the traditional four-engined piston-powered bomber was already reaching the end of its design potential. Despite the huge success and significance of the wartime Lancaster (together with the Halifax, Stirling and the twin-engined Wellington), subsequent development had only led to the Lincoln, a direct descendant of the Lancaster with an impressive range of some 2,250 miles, but still very much restricted in terms of speed and altitude performance, and certainly no match for the new fighter aircraft being produced by the Soviets. In many respects, the Lincoln was effectively obsolete by the time it entered service, and the RAF was already eager to look towards the jet age.

During 1950 the Lincoln was supplemented by a fleet of 87 Boeing B-29 bombers, which were supplied by America under a lease arrangement. Given the name Washington in RAF service, the B-29s were based at Marham and Coningsby and gave Bomber Command a significant boost in capability. With a range of more than 4,000 miles, the B-29 was obviously capable of carrying atomic bombs (having done so during the Second World War and during a variety of subsequent post-war test drops), but no official statement was ever made as to whether the Washingtons did in fact carry atomic bombs at any stage (which would have had to be provided by the United States), and it is unlikely that they ever did, although the very fact that the RAF was operating the B-29 must have given the Soviets a pretty clear indication of Britain's military ambitions. The Washingtons remained in RAF service until the end of 1954, when the new Canberra jet bombers began to emerge. Conveniently bridging a 'bomber gap' between the dwindling number of obsolescent Lincolns and the new Canberra, the Washingtons were a welcome addition to Britain's offensive capability, not least because they were effectively provided for free (as part of the post-war Military Assistance Programme), and placated growing political concerns about the situation in Korea.

With the arrival of the Canberra, the RAF had a new and very capable bomber aircraft, but of course it did not have the ability to carry a heavy and cumbersome atomic bomb all the way to the Soviet Union. During 1946 the Air Staff directly addressed the need to acquire a long-range bomber, essentially creating a truly effective and direct replacement for the Lincoln (and the Washingtons that had supplemented them in RAF service). Although at this stage there was no guarantee that the government would ultimately press ahead with the development of an atomic bomb, the Air Staff clearly believed that this was likely to happen, and even though the new aircraft would be designed to carry conventional weapons too, it was based around the assumption that its primary war load would be a single atomic bomb.

This Operational Requirement (OR230) called for a landplane capable of carrying one 10,000lb bomb (ie an atomic bomb) to a target 2,000 nautical miles from a base situated anywhere in the world. Because it would undoubtedly have to encounter enemy radar and defences, it would need to be capable of flying at high altitude (between 35,000 and 50,000 feet) and also be able to fly at high speed (some 500 knots). Although it would be expected to carry warning devices and defensive equipment, it would not (unlike the bombers that had previously been in RAF service) be armed with any self-defence guns, as this would add greatly to the all-up weight of the aircraft and therefore inhibit its speed and altitude performance. It was thought that the ability to fly both high and fast would be enough to evade enemy defences, and although this assumption ultimately proved to be incorrect, at the time that the specification was issued the decision to opt for an unarmed aircraft was made on current military appraisals of Soviet capabilities.

However, even without defensive armament the aircraft was expected to be heavy and cumbersome and, with an all-up weight not exceeding 200,000lb, was unlikely to be capable of operating from existing RAF runways, many of which were only 6,000 feet in length or even shorter (likewise, the bomber would probably have been incapable of fitting inside the RAF's wartime-era hangars). By December 1946 the draft requirement was eventually dropped when there was sufficient doubt that a bomber with such ambitious capabilities could either be successfully constructed, or successfully operated from existing facilities. But the basic concept of a fast and high-flying jet bomber had been established, and any notion of developing another slow and low-flying piston-engined bomber had finally gone for ever.

With the abandonment of OR230, another requirement was issued in draft form, this being OR229, which (although broadly similar to the earlier requirement) called for a bomber with a slightly less-ambitious performance. It would have to be capable of carrying the same bomb load (the 10,000lb 'special') but range would now be set at 1,500 nautical miles with a similar operational ceiling but an all-up weight not exceeding 100,000lb. As with the original requirement, the bomber would be capable of carrying conventional weapons if necessary, but fundamentally it would be designed in order to carry just one atomic bomb, the dimensions of which were still unclear but estimated to be roughly equivalent to the wartime 'Grand Slam' high-explosive weapon that had been carried by Lancasters. As with OR230, the aircraft would not carry defensive armament and (unlike earlier operational bombers) the crew of five would not be spread throughout the aircraft, but would be housed together in a pressurised compartment; even at this stage it was

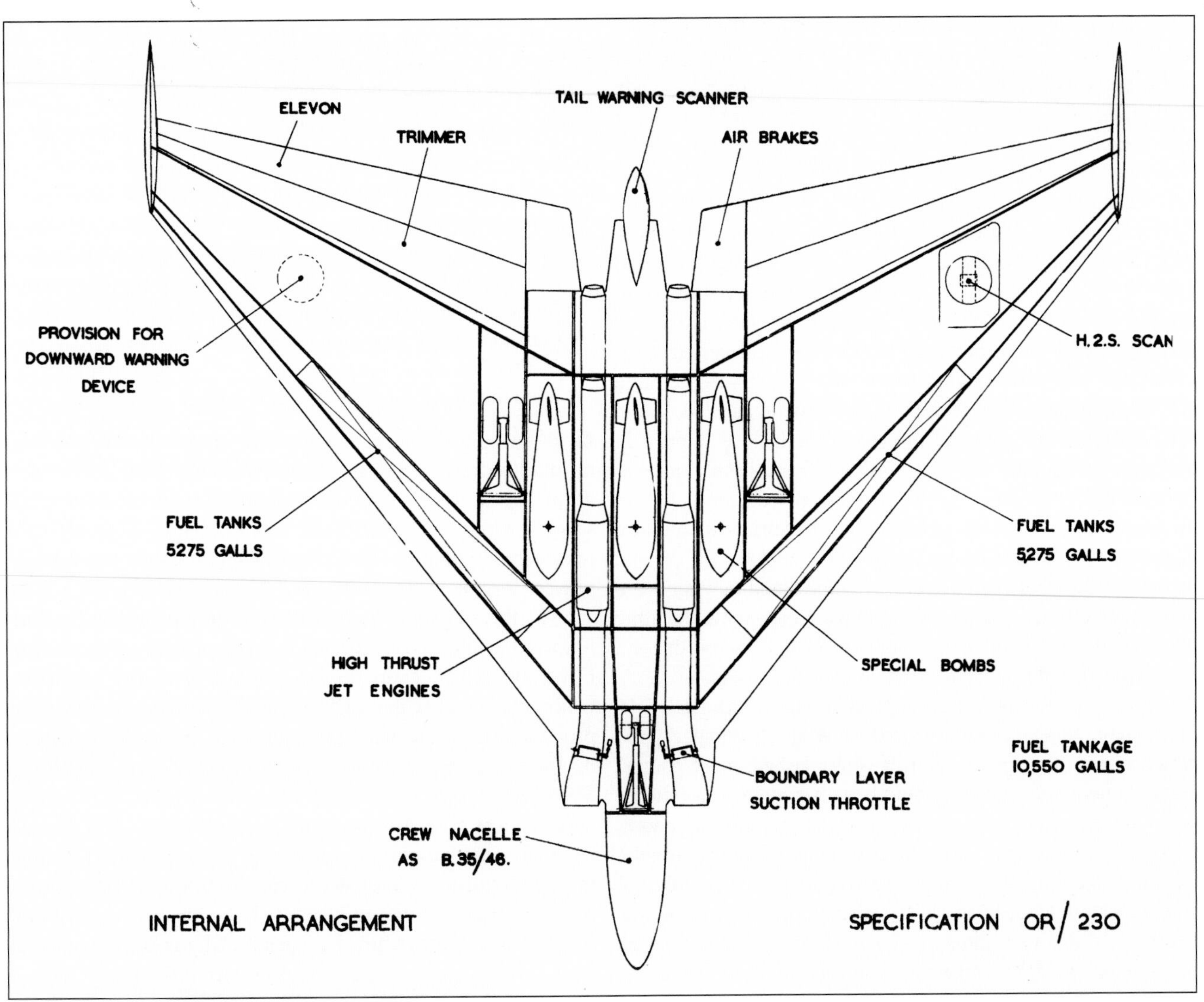

An Avro drawing of an early design for the aborted OR230 specification. In many respects the later 698 was a scaled-down version of this design.

proposed that this compartment should be jettisonable and fitted with parachutes for retardation during descent, although as an alternative it was specified that jettisonable seats could be fitted. The Chief Ministry of Supply representative, Stuart Scott-Hall, commented that

> '...the conclusion had been reached that the long-range bomber, the all-up weight of which would be in the region of 200,000lb, represented too great an advance in design to be entertained at the present juncture. Considerable research and development would be necessary – including, in all probability, the construction of half-scale flying models.'

He therefore recommended that 'consideration should be given to the medium-range aircraft, holding the long-range requirements in abeyance for a time'. He went on to propose a three-phase development: first, an 'insurance design', which would be a relatively unambitious aircraft created purely as a direct replacement for the Lincoln; second, the medium-range jet bomber; and third, the long-range jet bomber that would be regarded as a long-term project.

The approved OR229 was therefore issued on 7 January 1947, and the next day (when, by coincidence, the decision to authorise development of atomic weapons was also finally made), Scott-Hall began to send letters to Britain's aerospace manufacturers, setting out the terms of the requirement and inviting them to submit tenders. The first letter went to Handley Page, with others to Armstrong-Whitworth, Avro, Bristol and Shorts the next day, and finally English Electric some days later. On 24 January invitations were given to Avro, Armstrong-Whitworth, English Electric and Handley Page to submit formal design tenders meeting the new Specification B.35/46, although Shorts and Vickers-Armstrong also submitted designs. A Tender Design Conference was held on 28 July, and it was decided that an order should be placed for the Avro design, together with a smaller flying model. Additionally, it was agreed that either the Handley Page or Armstrong-Whitworth design should also be ordered (together with flying models), but only after further research had been completed on the designs by the Royal Aircraft Establishment (in its high-speed wind tunnel) at Farnborough.

Although all of the B.35/46 designs were ambitious and unusual, confidence seemed to rest primarily with Avro's submission, even though Sir Frederick Handley-Page had tried to persuade Scott-Hall to accept his design even before the conference, having sent a letter to him a few days previously explaining why he had turned down a delta design in favour of a crescent wing. Ultimately, after having given financial cover to Avro's design in November 1947, similar cover was given to the Handley Page design late in the following month. As for the fate of the other submissions, it seems that there was insufficient confidence that Armstrong-Whitworth would have the necessary design and manufacture capabilities to successfully complete its design, and both this submission and that supplied by Shorts were judged to be best suited to an unmanned and expendable design; however, given the huge cost of production, their fates were consequently sealed. While little evidence has emerged to explain the lack of interest in the Bristol design, the submissions from both Vickers-Armstrong and English Electric were deemed to be unimaginative, and of course English Electric was already heavily committed to production of the Canberra. Of course, it is also quite likely that as both Avro and Handley Page were unquestionably the 'big players' within the industry at the time, they had the distinct advantages of good design and production track records with which to back up their proposals.

Having established the very beginnings of the B.35/46 programme, attention also turned to Scott-Hall's 'insurance type', which resulted in Specification B.14/46, issued on 11 August 1947. Although this design would be rather more conventional with straight (instead of swept) wings and a significantly poorer performance (and therefore not much of an improvement over the existing Lincoln and Washington fleet), there was simply not sufficient confidence in any of the B.35/46 proposals to be able to afford the luxury of pinning all of the RAF's requirements on just one design. The worsening political situation created an increasing urgency for an aircraft capable of successfully delivering an atomic bomb, and although both the Avro and Handley Page designs looked promising, there was a very real fear that either (or both) might prove to be fundamentally flawed, or require a long period of design and developmental work during which Britain might ultimately be found in possession of an atomic bomb, but no aircraft capable of carrying it. Consequently, the 'insurance bomber' made good sense as it was almost guaranteed to provide the RAF with a functional bomber, even if it was one that was much less able that those proposed by Avro and Handley Page. The specification was based on the original requirements set out during the previous January, and stated that:

> 'The Air Staff require an additional aircraft built as nearly as possible to Requirement No OR229 but constructed on more or less conventional lines, so that it could go into service in the event of the more exacting requirement being held up or delayed an undue length of time.'

With less demanding requirements on both speed and altitude, but with weight kept down to 140,000lb or even 120,000lb if possible so that existing airfields could be used, the Advanced Bomber Project Group chose Belfast-based Shorts to build a suitable aircraft. However, even from the outset there was some doubt as to whether the Shorts design would even be capable of meeting the performance figures as set out in B.14/46. The Director of Military Aircraft Research and Development stated that:

> 'It has been apparent for some months now that the Short B.14/46 design will not quite meet the performance requirements written by DOR in OR.239 and incorporated by us as the Appendix B in Specification B.14/46. The advisory design conference on this aeroplane was held on 10 July and we are now fairly clear on the probable extent of the deficiency. I consider that Shorts have made the best job as they can of this design and it is no discredit to them that they have fallen a little short on performance.'

In essence, the Shorts design was based on the fuselage originally drawn up to meet the B.35/46 submission (albeit with the addition of a conventional tail), but with simple straight wings, each containing a pair of engines – in effect, a relatively simple Second World War-era bomber, but with jet power.

The three designs (two B.35/46 projects and one B.14/46 proposal) were given official ITP (Intention to Proceed) in December 1947, even though preliminary work had been taking place before then. However, although the procurement process already seemed less than sure, the situation was to become even more complicated following a visit made by Scott-Hall and colleagues to the Vickers-Armstrong headquarters at Weybridge.

Such was the lack of confidence in all three designs that there seemed to be a growing case for the production of a fourth design that would effectively be an 'interim-interim' project, fitting somewhere between the Shorts B.14/46 and the Avro and Handley Page B.35/46 aircraft. The Vickers-Armstrong Chief Designer, George Edwards, had been appraised of the situation and was asked to look at the possibility of creating an aircraft that would incorporate an all-up weight of 115,000lb, a tandem wheel arrangement, which would enable the aircraft to operate from existing runways, a total bomb load of 10,000lb, and the removal of the jettisonable cabin requirement. It was also agreed that English Electric should also be asked to look at revising its B.35/45 submission based on these new requirements. Consequently, the development of Britain's new atomic bomber looked even more uncertain than ever, as expressed by the Vice-Chief of the Air Staff (VCAS) Air Marshal Sir William Dickson:

> 'We have set the Ministry of Supply two main tasks in the production of replacements for the Lincoln. The first is a long-term replacement; a bomber which will have an approximate performance of 3,350nm range at 45,000ft at 500kt. In our specification we have said that it is desirable that the all-up weight of this type should not exceed 100,000lb and we have stressed that this aircraft should be able to operate from existing heavy bomber airfields. To meet these requirements it is inevitable that we must venture into revolutionary changes in aerodynamics. In other words the delta wing. At the current rate of research and development

A publicity photograph of the three 'V-Bomber' types that were operated by A&AEE (the Aeroplane and Armament Experimental Establishment, Boscombe Down) at the time the picture was taken.

it is unlikely that an aircraft of this performance will be ready for production inside eight years. As an insurance against the possibility that the firms in question will not be able to solve the aerodynamical problems involved in the production of this new type of bomber, we have asked the Ministry of Supply to build a bomber of conventional design with a reduced performance of not less than 3,350nm at a height of 40,000ft and a speed of 435kt. While this reduced requirement is less than we think to be essential, we cannot afford to have a replacement for the Lincoln which is already obsolescent if not obsolete. To meet our requirements for this "insurance" bomber, the Ministry of Supply have already placed an order with Shorts. We are not at all happy about this because from what we know, the Short design is very unimaginative and its estimated performance is already dropping below the Air Staff figures I have quoted above. From our knowledge of the work of this firm it is probable that the performance will drop still further, which is very serious bearing in mind that we do not expect to get even this "insurance" type into production inside six to seven years. We also know that since the Ministry of Supply have placed this order, two further designs have been submitted for this "insurance" specification; one from English Electric and the other from Vickers. From what the Air Staff know these designs are superior to that of Shorts. On the other hand these two alternative designs are based on a new jet engine, which is still on the drawing board, whereas the Short design employs an engine which is much further advanced in design.'

The increasingly complicated nature of the whole medium bomber programme was reviewed towards the end of 1947, and a few months later the introduction of a fourth design within the programme seemed to have been accepted, as indicated by an Air Staff report saying that

'…a complete review of the Bomber Programme has been made in view of its great importance. It has been decided that another type of bomber should be built to bridge the gap between the conventional medium-range bomber – the Shorts B.14/46 – and the two more advanced types which have been ordered from Handley Page and Avro – the B.35/46. Design studies were received from a number of firms and that of Vickers has been judged to be the most promising and a contract is about to be placed for prototypes of this aircraft. The Vickers medium-range bomber will have a still air range of 3,350nm carrying a bomb load of 10,000lb at a speed of about 465kt and height of about

45,000ft. It will weigh approximately 110,000lb and this will be distributed on a multi-wheeled undercarriage. The aircraft will be powered by four Rolls Royce Avon engines and will start with an initial sweep back of 20 degrees on the outer plane with the possibility of increasing this in future development to 30 degrees and later 42 degrees. The inner section of the wing is swept back to 42 degrees initially.'

The MoS gave Vickers an ITP notice in April 1948, followed by a contract for two prototypes of its Type 660 in February 1949, which were to be delivered to the RAF 'as early as possible'. Meanwhile, over in Belfast, work on the Short S.A.4 was already well under way. Shorts based much of the original design work on the hydraulic analogy tests performed in the company's seaplane tank at Rochester, not having the luxury of its own wind tunnel (the company normally relied on access to RAE and National Physics Laboratory facilities, but by the mid-1940s the waiting times for access had become prohibitive). In contrast to the company's earlier B.35/46 submission, the S.A.4 was extremely simple in layout, having a uniformly tapered wing with constant dihedral from root to tip. Likewise, the fuselage was of fairly straightforward construction and included a large Sunderland-style tail unit. In contrast, the engine layout was rather less orthodox, with a pair of Rolls Royce Avon turbojets mounted above and below each wing in huge one-piece nacelles, which, despite their bulky appearance, created surprisingly little drag. The advantage of this arrangement was that servicing could be completed easily and the wing construction could remain simple, thereby keeping the all-up weight fairly low.

Three production jigs were set up in Belfast, one for each prototype and a third for a static structural test specimen, and the first aircraft (VX158) was completed early in 1951. However, by this time the S.A.4 (by now named 'Sperrin') was already facing an early demise. The Air Staff continued to compare the predicted performance of the Shorts and Vickers designs, and as more knowledge of swept-wing design was established, it gradually became clear that work could go ahead on the Vickers design with relative confidence, thus making the Sperrin redundant. As AVM Pelly commented:

'At a meeting held at the Ministry of Supply on 11 October, I said that we could do without the B.14 for the following reasons. If the long-term planning dates to which the whole of our programme is aimed are still valid, there is every reason to hope that one of the B.35 designs will be available in time. We still need one earlier type with which to re-equip Bomber Command in order to practice the techniques involved in long-range operations at such high altitudes and to be ready at the same time as the special bomb. Nevertheless,

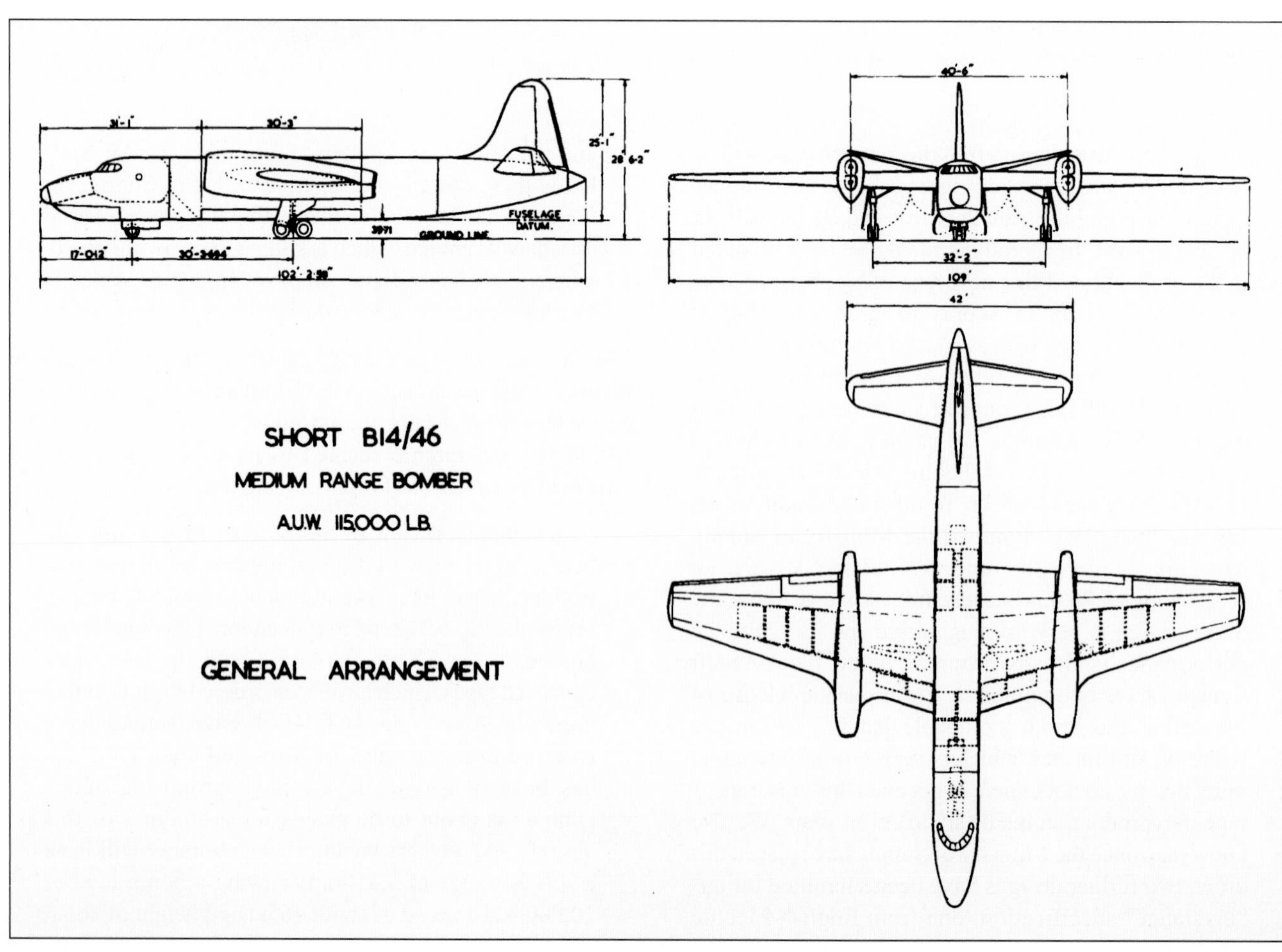

The Short Sperrin 'interim' bomber design.

only one type of aircraft would be required and I feel sure that the B.9, in view of its better performance, offers a far better solution to our problem, the only disadvantage being that it is six months behind the B.14. Although the B.9 is of more advanced design than the B.14, the increased knowledge gained lately on swept-back wings and other high-speed complications leads to the belief that no major troubles need to be expected with the B.9, and therefore production of that aircraft could start early in 1953 if need be and would, I understand, match up with the production of the special bomb.' (B.9 refers to Specification B.9/48 which was raised to cover the Vickers project.)

Subsequently it was agreed that the Sperrin should not go into production, but it was also agreed that the two prototypes should be completed and used for research and development work for the on-going bomber programme. VX158 made its maiden flight on 10 August 1951, some three months after the Vickers design had also taken to the air for the first time. After displaying at the 1951 SBAC show at Farnborough (resplendent in an unusual black, grey and red paint scheme), the Sperrin returned to Belfast and began a series of trials from Sydenham before moving to Farnborough, where it was employed on navigation and bombing development. The second Sperrin (VX161) flew during 1952 and became more directly involved in the B.35/45 programme as a host aircraft for bomb loading and drop trials with the AIEU (Armament & Instrument Experimental Unit) at Martlesham Heath, being based at nearby Woodbridge in company with the resident USAF F-84Gs, where it regularly operated to and from the Orfordness ranges. From here a variety of dummy bomb carcasses were dropped by the Sperrin, leading towards a definitive design of the 'Blue Danube' casing that would ultimately be carried by Valiants, Victors and Vulcans. Likewise, the Sperrin was also involved in development of the 'Blue Boar' guided bomb, which was also intended for the B.35/46 bombers, although this project was later abandoned (ultimately becoming 'Blue Steel').

The two Sperrins enjoyed a relatively short but varied flying career, providing valuable contributions towards the development of the Blue Danube bomb, the Blue Boar missile and the Gyron engine, as well as the design of landing gear eventually fitted to the Britannia airliner. It is certain that the Sperrin would never have matched the performance figures attained by the rather more advanced Vickers design, but given the relatively rapid development of engine technology and the eventual reductions in the predicted performance of the Vickers aircraft, there was certainly not much to separate them. But of course the Sperrins were never intended to be high-performance aircraft, and by definition they were designed to be simple, predictable and, most importantly, achievable designs on which the Air Staff could rely.

Perhaps the only irony is that the original design submitted by Shorts in response to OR.229 was possibly the most advanced design of all, based on the development of an unusual swept-wing layout that avoided the natural tendency of swept wings to twist under aerodynamic load by moving the wing's torsional box further aft. This 'isoclinic' wing was judged to be too imaginative and unpredictable for the MoS's taste, but despite being dropped from the original OR.229 selection process, Shorts embarked upon production of a one-third-scale glider employing the proposed wing layout. The completed glider was a great success and encouraged Shorts to construct a powered version, fitted with small Turbomeca Palais jet engines. This aircraft (the S.B.4 Sherpa) also served to prove that, despite the unusual nature of the Shorts wing design, it worked very well, and avoided many of the unusual aerodynamic difficulties created by the swept wing. With the movement of the wing's torsional box further aft along the fuselage datum, the wings were less inclined to flex and pull upwards towards the tips, thereby avoiding the usual problems with unpredictable (and unwanted) changes in the wing's lift properties in various configurations.

In some respects it may have been something of a shame that the MoS had not shown more faith in the original design, especially when so much interest was shown in the little Sherpa when it ultimately appeared at the 1954 SBAC display, and it was estimated that a full-scale version of the aircraft would probably have been more than capable of meeting (or even exceeding) the requirements set out by OR.229, but by then the future path of the B.35/46 programme was effectively fixed. However, the question of aeroelastic flexing on the swept wing was an issue that Avro had also been considering, and the company's interest in a 'flying wing' design also

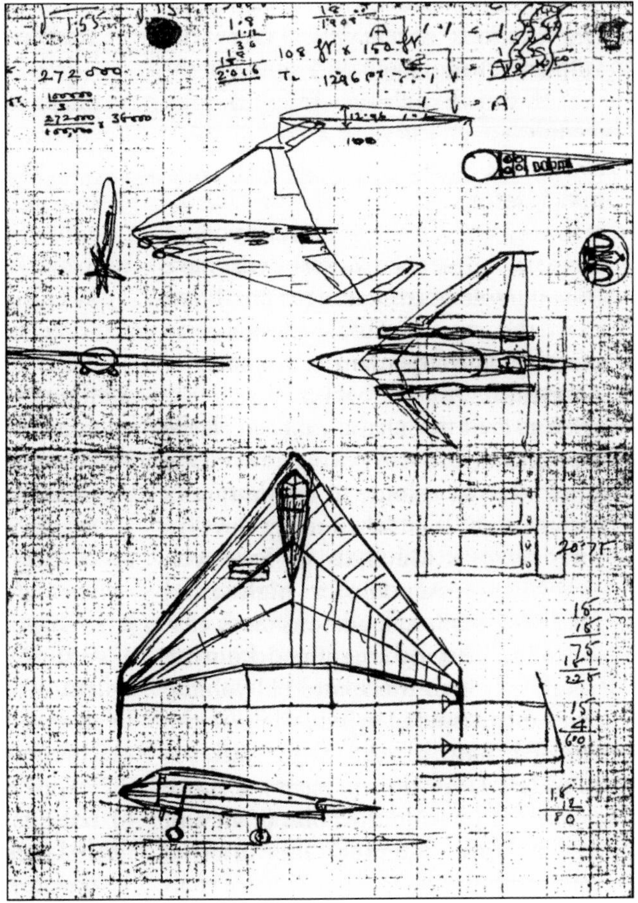

An undated sketch of the B.35/46 design project. The drawing is often credited to Roy Chadwick but the origins remain unclear; however, it illustrates Avro's early interest in a delta design.

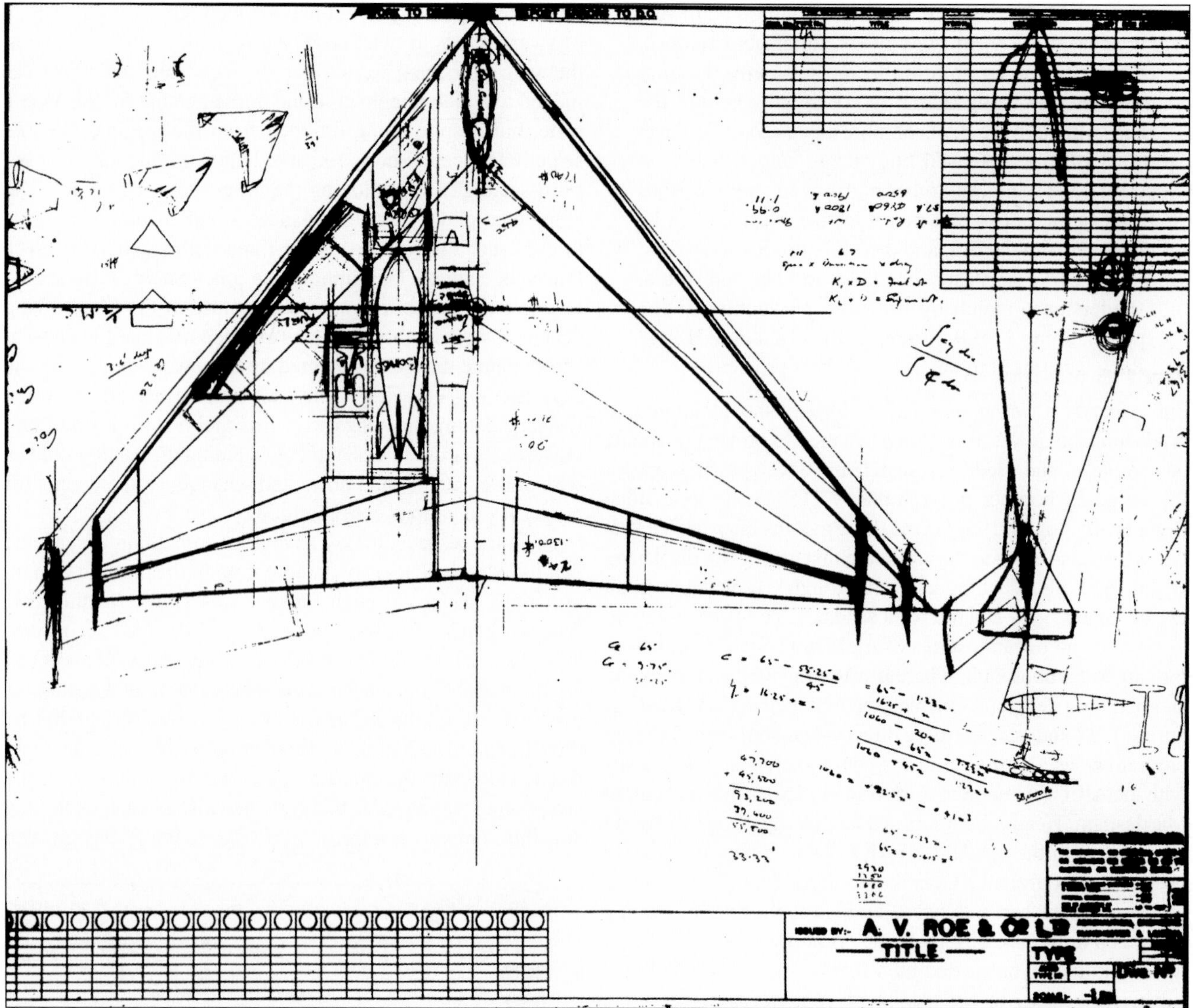

Certainly the best-known drawing to have survived from the Vulcan's design era, this illustration shows the simple delta wing design originally envisaged.

enabled it to address this potential problem, the layout of the delta wing being ideal for the creation of torsional strength across much of the wing surface.

Of course, it was the Vickers design that was ultimately adopted as the true 'interim' bomber. Although the original Vickers B.35/45 submission was regarded as possibly too simple (and therefore unlikely to perform as well as might be hoped), its relative simplicity was a distinct advantage when the Air Staff began to accept that basing all of its plans on two very ambitious designs might be unwise. Vickers already had an excellent history of aircraft production, having created types such as the Wellesley, Wellington and, more recently, the Viking and Viscount airliners, therefore it was unlikely that Chief Designer George Edwards would submit a design that was not capable of being translated into a reliable (and readily available) aircraft. Edwards argued that the relative simplicity of his revised B.35/46 design was its main virtue and it would enable the RAF to be equipped with an adequate (if less than ideal) aircraft in time for the arrival of the first atomic bombs, far in advance of the expected arrival of the advanced Avro and Handley Page designs.

The Vickers design was modern but fairly conventional, and did not require a long research and development programme, including the expensive and time-consuming creation of scale models. Edwards even went so far as to guarantee a production timescale for the Type 660 (eventually named 'Valiant' after a company competition to select a name, although 'Vimy' was another favoured option), promising that a flying prototype would be ready in 1951 followed by a production version in 1953 and quantity deliveries in 1955. Once the 'interim' Specification B.9/48 was issued and an ITP given to Vickers, development of the Valiant began, effectively ordered 'off the drawing-board' to save time (indeed, the airframe design was well established even before wind tunnel tests had been conducted). The first prototype made its first flight from the (then) grass airfield at Wisley on 18 May 1951, piloted by Joseph 'Mutt' Summers, who had previously conducted the first flight of the legendary Spitfire.

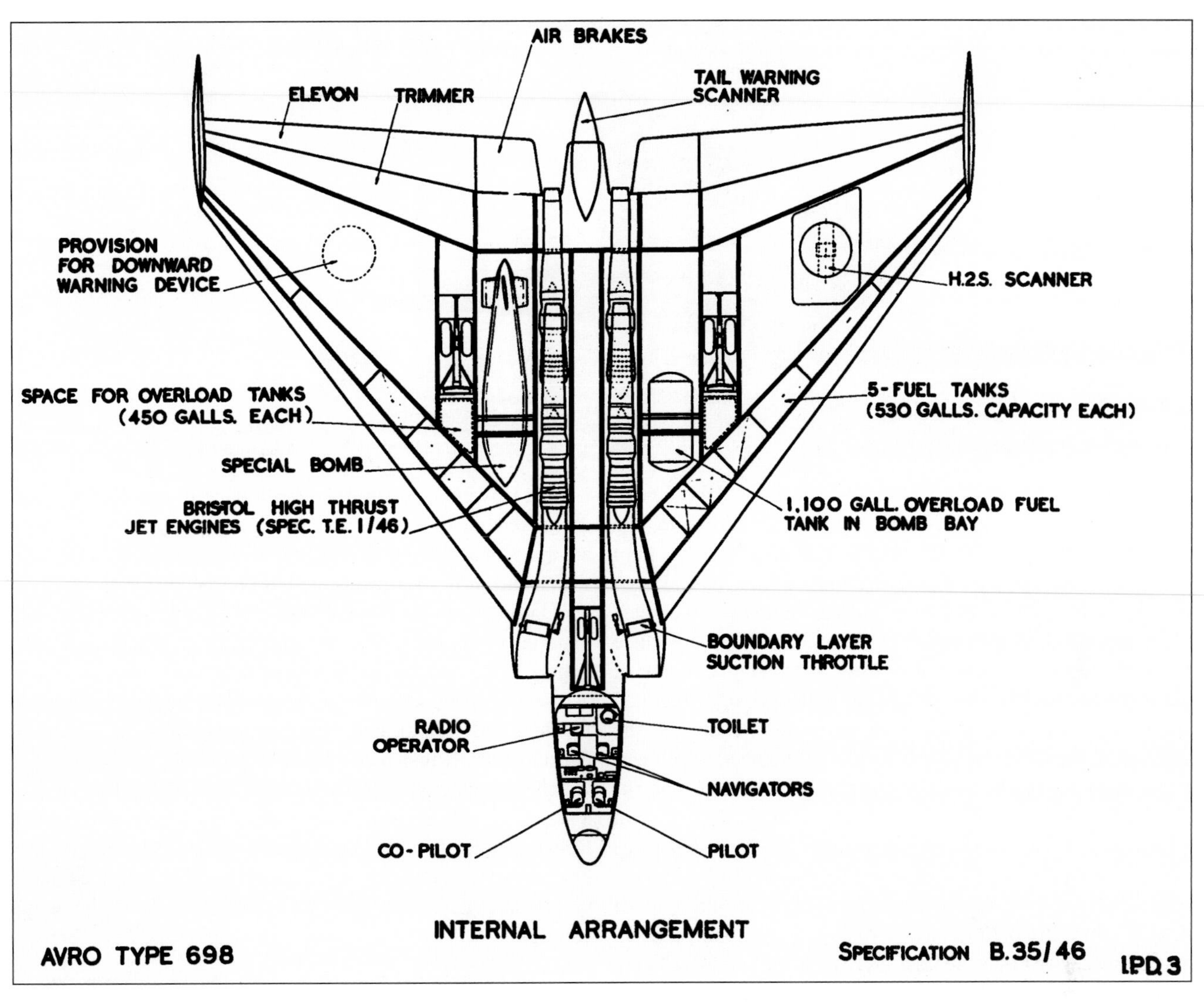

Above: An early Avro drawing of the initial B.35/46 submission.

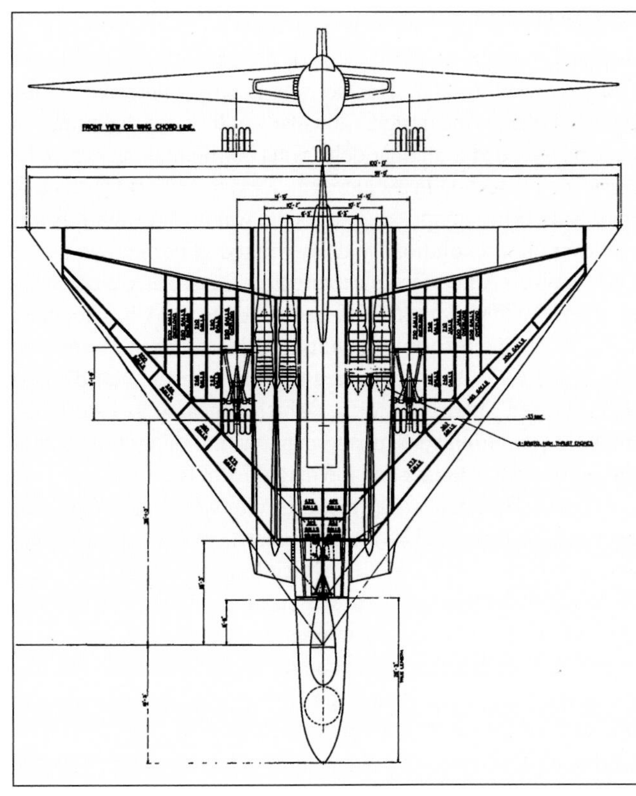

Right: An October 1948 drawing of the progressing B.35/46 design with revised intakes and a relatively large forward fuselage.

The proportions of the huge Valiant were dictated by the dimensions of the bomb bay (which, in turn, was proportioned to accommodate the Blue Danube bomb), with a huge 'backbone' member supporting the fuselage and bomb bay structures, onto which two right-angled branches formed the main wing spars. Although the basic structure was conventional in design, the Valiant did introduce some innovations, not least all-electrical systems, the only exceptions being the hydraulic brakes and steering. The overall design of the aircraft proved to be successful, and test flying continued without any difficulties, leading to the first deliveries to the RAF early in 1955, just as Edwards had promised.

With the introduction of the 'insurance bomber' now under way, attention turned to the two advanced designs that were still being developed, and which represented the ultimate requirements of the Air Staff. In Manchester, preliminary work on Avro's OR.229 design began in 1946, although the actual B.35/46 project did not actually get under way until January 1947. Under the leadership of Roy Chadwick (famous as the

The Lippisch LD-1, which significantly influenced the Vulcan's delta design.

creator of the legendary Lancaster bomber), the Avro design team looked at a variety of potential solutions to the requirements set out by the Air Staff, but ultimately kept returning to the concept of a 'flying wing' design, as had been developed by German designers in Nazi Germany.

A great deal of information had come from Germany since the end of the war, and there was a growing amount of aerodynamic data on the potential of tailless designs that had been pursued by Walter and Reimar Horten, leading to the creation of their H-1 glider and the Go229 bomber, which was about to go into production as the war ended. Although many of the Horten brothers' design drawings had been destroyed in anticipation of the Allies' advance, much still survived and it was estimated that the Go229 would have been capable of achieving almost 600mph at sea level and an initial climb rate of 4,000ft/min with a bomb load of 4,400lb and four 30mm cannon. It was impressive to say the least, and caught the interest of many British designers, not least the Avro team in Manchester.

Dr Alexander Lippisch was another leading German designer, famous as the creator of the Messerschmitt Me163 Komet rocket fighter. He too was deeply interested in the potential of all-wing designs, and many of his early experimental designs included triangular delta-wing configurations, which he succeeded in testing in a supersonic wind tunnel – a facility that only Nazi Germany possessed. As the war ended Lippisch was working on a revolutionary delta-winged glider known as the DM-1. When Allied forces captured his factory at Weiner Wald in 1945, the DM-1 glider was transferred to the USA, together with its designer. Lippisch continued his work in the United States and ultimately developed the XF-92A, the world's first delta-winged jet aircraft to fly (which appeared in 1948), and ultimately led to the design and manufacture of aircraft such as the F-102 Delta Dagger and the magnificent B-58 Hustler.

Over in Britain it was Armstrong-Whitworth who pioneered research in this field, with the construction of an all-wing research glider, the A.W.52G, which first flew in March 1945. Two larger designs were subsequently built, and Armstrong-Whitworth maintained an interest in the concept for many years. American interest also continued, and Northrop became particularly keen to develop the 'flying wing' concept into an operational bomber, eventually creating the huge XB-35, which, although successful, was ultimately abandoned by the USAF. However, Northrop's confidence in the 'flying wing' design remained, and after the passage of many years and many advances in aerodynamic design it re-emerged as the B-2 Spirit, which currently spearheads the USAF's offensive capability.

Unfortunately for Britain, most of the available aerodynamic material in Germany was secured by the Americans, thanks largely to the route taken by American forces as they advanced across central and southern Germany, where most of the research facilities had been located. However, a considerable amount of data did reach British hands together with some hardware, all of which was gathered by the RAE at Farnborough, culminating in a series of reports issued to British aerospace manufacturers. Robert Lindley (who later became Director of Engineering and Operations for manned space flight on the NASA Space Shuttle programme) describes Avro's early interest in the 'flying wing' concept:

> 'The Operational requirement for the aircraft was put into the Project Office in early 1947. Before we had managed to get very far the great fuel shortage hit us and the plant was closed down except for a few of us who huddled together in the offices along the front of the main building. If I remember correctly another upheaval was taking place at the time, and Chadwick was at home suffering from shingles. The performance requirements of the OR were rather startling to people nurtured on Lancasters and Tudors, and the only jet investigations we had made up to that time were for the Tudor 8 and the Brabazon 3 projects, and the latter was designed for a Mach number around 0.7. The original conception of the delta was not a result of spontaneous inspiration, but was arrived at by what seemed at the time to be an honest design study encompassing a whole series of aircraft, some with tails, some tailless, each type checked for a range of aspect ratios and weights. In retrospect I shudder to think just how much reliance was placed on the wing weight formula used, but the end product seems to justify the means. The first preliminary study was made for aircraft of aspect ratios not less than four and the result clearly showed that the aircraft required would be tailless and would give a much lower aspect ratio – probably about two. A second investigation covering the

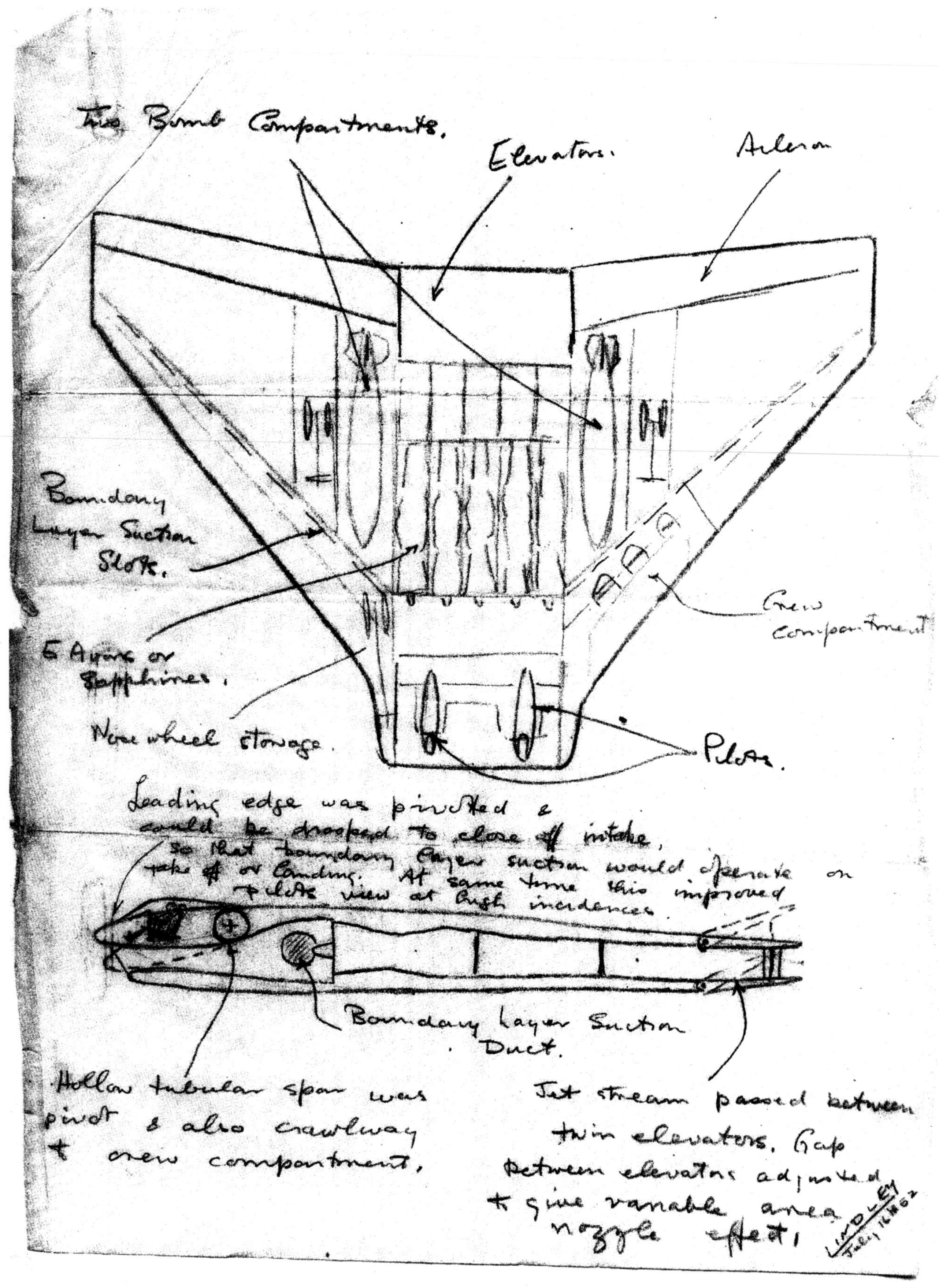

The scribbled sketch by Robert Lindley shows the original delta wing concept as first shown to Roy Chadwick. It is hardly surprising that Chadwick was initially rather surprised and almost dismissive of the radical design.

lower aspect ratios gave a solution 2.4, which was inevitably a delta wing. I knew that Lippisch had been working on a delta fighter – I had managed to see some reports on his coal-burning ram jet delta in Frankfurt during my trip to Germany in 1945 – and the possibility of making use of this configuration for a bomber was most intriguing. More elaborate checks were made, but only served to confirm the delta configuration.'

Lindley continues:

'The original arrangement of the aircraft was, of course, somewhat more advanced than that which was finally proposed in the original brochure; in fact I think it would still look advanced today. It had boundary layer suction combined with a movable cockpit so that the pilot could have good vision even when the aircraft was at 30 degrees incidence, and it had a very elaborate arrangement of combined elevator, air brakes and variable area jet pipe nozzle. Just about the time these first drawings were finished Chadwick recovered from his illness. I must say that he was considerably shaken to see the proposal – he had left the project as a sort of jet-propelled Lincoln and returned to find something apparently from a Buck Rogers comic strip in its place. He expressed his doubts very forcibly. I remember going home and sulking all weekend. I was very much in love with my project and couldn't stand the criticism. However, by Monday he had decided that it had its good points, and from there on he waded in with great enthusiasm and did much to make it a practical aeroplane. During this period of early development the aircraft underwent a number of changes. My original proposal had five Avons or Sapphires. In the interests of simplicity we decided to go for a twin-engined version, and for this we required an engine of around 20,000lb static thrust. Chadwick wrote around the engine industry for proposals for such an engine and the replies were very interesting, ranging from supreme optimism from Armstrong-Siddeley to complete pessimism from Metro-Vickers. I can recall Chadwick taking a 1/48th scale model of this twin-engined version up to London. The model was left with Air Marshall Boothman – Chadwick described how Boothman had "flown" it round the office, presumably making appropriate noises. At this stage we heard of the Olympus, and the four-engined version was investigated and adopted for the brochure. The other feature that was kicked around considerably was the crew accommodation. The first proposal had the crew compartment inside the wing, with the pilots under two fighter-type hoods. Then the requirement for a jettisonable crew compartment was emphasised, and we devoted much effort to getting the crew into the minimum nacelle, demountable just aft of the pressure bulkhead, and with a multiple parachute packed into the fairing aft of the canopy. This design was put forward in the brochure. In those days the radar scanner was installed inside the wing. The first issue of the brochure was finished, I think, in April 1947. After completing this we produced the drawings for the very beautiful 1/24th scale plastic model. We did some work on a civil version of the aircraft, employing a slightly higher wing loading and operating at a lower height than the bomber project. I remember it made a very attractive trans-Atlantic aircraft. Headwinds didn't seem to worry it too much.'

Avro's Roy Chadwick quickly became convinced that a delta-shaped design would prove to be the best solution to the Air Staff's requirement both in terms of aerodynamic efficiency (and therefore performance) and also in terms of structural integrity. However, he was also aware that the radical design would probably be a little too exotic for the tastes of the Air Staff to be considered as a serious contender. Certainly it was clear that, on the basis of German research and data, the delta wing was the right option, and the early Avro drawings also bore more than a passing resemblance to an earlier design submitted by Bristol ahead of OR.229. Likewise, Blackburn also submitted an early proposal, which may well have been the very first long-range jet bomber design to have been drawn up, and it too featured a delta design.

But although Avro's early design sketches were promising, the aircraft looked likely to exceed the maximum all-up weight figure by a considerable margin. Attempts to make the wing thinner simply pushed the weight still higher. With a specified cruising speed effectively fixing the basic design configuration, it was clear that the structure's weight would always be too high unless the payload was gradually reduced until a point where it no longer met the terms of OR.229. It was at this stage that the design team looked at the possibility of using the wingtips to provide longitudinal control, thereby making the cumbersome tail assembly redundant and saving a great deal of weight and aerodynamic drag. Thus the 'flying wing' concept began to look like the most logical solution to the Air Staff's requirement. With a 45-degree wing sweep, combined elevators and rudders (elevons), and small fins and rudders placed at the wingtips, the gross weight dropped to 137,000lb, and both Stuart Davies (who became Chief Designer) and Robert Lindley believed that they had finally identified the best design configuration, after having considered an endless variety of potential solutions over a period of more than a month.

Drawing on German data, Armstrong-Whitworth's experience, and Northrop's continuing development of the XB-35 (which was in the process of being converted into the jet-powered YB-49), the team determined that the all-wing design seemed practical but still rather too heavy. The obvious way to reduce weight would be to make the wing section deeper, but in order to achieve high subsonic speed the thickness/chord ratio had to be kept as low as possible, so the root chord was increased in order to provide compensation. Further attempts to reduce the wing's span while maintaining the same wing sweep, thickness and overall area meant that the 'missing' wing space had to be effectively relocated in the space between the wing's trailing edge and the fuselage. By March 1947 the design team had followed this process to its logical conclusion and concluded that the delta-shaped all-wing design was the logical choice, but even more weeks of research were devoted to consideration of how the air intakes, radar, crew compartment, fuel tanks, engines, undercarriage and weapon

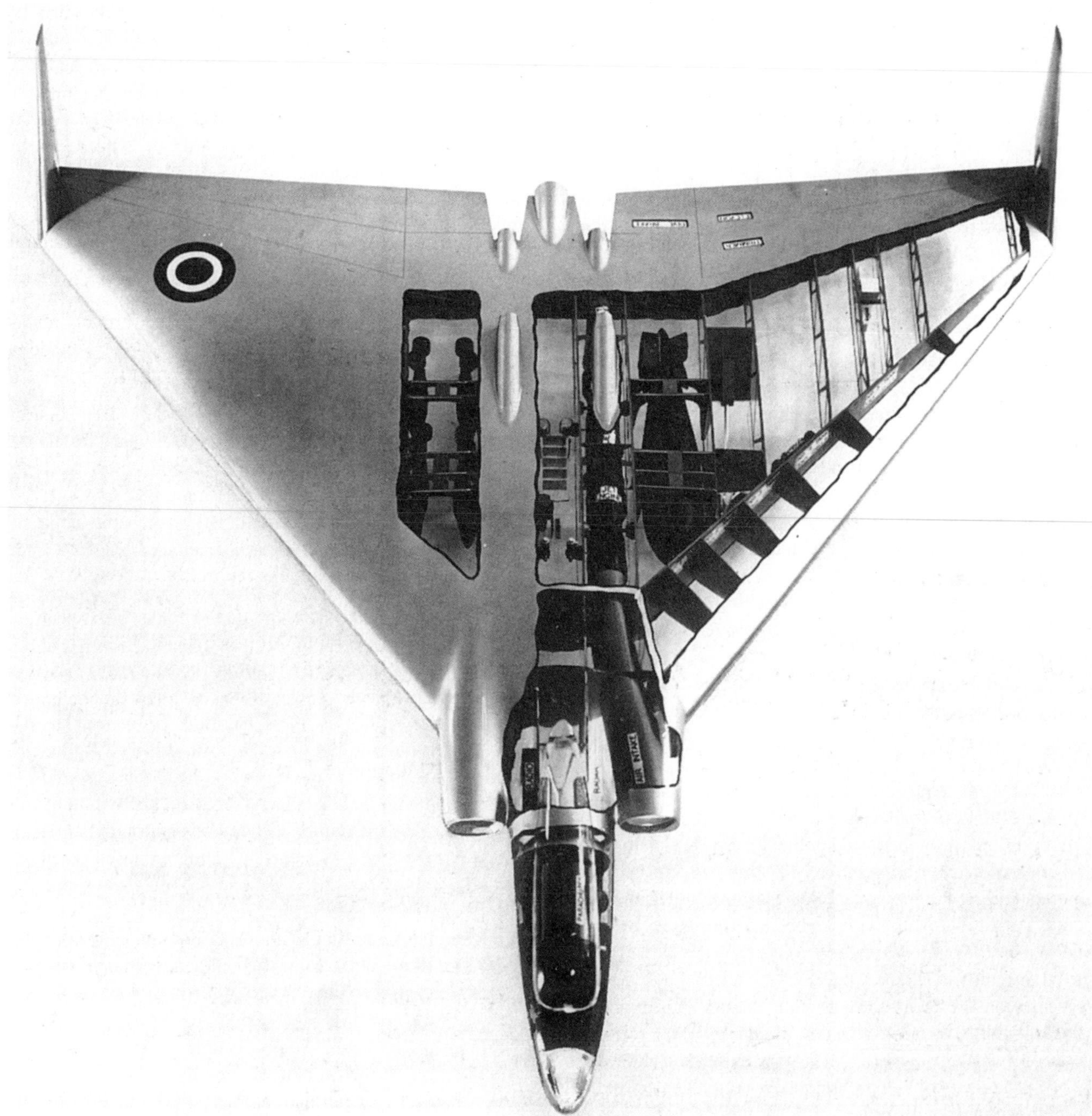

Avro's early 698 promotional model, which illustrates the very unusual twin bomb bay configuration that was originally favoured by the design team. The engine exhaust arrangement is also rather unconventional.

load could be incorporated into the design. Thanks to the relatively deep wing section (which was still sufficiently reduced to meet performance figures), all of the necessary equipment could be fitted neatly into an all-wing configuration, including the crew compartment (which resulted in two identical fighter-type blister canopies for Captain and Co-Pilot).

However, the Air Staff's subsequent specification for a jettisonable crew compartment required a fairly radical redesign of the bomber, with the crew being relocated in a pressurised nacelle positioned ahead of the wing leading edge so that it could be separated by means of a bulkhead. This effectively destroyed the 'flying wing' configuration and formed the basis of the design that eventually went into production, but it was with some irony that the plan to provide a jettisonable crew compartment was later abandoned, after it was established by Avro, Handley Page and Vickers that the complexity, time delays and overall coast of such a feature meant that it had to be dropped in favour of ejection seats for the two pilots, leaving the rear crew without any means of assisted evacuation. Consequently, the extended nose section, which became so characteristic of the Avro design, was probably never actually needed, and the design team could have (with hindsight) continued to develop a true 'flying wing' layout.

Having established the position of the crew compartment, the air intakes emerged ahead of the wing leading edge and were positioned each side of the short fuselage section. Each intake was routed through the wing and bifurcated into a pair of vertically stacked trunks, each feeding an engine with the upper engine exhaust emerging through the wing leading edge a few feet ahead of the lower exhaust. The huge bomb bay for the 10,000lb bomb was positioned within the inner section of the wing, but (unusually) was not positioned on the centreline, as has always been the case with previous bomber designs. Part of Avro's innovative approach was to create two identical bomb bays each side of the aircraft's centreline, although there must have been some debate as to the aircraft's predicted handling and trim qualities when a 10,000lb weapon would have been dropped asymmetrically.

However, when further research was conducted in Farnborough's wind tunnels it was determined that the wing section was still too thick and, after much discussion, the Avro team accepted that the thickness/chord ratio would have to be reduced still further. This meant that the engines and intakes could no longer be stacked vertically within the available space, and they were consequently repositioned in a side-by-side configuration. This meant that there was now insufficient space to accommodate twin bomb bays, so a more conventional single bomb bay was relocated to the aircraft's centreline. The undercarriage was initially envisaged as being similar to that employed by Boeing on its B-47, with main centreline-mounted gear and smaller outrigger wheels at the wingtips, but once the bomb bay had been fixed in a more conventional position, a traditional tricycle undercarriage could be comfortably incorporated into the design.

Much consideration was given to the idea of using moveable wingtips in place of traditional ailerons, moving the wingtip fins slightly in-board in order to accommodate power and drive units. By September 1947 this concept had been abandoned after it was concluded that individual elevators could be fitted to the wing's inboard trailing edge, with ailerons on the outboard sections. This effectively meant that the wingtip fins would now be better replaced by a larger one-piece fin on the centreline, especially after concerns were raised that the wingtip fins might not provide sufficient control. It was also agreed that a more traditional fin section would provide a convenient mount for an equally traditional tailplane unit, should further aerodynamic research indicate that one was needed. It is interesting to note that if this had been the case, the bomber would have looked like a scaled-up version on Gloster's distinctive Javelin fighter design, although the similarity was lost when the Avro team opted to remove the large, circular intakes and replace them with horizontal 'letterbox' intakes fitted within the extended wing leading edge, a move that seemed to make aerodynamic sense, even though little was known about the performance of intakes designed to this configuration.

During the summer of 1947 Chadwick had a series of meetings with Ministry of Supply and Air Staff officials in order to convince them that the revolutionary Avro design made sense and was being developed without undue difficulty. It was probably the fact that Chadwick had a well-deserved reputation within the aerospace industry that enabled him to maintain official confidence in the Avro design when it may well have been overlooked in other circumstances in favour of something rather more conventional. Tragically, Chadwick was killed when the Tudor 2 prototype crashed on take-off from Avro's airfield at Woodford on 23 August (it was later established that the aircraft's aileron control cables had been fitted the wrong way round). The shock of Chadwick's sudden death reverberated around the Woodford facility and the Chadderton factory, and there was a natural fear that the bomber programme would be abandoned without his leadership to push it along. Indeed, official interest in the design did begin to wane because there was a widespread belief that the design and development of the aircraft would inevitably suffer from Chadwick's absence, resulting in delays that would be unacceptable to the Air Staff. However, William Farren was appointed as Avro's new Technical Director, and as a former director of RAE Farnborough he had an equally formidable reputation, which he immediately used to restore faith in the Avro project. In order to explain the reasoning behind the Avro team's design conclusions, J. R. Ewans (who became Avro's Chief Aerodynamicist) wrote up a detailed account of the aerodynamic theories behind the choice of the delta wing early in 1951, and his report is as follows:

> 'So far as can be ascertained, the idea of using a triangular plan form for aircraft wings, now known as the delta wing, was first put forward in 1943 by Professor Lippisch, who will be remembered for his association with the Messerschmitt company. His studies led him to think that this form was most suited for flight at speeds in the region of the speed of sound where conventional designs were already known to be in trouble. By the end of the war he had a number of delta wing projects in hand, including an un-powered wooden glider intended to explore the low-speed properties of the wing. This was by then partly built and was later completed under United States orders. The idea of the delta wing was studied by many other aeronautical experts and a strong recommendation for its use was given, for instance, by Professor Von Karman at the 1947 Anglo American Aeronautical Conference. At the time of writing, three British delta aircraft and two American are known to have flown, and it is pretty certain that others are on the way. In the date order of their first flight these are: Consolidated-Vultee XF-92, Avro 707, Boulton Paul P.111, Douglas XF3D, and Fairey FD.1. With the exception of the last-named, which is fitted with a small fixed tailplane for the first flights, all the above aircraft are tailless.
>
> The following notes are intended to give a logical explanation of why there is this considerable interest in the delta wing, and just what advantages it promises the aircraft designer. But consideration must first be given to the type of aircraft the designer is trying to produce. The delta wing is of value only for very-high-speed aircraft, and at the present stage of engine development this implies the use of jet engines. When projecting his high-speed aircraft, the designer will attempt to produce something carrying the greatest payload over

the greatest distance at the highest speed, and for the least expenditure of power (ie using the least amount of fuel). This applies to all types of aircraft whether they be bombers, in which the payload is bombs, or civil aircraft, in which it is passengers or cargo, or fighters, in which it is guns and ammunition. The most fundamental factor determining the ultimate achievement is the height at which the aircraft flies. As height increases the density of air is reduced so that drag is less; it is possible to fly at a given speed at, say 40,000ft, for an expenditure of only one-quarter of the power required at sea level. The advent of the jet engine has enabled the aircraft designer to get his aircraft up to considerable altitudes and take advantage of the reduced drag, but a new factor is coming in to limit the speed of the aircraft. This is the speed of sound.

The speed of sound occupies a fundamental position in the speed range of aircraft. It is roughly 760mph at heights above 36,000ft. Because the speed of sound is of such importance, aircraft speeds are commonly related to the speed of sound using the term Mach number (the ratio of the speed of an aircraft to the speed of sound at the same height). As an aircraft approached the speed of sound – in fact for conventional aircraft when a speed of around 70 per cent of the speed of sound (Mach 0.7) is reached – the effects of compressibility become important, for the characteristics of airflow change fundamentally. There is a very large increase in the air resistance, or drag, and an excessive expenditure becomes necessary to increase the speed any further. For transport and bomber aircraft the speed at which the drag starts to increase (known as the "drag rise" Mach number) becomes the maximum cruising speed, because if the aircraft is flown at higher speeds the disproportionately higher thrust required from the engine means excessive fuel consumption and loss of range. At a rather higher Mach number there will be changes in the stability of the aircraft and in its response to the pilot's control, leading possibly even to the loss of control. In order to progress to higher speeds it is therefore necessary to design aircraft so as to postpone and/or overcome these effects.

We have noted that with an old-fashioned type of aircraft, ie that of jet propelled aircraft current in 1945, the limiting speed in steady cruising flight is likely to be a Mach number of 0.7 (higher speeds have of course already been achieved and a number of aircraft have exceeded the speed of sound, but only for short periods by either diving or by use of rocket power). However, from the knowledge now available, it appears possible, by careful aerodynamic design, to postpone the rise in drag until a Mach number in the region of 0.9 is reached, and this figure is likely to be the practical limit of cruising speed for transport aircraft of all types for many years to come. The designer of a civil aircraft, a bomber or a long-range fighter will, therefore, bend all his energies to achieving a Mach number of this order without any change in drag rise. In addition, he must pay attention to the changes in stability or lack of control which might occur in this region, and this will occupy his attention to the same extent as the purely performance aspect of the drag rise. It is quite easy to design a fuselage shape which is relatively immune from Mach number effects. It is the design of wings which is difficult, particularly since a wing that is suitable for high speed must also give satisfactory flying properties at low speeds for take-off and landing. As air flows past a wing its speed is increased over the upper surface to a considerable extent and over the lower surface to a lesser extent, so that there is greater suction on the upper surface than on the lower surface. Thus, at whatever speed an aircraft is flying, the speed of the air around the wing will, in fact, be higher. In the case of an aircraft flying at a Mach number of 0.8 the speed of the air around its upper surface may be equal to the speed of sound or may easily exceed it. At this stage the airflow pattern round the wing will be considerably changed and it is in fact this change which gives rise to the drag and stability effects mentioned above. It is essential, therefore, to keep the velocity above the wing as little in excess of the speed of the aircraft as possible. There are four ways of improving the behaviour of a wing. They are all different methods of keeping down the air velocities and can all be applied simultaneously: 1) sweep back 2) thinness 3) low wing loading 4) low aspect ratio.

1) Sweepback – The amount of sweep back is measured by the angle at by which the tip of the wing lies behind the centreline. The extent of the gains possible from sweep back is very considerable, and sweeping back a wing may easily lead to a postponement of the compressibility effects by a Mach number of 0.1.

2) Thinness – Keeping a wing thin leads to a reduction in the amount of air that must be pushed out of the way by the wing, and this helps the passage of the wing through the air. The thickness of a wing is measured by the thickness/chord ratio, which is the maximum depth of the wing divided by its length in the line of flight. In the past the thickness/chord ratios of aircraft wings have ranged from 21 per cent down to perhaps 12 per cent. Now values of 10 per cent down to 7 per cent are becoming common.

3) Low wing loading – The wing loading is the weight of aircraft carried by a unit area of wing, measured in pounds per square foot. Mach number effects are postponed by keeping the wing loading as low as possible, ie by supporting the weight of the aircraft with a large wing area. This is particularly important for flight at high altitudes where the low air density puts a premium on keeping the wing loading low. In fact, flight at high altitudes becomes virtually impossible unless this is done.

4) Low aspect ratio – Aspect ratio is the ratio of the span of a wing to the average chord. For moderate speeds, a high aspect ratio, ie a large span relative to the chord, gives greatest efficiency. At high Mach numbers this consideration is no longer important; in fact, some alleviation of compressibility effects is given by reducing aspect ratio.

There is another reason for choosing a low aspect ratio. One of the disadvantages of sweeping a wing back is that the flying characteristics at low speed become poor. A typical symptom is that the wingtip of a swept wing stalls, giving violent behaviour if the speed is allowed to fall too low. Research has, however, shown that this bad characteristic of swept wings may be overcome relatively easily. Although almost any aspect ratio can be accepted with an un-swept wing, for wings of 45 degrees sweep-back an aspect ratio of little over 3 is the most satisfactory. There is yet a third reason for choosing a low aspect ratio – the behaviour (as regards stability etc) in the high Mach number region. Compressibility effects are minimized and a transition from speeds below that of sound to the speed of sound and above is much more readily accomplished if the aspect ratio is low, say in the order of 2 to 4. Put the above requirements together, and the result is an aircraft highly swept back, with a thin wing, a moderately large wing area and a low aspect ratio. A little consideration of the geometrical properties and possible plan form of wings leads to the conclusion that the delta wing is the only form which satisfies these requirements. It possesses high sweep-back and low aspect ratio. The wing area will of necessity, be generous for the size of aircraft and, for reasons which will be detailed later, it is easy to build it with a low thickness/chord ratio.

Next, how does the delta plan form, indicated from considerations of aerodynamic performance, line up with practical design requirements and, in particular, the over-riding necessity for keeping weight and drag low in order to obtain maximum performance? A preliminary question is whether a tailplane is necessary. From the earliest days of flying, the question has been raised as to whether aircraft can be flown satisfactorily without a tailplane. Confining attention only to the case of the high-speed jet aircraft, each of the functions of a tailplane will be examined in turn, in relation to the delta wing aircraft. The functions are:

a) To trim out changes of centre of gravity position according to the load carried and the consumption of fuel. Investigation shows that a control surface at the trailing edge of the wing, provided that the latter has a large root chord (as has the delta), can cater for all but extreme centre of gravity movements.

b) To deal with trim changes due to landing flaps, etc. With the low wing loading associated with the delta wing, take-off and landing speeds are moderate without the use of flaps, and this question does not therefore arise.

c) To deal with loss of stability or control power consequent on distortion of the wing structure at high speed (Aerolastic Distortion). At very high speeds, all aircraft structures distort to a greater or lesser extent under the high loads imposed, and this distortion alters the aerodynamic form. In extreme cases this leads to loss of stability or control power, making the aircraft dangerous or impossible to fly at high speeds. An aircraft with a high aspect ratio, swept-back wing would need a tailplane to deal with this, but the shape of the delta wing makes it extremely stiff both in bending and in torsion, and a tailplane does not appear necessary.

d) To provide for spin recovery. Although this point has not been proved, it is expected that the controls on a delta wing would not be powerful enough to ensure recovery from a fully developed spin. A tailplane appears to be the only way of dealing with the problem. This restriction is of small significance for transport or bomber type aircraft. It can therefore be concluded that for a delta wing aircraft of the transport type, a tailplane is unnecessary. Its deletion leads immediately to a considerable saving of weight and drag, and to a major gain in performance.

We have now shown that, compared with a conventional aircraft, the delta wing aircraft will be simpler by the omission of the following items: the tailplane, the rear fuselage necessary in order to carry the tailplane, wing flaps, and other high-lift devices such as the drooped wing leading edge. There is a saving of weight, of design and manufacturing effort, and of maintenance when the aircraft is in service. These economies will have a considerable bearing on the initial cost and the manpower necessary to produce and maintain a number of aircraft. Because of its shape and the large root chord, the delta wing provides a large internal volume in relation to its surface area, even when using the thin sections, which, as noted above, are essential for high-speed aircraft. Simple calculations show that, for the same wing area, the delta wing has 33 per cent more internal volume than an un-tapered wing, while if the inboard half of the wing only is considered (as this represents a more practical case from the point of view of the aircraft designer) the internal volume of the delta wing is more than twice that of the corresponding tapered wing. It is found that without exceeding a wing thickness of as little as 8 to 10 per cent, it is possible on a moderate-sized delta wing aircraft to bury completely the engines, the undercarriage and sufficient fuel tanks for long range. The fuselage also has a tendency to disappear into the wing at the root. The result is the attainment of an aircraft consisting only of a wing, a fin and a rudimentary fuselage, representing a degree of aerodynamic cleanliness which has never before been reached. In fairness, it must be pointed out that this is achieved at the expense of a rather larger wing area than usual, but investigation shows that the drag of this area is less than that due to a conglomeration of items such as engine nacelles, tailplane, etc.

From the design point of view, the shape of the delta wing leads to an extremely stiff structure without the use of thick wing skins, and strength becomes the determining feature rather than structural stiffness. This avoids the inefficiency of conventional swept-back wings where the wing has to be made stronger than necessary in order that it shall be stiff enough. Summing up, it can be said that in order to meet the requirements of large loads for a long range at high speed, the high-performance transport or military aircraft of the future will cruise at a considerable altitude, at a speed not much below that of sound. The delta wing provides the only satisfactory solution to these requirements for the following reasons:

1) It meets the four features necessary for avoiding the drag rise near the speed of sound, ie it is highly swept back, it can be made very thin, the wing loading is low, and the aspect ratio is low.

2) Extensive wind tunnel and flight tests have shown that the low aspect ratio delta wing gives minimum changes in stability and control characteristics at speeds near the speed of sound.

3) In spite of the wing being thin, its internal volume is large, so that the engines, undercarriage, fuel and all the necessary equipment can be contained within the wing and a rudimentary fuselage.

4) Adequate control can be obtained by control surfaces on the wing, thus eliminating the need for a conventional tailplane. Together with item No 3, this leads to a considerable reduction in the drag of the aircraft and, therefore, to high performance.

5) Auxiliary devices such as flaps, nose flaps or slots, and the all-moving tailplane are unnecessary, thereby saving weight and design effort, and simplifying manufacture and maintenance.

6) The delta wing is very stiff and free from distortion troubles.'

Instruction to Proceed with the construction of two Avro 698 prototypes was given in January 1948, and although this obviously fell short of a full-scale contract for production aircraft, it was sufficient to allow the programme to continue. But before the prototype bomber could be completed, Avro began work on another aircraft design that ultimately became both a proverbial stepping stone and sometimes a hindrance on the path towards the creation of the mighty 698.

CHAPTER THREE

Baby Vulcans

SUCH was the concern that both the Handley Page and Avro bomber designs might ultimately prove to be too complex to actually complete, that the Ministry of Supply thought it prudent to construct smaller scaled-down versions of the two bombers in order to gain some real-time flight data on the basic design shapes, in advance of the actual full-sized airframes. Not content with the creation of two 'interim' bomber designs (in the shape of the Sperrin and Valiant), the MoS felt that there was a very real need for more research, which required the construction of yet another aircraft design. In order to explore the properties of the crescent wing design being developed by Handley Page for its bomber, a single research aircraft was ordered in the shape of the Handley Page HP.88. Constructed by Blackburn, the aircraft was in fact a Supermarine Attacker fitted with scaled-down versions of the bomber's crescent wings. It flew for the first time in June 1951 (from Carnaby), but crashed only two months later after a malfunction of a tailplane servo control. However, by this time the development of the full-size bomber was well advanced, and it was judged to be unnecessary to build another research aircraft.

Meanwhile Avro was asked to construct no fewer than three distinctly different research aircraft. The first – the Avro 707 – was to be a small and simple single-engined aircraft that would explore stability and handling of the delta wing at low speeds. The second aircraft – the Avro 710 – would be somewhat larger, powered by two engines, and would be assigned to research in the higher end of the predicted flight envelope, up to around Mach.0.95 and 60,000 feet. A third aircraft would be a full-scale, four-engined representation of the bomber but produced in a simplified form without the installation of all the necessary military equipment. It was by any standards a case of extreme over-caution, and the prospect of creating three very different airframes, while also endeavouring to continue development of the actual bomber, was too much to seriously contemplate, especially when Avro had originally been considering nothing more than a simple glider design on which low-speed delta research could have been conducted.

Eventually it was agreed that the full-scale 'stripped-down' 698 and the smaller Avro 710 were unnecessary, and would probably serve to draw resources and attention away from the main bomber programme, rather than constructively contribute towards it. As an alternative, the MoS opted for the construction of two complete 698 prototypes (which would provide all of the potential data that the simple 'stripped-down' version would have generated in any case) and a series of small (one-third scale), single-engined 707 aircraft under Specification E.15/48. In contrast to the Handley Page programme (which resulted in just one small-scale research vehicle), Avro agreed to build no fewer than three, with a pair assigned to low-speed research and another allocated to high-speed development.

Probably the best-known (and possibly the best-quality) photograph of VX784 pictured at Boscombe Down shortly after her first flight.

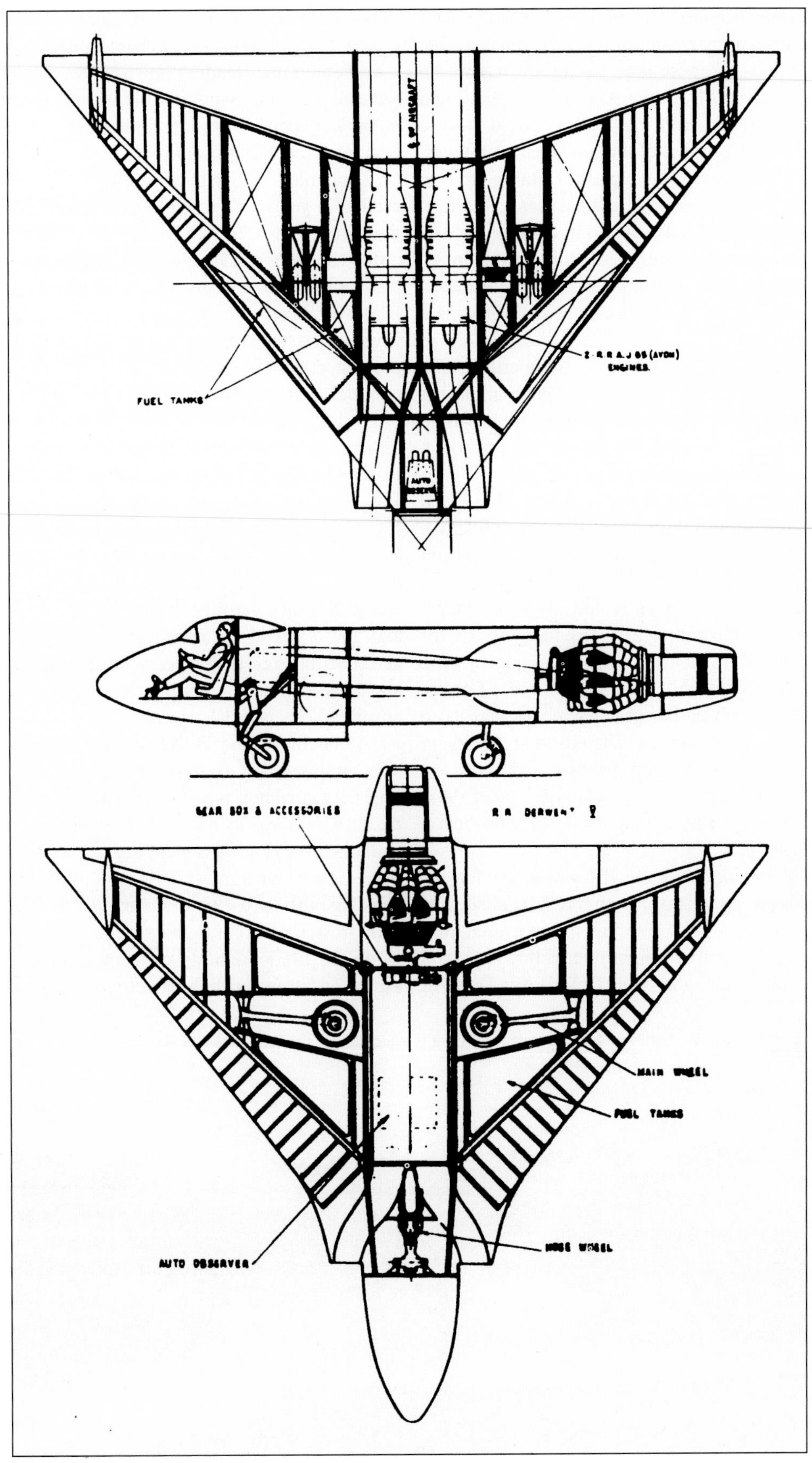

The original Avro 710 (top), which was subsequently abandoned, and the original Avro 707 design.

In principle, the decision to construct the 707s made good sense; they would be extremely simple in design and therefore relatively inexpensive. Likewise they would not require an undue amount of attention from the Avro design team, and would provide useful research data that could secure confidence in the ongoing design of the full-sized prototypes, and perhaps even speed up their development. But in order to realise the value of these research aircraft it was, of course, fundamentally important that they were (as much as physically possible) true scaled-down versions of the full-sized bomber, and would be available at the right time to allow the gathered data to be used in making any necessary changes to the machine tooling being set up for the full-sized 698. Unfortunately, in many respects the tiny 707s failed to meet these criteria and they ultimately became a separate development programme that continued in parallel to the main bomber programme, often diverting attention and resources away from the very project for which they had been designed to assist.

Specification E.15/48 (issued on 3 November 1948) called for a simple research aircraft with a top speed of 400 knots. Initially the Avro team proposed building the 707's wing from wood, but eventually it was decided that simple pressed sheet metal around two supporting spars would give the aircraft sufficient strength for the job, while providing space for internally housed fuel and test monitoring equipment. The canopy and nosewheel assembly were taken from a Gloster Meteor fighter, and the main landing gear assemblies were those designed for the Avro Athena trainer. The no-nonsense design was created at Avro's Chadderton factory, and work on manufacturing the aircraft commenced towards the end of 1948, with the new (scaled-down) delta wing inside its construction jig early the following year.

Although the main purpose of the 707 was to test the all-important wing design, the rest of the aircraft was created as simply as possible, with a short nose section manufactured from welded steel tubing covered with light metal alloy skin, and removable panels surrounding the single Rolls Royce Derwent turbojet engine, which was positioned below the swept fin and rudder. The engine intake was faired into the upper fuselage with a bifurcated duct feeding down above the wing spars. There were no powered flying controls and there was no ejection seat, but as a concession towards flight safety an anti-spin parachute was fitted in a fairing at the base of the fin assembly.

Unpainted, but with a prototype 'P' applied to the nose section in yellow (together with national insignia and serials), the diminutive 707 VX784 was assembled at Avro's Woodford airfield and subjected to a series of engine trials and short taxiing tests before being dismantled and transported to Boscombe Down in Wiltshire on 26 August 1949 for re-assembly. The first flight was scheduled for 3 September, but a 20-knot crosswind prevented the aircraft from flying that day. Having already completed some short 'hops' at Boscombe Down, test pilot Flight Lieutenant Eric Esler (Chief Test Pilot at the Aeroplane & Armament Experimental Establishment) taxied VX784 on to Boscombe's huge 10,000-foot runway at 19:30 on 4 September and proceeded to conduct a very satisfactory twenty-minute test flight in the local area. Esler was suitably pleased with the 707's handling, and Avro immediately issued a press release announcing the aircraft's first flight, claiming that the new delta would 'permit controlled flight at and above the speed of sound' and that it would be 'necessary to conduct a vast amount of aerodynamic research over a large range of speeds before application of the new configuration can be made to either civil or military operational aircraft'.

At the time of the 707's first flight, the new delta shape was seen as something exciting and experimental, and Avro often used the innovation as a promotional tool, even though the wing design was not really as radical as might have been implied. As explained previously, a great deal of work had already taken place on various aircraft designs in the United States, and by this stage the Avro 698 was making considerable progress, as was Gloster's Javelin fighter design, so in reality the new 707 would be predominantly concerned with only one specific aspect of the 698's flight envelope – its low-speed handling.

Because VX784's flying career was cut short, very few photographs exist of the aircraft, but this well-known image shows its simple construction and the Meteor-style nose section.

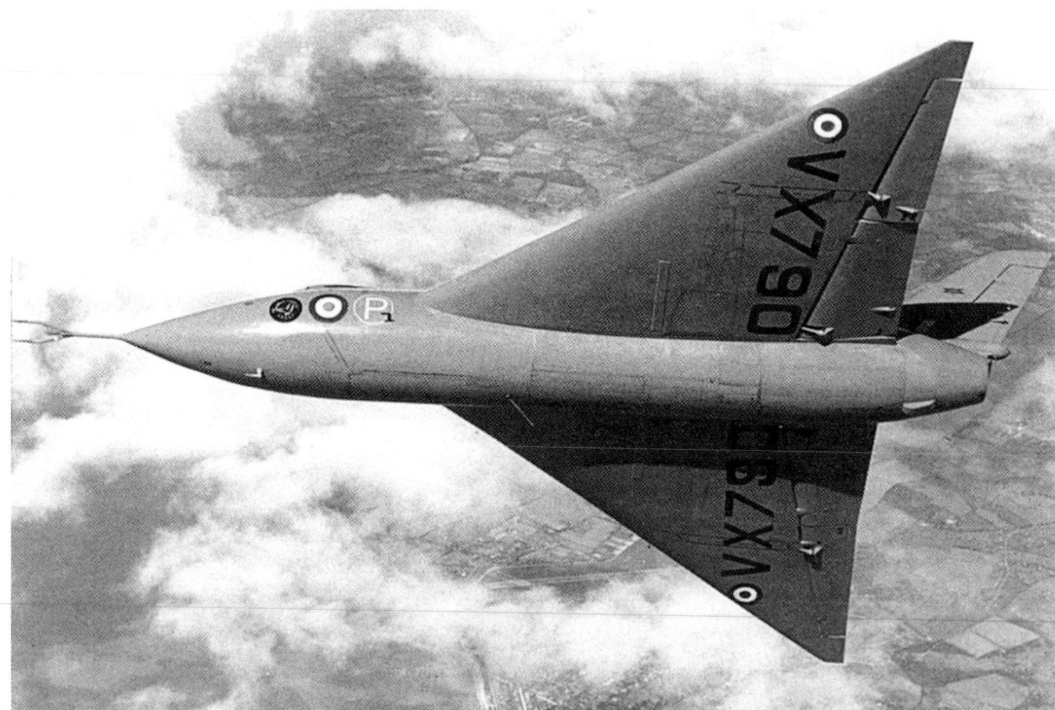

Right: VX790 breaking away from the camera aircraft during a publicity photography sortie.

Below: VX790 touching down at Boscombe Down, illustrating the alarming nose-high landing attitude peculiar to the 707 fleet.

Even in this respect, it was hoped that the aircraft would merely serve to prove the concept rather than highlight any deficiencies in the design that might require significant modifications.

The second and third test flights (totalling two and a half hours) effectively completed the 707's 'shake-down', and Esler then flew the aircraft directly to Farnborough where it was placed on static display at the 1949 SBAC air show. After returning from Farnborough, various items of test and data recording equipment were installed and the 707's flying programme resumed. The aircraft continued to handle remarkably well and Esler reported that the general characteristics were much the same as those found in more conventional aircraft with the possible exception of the take-off run, which required a considerable distance before the aircraft became airborne. All was proceeding well until 30 September when tragedy struck and VX784 crashed near Blackbushe, killing test pilot Esler. The cause of the disaster was never fully explained and many observers used the accident to cast doubt upon the delta design and the Avro 698 programme. However, investigation suggested that the cause was probably the failure of a control circuit, which may have locked the air brakes in their extended (open) position, causing a low-speed stall from which Esler (without an ejection seat) had no means of escape.

Two views of VX790 (above and below) illustrating the unusual dorsal air intake that was redesigned during the 707 programme. It was hardly representative of the wing-mounted 'letterbox' design ultimately used on the Vulcan, and added yet more complications and potential delays to the aircraft's development.

Work was suspended on the second 707 until more details of the accident were established, but there was much relief (despite the tragic loss of Esler) when the cause was determined to be something other than the design of the aircraft's wing. The Avro design team decided that the second 707 should have an ejection seat, and as the first Avro 707A (destined for high-speed research) was already under construction (in response to Specification E.10.49) it was decided that the most effective solution would be to take the nose from the 707A and fit it to the new 707 VX790, effectively lengthening the fuselage by some 12 feet.. In response to Esler's findings concerning the lengthy take-off run, a new (longer) nose gear assembly (from the Hawker P.1052) was installed, which the design team believed would lower the 707's 'unstick' speed, allowing the elevators to become effective rather sooner. This relatively simple act proved to be one of the most important contributions that the 707s made to the 698's development, as the full-size bomber's nose gear was lengthened, as a direct result of Esler's findings, while the design was still at the drawing-board stage. The wing air brakes were revised (although, rather oddly, not to the same design that had by now been approved for use on the 698) and the fuselage air brakes were deleted, but the revisions to the 707's design delayed its completion and it was not until September 1950 that the aircraft was ready to fly.

Emerging from the flight sheds at Avro's Woodford facility, the new 707B VX790 (painted in a bright blue colour scheme) was prepared for a first flight on 5 September. The pilot was to be Wing Commander R. J. 'Roly' Falk, Avro's recently appointed Chief Test Pilot, who had previously worked for Vickers-Armstrong and had also been RAE Farnborough's Chief Test Pilot. After completing a series of pre-flight checks, darkness was beginning to fall over Greater Manchester, so Falk elected to confine the day's activities to a series of hops along the runway, before making a successful first flight of some fifteen minutes' duration the next day. Following completion of the flight, Falk telephoned Avro's Managing Director Roy Dobson and Air Marshall Boothman, Controller of Supplies – Air, in order to obtain permission to immediately fly the aircraft to Farnborough, so that it could join the static line-up at the 1950 SBAC show. Arriving at the end of the day's flying, it was a testament to Falk's and Avro's faith in the design that VX790 made its public debut just a few hours after its very first flight.

After returning to Woodford, the 707B began test flying and was eventually flown up to a maximum speed of 350 knots. It was found that the air intake was suffering from a degree of air starvation because of the cockpit canopy immediately ahead of it, and modifications had to be made (following tests in the Rolls Royce wind tunnel at Hucknall) before the flight envelope could be extended further. Ultimately, the intake area was replaced with a completely new NACA Venturi design, which proved to be much more efficient. Minor oscillations in the pitching plane were also reported, but after investigation the cause was found to be out-of-phase movement of the elevators, and as the 698 would have powered flying controls, the problem was ignored. Although these minor problems were peculiar to the 707 (and therefore only served to distract attention from the 698 programme), the 707B did make a significant contribution to the 698's development when a series of flights aimed at investigating trim settings at different engine outputs led to the 698's engine exhausts being positioned downwards and outwards (in relation to the fuselage) to minimise trim changes. Perhaps most importantly, the 707B served to confirm that there were no fundamental flaws in the wing design, and there was no reason not to be fully confident in the 698's continuing development. It was probably just as well, because the 698 had by now established its own pace, and was making progress without any real input from the 707 test flying.

Capable of loops and vertical rolls, the sprightly 707B impressed Falk, who reported that the aircraft was very stable and showed no tendency to depart from normal flight characteristics; in fact, it was rather more stable in some respects, as Falk took the aircraft down to speeds below 100 knots, and up to angles of attack around 30 degrees, and it was clear that the aircraft showed no inclination to stall at angles that would have been well beyond those at which many other aircraft would have tumbled out of the sky. Without cockpit pressurisation, test pilot Falk was (according to an Avro press release) 'in the habit of taking a surfeit of oxygen for an hour or so before high-altitude flights, and then putting up with discomfort for short periods at upwards of 40,000 feet.' For reasons that no longer appear to be clear, some of VX790's test programme was conducted from Dunsfold, from where the aircraft continued to astound all who were lucky enough to fly it. The aircraft's rate of roll could exceed 200 degrees per second, and one pilot even managed to execute a complete inverted loop – not bad for a 1950 design.

When Air Marshal Boothman had an opportunity to fly the aircraft in September 1951 he was also greatly impressed and immediately issued an instruction that 'twenty-five selected pilots must fly it at once'. After completing around a hundred hours of test flying from Woodford and Dunsfold, the 707B suffered a landing accident and was withdrawn for repair, after which it was transferred to Boscombe Down to continue work on other research programmes. After suffering another landing accident at the hands of an ETPS student in September 1956, the aircraft was judged to be beyond economical repair and its flying career was brought to a premature end. The airframe was used as a spares source for other 707s and eventually the aircraft was moved to RAE Bedford, where it was dumped in 1960.

Attention was now redirected to the 707A, which was still under construction. There was some understandable doubt as to whether there was any point in attributing more resources and time to this (third) 707 when the 698 programme was so well advanced. Having proved the concept of the delta wing, there was clearly not much that the third aircraft could contribute other than further 'proof of concept'. After having discovered the deficiencies of VX790's air intake (and the time-consuming modifications that were necessary) it was finally agreed that the 707A (WD280) should be fitted with 'letterbox' intakes similar (but not identical) to those that had now been fixed into the design of the 698, although it was accepted that even if the 707's new intakes highlighted any necessary changes to their design, it would be too late to influence the construction of the 698 prototype.

Oddly, construction of the 707A continued, and WD280 was duly fitted with new wing-mounted air intakes, new true-to-scale

WD280 breaking away from the camera ship, revealing the delta wing layout and the 'letterbox'-style air intakes as eventually fitted to the Vulcan (albeit in a slightly revised shape).

elevators and ailerons, and servo tabs and balances to assist the manual flying controls. Painted in a very unusual 'salmon pink', the new 707A finally got airborne on 14 July 1951, by which time metal was already being cut on the 698 prototype. As predicted, despite some ninety-two hours of test flying, WD280 had no direct influence on the design of the 698; indeed, when Roy Ewans (who became Avro's Chief Designer) was asked just how much the design team had learned from the 707's, he replied, 'Not a great deal, apart from the reassurance that the thing would fly.'

Later in the 707A's test programme a series of flights was conducted to explore buffet boundaries, which required the pilot to apply a heavy amount of g-force at the greatest possible speed and height. It was quickly found that the airframe 'buzzed' (a high-frequency vibration) at speeds and heights that would easily be achieved by the 698. After a great deal of investigation, it was finally established that the 707's wing leading edge needed to be modified, reducing the angle of wing sweep inboard before increasing it outboard, producing a 'kinked' effect that would give the outer wing a greater chord and a mild leading edge droop. WD280 was suitably modified, but it was too late to make any changes to either the prototypes or the initial production 698s, as the unusual 'reverse envelope' jigs for the bomber's wing leading edge had already been constructed. Consequently, the wing modification had to be retro-fitted to the second 698 prototype, and sixteen sets of wing leading edges, which had already been constructed, had to be destroyed. It was this kind of absurd situation that plagued the 707 programme, which persistently lagged behind the 698's development instead of being ahead of it. At some stages the Avro team were forced to spend time making changes to the 707s in response to developments on the 698 – an absurd situation that clearly should have been the other way round – but the sad loss of the first 707 led to a mismatch of programme co-ordination that Avro was never able to rectify. Of course many of these difficulties were a direct result of the basic soundness of the 698's design, which never suffered from any catastrophic problems or long delays in the design process. With the benefit of hindsight it could be said that the 707s were an unnecessary and expensive diversion, but at the time that they were first ordered, they certainly seemed like a very sensible precautionary step.

Another view of WD280, illustrating its clean lines and an overall aerodynamic shape that was close to the ultimate Vulcan layout.

A second 707A was ordered in 1952 (under Issue Two of Specification E.10/49), but not as part of the 698's development programme. WZ736 (painted in a bright orange scheme) was constructed for the Royal Aircraft Establishment to conduct various trials that were not directly related to Avro's 707 programme. Flying from RAF Waddington for the first time on 20 February 1953, the aircraft spent most of its life at Farnborough and Bedford engaged on general tests duties, much of which centred on auto-throttle development, before being withdrawn in 1964. After spending some considerable time in storage at RAF Finningley, the aircraft now resides atthe Science and Industry Museum in Manchester.

The final 707 to be built was WZ744, an Avro 707C fitted with side-by-side seating under an enlarged fighter-type canopy. Four of these dual-control aircraft were initially required (two being part of Issue Two of E.10/49) as potential trainers for the RAF's bomber fleet, but as the 707 programme continued it became clear that the bomber would not suffer from any unusual handling vices and it would be perfectly practical to conduct conversion training on the actual aircraft rather than a scaled-down version, which would require its own manpower and technical back-up. Consequently, three of the aircraft that had yet to be constructed were subsequently cancelled, but WZ744 was completed at Avro's Bracebridge Heath factory, and upon completion the unpainted aircraft (carrying national insignia and serials) was towed the same half-mile along the A15 that WZ736 had followed to Waddington's northern perimeter gate, and prepared for flight to Woodford on 1 July 1953 in the hands of Squadron Leader J. B. Wales. Like the second 707A, WZ744 had no direct effect upon the 698 programme, but the aircraft enjoyed a useful career conducting a variety of tests flights (mostly connected with the development of fly-by-wire electrically signalled hydraulic flying controls) before being retired to storage at RAF Finningley in 1967. The aircraft now resides within the RAF Museum at Cosford, and although it was never actually operated in RAF service it continues to serve as an illustration of just how complex (and how important to the United Kingdom) the atomic bomber programme was.

After completing trials work for Avro, WD280 embarked upon a completely new career far from Manchester's rainy skies. Early in 1955, as a result of recommendations from the Commonwealth Aeronautical Advisory Research Council, the possibility of Australia obtaining an Avro 707 for research was first investigated. Discussions between senior government officials of Australia and the United Kingdom (which included input from Wing Commander Falk, test pilot Squadron Leader J. Harrison and Squadron Leader J. Rowland) were held on 21 March at Woodford, and in November it was decided that 707A WD280 would be provided for the programme. After having flown for more than a year on research duties for Avro (with

Left: WZ736 with elevators slightly raised as the aircraft slows behind the Lancaster camera ship for a photograph.

Below: WZ736 displays her simple but graceful lines at Woodford, with a slightly older Avro product visible in the background. Although much of the 707 programme served to merely draw resources from the 698 programme, the 707s did provide both Avro and other agencies with useful aerodynamic information.

A meeting of two closely related aircraft as WZ744 gets a hose-down ready for a public appearance at one of RAF Finningley's Battle of Britain 'At Home' days. *Joe L'Estrange*

A well-known publicity photograph of the twin-seat Avro 707C, illustrating the revised canopy and flat windscreen arrangement.

ninety-two hours of flying time 'on the clock'), the aircraft was withdrawn from flying to have powered controls fitted. Emergency equipment included a Martin Baker ejection seat, a jettisonable canopy hood, an emergency air system to lower the undercarriage, a Graviner fire extinguisher system for the engine, and an anti-spin parachute. There was also provision for reversion to manual control in the event of the hydraulic power system failing. At the time of delivery the aircraft had been painted pink, but it was subsequently repainted so that a suitably patriotic formation flypast could be made at the 1952 SBAC show, comprising the white-painted 698 prototype, flanked by the blue 707B and WD280 in bright red. As described, the need to overcome the high-speed and high-altitude buffeting had resulted in the complete redesign of the wing leading edge. In January 1955 the wing had been rebuilt by extending the leading edge forward by 19.5 per cent of the local chord from 80 per cent semi-span, going outboard to the wingtip. Going inboard, this forward extension washed out at 51 per cent semi-span. Flight testing confirmed the theoretical advantage of the design, and the new Phase Two wing was finalised.

Historian Denis O'Brien continues the story:

'A conference at Woodford was held in November with Avro chiefs, the Chief Test Pilot, representatives of the RAAF and Royal Australian Navy, the Department of Supply from Australia House and an officer from HMAS *Melbourne*. The main points discussed at this meeting related to transportation of the aircraft across the globe, the provision of spares, and arrangements for Squadron Leader F. Cousins to gather flying experience on the 707. With the RAN due to take delivery of their new aircraft carrier (the HMAS *Melbourne*), the transportation problems were easily solved. It was established that the 707 would just fit on to one of the carrier's aircraft lifts, which would enable the aircraft to be safely stowed in a hangar below deck. The question of engine spares didn't present a problem as the Rolls Royce Derwent Mk 8 was fitted to the Meteor aircraft that were still in service with the RAAF. It was during the conference that a rather more important point arose concerning some very evident handling and control characteristics that had become evident at high speeds. This was found to be due to the remote location of the jack that operated the system of rods that moved the control surfaces. The problem had been overcome in the dual-control 707C and the other 707A by the use of double control jacks that were closer to the control surfaces. This was yet another example of a design modification that was peculiar to the 707, which had no bearing on the 698, and simply created another headache for the Avro team to solve. WD280 was finally transferred to Australian ownership on 3 March 1956 after having completed just over 283 hours of flying time. The aircraft was flown from Woodford to Renfrew, from where it was transported to the King George V Docks at Glasgow. Loading on to HMAS *Melbourne* went smoothly and on 9 March the aircraft joined twenty-one Fairey Gannets, thirty-nine De Havilland Sea Venoms, two Bristol Sycamores and a single Meteor. Safely stored in C

WD280 pictured during her years of research flying in Australia. *Denis O'Brien*

A well-known but nonetheless impressive portrait of the 698 prototype shortly after her first flight in 1952.

Hangar, the long sail south began on 12 March via Naples, Malta, the Suez Canal, Aden and Colombo, then on to Fremantle and Melbourne before arriving off Sydney on 11 May 1956. A great deal of publicity surrounded the arrival but rather perversely the Prime Minister's Department advised the Aircraft Research & Development Unit that photography of the aircraft would be restricted, and that the interior of the cockpit and any exterior areas where access panels were open were strictly off-limits. More general photography would be permitted but not if any part of the aircraft's interior was visible – a strange decision considering the extreme simplicity of the 707's construction, which was geared towards the design of the exterior.

After the fuss of the official arrival in Sydney, the 707 was off-loaded onto HMAS *Sydney* to be transported back to Melbourne where disaster almost struck as the aircraft was finally taken on shore. As the aircraft was hoisted on to a low-loader, a metal support failed under the weight of the aircraft and a great deal of fuss ensued (ultimately resulting in parliamentary questions) until a second loader was provided with which to take the aircraft to RAAF Laverton. As a dedicated research aircraft, WD280 had several important features not normally found on conventional aircraft. In the starboard wing were five chord-wise rows of static pressure holes flush with the wing skin. These were designated chord A to E from the wing root to the tip. These pressure points (on the upper and lower surfaces) were connected inside the wing to a tube that ran to a pressure gauge. Whilst there were some 178 pressure points, most recordings were achieved using only 33 from each of the chords. The chord-wise position of the pressure points was such that the span-wise rows formed a specific percentage of the local chord as measured from the leading edge. Situated in the aft portion of the nosewheel bay was a large auto-observer unit. The main section consisted of an instrument panel, two cameras and a mirror in an open tubular box structure. The cameras, under the control of the pilot, were mounted on the rear side of the instrument panel. The mirror enabled the cameras to photograph the reflected image of the instruments. These cameras, running in sequence, could be operated continuously or in short bursts as required. For pressure plotting, a pressure auto-observer was fitted in place of the handling observer, which was replaced by a smaller unit fitted in the roof of the nosewheel bay. An SFIM A20 recorder was installed to record accelerations and aircraft attitude, monitoring aileron, elevator and trim tab positions. The centre of gravity datum was 50 inches forward of the datum face of the front spar former, and 4 inches below the aircraft datum. To enable varying centre of gravity positions to be obtained, or to allow for weight and balance changes resulting from varying test instrument configurations, ballast storage areas were provided in the nose and in the wing as required. Provision was also made for two aircraft accumulators to be fitted either in the nose or wing. Maximum ballast permitted was normally 418lb in the nose and up to 350lb at the rear. A short mast attached to the fuselage nose cap carried an Avimo high-speed pitot-static head to provide pitot and static pressure readings direct to instruments in the auto-observer. Provision was also made for a yaw vane on a horizontal mast situated at the extremity of the starboard wingtip.

The first flight of the 707 in Australia was made on 13 July 1956, and a few days later the aircraft made its first public display with Squadron Leader Cousins at the controls. The five weeks prior to this event were spent on a careful servicing programme, preparing the aircraft for its test programme. The auto-observer was removed and re-wired so that it could be used for pressure plotting. The main auto-observer carried spaces for thirty-eight dial instruments, and after essential instruments such as the Air Speed Indicator and altimeters had been installed, there remained thirty-three places for pressure plotting instruments. To enable the aircraft weight to be accurately assessed, a "Kent"-type fuel flow meter was installed to record "gallons gone". The serviceable pressure points in the starboard wing were connected to a pressure switch designed and built in the Flight Group's workshop (it proved impossible to connect up the mid-chord pressure holes at chord A). To minimise pressure errors, a swivelling pitot-static head was mounted ahead of the starboard wing. This was necessary as the wing angle of incidence approached 16 degrees with the aircraft in level flight at its highest Coefficient of Lift. Following initial flight tests and pilot familiarisation, flights were conducted to obtain pressure error calibrations and to establish lift and trim curves. Flights were also conducted to perform pressure plotting and to establish high-speed buffet boundaries, and by April 1957 some fifty-one flights had been conducted, totalling thirty hours. During the pressure error calibrations it was established that pressure errors over the entire speed range were less at the wingtip than at the nose boom. The pressure error in the mid-speed range between 140 and 250 knots was determined by the tower aneroid method. The error in the low-speed range between 99 and 140 knots was determined by a formation technique involving the use of a specially calibrated Winjeel aircraft. During the course of this low-speed flying conducted to establish lift curves, it was found to be extremely difficult to accurately determine a lift value for the higher values of lift coefficient. Subsequent flying confirmed the existence of two different flight conditions when flying in the high lift coefficient region. It was found that varying amounts of engine power were necessary to maintain altitude at the same indicated airspeed and it was concluded that two distinct drag conditions existed between indicated airspeeds of 87 and 97 knots, and these different drag states were associated with very definite alterations in the aircraft's handling characteristics. The "high" drag state was accompanied by considerable wing dropping, pitch changes and to a lesser degree some yawing. Studies aimed at determining the buffet boundary were undertaken during this period. Ten flights were made employing the spiral dive technique to determine where the buffet boundary was. In the following year, owing to fatigue limitations on the aircraft structure, the high-speed buffet research was brought to a conclusion.

Simultaneously with the flight tests of the Avro 707, wind tunnel investigations were being conducted to supplement the flight tests. The low-speed wind tunnel tests were conducted using the "nine by seven" facility, using a one-eighth scale model of the aircraft that had been supplied by Avro, complete with adjustable elevators and trim flaps. Surface pressures were measured at low speed on both the 707 in flight and on the model in the wind tunnel. The first phase of the research programme using WD280 consisted of a detailed evaluation of the development of flow separation, and comparisons of this with the model. The tests showed that tip separation commenced and spread inboard at lift coefficients of 0.29 (roughly 5 degrees incidence). This condition was associated with large increases of drag and deterioration of both lateral and longitudinal control due to poor aileron and elevator control. Following on from these studies, the two different drag states that were established during the research into lift curves were more fully investigated and it illustrated that it was possible to enter the high drag state from the low drag state at the same value of lift coefficient, although the reverse was not true. These investigations called for a high degree of skilled flying on the part of Flight Lieutenant R. R. Green from the ARDU (the Royal Australian Air Force's Aircraft Research and Development Unit based at Laverton). The "high" and "low" drag conditions were defined by the engine rpm required to maintain height and speed. Qualitative assessment was provided by the pilot and his recognition of handling changes in the aircraft. The flying technique involved the careful deceleration from 180 knots with the dive recovery flaps (elevator trim flaps) fully extended to achieve the test airspeed and maintain height. Flying was conducted with the powered controls and no attempt was made to trim out stick forces during each run. Precise flying was necessary to ensure that no excessive incidence was applied during the speed reduction as this would immediately produce the high drag condition. In the speed range 87-97 knots IAS the pilot could change from low to high drag maintaining the same airspeed and altitude by co-ordinating the increase of incidence with the engine's power.

A series of flight tests was undertaken to evaluate low-speed handling. Carborundum dust was applied to the under surface of the wing leading edge using an acetate adhesive. Varying the sizes of grit and varying the span-wise extent of application gave rise to four different test conditions. Detailed measurements were only made when the pilot considered that a particular test condition had produced a significant alteration in the aircraft's handling characteristics or performance. Generally, it was found that with leading edge roughness the aircraft was much better and easier to control at low speeds than with a clean aircraft and there was also more aileron and elevator effectiveness. The improvement in aircraft handling with the application of carborundum dust prompted a series of tests employing turbulators. These tests involved the placement of small vane-type protuberances on the wing. Four test conditions were created by varying the height and span-wise extent. With the first turbulator test condition a greatly improved aileron and elevator effectiveness was apparent; in fact, 10-degree banked turns executed at 107 knots were executed comfortably – something not recommended with a clean aircraft. The first sign of wing drop became apparent at 92 knots IAS but was unassociated with any yaw movement and thus no wallowing. The latter effect was one of the most undesirable aspects of the clean aircraft. The minimum speed obtainable for 14,000rpm was 91.5 knots as compared with 89 knots for the clean aircraft. In the period July 1959 to February 1960 some extra instrumentation was fitted to the aircraft in preparation for more extensive investigation of the aerodynamic hysteresis. The discovery of differences in total drag as great as 40 percent at high values of lift coefficient prompted this further investigation and the extra instrumentation permitted a more accurate and detailed appreciation of the handling characteristics in the two different drag states. The auto-observer panel was photographed every 4 seconds by a robot recorder camera with a 2,000-shot magazine and the results were processed by an IBM 650 computer developed by A. A. Keeler to trace the path of the boundary layer on the wing in flight. The pattern was developed by flowing kerosene over a specially prepared portion of the port wing, which was photographed by a formating aircraft. The wing was meticulously prepared for the tests by carefully filling all areas of uneven panelling, and careful rubbing down to ensure as smooth a surface as possible. The wing was then painted with a coat of black shellac as an undercoat for the china clay top surface. The upper surface of the wing was sprayed with a fine layer of white china clay (with a mixture of kaolin, aircraft clear dope and dope thinners) and, because of the clay's porous qualities, it became translucent when wetted by the kerosene. The flow pattern within the boundary layer was thus visualised as the underlying black shellac was revealed beneath the china clay. The starboard wing was spray-painted with a grey undercoat and although no attempt was made to polish the surface, it still wasn't as rough as the china clay-treated port wing. A 5-gallon tank was fitted in the fuselage aft of the cockpit to house the kerosene, which was pressurised from the cockpit's air bottle via a reducing valve. A remotely controlled pressure cock enabled the pilot to control the flow of kerosene, which was fed into a nozzle situated on the under surface of the wing, some 8 inches from the leading edge at 50 per cent semi-span. The delivery pressure was adjusted on the ground so that the spray extended some 5 feet ahead of the leading edge, and it was found that this pressure was sufficient to carry the kerosene around the leading edge to flow onto the upper surface at speeds between 80 and 120 knots. These tests were flown by Flight Lieutenant Cameron from the RAAF and a Squadron Leader Blake from the RAF.

Following these tests, the aircraft was used to determine dynamic stability characteristics at very low speeds. Initially, longitudinal motion was investigated by recording the decay of a short period of oscillation resulting from a single rapid fore and aft stick movement. The need for great precision in measuring the overall drag of the 707 led to the development of a most successful thrust-calibration technique, which enabled the relationship between the installed thrust and the jet pipe temperature to be established. That last flight of WD280 was made in 1963, and during its seven years of research flying for the Aeronautical Research Laboratories the aircraft flew 203 hours and 30 minutes, with each flight lasting between twenty and thirty minutes. It is perhaps ironic that an aircraft designed for high-speed research should have spent so much of its operational life on low-speed research. Almost all of the flying was devoted to aerodynamic research although some demonstration flights were also performed and the aircraft participated in a flying display during Air Force Week in September 1958. With a total of just over 486 hours to its credit, WD280 was placed into long-term storage at Avalon until 1967, when it was sold by auction through the Department of Supply to a Mr Geoffrey Mallet for 1,000 dollars. On 12 June the aircraft was loaded onto a trailer and transported to Williamstown with police escort during the early hours of the morning. The delivery went well, encountering only one minor skirmish with a traffic pole. Upon arrival at Mr Mallet's house the aircraft was lifted by crane into his 44 foot-long garden, which was a tight fit for a 42ft 9in-long aircraft. Subsequently, the aircraft remained in Mr Mallet's garden, maintained in near-pristine condition. Today WD280 is preserved at the Royal Australian Air Force Museum at Point Cook in Victoria.'

Actual construction of the 698 prototype began in 1951 after a three-month delay, during which time the basic design had altered quite considerably in response to testing and research that was still continuing. Most significantly, the RAE at Farnborough had been conducting wind tunnel investigations that had revealed some predicted performance deficiencies caused by the distribution of air pressure over the wing surfaces and air intakes. The results contradicted Avro's predictions, and although there was no conclusive proof as yet (but may well have been, had the 707s been flying sooner), the data suggested that the onset of compressibility drag rise (which effectively capped the 698's maximum speed) would take place at a lower altitude and lower speed than had previously been expected. Clearly the design would have to be revised, and from December 1949 until May 1950 the Avro team worked long and hard to revise the wing shape, moving the thickest section forward to the leading edge, instead of being close to the root chord centre. The result was a wing root that was almost as deep as the short fuselage to which it was attached (this being of some 9 feet diameter) and in many respects it shifted the design back to more of a blended 'flying wing' layout from which the basic design had first emerged. There were other advantages in addition to the aerodynamic improvement, not least the ability to improve the shape, size and efficiency of the air intakes, but also the opportunity to create internal space for bigger engines whenever they became available. In fact, it was the projected need for more powerful engines that ultimately convinced Avro to revise the wing shape, rather than the theoretical advice from the RAE. A total of 190 draughtsmen worked on the 698 together with thousands of engineers both with Avro and various sub-contractors around the country. Companies such as Dowty, which produced the multi-wheel bogie undercarriage, and Boulton Paul, which designed the power controls, produced component designs for the 698 in response to Avro's often vague and constantly changing requirements, long before a proper contract was issued by the MoS.

Finally, construction of the 698 resumed with the huge outer wings gradually taking shape in the factory at Woodford, while the larger centre sections and nose assemblies were produced at Chadderton on the other side of Manchester. In June 1952 notification was received that a contract would be placed for twenty-five aircraft, and this was welcome news for Avro, although there was some concern that a contract had also been placed with Handley Page for twenty-five examples of their own design. Worse still, news came in that the new Handley Page prototype had been delivered by road to Boscombe Down (rather comically disguised as a ship), and was already being re-assembled in anticipation of its first flight. Until this stage, everyone in the aerospace industry and within the RAF had expected a final decision to be made in favour of either the Avro or Handley Page design, and it came as something of a surprise that an initial batch of both aircraft had been ordered before the respective prototypes had even flown.

Years later it was revealed that the original plan had indeed been for just one design to be chosen, for which an initial batch of fifty would have been ordered, but as was so often the case with the bomber programme there was insufficient confidence that one design would prove to be superior to the other, so there was only one choice to be made: either proceed with both aircraft or wait until one emerged as the better design. But Britain's urgent need for a capable atomic bomber meant that waiting was just not an option, so, despite the expense and wastefulness, both aircraft were ordered into production. Some observers commented that the real reason why no choice had been made was because nobody wanted the thankless task of informing the losing company that their design had been abandoned, and although aircraft procurement decisions obviously could not be clouded by such considerations, it is fair to say that even in the early 1950s there were political aspects to the programme that had to be considered, and had either Avro or Handley Page been forced to scrap its designs, one of the companies (each employing thousands of workers) would have faced a very bleak future.

Spurred on by news of Handley Page's progress, the Avro team worked around the clock to hasten the completion of the 698 in the hope that it would be ready in time to appear at the 1952 SBAC show. When the first centre section was completed it was moved in one piece on a low-loader through the streets of Manchester from the Chadderton factory to be mated with the remaining components at Woodford. The 17-mile road journey was a particularly tedious and difficult task as the massive combined inner wing and fuselage section proved to be almost too big to be successfully transported through some disturbingly narrow Manchester streets. As it was, the route had to be carefully planned, various signs and posts were repositioned, and some street lamps were redesigned so that they could be hinged downwards to enable the huge structure to pass by. Many subsequent centre sections made the same journey to Woodford, and the sight of the bomber's carcass became familiar to locals as each one emerged from the factory and made the ponderous journey through Manchester overnight.

Final assembly of the prototype was completed during August and the aircraft was towed across the airfield to Woodford's flight sheds, from where engine running trials began. On 30 August Roly Falk taxied the fully functional prototype (VX770) on to Woodford's runway, back-tracked to the turning circle at the eastern end, and opened up the four Avon engines. This first run along the runway provided Falk with an indication of the speed at which the nosewheel would 'unstick', and once satisfied that the aircraft would be capable of safely getting airborne from Woodford's relatively short runway, he concluded that further test runs would be unnecessary (and would avoid the risk of overheating the wheel

Left: Transporting the massive centre sections of the Vulcan's airframe from Chadderton to Woodford was far from easy. As can be seen from this 1952 image of the prototype 698's journey, the presence of tunnels and street lamps made the task particularly difficult.

Below: An excellent in-flight view of VX770 during her first publicity photography shoot.

VX770 pictured shortly after her maiden flight in August 1952.

brakes). He therefore taxied VX770 back to the runway threshold to position for take-off. After a brief interlude while a flock of seagulls was cleared from the area, the four Rolls Royce Avon RA.3s were opened up again to full power and, with a deafening roar, the glossy white-painted bomber lurched forwards, jumping upwards slightly as the pressure came off the brakes. With a relatively modest combined power of 26,000lb, the 698 might have been expected to require the full length of Woodford's runway to become airborne, but as the aircraft thundered towards the centre of the airfield Falk gently raised the nose and VX770 lifted cleanly into the air and settled into a steady climb. At long last the mighty delta was airborne.

Once safely off the runway, the undercarriage was retracted and the aircraft was slowly taken up to 10,000 feet, where Falk proceeded to make a series of gentle manoeuvres to establish a feel for the controls and to establish that everything handled as predicted. Once satisfied that the aircraft was not hiding any unpleasant characteristics, Falk completed his preliminary handling and began a long, gentle descent back to Woodford. Just thirty minutes later the prototype was back in Woodford's airfield circuit, generating more than a little attention from observers in the area who, quite literally, had never seen anything quite so unusual in the skies before. Falk lowered the undercarriage and at this stage the control tower staff at Woodford noticed something falling from the aircraft's underside. Falk reported that everything appeared normal from inside the cockpit and that the undercarriage had safely locked in the extended position, but rather than risk a catastrophe, the aircraft was held in the circuit while two other test pilots frantically got airborne in a Vampire and Avro 707 to made an airborne rendezvous with VX770 and examine the aircraft for signs of damage. It did not take much scrutiny to notice that the objects that had fallen from the aircraft were the rear doors attached to the main undercarriage, which had been broken clean off as they emerged into the airflow. Having established that the incident was a minor one, the chase planes cleared the area while Falk brought the 698 back on to final approach and gently settled the aircraft back on to Woodford's runway before popping the huge brake parachute.

The first flight had been a great success and everyone was delighted, not least Falk, who commented that the aircraft had been easier to handle than the Avro Anson that he had flown many times before. Two more test flights were then made before the aircraft was flown to Boscombe Down, from where another three hours of initial handling trials were completed. While

VX770 recovering to Woodford on a wintry day during the Vulcan's development programme. The brake parachute housing was fitted to the starboard side of the tail cone, unlike the more familiar position on top of the enlarged ECM tail fitted to later Vulcans.

flying from Boscombe Down, VX770 made the short hop to Farnborough to make daily appearances at the SBAC show. It was decided that for both security and technical reasons it would be unwise to allow the bomber to land at Farnborough, so the sparkling white aircraft was flown in each day to make a series of fly-bys, flanked by the red 707A (flown by Jimmy Nelson) and the blue 707B (flown by Jimmy Orrell). The patriotic sight of the huge bomber in company with the diminutive 707s was a sight that stirred the emotions of countless spectators and brought great satisfaction to the many industry and service chiefs present, who were eager to see some tangible results from the protracted and hugely expensive bomber programme.

Five displays were made at Farnborough in total, and because there had been no time to replace the missing undercarriage doors, Avro's personnel at Boscombe Down made daily checks on the various micro-switches and other components in the exposed undercarriage bays, making strengthening modifications where necessary in response to Falk's display, which got a little faster each day. After the SBAC show media interest in the 698 was widespread, and everyone with an interest in aviation wanted to know more about the unusual and futuristic bomber prototype.

Speculation began to mount as to what name would be applied to the aircraft, and following the appearance of the Vickers Valiant a variety of suitably alliterative names were put forward, including 'Albion' and (probably the most popular choice) 'Avenger'. But the matter was finally resolved when the Chief of the Air Staff announced his preference for a class of 'V-Bombers', which prompted Handley Page to give the name 'Victor' to their design, and Avro's 698 became the 'Vulcan'.

Once the SBAC show appearances had been completed, the undercarriage doors were replaced (more securely), some instrument positioning was revised, and the second pilot's seat was installed in the cockpit. Roly Falk had been heavily involved in the design of the cockpit (particularly the instrument layout), and it was he who suggested the installation of a fighter-type control stick rather than the more traditional 'spectacle' wheel, which would normally be found in the cockpit of a four-engined bomber. Although there was certainly a case for using more conventional controls, Falk was keen to create an environment where the pilot would immediately appreciate the light, fighter-style handling qualities of the Vulcan, and the control joystick certainly gave the impression that the Vulcan was not just another cumbersome heavy bomber, and actually possessed a performance capability that outclassed many contemporary fighters. Back at Woodford, the second pilot's seat was finally fitted, and there was some media speculation (which is sometimes perpetuated even to this day) that the second pilot's seat was literally crammed under the small canopy hood into what was essentially a single-seat cockpit. In reality this was not the case, as the 698 had been designed from the outset to have seating for two pilots. Clearly, no matter how impressive the aircraft's performance may be, it would be unwise to entrust control of a four-engined jet bomber armed with an atomic bomb to just one pilot.

By the end of January 1953 VX770 had completed thirty-two hours of test flying and another period of modification was undertaken, which included installation of fuel cells in the wings, as until this time only a temporary fuel system had been used, incorporated inside the bomb bay. The cockpit pressurisation system was made operational and the Avon engines were replaced by Armstrong Siddeley Sapphire ASSa.6s, each rated at 7,500lb thrust. This gave the Vulcan the same engine power as Handley Page's Victor prototype, and enabled the aircraft to reach higher speeds and altitudes. It had been the design team's intention to fit Bristol BE.10 Olympus engines (each rated at 11,000lb) right from the start, but jet engine development was a relatively new science and progress with new engines continually lagged behind development of the aircraft that required them. The Sapphires made a suitable interim replacement for the Avons, and flying was resumed in July 1953.

By this stage the second prototype (VX777) was nearing completion, and although some initial ground trials were completed with Olympus Mk 99 engines fitted, they had been replaced with 9,750lb Olympus 100 engines in time for the second prototype's first flight on 3 September 1953. In response to data from the 707 programme, VX770's nosewheel leg was extended by a few inches during construction in order to give the aircraft a 3.5 degree angle of incidence in relation to the ground, thereby allowing it to 'unstick' at a lower speed (using less runway length). This design modification was built into VX777 from the start, and rather than relying on a telescoping mechanism to allow the gear to fit into the wheel bay (as had been the case with VX770) a one-piece leg was fitted to VX777 together with a correspondingly larger nosewheel bay to accommodate it. As a result, the entire nose section of VX777 was slightly longer and also included a bomb-aimer's blister under the nose, complete with a clear forward panel.

Less than a week after its first flight, the second prototype appeared over Farnborough at the 1953 SBAC show. It was at this event that the aircraft created what has become one of the most iconic moments in British aviation history, when VX777 led an enormous 'delta of deltas' formation over Farnborough, the two Vulcan prototypes being flanked by four Avro 707s, creating a sight that will never be forgotten by those who were lucky enough to witness it. After Farnborough week, VX777 was delivered to Boscombe Down in preparation for high-speed and high-altitude trials, but before they could begin a series of modifications was made to the Olympus engines, together with their control and fuel systems, and almost six months were to pass before flight trials could commence. Unfortunately VX777 suffered a heavy landing while conducting flight trials at Farnborough and this caused significant damage to the airframe. This created another delay in the flight test programme (particularly in terms of engine development), and another six months went by before flying could continue, much to the frustration of the Avro design team. During the time that VX777 was laid up, new Olympus 101 engines were installed, each rated at 10,000lb, and the whole airframe was structurally reworked, using data from a static test specimen that had also been constructed. While VX777 remained grounded, the first prototype remained active on flight trials, albeit within the more limited flight envelope in which it could safely perform.

When VX777 eventually resumed flying duties in 1955, the exploration of the high-speed and high-altitude characteristics of the aircraft quickly confirmed the mild buffeting that was predicted to occur when pulling 'g' at speeds of Mach 0.8. This problem had been highlighted during the 707 test programme, but only at a fairly late stage, and with 9,000lb engines it was not a significant difficulty. However, with projected improvements in engine thrust likely to result is significantly more powerful engines, the high-altitude buffet would probably be reached with only the slightest application of engine power. This would lead to problems with bomb-aiming accuracy and would also create a significant risk that the outer wing sections could ultimately fail through fatigue stress. Consequently the Phase Two wing design emerged with a 'kinked' leading edge that neatly solved the problem. However, because the modification was made at such a late stage in the development

Assembled in a secured compound at Farnborough, the two Vulcan prototypes prepare to make an overnight stay with their Handley Page competitor at the 1953 SBAC Show.

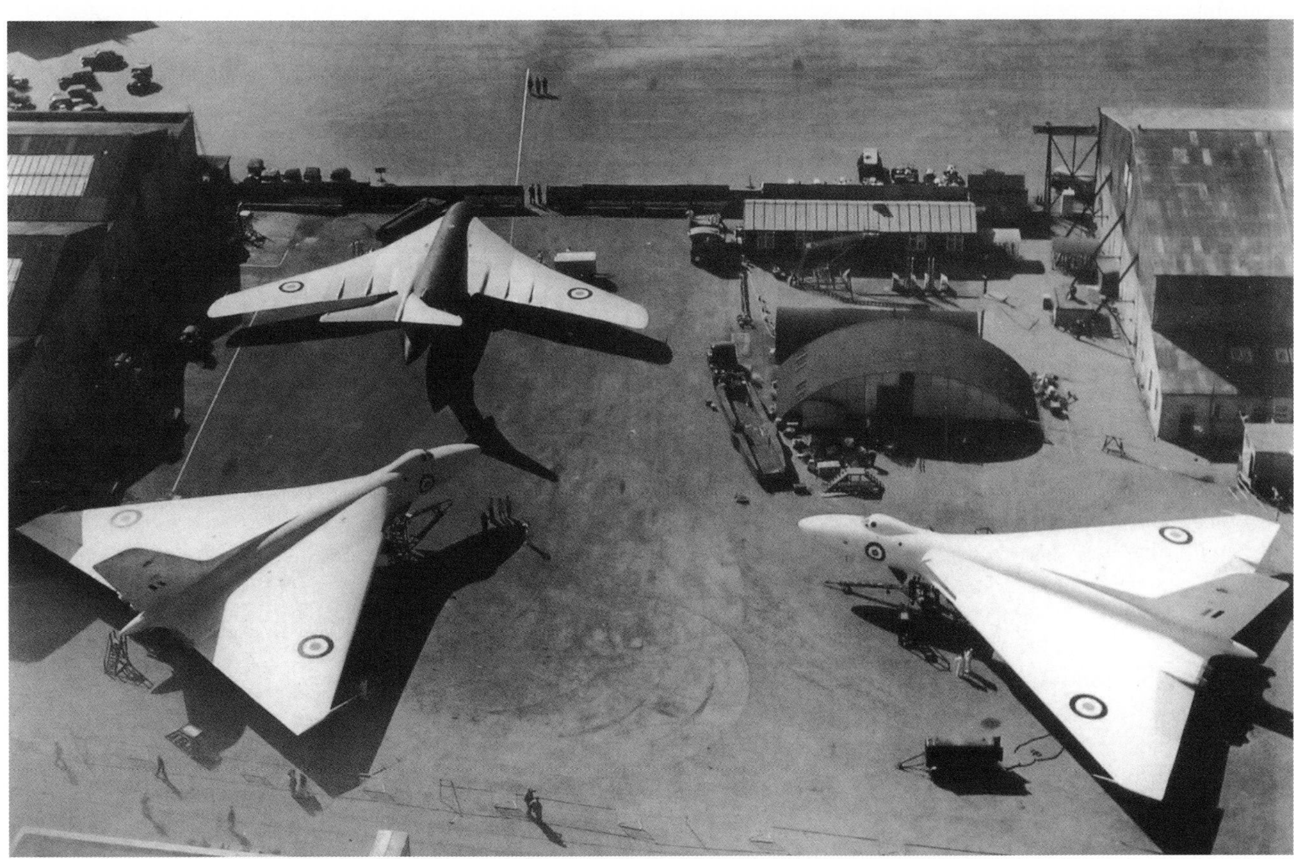

A famous (but nonetheless magnificent) picture of the two Vulcan prototypes in company with the 707 test fleet, passing RAF Thorney Island en route for the 1953 SBAC show.

programme, the first production Vulcan (XA889) was rapidly nearing completion and it was too late to incorporate the new wing modification. Consequently, while VX777 re-emerged with a revised wing leading edge, the first production Vulcan made its first flight on 4 February 1955 with the original straight wing leading edge still installed.

Painted silver, complete with national insignia and a striking dark-coloured radome made from a glass-fibre/Hycar sandwich, XA889 had successfully become airborne a whole 12 months ahead of Handley Page's first production Victor, much to the delight of Avro, which was still very keen to be visibly ahead of Handley Page in terms of development progress while there was still any doubt as to which bomber design might ultimately receive the most orders. The second production Vulcan (XA890) joined the flight test programme later in 1955, and although it was also not fitted with the Phase Two wing, the first production aircraft was eventually retro-fitted with the new wing design, complete with vortex generators along the upper surface, which speeded up boundary layer airflow across the wing. Roly Falk took XA890 to the 1955 SBAC show and once again stole the show, this time by executing a neat barrel roll at a surprisingly low altitude in front of an assembled mass of astonished spectators.

The SBAC President, Sir Arnold Hall, was just as astonished as every other observer, and immediately forbade Falk from repeating the performance on subsequent show days. Falk explained that the manoeuvre was well within the aircraft's capabilities providing that the roll was kept within a steady 1g throughout, but Hall remained unconvinced. His only plausible excuse for stopping what would have easily been the highlight of the show was that the manoeuvre would ultimately set a bad example to service pilots, and in some respects his comments did prove to have some prophetic value, as other pilots did eventually try to repeat the manoeuvre (some without the same degree of success), until it was ultimately forbidden by senior industry and RAF chiefs. On 7 September Prime Minister Anthony Eden arrived at Farnborough in a CFS Dragonfly and, after watching Falk's Vulcan display (minus the barrel roll), was invited on board and treated to a short flight, during which he occupied the co-pilot's seat and briefly took control of the aircraft before Falk brought it back to nearby Blackbushe. After the flight Falk received a hand-written note of thanks from Eden.

In March 1956 the first production Vulcan (XA889) was delivered to Boscombe Down to begin acceptance trials with the RAF. The initial CA (Controller of Aircraft) release was subsequently made on 28 May as follows:

> 'Tests have been made on the first production Vulcan B Mk 1 to assess the type for use by the Royal Air Force, in the medium bomber role. The trials programme was completed in 26 sorties, totalling 48 hours 15 minutes flying time. During these tests the aircraft was flown over the full centre of gravity range, and at take-off weights up to a maximum of 165,220lb.

Above: VX777 pictured shortly after having been re-manufactured to carry the Phase 2C wing, becoming the B2 aerodynamic prototype.

Right: B.1 XA890 has early non-production-standard double airbrakes in the wing lower surfaces. Unusually, the aircraft remained unmodified and retained its straight wing leading edge throughout its life as a developmental aircraft, before ending her days at Boscombe Down as a fire crash rescue airframe in 1969.

Above: XA890 inside the EMI test rig at Boscombe Down. Despite the poor quality of this image, it appears that white (or possibly fluorescent orange) patches have been applied to the tail and wing surfaces.

Left: XH478 making a public appearance at Coltishall's Battle of Britain 'At Home' Day. This image clearly shows the vortex generators attached to the B.1A's wing upper surface. *Ken Elliott*

The first production Vulcan XA889 was representative of service aircraft in all respects save those of operational equipment, automatic pilot, the rear crew stations and certain items of cockpit layout. The aircraft incorporated the drooped leading edge outer wing with vortex generators, the longitudinal auto-Mach trimmer, the pitch damper and revised airbrake configuration. These modifications have successfully overcome the unacceptable flying characteristics exhibited by the second prototype in the preliminary assessment carried out by this establishment, and when all stability aids are functioning, the Vulcan has safe and adequate flying qualities for its primary role as a medium bomber.'

The CA release reflected early concerns over the Vulcan's flying characteristics following A&AEE trials with the second prototype, VX777, after which the following comments had been made:

> 'A preliminary flight assessment has been made on the second prototype Vulcan in 17 sorties totalling twenty-seven flying hours. During these tests the aircraft was flown at a mid centre of gravity and take-off weights of 119,000lb and 130,000lb. The expected operational take-off weight of production aircraft is about 165,000lb. The expected cruising Mach number is 0.87M (500 knots) and the design Mach number is

0.95M. Above 0.86M a nose-down change of trim occurred which became pronounced with increase of Mach number towards the limit, making the aircraft difficult to fly accurately and requiring great care on the part of the pilot to avoid exceeding the maximum permitted Mach number. This characteristic is unacceptable; the Firm propose to eliminate it in production aircraft by the introduction of an artificial stability device (a Mach trimmer).

With increase of Mach number above 0.89 the damping in pitch decreased to an unacceptably low level, particularly near the maximum permitted Mach number, and the aircraft was difficult to fly steadily. The Firm propose installing a pitch damper in production aircraft. As tested the Mach number/buffet characteristics were unacceptable for a high-altitude bomber, but considerable improvement is hoped for with the drooped leading edge and vortex generators. Associated with the buffet were oscillating aileron hinge movements which in these tests imposed severe manoeuvre limitations from considerations of structural safety. Making allowances for the differences in engine thrust and aircraft weight between the aircraft tested and the production version, the performance in terms of attainable altitude was not outstanding. The likely target height with a 10,000lb bomb will only be about 43,000 feet with 11,000lb thrust engines, and the high-altitude performance will be poor. The level of performance is considered to be inadequate for an unarmed subsonic bomber, even under cover of darkness. In summary, although the aircraft has certain outstanding features, serious deficiencies are present, particularly in and above the cruising Mach number range, and until these are rectified the Vulcan cannot be considered satisfactory for service use.'

It is interesting to compare this less than enthusiastic appraisal of VX777 with the shorter (but generally praiseworthy) comments made about XA889. Although such reports are primarily concerned with highlighting deficiencies, it is clear that the prototype (with an unmodified wing) was never going to match the performance required by the RAF. Despite being extremely manoeuvrable and easy to handle, the early Vulcan was simply not suited to the role for which is was designed, but thanks to the extensive research and testing conducted by both Avro and the RAE, and more than a little help from the diminutive 707s, the Vulcan was modified and refined to become a truly outstanding jet bomber.

CHAPTER FOUR

Design and Development

WORKING within Avro's design and manufacturing business in the early 1950s was certainly a fascinating experience. Peter Rivers was one of many Avro employees during the years when the Vulcan was being developed, but unlike most of his colleagues he had also previously worked on Handley Page's competing design, the Victor, which gave him a fascinating insight into the creation of both advanced bomber designs. In this chapter he recalls some of his experiences.

'The design of the H.P.80 which later became the Victor had been progressing for just over a year when I arrived at Handley Page, and the aircraft's shape had not yet settled into the final configuration, the tail in particular being much smaller and with the tailplane wandering up and down the fin as the design continually changed. The department that I joined was small but expanding, and was responsible for the theoretical design calculations for all mechanical systems such as hydraulics, cabin systems, de-icing and anti-icing, fuel supply, etc. We shared a small area in the open-plan design office with the electrical engineers with whom we worked closely, but we had little contact with other technical departments or the general drawing office, which was in another part of the building. Maybe this was because we were not really involved with design detail at this stage, but there was a general practise at Handley Page of channelling our results to the Chief Designers and the heads of the other departments, via the head of our own department. This was a system that contrasted dramatically with my experience at Avro. Compared to the stress and aerodynamic departments at Handley Page, our calculations did not involve any higher mathematics. Even differential equations were very rare, but the range of environmental

Above and opposite: Two views of the Vulcan assembly production lines at Woodford (B.2s below and B.1s overleaf). Although the upward-hinged canopy suggests that there may have originally been some design facility for fighter-style cockpit access, this was never actively considered and the canopy was only designed for removal during servicing or crew abandonment.

conditions which we had to investigate was vast, especially for the anti-icing systems. Endless computations of complex arithmetical relationships had to be carried out, and the discovery of a simple mistake could mean the scrapping of work that might have taken a month or two to complete. We didn't have the luxury of computers, as they didn't arrive in the industry until the late 1950s, but the more favoured departments had electronic calculators, each one being about the size of a typewriter. We had hand-wound mechanical versions, rather like old- fashioned cash registers, and they always jammed if one tried to work too quickly.

Thanks to Handley Page's "economical" ways, even pencils were rationed to one per month, so if one had a heavy spell of writing one had to go begging to other people who hadn't used up their rations! Our main difficulty was that there was virtually no relevant input data to feed into our calculations, and whereas work was being done at the RAE and elsewhere into the radically new aerodynamic and structural aspects of the design, nothing was being fed into our side of things, or at least some work was being done, but tended to appear only after we had finished! This was especially true of icing work, which depended crucially on the proportion of water droplets in an icing cloud that would be caught on the wing or deflected around it. That required detailed mapping of the streamlines around the nose of the aerofoil sections, and the only highly theoretical plots that had been made in the US were for aerofoils that were most unlike the Victor's wing. All we could do was modify the results by factors that we hoped would be realistic. At least we knew that nobody else knew any better. The figures for the actual icing conditions were a little more reliable but still subject to annual changes. They came from NACA (the predecessor to NASA), who had an ageing C-46 that spent each icing season plodding around, flying into freezing clouds, with a variety of weird instruments sticking out of the windows to measure things like the total water content in each cubic metre of air. When we had spent a year doing our complicated step-by-step "catch and heat" calculations, NACA would issue the next annual report, saying that last year's was wrong, so we had to start all over again! Another area that caused us headaches was the design of jet pumps to mix high-pressure air, bled from the engine compressors (which was too hot to pass directly through the light alloy wing structure) with cold ram air, in order to achieve the maximum usable air temperature while still delivering a high enough pressure to force the air through the heat exchange passages in the wing and tail leading edges, although railway engine designers had been using the same principle for the better part of a century.

There was no reliable theory for the mixing of air streams in the kind of proportions necessary to achieve the pressure performance that we needed. We adapted the few theoretical analyses that we could find, and pounced upon the odd wartime German paper that came through from the post-war translators, but the results, which had to be turned into some massive pieces of hardware, were based upon guesswork to an even greater extent than our usual work. Ironically, just after all this work was complete and the detailed design finalised, a classic paper on the subject was published, and although I was working for Avro by that time, I was able to check that our answers had been pretty accurate. I've described this work in some detail because the same things were going on at Avro at the same time, and when I joined them I found that very similar conclusions and guesses had been made there. It was interesting to have worked on the Victor during the paper stage and then the Vulcan, when the theory was being put into practice. One question did arise concerning the Victor that didn't affect the Vulcan, and that was heat loss from the fuel during long flights at high altitude, which might cause the fuel to freeze or at least turn waxy. The Vulcan's tanks were separate items within the deep wing structure, whereas the Victor's were integral, with the inner metal skin separated from the outer one only by stringers and heavy spar booms at the corners of each span-wise bay. The latter feature meant that as well as encountering heat flow directly from one skin to the other, some would flow along the inner skin to the conductive boom. A rather unusual adaptation of some standard school physics was needed to solve the problem, and a further surprising extension of my calculations was used years later, when we were trying to establish the fuel temperature variations through day and night for the Avro 730 supersonic reconnaissance bomber, for which the skin temperature was expected to be around 225 degrees Celsius. In this design the fuel was to be the heat sink for all cooling purposes. The same system was later used in the Concorde.

Towards the end of 1951 I was getting tired of Handley Page, and I needed some practical experience away from a desk, so I joined the test section of Fairey's missile division at Heston. They were developing the beam-riding Blue Sky missile at the time, and I quickly determined that missile work, especially trials, would not be much fun, so after a few months I was looking for a way out. In September of the same year I saw the Avro 698 prototype performing at Farnborough, never imagining that within three months I would be part of the Vulcan design team. I attended a lecture at the RAeS in London, presented by Dr Still, the Technical Director of Teddington Controls, who were to supply much of the control equipment for the V-Bomber hot air systems. At the lecture I met a former colleague from Handley Page, who had joined Avro a couple of years previously. He was about to leave for Canada, and after introducing me to Avro's Chief Designer and the head of his own department, he virtually left me his job to step into.

It's probably worth describing the design organisation at Avro as it was somewhat unusual for a British company. They had adopted a more American practice, with an individual Project Designer for each type of aircraft, all reporting to the Chief Designer, as did the heads of the separate technical departments, these being Stress, Aerodynamics, which included New Projects, Weight Control, and the Drawing Office. The job of each Project Designer was to push through his machine, regardless of the efforts of the other Project Designers, and this form of organisation ensured that one man at least would be committed to the success of each project, ensuring that maximum effort was provided by every department that worked for him. Compared to Handley Page, the Drawing Office was enormous, as was the whole Avro organisation, especially the production side. There was an old saying that when it was foggy outside, and it often was around Manchester, the far end of the Drawing Office was pretty hazy too! Each project was allocated a numbered Drawing Office section to carry out all the detail design, and only the Project Designer could issue instructions to them, and he had six Project Engineers to do the necessary work.

The group which I joined at Avro had grown up during the post-war period and expanded with the Vulcan design requirements, but, unlike my department at Handley Page, it had remained as a group of individual specialists, each reporting to the Chief Designer, in parallel with the Project Designers and heads of the major technical departments. We were called Specialists but we were much less so than the people in the Stress and Aerodynamic offices. Our position was rather odd in that we had direct access to the top but we couldn't actually instruct any department to do anything without the authority of the Project Designers. Because the final assembly, test flying and work on the aircraft was done at Woodford, we travelled between there and the main offices at Chadderton, roughly 15 miles away, several times each week. Because we were involved with the cabin and control system, we dealt directly with the test pilots and flight crews. The only person we didn't talk to was Sir Roy Dobson himself, although we did sit in on post-flight inquests that he chaired. With all of the other directors we were on first-name terms, with the people who had produced thousands of Lancasters.

During 1954 the head of my department left for Woodford to take charge of a new department that would produce our own control equipment, replacing the items previously supplied by companies such as Normalair and Teddington, as their supplies were often underdeveloped, and delivered late. I then took charge of the Systems group, essentially a bunch of individuals, as we didn't actually call ourselves a Department as such. Throughout the time I spent with Avro from December 1952 to December 1960, the Vulcan was only one of several projects that we were working on simultaneously, and we were involved in every stage of development from initial project studies to sorting out problems in service, as nobody else within the company had a detailed knowledge of our systems. Before my arrival the company had started a practice of writing simple descriptions of each system, describing the design and operation for the benefit of our Senior Designers, inspection staff, pilots, flight test crews, and anyone else who needed to understand. We called them Children's Guides, but they also helped us to understand better what we were doing, and they were sufficiently well written to be used by the Technical Publications Department as the basis for Operating Manuals for the type.

Apart from the very open form of organisation and the direct personal friendship at all levels, there were other noticeable differences between Avro and Handley Page. At the latter company, design had the upper hand over production, and no matter how complex or difficult the structural or detail design might be, the production side had to find a way of making it. At Avro, however, production was dominant and if some feature was judged to be too difficult to produce economically or easily, it was simply redesigned. This resulted in more simple and therefore more reliable and easily maintained machines, and whenever we found ourselves thinking in too complicated a way, we'd say that it was "becoming a Handley Page design" and then reconsider the subject in an effort to simplify it. At the same time as trying to keep the design simple and aiming to use established production methods, the Avro production "machine" was capable of thinking big in a way that Handley Page never was. If a major alteration was found to be necessary, such as the Vulcan's extended and kinked leading edges, a modified production facility would be set up without delay, and massive jigs would often appear in the shops every few weeks.

When I joined Avro the Vulcan prototype VX770 had been flying for about three months. It was without most of the systems with which I was concerned. There was a hydraulics system for the undercarriage and bomb doors, electrical power for the airframes and the electro-hydraulic-powered flying controls, but the cabin was unpressurised and had only a rudimentary ventilation system. The aircraft was a long way from being fitted with production-standard engines, but even then the engine bays were large enough to accommodate the ultimate Olympus variants. Handley Page of course had to redesign and enlarge the entire Victor centre section when they changed to Conway engines.

Also, when I joined Avro they were working on the 720 rocket-propelled fighter, early schemes for the Atlantic, which was an airliner derivative of the Vulcan, some Shackleton developments leading to the Mark Three, and the clearing of the last of six Ashton high-altitude research machines for delivery. The main activities in 1953 were the testing of hot air systems on ground rigs, analysing flight results on the first prototype's flying controls, and preparing for the fitting and operation of the fully equipped and pressurised second prototype, which flew in time for the Farnborough show in September, but wasn't pressurised until the following year.

The contracts for the V-Bombers called for all mechanical systems to be tested on realistic ground rigs before they were operated in flight. Before the first flight, the hydraulic system had to be tested with all components and pipework in their correct dimensional positions. A separate power plant and engine intake rig was set up in a building behind the flight sheds. This had the wing root intake set at the correct height and angle above the ground. The rig that simulated the hot air system that would feed the cabin and icing systems caused large problems. A source of air at a temperature and pressure corresponding to that of a jet engine compressor was not available, and a supply had been devised from standard ground-type compressors and storage vessels, with heat from combustion heaters. These were specified to burn with a clean flame so that their output when mixed with the main air flow would not contaminate the aircraft equipment. The ducting was supported on steel framework in the exact configuration of the aircraft system, with a used boiler of the correct volume to represent the cabin. When the rig was started up, control of the flow and heat inputs seemed to be excessively tricky, and required the entire test department staff, standing at different valves and regulators around the system, frantically twiddling them in an attempt to get stable operation. After a few hours of this, without any useful results, the electro-pneumatic controllers started to misbehave, and they were found to be clogged with soot and corrosion from the supposedly clean burners. The whole concept of the rig was seen to be a disaster, but the difficulty was solved by the quick and forceful decision made by the Chief Designer, who ordered a complete ducting system to be removed from the power plant test house. It was laid beneath the engine intake, and connected to the engine compressor. With this realistic supply we began to get usable results for temperate conditions, although we were unable to simulate the maximum engine output temperatures. The engine test crew controlled the power plant from inside their specially designed soundproof control room, but the mechanical test people had to stand in the open, below the engine bay, wearing what were obviously rather inadequate ear protectors to protect them from the terrific noise.

The pressure cabins were built at Chadderton, and in their boiler-like form, with a blanking plate over the pilot's canopy, they were pressure-tested to 1.5 times their maximum operating differential of 8psi and leak-tested to a schedule of pressure against time. Following the problems with the de Havilland Comet, which occurred at the same time, structural testing to the point of failure was carried out in water tanks, in order to avoid the explosive effects of the rupture of an air-filled cabin, but obviously this could not be done with usable cabins, and in any case we hardly expected a failure at what was a normal proof pressure by industrial standards. All the same, the area at Chadderton where the tests were done was roped off to keep bystanders away, and the inspector making the test had a heavy armoured screen to stand behind. The cabin of the first prototype had been completed and had flown before I arrived, and I doubt if it was ever pressure tested.

The second prototype's cabin was tested, and the next one, destined for the first production aircraft, surprised us all by blowing up. This is how it happened: the front bulkhead of the cabin was domed, the usual way to carry pressure economically on the end of a cylinder. However, the rear bulkhead was flat because it could not take up space in the nosewheel bay, which was full of equipment. It was strengthened by massive vertical beams, which carried the nosewheel leg and operating jacks. When the cabin was pressure-tested the support was missing, as the cabin couldn't be tested once it was mated to the fuselage. So a dummy structural bay was bolted to the rear bulkhead to give it the necessary strength. When the prototype cabin was tested, the holes in the two pieces of structure didn't line up very well and they were opened up and aligned to take the temporary connecting bolts. With the next cabin the holes were again out of line; the bolts were consequently a little slack, and the stiffening effect was inadequate, so the bulkhead blew out. The effect was like a sizeable bomb going off and impressed all concerned, especially the inspector, who had only just retreated behind his screen after checking the manometers attached to the bulkhead, which were fundamentally more accurate than the gauges. So in total, one more cabin was completed than the actual number of airframes.

With the first flights of the second prototype with full pressurisation scheduled for early 1954, it was decided that I should go on all of the early flights to take charge of the pressure and air conditioning controls, which were on the panel of the port-side rear crew member. It was evidently thought that if the system misbehaved, I would understand and know what to do. In case such a failure resulted in cabin pressure being lost at more than 40,000 feet, I had to wear a pressure suit. It was an American-designed, close-fitting, nylon partial pressure suit that pulled tight when required through laces and pneumatic tubes along the arms and legs. Unlike the original American version, which had a helmet with opening visor, the British version had a cloth headpiece with a hard leather inner shield, and a fixed visor. Each one was custom-fitted and all our test crews were provided with one, but the Chief Test Pilot Roly Falk refused to wear one as he felt it would hamper his movements and create a potential danger. To take account of possible decompression, our Bomber Command liaison pilot, Wing Commander C.C. Calders, rode in the second pilot's seat, wearing a pressure suit, and really did little other than wait for an emergency. If a failure had occurred, his suit and my suit would have inflated, and we would have brought the aircraft down to a level where the remaining three crew would hopefully have regained consciousness.

Late in 1953, therefore, I was fitted with my suit at the appropriately named Frankenstein's works in Manchester, and had a trial inflation while seated on the edge of a table. It was shaped to a seated position, and when pressurised it took all the weight off one's arms and legs and you felt like you were going to float up to the ceiling. I then went to the Institute of Aviation Medicine at Farnborough to experience a proper inflation of the suit in their decompression chamber, where pressure was explosively changed from 8,000 feet to 70,000 feet. I shared the chamber with a representative from an equipment company, who passed out and flopped over, but I couldn't do anything about it as I was held firmly on my chair by my suit.

After all the preliminary testing, the actual flying seemed quite tame, but it was interesting to be amongst the first people in Britain to fly regularly at 50,000 feet and near to Mach 1. Most of the flights I participated in were around February-March 1954, and after five or six flights it was clear that the system was not going to fail or do anything untoward, so I stopped flying. The usual routine for each flight was to take off from Woodford, climb through what seemed to be a permanent overcast heading south, and fly straight ahead for about thirty minutes. That would place us over the Isle of Wight. I could see the ground through the small porthole window above the rear crew's bench when we were well banked over. We would then turn north, flying for an hour until we were over the central Highlands of Scotland, before turning south again. Heading back to Lancashire, we would finally home in on the voice transmissions from controllers at Woodford and sometimes with the help of the radar controllers at Manchester Airport. The last flight I flew on included the Minister of Supply, the infamous Duncan Sandys, who was permitted to fly the Vulcan himself for a few minutes, making the customary flattering remarks afterwards. The flight was reported in many newspapers and I was listed as part of that crew. One particular feature of each flight was the buffeting that occurred when flying at altitudes in the high 40,000s at high Mach numbers. It seemed quite gentle to me, just a low-frequency shaking like driving along a bumpy road, but of course it led to the extended and drooped wing leading edges that were retro-fitted to early Vulcans.

From early 1954 onwards we were concerned with production machines, of which the first few were used for trials, and I remember XA894 and XA897 in particular, as they were used for cabin systems and anti-icing development. Perhaps it would be helpful if I described the arrangement of the hot-air systems on the Vulcan. These were fed by a tapping at the high-pressure end of each main engine compressor, and were designed for a maximum pressure of around 200psi and a maximum temperature of 400 degrees Celsius – figures which were unprecedented at that time. The ducting was all stainless steel 0.28 inches thick, which led to some tricky welding problems. Simple things like joints, clamps and sealing rings took on a great deal of our attention, and the long runs of ducting, typically 4 inches in diameter, had to be supported from the aircraft structure with flexible steel bellows at the anchor points, to allow for differential expansion of the ducts and structure, with the wide temperature ranges that went through. This is an area that troubles the designers of ground-based pipe systems even today, but we must have got things pretty well right, as I never heard of any major failures in RAF service. The ducts in each pair of engine bays ran forward, joined up into one from each side, and met on the rear wall of the nosewheel bay (in each wing root, a branch went off to supply the anti-icing jet pump, and as described for the Victor, each one was a hefty device about 8 inches in diameter and 6 feet long). From the junction, a single duct ran forward along the port side of the nosewheel bay, to the air conditioning pack (consisting of a cold-air expansion turbine, usually called the CAU), heat exchangers (supplied with cooling air from that unobtrusive intake inboard of the engine intake fence) and control valves. From the pack, the air entered the cabin through a non-return valve on the rear bulkhead (so that the cabin would hold pressure if the ducting or supply failed). Cabin pressure was controlled, as always, by an outlet valve on the front bulkhead (so that leakage from the cabin structure did not affect the control), ventilated the radar in the nose, and finally left through the small grilles low down on the nose.

An excellent in-flight view of XA894 carrying the Olympus 22R, which was destined for the BAC TSR2.

A magnificent Avro publicity photograph of an early trials fleet Vulcan B2 carrying a Blue Steel missile.

The flow of air to the cabin was far greater than necessary to keep the crew comfortable because the requirement for cooling the radar and navigational sets was overpowering. These sets were in sealed drums about the same size as a typical dustbin, and what was inside was so secret that we were not allowed to discuss the cooling load, only supply what the Ministry specification called for. Years later, when we did find out what was inside, it confirmed our suspicions – the designers may have known plenty about electronics, but they hadn't a clue about heat and all our efforts were being devoted to the cooling of a few hot spots, such as sensitive valves placed right above heat-generating components. This was the pre-transistor era, and the valves in this case were of the electronic kind, unlike those mentioned previously!

The pressure control side of things, manufactured by Normalair, didn't give us any undue trouble, but the flow control side, made by Teddington, turned out to be a major headache, both on technical grounds and unreliability. At the time they were the only suppliers of high-temperature valves with sliding carbon gapes, which were used for shutting off airflows from the engines, and initially for regulating the flows to the different parts of the system. The valves were moved by electric motors, and it was therefore logical for Teddington to develop electrical flow sensing control units, to operate the valves. Unfortunately the controllers were underdeveloped when they were installed in production Vulcans and I personally felt that they were designed to unrealistic specifications, typical of the time, especially in the case of equipment with which the Ministry was concerned. For example, a requirement for airflow to be controlled at 40lb/min with a tolerance of plus/minus 11lb/min led to complicated and unreliable gadgetry. To my way of thinking it didn't matter if the flow was 40 or 45 as long as it was steady, and I later changed the systems to this standard.

Our early tests, starting with the rig, showed that the flow controllers were not stable and could not cope with various aspects of the special requirements that we considered unrealistic, but which the Ministry insisted upon. Teddington kept adding more fiddle units to back up the ones that were misbehaving, without doing much good. At one stage they turned up with yet another gadget, saying that it had got Handley Page out of trouble, and it would do the same for us.

Of course, Handley Page was regarded as a major competitor, and during the war it was often said in Avro circles that the company had three enemies: Handley Page, the Ministry and the Germans in that order! However, having previously worked for Handley Page, I telephoned my old colleagues to enquire how they were getting on with Teddington's new panacea. They said that Teddington had told them that the gadget had worked for Avro, so it would work for them too, just as they had told us. At one stage I visited Teddington's factory, and found that the inspection department, to save themselves undue effort, were leaving ready-signed clearance forms for items that had yet to be tested. That didn't improve my opinion of them, and when their deliveries turned out to be late and unreliable, Avro, in typical ruthless style, virtually took over their test and inspection department with our own people, ensuring that whatever happened to everyone else's orders, the Avro units would be completed on time and work properly. In the end I solved our flow problems by throwing out all the Teddington equipment, except for the shut-off valves for which there were no alternatives, and putting in Normalair controllers of an extremely simple kind, based on the ones that had been used in smaller aircraft since the Second World War.

One thing we could not do with the Vulcan was to keep the aircrew comfortable on the ground in hot conditions, although neither could any other military or civil aircraft design at that time. By the time we had reached the Vulcan Mk 2 the operational procedure had changed, making crews spend long periods on board the aircraft on standby, and the heat situation could not be tolerated. The RAF's answer was to put the crews in ventilated suits, which were a light nylon overall design, with fine tubes directing conditioned air to evaporate sweat from appropriate areas of the aircrew's bodies. The additional airflow, drawn from idling engines or a ground supply, and the tight temperature control needed (too hot or too cold either cooked or chilled the unfortunate crewman) meant that we had to install a separate conditioning pack, which was somehow squeezed in next to the existing one. This was again a Normalair unit, with a fighter-sized air turbine running it at over 100,000rpm, and we did not have any particular troubles with it. I should add that there was no connection between the fact that I'd worked for Normalair and the way that they took over many of the Vulcan's systems; it was just that they were slightly less unreliable than the other suppliers!

I cannot recall doing any useful testing of the icing system as installed. All we could check was that the surfaces were being evenly heated, as it would have been impossible to fly in real icing conditions as the RAE's attempts to find such conditions with an instrumented Valetta had indicated how difficult it would have been, and we couldn't have measured the complicated icing factors anyway. As the extended and drooped leading edges were introduced on the Vulcan Mk 1, and extended further on the Mk 2, our heating passages were getting hacked about and restricted by the changes in internal structure, and there was nothing we could do about it. Airframe icing was becoming less of a concern, and I presented a paper at a Napier icing symposium to show that with the fast climbing and descending flight profile for which the V-Bombers were designed, they would go through the icing layers in a matter of seconds, so that even with no heating system the amount of ice picked up would be unimportant. Of course I did not anticipate that in a few years' time virtually all operations would be conducted at low level, but even so I never heard of any icing problems with the Vulcan.

I made myself pretty unpopular at one stage by pointing out that years of calculating anti-icing performance, and agonising over the assumptions we had to make, simply resulted in our demanding more bleed air from the engines than we were allowed to take, so that all we were able to do was to take as much air as we could and spread it around evenly. Nevertheless, the Ministry insisted that we certify that the system would meet their requirements on paper, as we were still calculating and refining our figures when Avro installed the first computer in the industry. It was about the same size as a bungalow, and we jumped at the chance to have many more points in the icing envelope calculated.

Of course, the occasional accidents that the Vulcan suffered often affected us on the design side and such events have only been briefly mentioned in other historical accounts. Aircraft accidents often have a funny side when nobody gets hurt, and the first one we had was like that. One of the prototypes, VX777, was being demonstrated to one of the RAE pilots at Farnborough in 1954; at that time the early Vulcans were being thrown about in an almost carefree way, and on this occasion the crew had made a rather snappy yawing manoeuvre, probably a sort of stall turn, as the rudder was locked hard over but stayed there. At this stage the aircraft didn't have any periscopes, so the pilots could only see the wing leading edges as far as the wingtips, and no further aft. Periscopes were later fitted for navigation and were essential to enable the crew to see what was going on behind them, such as the undercarriage position, bomb doors position, or, in this case, whether the fin was still attached. The crew flew VX777 past Farnborough's control tower and received confirmation that the fin was still there. They then landed, using asymmetric power to counter the jammed rudder, but this meant that the brakes were unable to stop them before the aircraft rolled off the end of the runway onto soft ground, where the undercarriage collapsed. As there was no great danger of a fire, there wasn't any hurry to abandon the aircraft, but the question on the crew's minds was how to actually get out, with the entry door firmly wedged into the ground. Oddly enough the situation had been discussed in our design office a few weeks previously, and it had been concluded that the pilot would have to hold up the canopy while everyone else climbed over the side. The canopy could be jettisoned in flight and sucked away by the airflow, but there was no way of getting rid of it on the ground, other than by unlocking it and tipping it over the side. The crew members were unwilling to do this, feeling that they'd caused enough damage already, so Roly Falk, who was quite tall, held the canopy on his shoulders while everyone got out. After that, explosive jacks were fitted to the canopy for ground jettison. The next job was to raise the aircraft, by digging holes under the wings for lifting jacks, and it was then revealed that the ground had been used by cavalry regiments for a century or so of Army occupation, and they were digging into feet-deep layers of manure! When the Vulcan was finally lifted, the undercarriage was fixed down with structural steel, and VX777 was flown back to Woodford.

The cause of the rudder failure was a fine example of how deadly tiny details can be. The power controls were all Boulton Paul electro-hydraulic units, in which an electric motor drove a swashplate pump (one with an angled driving block that could be swung in either direction to give flow either way) to a jack, which moved the surface. The pilot's control moved the swashplate of each unit, and for the rudder there was a standby unit that idled while the normal unit did the work, with a spring strut arrangement to keep the standby out of action until the normal unit stopped. On the wing units, four of which operated elevators and four on the ailerons (the Mk 2 changing to elevons), the philosophy was that if one stopped, there would still be three to keep going, and even two failures left half power available. In the case of the VX777 accident, the rudder had been kicked over so hard that the spring in the telescopic strut had expanded more than was intended, so that it jammed inside its tube. A simple problem really, and one that didn't take much of a modification to prevent it from happening again.

The next accident that affected us was the infamous Heathrow tragedy, at the end of a very successful tour of Australia and New Zealand, with Air Marshal Harry Broadhurst as second pilot. There was no technical failure of the aircraft this time, but the fact that the pilots were able to eject and the rear crew could not led to a media and political outcry about ejection seats for navigators and other crew, creating a great deal of design investigation for us. The basic cause of the crash was as old as flying itself: we must land at place A because the welcoming VIPs and brass bands are there, never mind the weather, even though our alternative place B is basking in sunshine. The inquiry into the affair was revealing and was conducted by an electronics pundit, and the Vulcan's final flight path was a perfect example of divergent oscillation due to the time lag between the GCA controller and the pilot. You could have reproduced it exactly on a cathode ray tube with appropriate time resistances for the two participants. The controller hadn't handled a fast jet before, and Podge Howard, the pilot, had not used civilian GCA before. The delay in the pilot's response to the controller's instructions, and the delay in the controller following the radar plot, led to the flight path swinging further and further above and below the correct one until the final low point caused the aircraft to hit the ground, pushing the undercarriage through the wing's flying controls.

When we came to investigate the possible fitting of rear crew ejection seats, the problems became quickly apparent. Apart from having to remove virtually the complete top of the pressure cabin, which would have been a structural nightmare, there was no way that we could get the rear crews out safely. As they sat facing aft, the seats would come out on a forward trajectory, and probably hit the tail, which the pilots' seats would probably clear. To enable the rear seats to fire on a

Factory-fresh XA891 pictured over the Mersey during a publicity photo flight from Woodford.

rearward trajectory meant a complicated drill of turning one round at a time, because of the cramped width, and the procedure would take so long they would never get out in time at low level. At high altitude there was no problem in getting out by the normal crew entry door as already designed. Although Martin-Baker did later offer a suitable escape system, it was never adopted, probably because the huge expense was felt to be wasteful for an aircraft that wouldn't be in service all that long, never realising just how long the Vulcan's service career would be.

Once the RAF got their hands on the Vulcan, they started touring all around the world, and one aircraft, XA908, suffered a major electrical failure while flying over Canada. The Vulcan B.1s had DC electrical systems with batteries to supply reserve power. The machine suffered a progressive generator failure in which the load cascaded in the main distribution system so that an initial failure of one generator led to all four cutting out. Following the accident all B.1s were modified, including our test aircraft at Woodford, but before our last aircraft (XA891) was modified it suffered the same fate soon after taking off from Woodford. Fortunately our Chief Test Pilot, Jimmy Harrison, and his crew were able to aim for open countryside and bale out close to Hull. Jimmy later remarked that he hadn't previously realised what a beautiful machine the Vulcan was until he saw it from above, presumably after he had got rid of the ejection seat face blind, and disengaged from his seat.

The last Vulcan accident for which we provided a design explanation, although there was no fault with the aircraft as such, was the crash of the 698 prototype, VX770, at Syerston in 1958. Our old friend had been passed to Rolls Royce at Hucknall for engine test work, and it was being flown at a Battle of Britain display by a Rolls Royce crew, when it broke up during a low flypast. The visual evidence was primarily drawn from an amateur cine film, from which was taken the famous picture that has been featured in many books and articles. We took the film to Chadderton and ran it through many times, but major changes took place between frames and the complete disintegration took place over no more than four or five frames. However, we had some solid technical evidence to work on. The accelerometer on the normal flight panel showed the maximum g-forces that had been applied, and we were also able to read the airspeed at which electrical power was cut off from the artificial feel units, and we quickly concluded that the combination was outside the aircraft's safe flight envelope. Once power flying controls were introduced, pilots had to be given some sort of artificial feel, pushing against springs in effect, to stop them from breaking the aeroplanes. The Vulcan's feel units were "q" feel, in which the leverage against the spring increased in proportion to the pitot pressure and was thus the square of the speed, which was the way manual control forces normally behaved. The actual mechanism was a lever and roller moving in a curved slot, known as the Banana Lever, shifted by an electric motor responding to a signal from a pitot

The elegant and clean lines of the Vulcan B.1 are dramatically illustrated in this image of Waddington Wing aircraft, pictured over Lincolnshire prior to an overseas deployment.

A beautiful image of XJ824 high above the clouds, proudly wearing the white rose marking applied to Finningley's OCU aircraft during the 1960s.

The tragic loss of VX770 at Syerston in September 1958 – caused by a simple case of over-exuberance, flying the aircraft beyond its design limitations.

pressure capsule, as in an airspeed indicator. So the position at which the levers stopped, when the aircraft disintegrated, gave us the speed at which it happened.

Power controls and the associated technology were taken out of my care early in 1956, but my group was still involved with cooling the controls and everything else too. On the Mk 1 the Boulton Paul units were inside the wing, which was relatively deep all the way along, and one unit was positioned in front of each section of control surface. Each had a small NACA-type intake in the lower wing surface ahead of it, and the outlet air spilled into the wing interior, escaping through gaps around the control surfaces. With the Mk 2 the two inboard units on each side were the same, but the new outer wing was too thin for the outboard units to fit inside, so they were mounted underneath the wing in long blister fairings. There was no fundamental difference from our point of view, and as we were busy with other work we let the Drawing Office get on with the new installation, without paying attention to what they were really doing. So on the Mk 2's first flight these outboard control units overheated drastically and the Drawing Office, backed up by the Project Designer's people, quickly descended upon us to find out what we had allowed them to do Wrong! I had a look at the drawings and pointed out that the units were in sealed fairings with air intakes, which they had faithfully copied, but there was no outlet, because there had been no obvious one before, so how could they expect air to go in and around the power unit, if it could not get out? There was a great fuss about needing to fly the next day, so I said that they simply needed a backwards-facing hole and the easiest way to get one would be to cut the back end of the fairing blisters. A quick 'guesstimation' of how forward to cut and off they went. The cooling was adequate thereafter and that is why, if you look at the Mk 2's control blisters, you'll see that they are cut off a few inches short of the trailing edge, even though the riveted flanges continue all the way. Designing with a hacksaw was one of the things that the great Roy Chadwick was noted for, so I felt like I was following a great tradition.

My last flight in a Vulcan was in the first Mk 2, as I claimed a flight every now and then just to see how things were feeling inside the aircraft, while in flight. This flight was also the only one where I rode in the second pilot's seat and was given a chance to handle the machine at altitude, not doing anything fancy, just gentle turns and speed variations. However, it was a notable flight as Jimmy Harrison did a beat-up over Woodford, culminating in a climbing roll, so I was able to experience one of the famous Vulcan manoeuvres that had been a feature of many displays up until that time. It was just as well that I did have a good forward view, as if I'd been in the back I wouldn't have known that we were doing anything more than a gentle turn. The trick of doing these rolls was to keep the manoeuvre barrelled just enough to keep a little over 1g all the way around.

I called in at Woodford during 1984 in connection with some work I was doing at the time, and in the reception hall I met a member of the flight shed staff from the late 1950s, and without introduction he immediately recognised me and recalled the day when Jimmy Harrison had rolled me in the Mk 2. Eventually the RAF asked Avro to stop rolling the Vulcan as some service pilots were trying it, but hadn't quite got the knack, so the airframes were in danger of being overstressed. One funny incident was when Jimmy visited Finningley for some reason, departed with the usual beat-up and roll, only to return to Woodford to find a letter from Bomber Command asking him not to do it any more – but the letter had arrived after he'd left!

Navigation on these test flights had some amusing moments too. Having described the typical flight plan of the early tests, I recall that the later flights tended to amble around the coastline, as even with extensive weather systems and low-level cloud there was normally some recognisable part of the coastline visible. If we could only see a small area of the ground, the first thing we checked was the colour of the soil, which tended to establish the region we were flying over pretty well, with red soil in Devon, chalky soil in Sussex, black in the Fenlands, and so on. If an aerodrome was visible the type of aircraft on it gave us a second clue, the RAF having a huge variety of aircraft in those days. One day, though, the crew had been flying on a steady course for some time while taking performance measurements that required steady conditions. We were below 40,000 feet when a flight of Meteors shot past, and one of the pilots commented that he'd never seen Meteors carrying red, yellow and black roundels before. It turned out that we were over Belgium! After that, the first visual check we made was to see which side of the road the cars were driving on!

In the summer of 1958 I attended a meeting at White Waltham where we discussed the whole question of escape from the Vulcan's cabin. To start with there was a move to co-ordinate the various connections that the crews, pilots in particular, had to make on getting into their seats, and therefore had to break on leaving to bale out, such as intercom, oxygen, suit air and so on. Each item had been developed by small departments in the Ministry, and a crew member had many separate pipes and plugs to undo every time he got out of the aircraft. Eventually two companies were given contracts to do what the Ministry apparently could not, to bring all the systems to a common point and develop a multi-way connector, so the crew member had only one fitting to connect. M. L. Aviation did the job for bomber aircraft, and they also had a contract to improve escape systems generally, so they built a full-size mock-up of the cabin, with no skin, so that one could see what was going on inside. In theory, every piece of equipment, or bracket on which one could get snagged, was in place. We gave them all the drawings and they gathered equipment from various sources, but I doubt if every item could have been fitted.

However, a meeting was set up at White Waltham, with myself and a deputy from Avro, representatives from the Ministry, the Air Staff, Institute of Aviation Medicine and so on, plus an RAF crew from Waddington to act as guinea-pigs in the cabin. The mock-up cabin was in a large room, with everyone seated on benches either side. It was an incredibly hot day, and we sweated away until M. L. Aviation brought in an air conditioning van they had built, and poked a large hose through the room's window to provide cooling air. It made so much noise that nobody could hear what was being said, so we opted to go back to the heat. The crew, of course, played up and pretended to be as clumsy and awkward as possible, so it was a lengthy performance. When it came to looking at belly landings, the pilots climbed over the side, having disposed of the canopy, but the rear crew pointed out that the narrow gap between the ejection seats would delay their escape. They suggested that they should have a switch to blow the pilots' seats out, but the pilots were obviously not very enthusiastic about the rear crew having the power to shoot them out if they had a disagreement or, more seriously, they might trigger the switch by mistake. This was the tone of the meeting, and it continued throughout the day, finally ending at about seven with no real conclusions having been reached.

I left Avro at the end of 1960 to try my luck in general industry, and it was the only company that I was ever really sorry to leave. The last year or so had been darkened by the shadows of missiles. Avro set up their missile division at Woodford and introduced a bunch of rather high-toned individuals from Farnborough and places like that. Their attitude was that their work was far beyond anything that we had being doing, and our old-fashioned ideas about sound engineering didn't really apply to missiles. However unprecedented the performance we were designing for, we always had been very careful to stick to sound principles of reliable engineering. The missile work was regarded as being so advanced and secret that the service requirements couldn't be discussed with us, and we just had to provide the airflows or whatever else was called for, and not ask any questions. It reminded me of the radar cans mentioned earlier, and no matter what we always thought of the Vulcan as "our" aeroplane.'

An early publicity photograph of a 230 OCU Vulcan B.1 at Waddington, clearly illustrating the revised 'cranked' wing leading edge.

CHAPTER FIVE

Into Service

FOLLOWING the issuing of Service release on 31 May 1956, the first Vulcan delivery to the RAF was made on 20 July 1956 when XA897, the ninth production aircraft, flew to RAF Waddington, joining the newly formed No 230 Operational Conversion Unit. The arrival was essentially symbolic, however, as XA897 soon returned to Woodford to undergo a series of minor modifications in preparation for a long-range 'goodwill' flight to New Zealand. The next arrival at Waddington was XA895, which was delivered to No 230 OCU on 16 August as a substitute, although this aircraft was also only a temporary resident, spending most of its time with the A&AEE on Operational Reliability Trials.

Vulcan XA897's flight to New Zealand was primarily a diplomatic 'flag-waving' exercise, although it also presented an opportunity to test the Vulcan's long-range capabilities. The aircraft captain was Squadron Leader Donald Howard, with a very distinguished VIP taking the co-pilot's seat, the Commander-in-Chief of Bomber Command, Air Marshal Sir Harry Broadhurst. The rear crew comprised Squadron Leaders Albert Gamble, Edward Eames AFC and James Stroud, who was also a qualified Vulcan pilot and who would be able to take turns on the flight deck with Broadhurst; they were joined by Fredrick Bassett, a technical representative from Avro. Although the Vulcan had only just entered RAF service, the outbound flight went remarkably smoothly, via Aden and Singapore and on to Melbourne, with a flying time of 47 hours and 26 minutes including stop-overs. After visiting Sydney and Adelaide, the crew flew on to Christchurch on 18 September to complete what was a satisfactory and enjoyable journey. For the return trip the aircraft was flown via Brisbane, Darwin, Singapore, Ceylon and Aden, finally departing for the UK at 02:50 on the morning of 1 October 1956. The seven-hour flight back to England was scheduled to culminate with a VIP reception at London's Heathrow Airport, where representatives of the RAF, Ministry of Aviation and Avro were joined by the crew's families and the media. Unfortunately the weather conditions were very poor, and from inside the Queen's Building the airfield was completely invisible, shrouded in cloud and heavy rain. As XA897

Another suitably posed publicity photograph of an early Vulcan B.1 with 230 OCU at Waddington.

The OCU flight line at Waddington shortly after the first Vulcans were delivered to the RAF. As can be seen, RAF crews were already gaining a great deal of jet bomber experience flying the Canberra.

Above: Vulcan B.1 XA895 spent a great deal of time on developmental duties and was assigned to the Bomber Command Development Unit at Finningley in the early 1960s. Unusually, the aircraft carries BCDU titles on the tail fin. *Ray Deacon*

XA907 completing a landing run at Waddington, air brakes out and brake parachute streaming. The aircraft was subsequently transferred to the BCDU at Finningley before being withdrawn late in 1966. *Robin A. Walker*

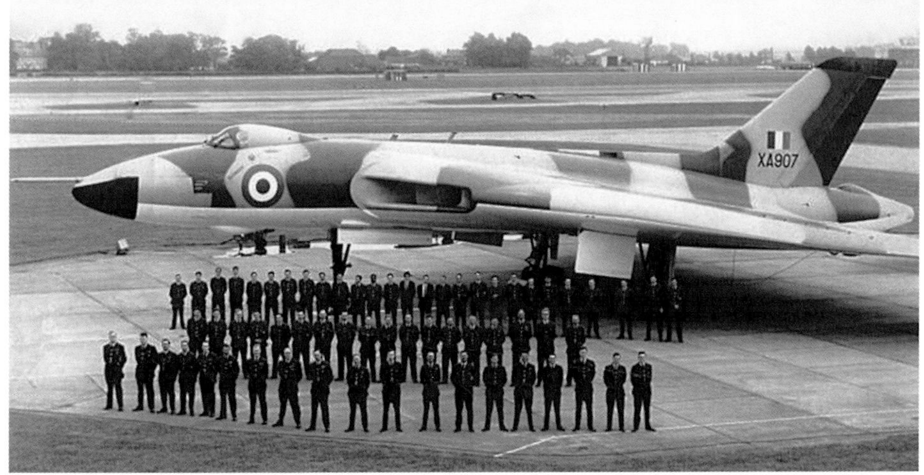

After transfer from Waddington to the BCDU at Finningley, XA907 received low-level tactical camouflage. This image shows her on a single-aircraft dispersal, with the ORP adjacent to the runway visible in the distance.

approached the English Channel, Howard called Bomber Command Operations at High Wycombe and received the bad news that Heathrow's current weather status was eight-eighths cloud at 700 feet, with two-eighths cover at 300 feet. The airfield was in rain and visibility was just over 3,000 feet.

Although the VIP spectators were somewhat disappointed by the prevailing conditions, there was no reason why XA897 could not land at Heathrow, as normal airliner operations were continuing without difficulty. However, Howard did have the option of diverting to Waddington, where the weather was much better, but the decision was his, and in view of his extensive flying experience he saw no reason to divert, and Broadhurst agreed. The aircraft entered an instrument approach at 5 nautical miles from touchdown, at 1,500 feet, and the hardy souls on top of the Queen's Building stepped out into the rain, determined to brave the elements and watch for the Vulcan's arrival. The GCA talk-down began as normal, with glide path and centre-line corrections being given until the aircraft was three-quarters of a mile from the runway, when the ground controller instructed the captain that he was 80 feet above the glide path. This was the last elevation advice received from the controller, and a few seconds later the Vulcan suddenly hit the ground. The first indication the assembled spectators had of the Vulcan's proximity was a roar of engine noise as all four engines wound up to full power. XA897 came into view, climbing steeply from the runway, with its landing gear extended. The noise quickly subsided and at around 800 feet the canopy was jettisoned with both pilots ejecting, leaving the Vulcan to turn sharply to starboard before making a 30-degree descent back to the runway, where it instantly exploded on impact, killing all four rear crew members.

The VIPs inside the Queen's Building were informed that the aircraft had crashed. As the crash rescue crews battled to extinguish the burning wreckage (which continued to burn for almost an hour) it quickly became clear that the aircraft's main wheels were missing. Beyond the 300-yard scattering of wreckage, there was still no sign of the missing undercarriage, and Avro's Chief Aerodynamicist, Roy Ewans (who had been waiting for the Vulcan's arrival at Heathrow), drove along the runway to the airport boundary, where still nothing could be seen. He then drove out from the airport and proceeded to search the fields surrounding the runway approach. He soon found two deep holes that were rapidly filling with rainwater. The holes had obviously been created by the Vulcan's main gear as it had briefly hit the ground, short of the runway by 1,988 feet (displaced to the north by 250 feet). Ahead of the holes were two swathes of Brussels sprouts that had been flattened by XA897's exhausts as the aircraft had begun to climb away. The main wheels were scattered about the field.

Few aircraft have landing gear designed for landing on soft ground, but the loss of the Vulcan's lower leg assembly would not in itself have been too serious. However, the excessive backwards force created by hitting the ground had caused the drag struts to fail, allowing the legs to swing back on their main hinges to hit the lower surface of the wing, ahead of the trailing edge control surfaces. The gear assembly had penetrated the wing at a point where the aileron control rods were located, damaging them to such an extent that all lateral control was lost.

The following day an RAF Court of Inquiry was opened, headed by Air Chief Marshal Sir Donald Hardman. On 26 October the Minister of Transport and Civil Aviation requested that Dr A. G. Touch, Director of Electronics Research and Development at the Ministry of Supply, should carry out an independent investigation into the accident, particularly the GCA (Ground Controlled Approach) arrangements at Heathrow. The conclusions reached by Dr Touch's report were as follows:

> 'The GCA equipment was correctly set up and calibrated. There is no evidence of malfunctioning or failure. The Controller failed to warn the pilot of his closeness to the ground. During the last ten seconds of the approach the aircraft made a steep descent to the ground. The cause for this descent was probably due to the build-up of oscillations about the glide path. Poor talk-down by the Controller contributed to things, but as the approach was subject to the overriding judgment of the pilot, the Controller was not to blame for events

XA911 was one of many early Vulcan B.1s that enjoyed a relatively brief but successful service life, mostly with the Waddington Wing, before being withdrawn and scrapped around 1968. *Robin A. Walker*

The ill-fated XA897 pictured shortly before her final journey, which ended in disaster at Heathrow.

arising from the control. The critical phase was the first four seconds after the descent steepened, during which no height guidance was given to the pilot. It is very difficult to pass judgment on this matter, but in view of all the circumstances I do not think the Controller should be blamed. No warning was given during the final five or six seconds. It should have been, although it would have been too late. Although it cannot definitely be proved, the most likely theory is that the Controller made an error of judgment, concentrating too much on azimuthal correctional and paying insufficient attention to the elevation error meter. Human errors are more likely to occur under stress or unusual circumstances. In my opinion, evidence exists to show that all the elements in the GCA "servo-chain" were strained.'

Dr Touch later explained that the term 'servo-chain' referred to the way in which the GCA information was acted upon by the pilot, in a kind of 'servo-loop' between the pilot and the ground controller. The post-accident technical investigation – part of the RAF Court of Inquiry – confirmed that the aircraft had not suffered any technical failures. The Inquiry did conclude that the captain of the Vulcan had made an error of judgment in selecting a break-off height of 300 feet and also in going below that height. The ground controller had advised the captain that he was 80 feet above the glide path just seven seconds before hitting the ground, but no subsequent warning was given when the aircraft rapidly went below the glide path; indeed, the talk-down continued after the Vulcan had hit the ground, as if the approach was still continuing normally. Consequently the RAF Inquiry concluded that this failure to warn the pilot that he was below the glide path was the principal cause of the accident. This was a slightly different conclusion from that made by Dr Touch, but as it is RAF policy not to make Court of Inquiry evidence public, Dr Touch remained unaware of the RAF's views.

At the inquest into the deaths of the rear crew members, another contributory cause became evident – altimeter error. In his summing-up the Coroner, Dr H. G. Broadbridge, stated that there was nothing in the evidence to show criminal negligence on the part of anyone in the aircraft or on the ground, and everyone seemed to be doing their duty as they thought right at the time. Squadron Leader Howard said that before leaving Aden he had received a signal from Bomber Command saying that he was to land at Heathrow.

'I was going to make an attempt to land, in view of the weather. If I could not I was going to overshoot and go to Waddington, where it was promised that the weather would be very good. I decided to come down to 300 feet on my altimeter, which represented to me a minimum approach altitude for London Airport of 150 feet over the ground.'

Howard continued, saying that the talk-down from the ground controller was normal, and he acted on the information as soon as he received it. Before the talk-down was completed he hit the ground. Although he did not know it at the time, he now knew that he had been ahead of the runway when he hit the ground, and the last instruction he remembered hearing was that he was three-quarters of a mile from touchdown at 80 feet above the glide path. He increased the rate of descent. Continuing his account, Howard said:

'I asked the co-pilot to look for the high-intensity lighting which I was going to use for the landing. He told me he could see the lights over to starboard, and all this time I was looking at instruments and not looking out. I looked at the lights as he told me, and I did not recognise the pattern. They were not what I expected to see. Immediately I had looked I went back on instruments and he then told me I was very low and to pull up and so I did. At that precise time the aeroplane touched the ground and I decided to overshoot. This I tried to do, but as the aircraft accelerated it became obvious that I could not control it any more. It wanted to roll over to the right. I used all the control I had but I could not stop it and I realised I could do no more. My altimeter was showing slightly below 300 feet. I shouted to the crew to get out and when it was apparent that the aircraft was going to roll into the ground, I decided to eject.'

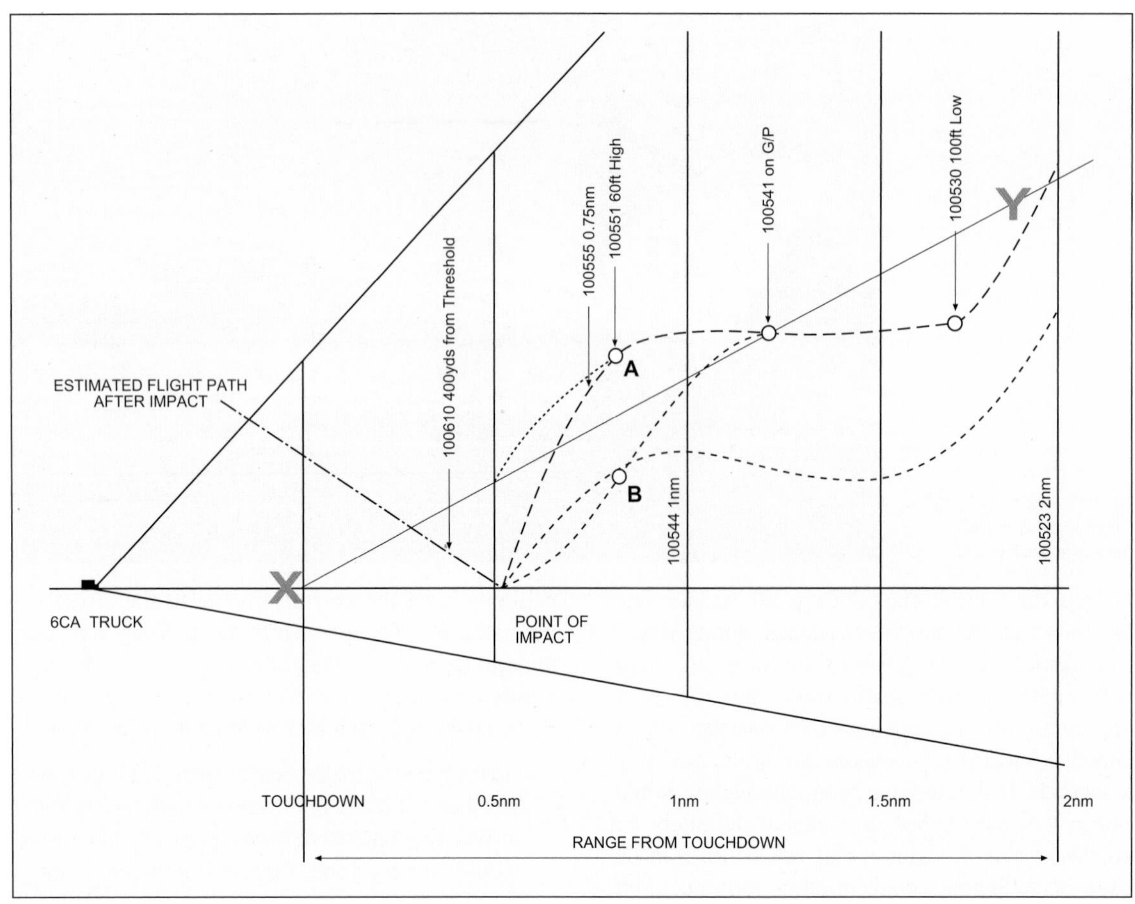

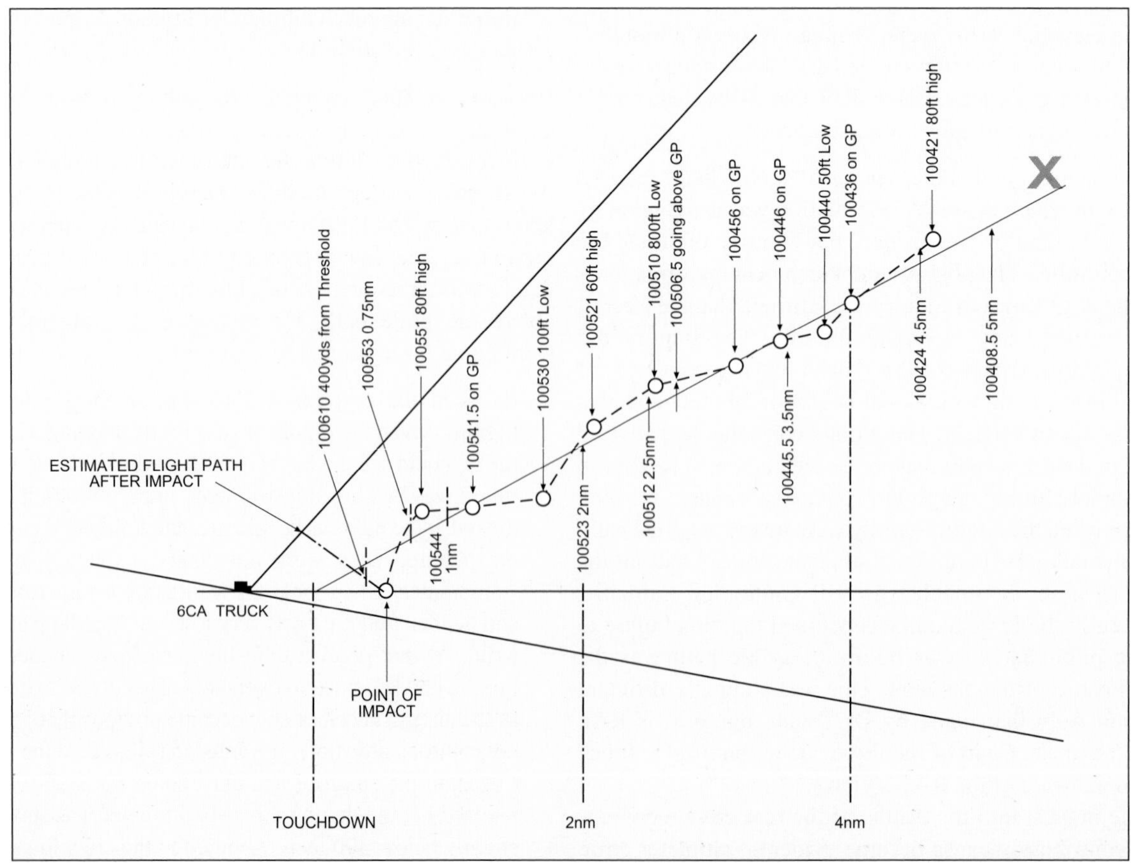

Diagrams illustrating the path of XA897 as it approached Heathrow.

Squadron Leader Howard was asked if the Vulcan's altimeter was gravely misleading him. He believed that it was, but he did not know how this could be accounted for. He stated that he held a Master Green instrument rating, the highest that an RAF pilot could have. Continuing, he said that his altimeter had a known error of 70 feet, and he set 80 feet as the height of London Airport above sea-level. He agreed that if he had been on the glide path at three-quarters of a mile from touchdown, he should have been at about 260 feet, and if the altimeter had been functioning correctly he would have aborted his attempt to land at that stage. Air Marshal Sir Harry Broadhurst was also questioned by the Coroner, and he commented that the talk-down was normal: 'There was nothing in it to alarm you. It seemed perfectly safe.' After hitting the ground with a glancing blow, he was convinced that no damage had been done, and had even said, 'If we turn slightly left we can still make it.' However, the captain had answered, 'No. I'm going to overshoot,' but quickly added, 'I think we've had it.' Broadhurst tried the controls as Howard ejected, but failed to get any response, so he also ejected.

Returning to the altimeters, Broadhurst said that there was an error of between 50 and 80 feet between his and the captain's. The ground control tracker, Miss A. C. Maley, said that as she watched and advised on height and range, she could not recollect any rapid descent. When the Vulcan was about 2 miles from touchdown it appeared to drop roughly 100 feet below the glide path, but she agreed that it must have recovered height, as later in the approach it was above the glide path again. Squadron Leader Howard was asked if it would be an abnormal rate of descent to drop 300 feet in 500 yards. Howard said this would be 'fantastic, about 4,000 feet per minute' and that he would certainly have known if they had been going down at that rate.

Finally, as if to confirm the underlying cause of the accident, Wing Commander C. K. Saxelby from the A&AEE said that it had been discovered that the 70-feet altimeter error could become as much as 130 feet when close to the ground, added to which there was the 80 feet for Heathrow's height above sea-level. In addition there could have been a further error of 70 feet because of friction phenomena. Despite numerous investigations and a variety of theories, no single factor was ever identified as the principal cause of XA897's tragic accident. Clearly the crew members were not to blame, and, as Dr Touch's report indicated, neither was the ground controller. The crash seemed to be the result of a culmination of errors.

Sir Harry Broadhurst commented on the accident:

> 'The whole thing is a puzzle to me. It seemed to me an absolutely normal glide approach until the ground appeared in the wrong place. If we had been coming down at an unprecedented rate we would have hit the ground, and the undercarriage would have been forced up into the wings. As it was we touched so lightly we merely thought the aircraft had burst a tyre or something. We had no idea that the undercarriage had been ripped off. Until then the captain, obviously very experienced, imagined he was being talked-down normally. The fact is, they were still talking him down normally after he had gone up again. Obviously something went wrong. We cannot supply the missing links.'

Although the rear crew lost their lives, it was something of a miracle that the two pilots escaped. When XA897 returned to Woodford for modification, prior to embarking on the New Zealand trip, Avro had just designed a new system to interconnect the canopy release with the operation of the seat pull-down blinds that actuated the ejection sequence. Instead of requiring the pilot to make two separate actions – one to release the canopy and the second to pull the seat handle – the single operation of either seat blind handle would now blow off the canopy and fire the seat in automatic sequence. Although the aircraft was scheduled to be delivered without this modification, the change was made prior to the overseas tour, and without it there is some doubt as to whether the pilots would have had time to eject safely from the Vulcan. It is also interesting to note that Sir Harry Broadhurst later went on to become Managing Director of A. V. Roe, after retiring from the RAF.

It was not until January 1957 that No 230 OCU finally received two Vulcans (XA895 and XA898) on a more permanent basis, and following a period of intensive flying trials the task of training Vulcan crews began on 21 February. During March, April and May three more aircraft were delivered (XA900, XA901 and XA902) and it was on these machines that the first OCU course qualified, graduating on 21 May 1957. No 1 Course then re-formed as 'A' Flight of No 83 Squadron, the first RAF Vulcan squadron, and after a brief spell during which the unit borrowed aircraft from the OCU, No 83 Squadron's first aircraft, XA905, was handed over on 11 July, the day on which the squadron was commissioned. Although the OCU Vulcan B.1s had been fitted with Olympus 101s, the five aircraft delivered to No 83 Squadron were powered by

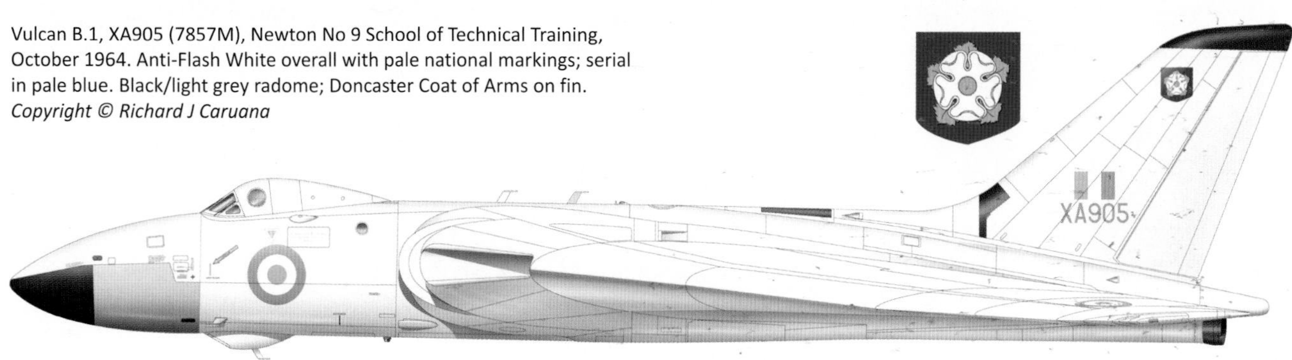

Vulcan B.1, XA905 (7857M), Newton No 9 School of Technical Training, October 1964. Anti-Flash White overall with pale national markings; serial in pale blue. Black/light grey radome; Doncaster Coat of Arms on fin.
Copyright © Richard J Caruana

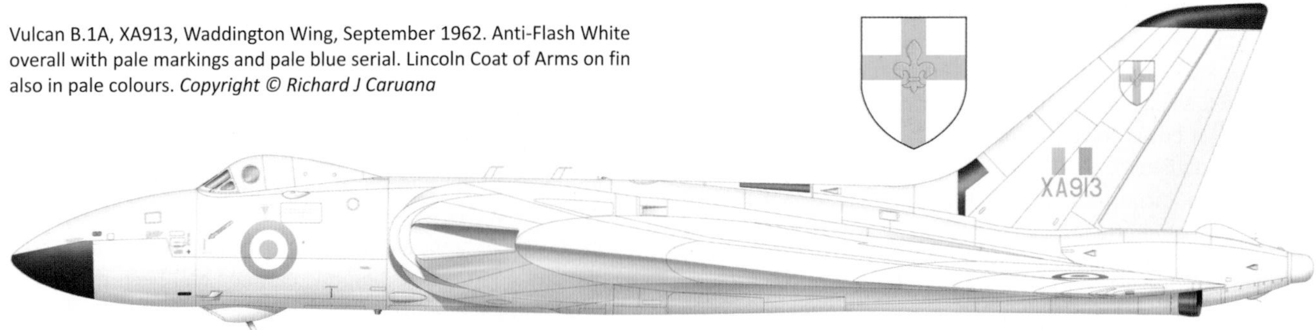

Vulcan B.1A, XA913, Waddington Wing, September 1962. Anti-Flash White overall with pale markings and pale blue serial. Lincoln Coat of Arms on fin also in pale colours. *Copyright © Richard J Caruana*

12,000lb-thrust Olympus 102s (the B.1 fleet eventually standardised on 13,500lb Olympus 104s). The second OCU course also became a Flight of No 83 Squadron, and subsequent courses, upon graduation, transferred to Finningley, where No 101 Squadron was formed on 15 October 1957. By the end of the year four aircraft had been delivered to Finningley, and the last aircraft from the 1952 order (XA913) had finally been handed over to the RAF.

At the start of the new year the first of a new batch of B.1s was completed, as part of a thirty-seven-aircraft order that had been placed in September 1954. Initial OCU and squadron flying proceeded surprisingly smoothly and no major problems were encountered during the Vulcan's introduction into RAF service. In fact, comments made in later years suggest that the Vulcan enjoyed one of the most trouble-free entries ever into RAF service. The aircraft were immediately operated far and

A suitably posed RAF PR photograph showing a factory-fresh 83 Squadron Vulcan. The Vulcan's tall undercarriage units made servicing access relatively simple, as can be seen by the accompanying groundcrew. Most access doors were located under the wings and fuselage.

XA900 was one of the first Vulcan B.1s to join the Waddington Wing, and after retirement the aircraft was housed at Cosford's Aerospace Museum. Despite becoming the sole surviving Vulcan B.1 airframe, the museum subsequently scrapped the aircraft after claiming that the cost of restoring it would be prohibitive. It was with some irony that huge sums of money were later spent on creating a new museum building at the site.

wide across the globe, participating in exercises and goodwill tours to exotic locations such as the United States, New Zealand, Libya, Kenya, Rhodesia, Brazil and Argentina.

Two Vulcans entered the Strategic Air Command Bombing Competition at Pinecastle AFB in Florida during October, the RAF returning after an absence of several years. In company with two Valiants, the Vulcans made quite an impression on their American audience, but their crews did not enjoy any great success in the competition, being placed 44th overall. The results were disappointing, but, considering the fact that the Vulcan had only just entered service, it was probably something of an achievement just to have taken part. AOCinC Bomber Command Sir Harry Broadhurst said that, 'The results have rather hinged on the experience of the ground crews in maintaining the equipment rather than the aircrews using it.' Other factors also conspired to create difficulties, such as the hot and humid conditions, which affected the Vulcan's electronic equipment, the operational heights, which were lower than the altitudes at which the Vulcan crews had been training, and the lack of experience that both the air and ground crews had with their new aircraft. But the SAC competition was just the start of a very long relationship between the RAF and USAF's Strategic Air Command, which marked a degree of integration between America's and Britain's forces that had not been experienced since the Second World War; as Sir Harry commented, it also demonstrated that Britain was 'back in the nuclear club'.

Sadly, one tragic accident marred the success of the Vulcan's early years – the loss of 83 Squadron's XA908 on 24 October 1958. The aircraft was flying on a 'Lone Ranger' exercise from Goose Bay in Labrador to Lincoln AFB in Nebraska when, approximately 60 miles north-east of Detroit, the Vulcan's main power supply failed, and although the engines continued to feed the electrical generators, a short circuit in the main busbar blocked the power supply. This should, in theory, have presented few problems for the crew, as the aircraft had battery standby power for some twenty minutes of further flying, so the captain requested an emergency descent to Kellogg Field in Michigan. Unfortunately the batteries supplied power for only three minutes, after which the powered flying controls ceased to operate. The captain immediately requested directions to the nearest landing field, but without any power the Vulcan' s control surfaces were useless, and the aircraft flew into the ground at a 60-degree angle, killing all but one of the crew. The co-pilot managed to eject, but tragically he landed in Lake St Clair, and as the only member of No.83 Squadron who could not swim (and without a lifejacket) he drowned.

Following this accident, the Vulcan B.1 fleet was quickly modified – the main busbar being divided in two – to prevent any such disaster happening again. Unfortunately, however, it did happen again when XA891 suffered an electrical failure on 24 July 1959. This Vulcan had yet to receive the electrical system modification, and shortly after take-off from Woodford (where the aircraft was flying trials with Olympus 200 engines) the same problem arose. This time, however, the batteries provided a greater reserve of power and the entire crew was able to parachute to safety. Despite such tragic instances, the Vulcan proved to be a very reliable aircraft, enjoying an excellent safety record, probably as a direct result of Avro's adherence to design simplicity and reliability.

Some doubt had certainly been cast over the Vulcan's safety when the prototype 698 (VX770) was destroyed at Syerston near Nottingham on 20 September 1958. The airframe broke up during a fast pass over the airfield during a Battle of Britain display in front of a crowd of horrified spectators. Investigation revealed that structural failure of the wing had been the cause of the accident, but more detailed analysis confirmed that the aircraft had been flying outside the safe speed and 'g' flight envelope, so the aircraft's design was clearly not at fault and the accident was a classic case of over-exuberance on the part of the pilot. Although the Vulcan was an astonishingly manoeuvrable machine it was often all too easy to forget that it was also a very big and bulky four-engined bomber, not a fighter, and like any other aircraft it could only be operated safely within clearly defined limits.

While deliveries of the Vulcan B.1 were still being made, Avro was already looking towards a second-generation Vulcan design, based on the promise of even greater thrust output from Bristol's Olympus engine series. During the first half of 1955 sufficient data was available to suggest that the Bristol B.01.6 engine would be capable of delivering 16,000lb thrust, with the prospect of even more powerful versions at a later stage. Design analysis indicated that, without suitable modifications, a Vulcan fitted with the more powerful engines would suffer from the same high-speed and high-altitude buffet problems that the Phase Two wing had been designed to counter. Clearly the wing would have to be redesigned again, and in September 1955 Avro submitted a brochure to the Ministry of Supply describing what would ultimately become the Vulcan B Mk 2.

Avro had been discussing the possibility of a Vulcan Mk 2 with the MoS for some time and, having gained a great deal of encouragement from the Ministry, funds had been made available within the company to initiate development of the new design. Although the revised wing layout was the most obvious external difference between the B.1 and B.2, many other changes were made to the Vulcan's design. Later it was switched from high-altitude to low-level operations and it was these less noticeable modifications that did much to improve what was already an outstanding aircraft. The airframe was re-stressed to a gross weight of over 200,000lb, which was more than double the figure originally set by Specification B.35/46. The landing gear was redesigned and strengthened to withstand much greater take-off and landing weights. The electrical system was changed to 200V AC, and instead of direct-drive 112V generators each engine had constant-speed drive and alternators. A gas turbine auxiliary power unit was installed, driven by a Rover 2S/150 engine, and the flying controls were changed, the ailerons and elevators being replaced by inner and outer elevons (combined ailerons and elevators), each with independent power control units. The new Phase 2C wing design extended the span by 12 feet (to 111 feet), while the thickness/chord ratio changed from 7.92 per cent to 4.25, and the wing area increased from 3,446 to 3,965 square feet. The compound taper of the leading edge was increased further, and the trailing edge was also now swept.

Construction of the B.2 was not, however, a simple continuation of the Vulcan's development process and, had circumstances been any different, the B.2 may never have even reached the production stage. Governmental discussions on the projected size of the V-Force (the Medium Bomber Force) had been continuing since 1954, and the ultimate size of the force was the subject of much debate. It was not until 31 May 1956 that Ministers agreed to proceed with the construction of more Vulcans and Victors, both of which would be of an improved 'second generation' design. Initially the plan was to order both types, but both with Rolls Royce Conway engines. There was little enthusiasm for proceeding with Bristol's Olympus engine and the MoS was invited to investigate whether the engine should be supported only as a civil programme. However, Bristol, when faced with the cancellation of the Olympus Mk 6, offered to carry all further development costs itself, and would also offer to provide completed engines at the same price as the Conways if an order for 200 was placed. It was this offer that ultimately saved the Olympus programme and resulted in further developments that went on to power the Vulcan, TSR2 and, eventually, Concorde.

The Minister of Supply, Reginald Maudling, commented that:

> '...there are at present 99 Vulcans on order and I am proceeding on the assumption that there is no suggestion that this number should be reduced. We have been re-examining the Vulcan programme and it now looks as if the Mk 2 version can be introduced at about the 45th aircraft. This means that 54 Vulcans should be the new and greatly improved Mk 2 version.'

Despite this statement there was still a great deal of doubt as to precisely how many medium bombers could be comfortably afforded. Even the Admiralty stepped in and stated that the force should not 'cost so much inside the total resources set aside for defence as to make it impossible to finance the other forces essential to ensure the cohesion of Commonwealth, dependencies and alliances'. Considering the way in which the Navy's Trident programme now takes up a significant proportion of modern defence spending, this statement has more than a little irony attached to it. Finally it was decided that the MBF should comprise a front-line strength of 144 aircraft, of which some 104 aircraft would be Mk 2 Vulcans and Victors.

A contract for one Vulcan B.2 prototype was received in March 1956. A further contract, issued in June, converted the order for the final seventeen B.1s into B.2s and added eight more machines. Interestingly, these aircraft were covered by Specification B.129P Issue 2, which referred to this new batch of Vulcans as being B(K) Mk 2 versions, suggesting that the aircraft would also have an in-flight refuelling (tanker) capability. This option appears to have quickly been abandoned, although design drawings for a tanker configuration certainly existed. It was, of course, a quarter of a century later that the Vulcan was introduced into the tanker role.

The second B.1 prototype (VX777) was selected for conversion into the B.2 prototype, and on 31 August 1957 it made its first flight in the revised configuration. As the B.2 development programme continued, Avro also received instructions to equip the Vulcan fleet with new ECM (electronic counter-measures) gear, reflecting the ever-increasing capabilities of Soviet forces. This required a fairly radical redesign of the rear fuselage in order to accommodate the bulky equipment. Until this stage in the Vulcan's career there had been little need for an advanced ECM fit, thanks largely to the Soviet Air Force's rigid system of fighter control, which was well monitored by the West. Only a limited number of VHF channels were used to control Soviet fighters, and just a single piece of equipment (code-named 'Green Palm') was capable of jamming all of them with a high-pitched wail. However, post-1960 developments suggested that something rather more sophisticated was now required, and a tail warning radar ('Red Steer') was to be fitted into the Vulcan, together with other ECM equipment, including a flat-plate aerial installation ('Red Shrimp'), which was accommodated between the starboard jet pipes (and between both sets of jet pipes on later B.2s). Although the new ECM fit was intended primarily for the B.2, a contract was also received to convert part of the B.1 fleet to B.1A standard by installing the new tail cone and plate aerials.

Above: This pre-scrapping image of XL444 shows the open ECM bay doors covering the bay in which the transmitter and power units were housed. The brake parachute housing is also popped open. *Joe L'Estrange*

Right: XL317 illustrating the semi-recessed Blue Steel missile and 'Red Shrimp' ECM under the starboard jet pipes.

Consideration was also given to the rather more ambitious idea of converting the B.1 fleet to the full B.2 standard, but as the cost of each aircraft conversion was estimated at roughly two-thirds of the cost of a new-build B.2, the idea was too expensive to seriously consider. However, some twenty Vulcan B.1s from the 1954 contract were eventually converted to B.1A standard, together with nine machines (those in the best condition) from the original 1952 contract. With VX777 back in the air following modifications, the new Phase 2C wing was displayed at the 1957 Farnborough show, where the prototype B.2 performed, powered by 12,000lb Olympus 102 engines. After the show the aircraft was put through a series of aerodynamic trials, while a further seven B.1s were brought into the B.2 development programme, ensuring that every

The last four-ship Vulcan scramble, at Finningley in September 1981. Each aircraft carries a different squadron marking; from right, XM575 of 44 Sqn, XL359 of 35 Sqn, XL444 of 9 Sqn and XM648 of 101 Sqn. *Fred Martin*

aspect of the redesigned Vulcan could be properly explored. Early in 1958 XA891 was fitted with the first Olympus 200-series engines, each rated at 16,000lb. Meanwhile XA890 was employed on avionics work and XA892 on weapons research. XA893 was fitted with the B.2's electrical system, XA894 was used for development of the B.2/B.1A ECM fit, while XA899 was also used for avionics research, and XA903 was later employed as a Blue Steel missile trials aircraft.

Flight trials with the B.2 prototype revealed a 25-30 per cent increase in range over the B.1, effectively extending Bomber Command's target range by a similar amount. Certainly the high-speed and high-altitude buffet problem had been cured, and on 4 March 1959 the first production Vulcan B.2 (XH533) reached 61,500 feet during a test flight. XH533 made its first flight on 19 August 1958 powered by 16,000lb Olympus 200s and still fitted with the original unmodified tail cone. It is interesting to note that the aircraft first flew before production of Vulcan B.1s had even been completed, with the last B.1 making its first flight in February 1959. Likewise, the second production B.2 (XH534) took to the air ahead of the last B.1 in January 1959, and was the first aircraft to fly with production-standard Olympus 201s rated at 17,0001b, together with a full ECM fit; however, the next three aircraft from the production line first flew with non-ECM tail cones.

As production of the Vulcan B.2 got under way, the first seven aircraft (including the prototype) were all used for trials work, XH534 being flown on CA release trials at Boscombe Down. The CA release was given in May 1960, and on 1 July 1960 XH558 was delivered to No 230 OCU at Waddington, becoming the first Vulcan B.2 to enter RAF service; many years later this particular aircraft also became the last Vulcan B.2 to leave RAF service. XH559 was delivered in August, with XH560, XH561 and XH562 arriving before the end of the year. XH557 was loaned to Bristol Siddeley for further engine trials, and was fitted with what was to be the ultimate Vulcan power plant, the Olympus 301 engine rated at 20,000lb thrust. Two engines were fitted, one in each of the outer nacelles, flying for the first time on 19 May 1961. Although these immensely powerful engines required a greatly increased amount of intake airflow when compared with the early Vulcan B.1 power plants, the first production Vulcan B.2s retained the same air intakes as the B.1, but on later aircraft the lower lip

B.2 XL317 of 617 Sqn, a Blue Steel trial veteran, at Mildenhall in September 1978. *Fred Martin*

of the intake was deepened (starting with XH557) and all subsequent B.2s were manufactured accordingly to accommodate the increased airflow mass. XH557 was later fitted with four Olympus 301s before entering RAF service, and approximately half the Vulcan B.2 fleet was completed or retro-fitted to this standard, the remainder utilising 17,000lb Olympus 201s. Contrary to popular belief, the Olympus 300-series aircraft were still designated Vulcan B.2s, just like their 200-series-engined counterparts. The much-quoted 'B.2A' designation never actually existed, even though it has persisted in books and magazines throughout the aircraft's history.

Although continued Vulcan production enabled the RAF to form more squadrons, policy dictated that the B.2s should first be delivered to an established unit, which would pass on its B.1s to newly converted squadrons. Consequently the first crews to leave B Flight of No 230 OCU went to No 83 Squadron at Scampton, where transition to the B.2 began in November 1960. On 1 April 1961 No 27 Squadron formed at Scampton on the B.2, and the Scampton Wing was completed in September 1961 when No 617 Squadron (the legendary 'Dambusters') started converting from B.1s on to B.2s. Waddington standardised on the B.1A, with No 44 Squadron forming on 10 August 1960 with aircraft that had previously been operated by No 83 Squadron (the first B.1A being delivered to No 44 in January 1961). No 101 Squadron relocated from Finningley in June, and No 50 Squadron formed on 1 August (using former No 617 Squadron aircraft). The Vulcan OCU relocated to Finningley, where it was divided into A Flight, with Vulcan B.1/B.1As, and B Flight, with Vulcan B.2s. Finningley also provided a base for the Bomber Command Development Unit, which operated a mixed fleet of Vulcans, Valiants and Victors for various trials operations. The second Vulcan B.2 Wing was established at Coningsby, where IX Squadron formed on 1 March 1962, followed by No 12 Squadron on 1 July and No 35 Squadron on 1 December. The Wing later moved to Cottesmore in November 1964, and the last production Vulcan B.2 (XM657) was delivered there on 15 January 1965, where it joined No 35 Squadron.

Following the initial order for twenty-five B.2s, a further contract had been issued for an additional twenty-four aircraft, followed by a final order for another forty machines. The final Vulcan B.1A conversion was XH503, which was delivered to Waddington in March 1963, after which the remaining B.1s

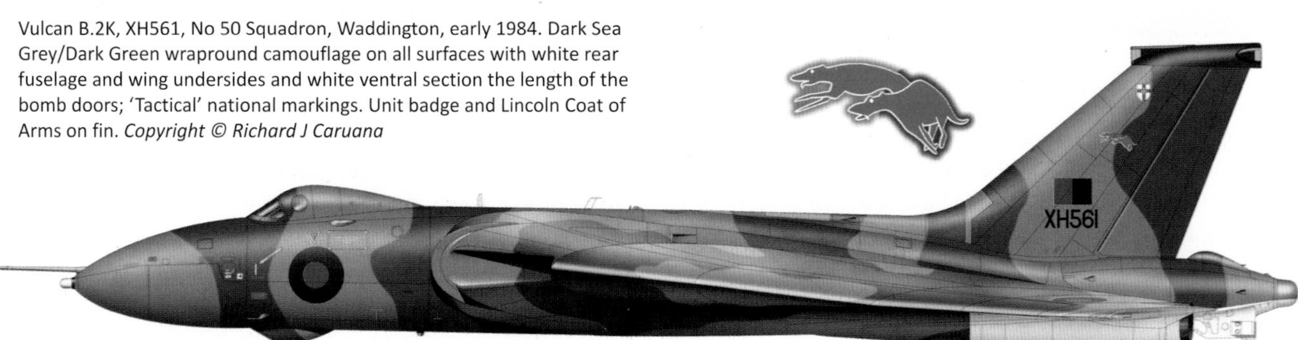

Vulcan B.2K, XH561, No 50 Squadron, Waddington, early 1984. Dark Sea Grey/Dark Green wrapround camouflage on all surfaces with white rear fuselage and wing undersides and white ventral section the length of the bomb doors; 'Tactical' national markings. Unit badge and Lincoln Coat of Arms on fin. *Copyright © Richard J Caruana*

Flight testing of later versions of the Olympus engine (destined for the TSR2) was conducted by Bristol Siddeley from Filton, using Vulcan XA894 fitted with an under-fuselage pod. The aircraft undertook almost eighty hours of testing until 8 December 1962 when, during a static ground run, an uncontained turbine failure led to the complete destruction of the aircraft (below). Subsequent engine flight testing (mostly connected with the Concorde programme) was conducted using Vulcan B.1 XA903 (above).

were either withdrawn from use or transferred to the OCU and the Bomber Command Development Unit at Finningley. Starting in 1966, the Waddington Wing also began converting to Vulcan B.2s, completing the transition at the end of 1967. The OCU then relinquished its B.1s and the RAF standardised on the Vulcan B.2 from 1968 onwards, with just a handful of Mk 1 Vulcans remaining active on test duties, the last such machine being XA903, which continued flying on engine development work until 1 March 1979, when it was grounded (and subsequently cut up) at Farnborough.

One direct result of Britain's possession of both fission (and eventually fusion) weapons was, as indicated, a much greater level of nuclear co-operation between the USA and Britain. America's reluctance to share nuclear secrets with Britain was largely due to pressure from politicians who either believed that America should endeavour to remain a nuclear monopoly (even though the Soviet Union had begun manufacturing her own thermonuclear bombs), or that Britain simply could not be trusted to share nuclear knowledge because of continual spy revelations, or because Britain could even be invaded by the

Soviets at some stage. Despite the fact that many American officials, including a succession of Presidents, believed in the Special Relationship, it was not until the RAF received atomic weapons (and demonstrated that Britain would therefore be a nuclear power either with or without American co-operation) that the transatlantic friendship was restored.

The strong working relationship between the RAF's Bomber Command and the USAF'S Strategic Air Command had survived the political ups and downs of the early post-war years and, as the climate improved, so the possibility of co-ordinating nuclear strike plans became a reality. Moves towards co-operation in this field were first made by the British, and in September 1955 the CAS visited the USA with a briefing paper that set out Britain's objectives. It began by stating, 'The primary aim of the defence policy of the United Kingdom is to prevent war,' and went on to say:

> 'The main instrument for achieving this aim lies in the nuclear capability together with the means of delivery, which is possessed by the United Kingdom and the United States alone. We should achieve a closer association with the United States world-wide in the field of defence strategy. This is particularly important in strategic air operations, where Bomber Command and the Strategic Air Command will be attacking components of the same vast target complex. It follows that unless there is a full exchange of information and a co-ordinated plan of attack, wasteful overlapping and dangerous omissions will result.'

The Chiefs of Staff had tried unsuccessfully to persuade the Americans to begin joint planning for some years, but little progress was made until the USA accepted that the V-Force had become a reality. Finally, agreement was achieved in August 1956 and a team of senior USAF officers was sent to London to discuss the co-ordination of strike plans with Bomber Command. Additionally, in a remarkable volte-face, the USA had also agreed to supply the RAF with atomic weapons as a stop-gap measure, until sufficient British-made bombs could be completed. Because the McMahon Act was still in force, the bombs would be kept in

The first production Vulcan B.2, XH533, on an early test flight. The aircraft was used on various trials and spent more than eleven years with the A&AEE, eventually being fitted with a camera fairing on the nose (in the refuelling probe attachment point). The aircraft was ultimately scrapped at St Athan.

XH557 pictured on final approach to Filton during trials with Rolls Royce. Although the starboard inboard engine is an Olympus 201, the larger port engine caps indicate that Olympus 300 series engines have been fitted. XH557 was first fitted with one such engine for trials, eventually taking two, before being fitted with a full set of four. *Rolls Royce*

This unusual rear view of XM597 illustrates the large-diameter 301 series engine exhausts and the square-shaped RWR fin cap modifications.

Above: XH557 is seen during 301 series engine trials. Just visible is the enlarged exhaust fitted to the inboard starboard engine, signifying installation of a 301 power plant.

Right: A very early shot of XM570, wearing the markings of No 27 Squadron, the first unit with which the aircraft served. *Ken Elliott*

USAF custody on RAF bases, from where they would be released to the RAF in a wartime emergency. The detailed agreements made in London included references to the general concept of Allied atomic air operations, stating that, in a general war, atomic weapons would be used right from the outset. It was established that an atomic war would probably begin with an initial phase characterised by an intensive exchange of atomic blows, followed by a subsequent exchange of indeterminate duration at a reduced atomic intensity. The 'Brief Plan of Action' stated that the Allied counter air offensive would begin with heavy co-ordinated attacks against airfields, logistic facilities, control centres and command headquarters, creating a contraction of forces that would concentrate the surviving enemy aircraft on remaining airfields, enabling SAC and Bomber Command to exploit the vulnerability of such concentrations.

In January 1956 the British Defence Minister, Duncan Sandys, wrote to Charles Wilson, his American counterpart, with a detailed set of proposals. The reply from Wilson was positive, and agreed that arrangements should be made to furnish the Royal Air Force with United States atomic bombs in the event of general war, and to co- ordinate the atomic strike plans of the United States Air Force with the Royal Air Force. However, Wilson also added that 'the provisions of United States legislation must govern and that the United States cannot engage in a commitment to transfer custody of such weapons to the Royal Air Force other than by Presidential decision in strict accordance with his constitutional and legislative authority'. The supply of American atomic bombs (Project E) was discussed by President Eisenhower and Prime Minister Macmillan when they met at the Bermuda Conference in March 1957. The 1956 Suez Crisis had been a turning point for Anglo-American relations, and the Bermuda meeting marked a significant improvement in mutual confidence between the two countries.

XH557 was delivered with Olympus 201s but the inlet lip was deepened for trials with 301s. Here she is seen when with 50 Sqn at Waddington in March 1982. *Fred Martin*

B.2 XM570 of the Scampton Wing participated at the 1964 Waddington Battle of Britain 'At Home' day. *Martin Derry*

XL320 enjoyed a relatively uneventful service life based only at Scampton, where she is pictured during the late 1970s while serving with 230 OCU. The aircraft was finally retired to St Athan where it was later scrapped. *Joe L'Estrange*

XL359 was another aircraft that operated only from Scampton during its service life, ultimately ending its days with the OCU, resplendent in wrap-around tactical camouflage. *Joe L'Estrange*

Below: Vulcan B2 XL446 served with almost every Vulcan unit during a service life, which saw it shift between the Waddington and Scampton Wings.

Looking pristine, B.2 XL359 of 230 OCU taxiing at Mildenhall in May 1977. *Fred Martin*

After the conference, Eisenhower outlined his views, stating:

> 'The United States Government welcome the agreement to co-ordinate the strike plans of the United States and United Kingdom bomber forces, and to store United States nuclear weapons on RAF airfields under United States custody for release subject to decision by the President in an emergency. We understand that for the present at least these weapons will be in the kiloton range. The United Kingdom forces could obviously play a much more effective part in joint strikes if the United States weapons made available to them in emergency were in the megaton range, and it is suggested that this possibility might be examined at the appropriate time.'

Interestingly, America's willingness not only to co-operate with Britain but actually to supply weapons was a fairly well-kept secret at the time, as the President indicated by saying in his communication to Macmillan:

> 'With respect to the item "Nuclear bomb release gear for RAF bombers", I agree of course that you shall probably have to make some statement in order to prevent speculation in the press that might prove not only inaccurate but damaging. However, as I explained to you verbally, the United States would prefer not to be a party to a public statement which might give rise to demands upon us by other Governments where we should not be in a position to meet the requests.
>
> Consequently, I suggest the possible adequacy of a unilateral statement by yourself or by the British Defence Minister to the effect that Canberras are now being equipped to carry atomic bombs.'

In fact, Project E covered the supply of atomic bombs not only to the RAF's Canberra tactical bomber force, but also to the V-Force itself. An Air Ministry meeting in August 1956 reported that the US Government had not given any indication of the numbers of weapons to be supplied, nor were they likely to, but it was believed to be almost certain that the numbers would exceed any delivery capacity that the UK was capable of developing. Records do not indicate when the first E weapons arrived in the UK, but US transport flights began towards the end of 1958, and both Honington and Waddington received the first suitably modified V-Bombers in October of that year. Approval was given for modifications to be made to seventy-two aircraft at Honington, Waddington and Marham to carry American Mk 5 (6,000lb) bombs (and subsequently the Mk 7, Mk 15/39 and Mk 28), and Nos 90 and 57 Squadrons at Honington, with Valiants and Victors respectively, were the first units to be assigned to Project E during the spring of 1959, followed by Nos 148, 207 and 214 Squadrons (Valiants) at Marham, and 83 Squadron (Vulcans) at Waddington. In 1960 No 7 Squadron (Valiants) and No 55 Squadron (Victors) at Honington, together with No 44 Squadron (Vulcans) at Waddington, joined Project E, and finally Nos 101 and 50 Squadrons (Vulcans) at Waddington were assigned in 1961.

The supply of American bombs enabled Bomber Command to equip the V-Force with atomic weapons at an earlier date than would have been possible with British weapons, and before 1961 there would otherwise have been an absurd situation in which the RAF possessed more bombers

A short-lived variant of the 35 Sqn insignia adorns B.2 XL446 at the Binbrook Open Day in July 1975, adjacent to a 43 Sqn Phantom FG.1. *Fred Martin*

than bombs. However, Project E was a less than perfect solution to Britain's early problems, chiefly because of American restrictions on the deployment of the weapons. The E weapons could not be distributed throughout Bomber Command, nor could they be dispersed to remote sites from which clutches of V-Bombers would operate during wartime, in order to evade enemy attack.

Consequently Project E was relatively short-lived, and the decision to begin a phase-out was taken on 7 July 1960. Although America was reluctant to disclose how many nuclear bombs were stored at RAF bases, they were equally ambiguous when revealing the nominal yield of each weapon. Eventually it was established that the bombs were largely of a type that would deliver only half the yield that they had previously been thought to possess, which meant that, once British equivalents were available, the greater flexibility they afforded (not least in being 'dispersible') enabled the American weapons to be withdrawn, although the Valiants assigned to Supreme Allied Commander Europe (Saceur) and RAF Germany's Canberras would continue to carry E weapons.

On finals to Waddington, XL321 in March 1982. She achieved the highest Vulcan flying hours but ended up being burnt at Catterick. *Fred Martin*

While Project E progressed, the co-operation between SAC and Bomber Command became much more intimate, as the CAS's comments relating to a November 1957 meeting indicate:

> 'Examination of the separate Bomber Command and SAC plans has shown that every Bomber Command target was, understandably, also on SAC's list for attack and that both Commands had doubled-up strikes on their selected targets to ensure success. A fully integrated plan has now been produced, taking into account Bomber Command's ability to be on target in the first wave several hours in advance of the main SAC forces from bases in the US. Under the combined plan, the total strategic air forces disposed by the Allies are sufficient to cover all Soviet targets, including airfields and air defence. Bomber Command's contribution has been given as 92 aircraft by October 1958, increasing to 108 aircraft by June 1959. 106 targets have been allocated to Bomber Command as follows:
>
> a) 69 cities which are centres of government or of other military significance.
>
> b) 17 long-range air force airfields which constitute part of the nuclear threat.
>
> c) 20 elements of the Soviet air defence system.'

Although SAC had 380 B-52s and 1,367 B-47s by the end of 1958, compared with Britain's three V-Bombers, it is worth noting that Bomber Command's contribution to the Allied counter air offensive plans was out of all proportion to the relative size of the SAC and RAF forces. The V-Bombers possessed superior speed and altitude performance, and the RAF'S training and technical expertise was at least as good.

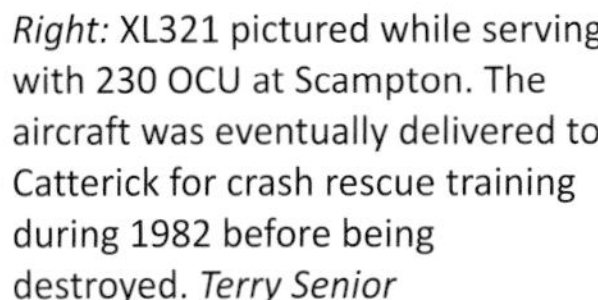

Right: XL321 pictured while serving with 230 OCU at Scampton. The aircraft was eventually delivered to Catterick for crash rescue training during 1982 before being destroyed. *Terry Senior*

Below: XA903 over the Severn Estuary, displaying the Olympus 593 engine test pod and with the icing rig attached to the nose, during pre-Concorde trials conducted from Filton.

Vulcan B.1 XA903 was used throughout her life as a trials aircraft. Prior to her role as a test-bed for Concorde's Olympus 593 test-bed as seen here (first flying as such in September 1966), she was a test aircraft for Blue Steel development; in November 1960 she carried a test round to Edinburgh Field in South Australia, returning after three months and seven test flights. Her final trials were as a test-bed for the MRCA/Tornado RB199 engine with Rolls-Royce, then at A&AEE investigating the effect of ground-firing the Mauser 27mm cannon. The nose section survives at Stranraer. *Martin Derry*

More importantly, although the RAF forces were obviously more vulnerable to air attack, Bomber Command had introduced a dispersal plan that would enable the V-Force to deploy in groups of four or two aircraft to a total of thirty-six airfields scattered around the UK, from where the combined force would be capable of becoming airborne in less than ten minutes. The performance of the V-Bombers, together with Britain's geographical location, meant that Bomber Command's aircraft would reach their targets long before SAC's aircraft would arrive, so the V-Force was, quite literally, very much at the forefront of the West's nuclear strike capability.

As previously mentioned, the 1957 Bermuda Conference was an historical turning point in Anglo-American relations. After the political embarrassment of the Suez Crisis, which had highlighted Britain's impotence in the face of American economic and political pressure, Macmillan was eager to restore the Special Relationship. Fortunately Eisenhower shared the same sentiments, as highlighted in a letter to Churchill, in which he wrote, 'I shall never be happy until our old-time closeness has been restored.' Consequently America agreed, as described, to joint strike plans, the supply of nuclear weapons and, as first mentioned in July 1956, the supply of the Thor intermediate-

XH478 also conducted a variety of development trials during her service life. This image shows a February 1968 sortie from Boscombe Down during which the Vulcan was used to determine the necessary spray pattern for Concorde icing trails, which were to be conducted by the Canberra WV787. *Shaun Churchill collection*

XH478 is seen again at St Mawgan while operating from Boscombe Down on the pre-Concorde trials. Dayglow orange stripes have been applied to the nose and tail areas.

range ballistic missile (IRBM) system to the UK. In essence, these nuclear-armed rockets would be all-American, but they would be jointly operated by RAF personnel and located on RAF bases. Thor would ease American worries of a growing 'missile gap' between the USA and the Soviet Union by placing missiles within striking distance of key Soviet targets, and at the same time it would 'give Britain a megaton rocket deterrent at least five years before we could provide it ourselves', as Macmillan commented. He also added:

> 'United States would provide weapons and specialised equipment, including anything costing dollars. Nuclear warheads would be held under same conditions as nuclear bombs for British bombers. We would undertake site works and would provide general supporting equipment. United States estimate of the cost to us for the four sites [eventually there were, in fact, five sites] is £10 million, apart from the costs of personnel and their training and housing.'

Britain had already addressed the possibility of manufacturing a ballistic missile, based on Air Staff Requirement OR.1139 of 8 August 1955. The Blue Streak missile was to be housed in a series of underground silos (a British concept that was ultimately adopted in other countries) scattered around the UK, whereas Thor would be a surface-housed weapon, stored horizontally in shelters before being erected for fuelling and launch. It was considered that Thor might provide some useful experience in advance of Blue Streak, which was expected to enter service in 1964, whereas Thor would be operational by 1958. In January of the latter year agreements were made for the establishment of SM-75 Thors in the UK, with the first missiles arriving by July. Each squadron would be deployed on five sites, and each site would house three missiles. A White Paper issued on 25 February 1958 stated that the missiles would be 'manned and operated by United Kingdom personnel', and that any decision to launch the Thors would be a joint one between the two governments, that the 1-megaton warheads would remain in full United States custody, and that the Thor agreement would remain in force for not less than five years. The first missile was delivered to No 77 Squadron at Feltwell on 19 September 1958, and the first live firing of a Thor by an RAF crew took place at Vandenberg AFB on 18 April 1959. Deployment was completed by March 1960 and a total of sixty Thors was ultimately deployed to the UK, all arriving on board transport aircraft from the USA.

Although the Thor missile could be accommodated easily inside Military Airlift Command's huge Douglas Globemaster and Cargomaster aircraft, the task of delivering the missiles, and transporting some back to the USA for RAF live firings at Vandenberg, was far from easy. The transport aircraft had to operate within strict rate of ascent and descent figures to avoid pressure damage to the Thor's fuel tanks. Even more exacting was the need to transport the hugely expensive guidance unit gyroscopes, which were suspended in a lubricant that had to be maintained at a constant temperature. Control of the temperature required direct power from the transport aircraft's engines, so the crews had to keep the outboard engines throttled up to high power, even while taxiing, brake power being applied against the thrust.

However, once the missile was installed it was regarded as being a very effective and reliable weapon. Its downfall was the fact that it was not mobile and could not be housed underground, rendering the entire Thor force vulnerable to a Soviet attack. Much consideration was given to the possibility of extending Thor's service life beyond the planned period, but like Britain's Blue Streak it was accepted that a fixed IRBM system could not remain viable as a credible nuclear deterrent. The last RAF Thor complex to close was at North Luffenham, and on 23 August 1963 the component squadrons, Nos 144, 130, 218, 223 and 254, disbanded, marking the end of Bomber Command's IRBM era.

Above: Pictured at the moment of rotation, a 44 Squadron Vulcan leaps into the air on a 'roller' just 2,000 feet down Waddington's runway.

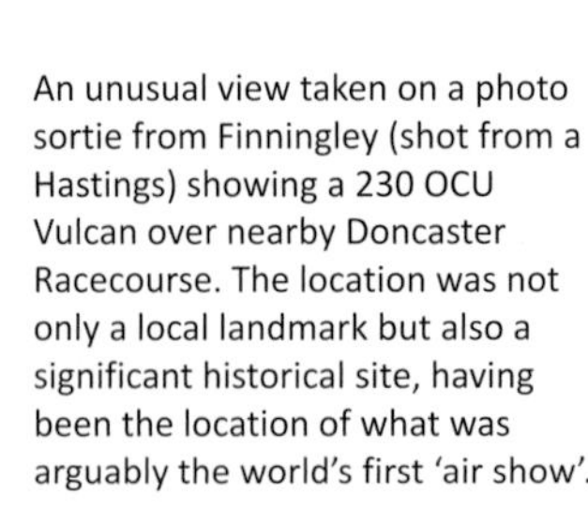

An unusual view taken on a photo sortie from Finningley (shot from a Hastings) showing a 230 OCU Vulcan over nearby Doncaster Racecourse. The location was not only a local landmark but also a significant historical site, having been the location of what was arguably the world's first 'air show'.

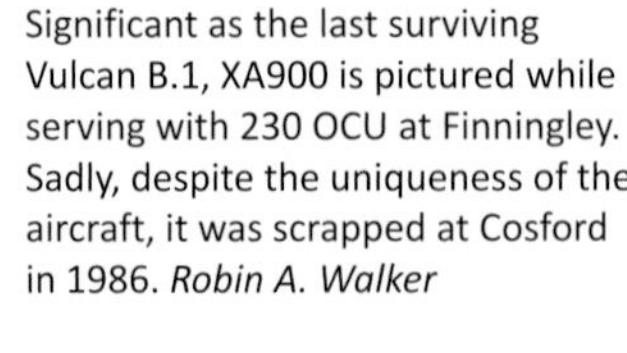

Significant as the last surviving Vulcan B.1, XA900 is pictured while serving with 230 OCU at Finningley. Sadly, despite the uniqueness of the aircraft, it was scrapped at Cosford in 1986. *Robin A. Walker*

A late-1960s publicity photograph of a preserved Spitfire in company with a 230 OCU Vulcan over Finningley. The unusual layout of Finningley's 'flight line' dispersals (which was unique to that airfield) can be seen below.

230 OCU's XH563 passing Conisbrough Keep during a photo sortie from nearby Finningley.

Basking in sunshine on one of Finningley's dispersals, XA905 served with 230 OCU until September 1964, when the aircraft was transferred to the Waddington Wing. *Robin A. Walker*

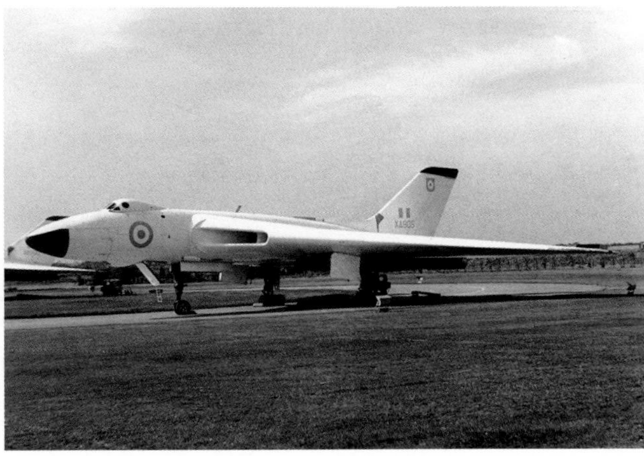

Coningsby Wing Vulcans pictured on a dispersal exercise on Wittering's Operational Readiness Platform (ORP). *Joe L'Estrange*

A nostalgic early-1960s view of RAF Coningsby showing a variety of white-painted Vulcans on the ORP and more distant dispersals. *Joe L'Estrange*

This second early-1960s view of white-painted Vulcans at Coningsby's ORP (above) makes a striking contrast to the Vulcan's late-1960s low-level camouflage as seen in the second picture (below), taken just a few years later. *Both Joe L'Estrange*

The CAS commented:

'When Thor came into service we knew that we would be faced with many new and complex technical and administrative problems and we fully expected that one of the greatest of these problems would be the task of maintaining the morale of the officers and men allocated to the missile sites. In the event, the problems were met and solved with a degree of enthusiasm, skill and resourcefulness which was in the finest traditions of Bomber Command and the Royal Air Force. The high morale which was a feature of the Force from its inception has never flagged, and Thor's fine record of serviceability and state of readiness over the years is a remarkable tribute to the loyalty and sense of duty of all the personnel who played a part. They will be able to look back with pride on a most valuable contribution to our deterrent force.'

As the Thor era ended in 1963, the Blue Streak programme had also ended in 1960, less than five years after the Air Staff Requirement for it had been issued. A Commons statement in April 1960 said:

'The Government have been considering the future of the project for developing the long-range ballistic missile Blue Streak and have been in touch with the Australian Government about it, in view of their interest in the joint project and the operation of the Woomera range. The technique of controlling ballistic missiles has rapidly advanced. The vulnerability of missiles launched from static sites, and the practicability of launching missiles of considerable range from mobile platforms, has now been established.

In the light of our military advice to this effect, and of the importance of reinforcing the effectiveness of the deterrent, we have concluded and the Australian Government have fully accepted that we ought not to continue to develop, as a military weapon, a missile that can be launched only from a fixed site. Today, our strategic nuclear force is an effective and significant contribution to the deterrent power of the free world. The Government does not intend to give up this independent contribution and, therefore, some other vehicle will in due course be needed in place of Blue Streak to carry British-manufactured warheads. The need for this is not immediately urgent, since the effectiveness of the V-Bomber force as the vehicle for these warheads will remain unimpaired for several years to come, nor is it possible at the moment to say with certainty which of several possibilities or combinations of them would be technically the most suitable. On present information, there appears to be much to be said for prolonging the effectiveness of the V-Bombers by buying supplies of the airborne ballistic missile Skybolt, which is being developed in the United States. Her Majesty's Government understands that the United States Government will be favourably disposed to the purchase by the United Kingdom at the appropriate time of supplies of this vehicle. The

The expensive, complicated and cumbersome Blue Streak rocket – ultimately abandoned in favour of the Thor programme.

An unusual view of the last production Vulcan B.2, XM657, about to touch down at Waddington while a queue of traffic forms at the runway traffic lights on the nearby A15. *Mike Jenvey*

XJ781 (pictured during 1962 while serving with 12 Squadron) was written off in a landing accident at Shiraz in Iran on 23 May 1973. The wreckage remained in Iran. *Ray Deacon*

> Government will now consider with the firms and other interests concerned as a matter of urgency, whether the Blue Streak programme could be adapted for the development of a launcher for space satellites.'

In fact, Blue Streak did eventually become part of the European Space Agency's rocket design process, but as a military weapon the fixed-site IRBM was obsolete. Britain was already looking towards American developments, most notably the aforementioned Skybolt and the sea-launched Polaris missiles, as possible future means of delivering British warheads to their targets.

Above: XH560 enjoyed a particularly varied career. After serving with 230 OCU and 12 Squadron (as illustrated), the aircraft was later converted to B.2 (MRR) standard and later re-converted to K.2 tanker standard before being used briefly as one of three dedicated air display aircraft. Her final days were spent at Marham, slowly rotting close to the control tower. *Ray Deacon*

A fairly rare image of XJ825 pictured during 1963 when the aircraft was in service with No 35 Squadron.

CHAPTER SIX

Bombs and Bombers

Having successfully proved the concept of atomic weaponry, and having ordered a fleet of medium-range and high-speed bombers with which to carry the weapons, the next development was to mate the atomic bomb with its carrier aircraft. Early ballistic trials of various bomb casings were undertaken using an Avro Lincoln (as mentioned previously, later casings were dropped from the Sperrin). The Lincoln was an appropriate choice, not only because the aircraft was readily available (and the only aircraft suitable for the job), but because the type would have been modified to carry operational atomic bombs if the V-Bomber programme failed – a far cry from the ambitions of the Air Staff, but a suitable alternative if necessary. In 1948 Lord Tedder suggested to the Chiefs of Staff that the RAF should begin training for the handling and storage of nuclear weapons, and subsequently a committee was set up to explore all matters relating to this subject. It was known as the Herod Committee, an acronym of High Explosives Research Operational Distribution, and the first meeting took place on 22 November.

XM605 is literally out in the cold at Canadian Forces Base (CFB) Goose Bay in Canada. Goose Bay was a regular destination for Vulcan crews either as a staging post on flights to SAC bases in the US, or as a detachment destination from where low-level flights could be conducted over the sparsely populated Canadian countryside.

The committee made a number of key decisions on operational activities, including planning for the acceptance of nuclear weapons at Wittering and Marham, and at a later meeting it agreed that the first in-service bombs should be designed for in-flight fusing, enabling the tube containing the fissile components and corresponding outer layers of the bomb to be inserted after take-off, as a safety measure. In November 1951 the Herod Committee decided that the RAF Atomic Weapons School should be located at the first RAF station where nuclear weapons would be stored (Wittering), and it was agreed that the Armament Training School should also be based there, after having first considered Honington, which was originally scheduled to be the first Valiant station. The first operational Valiant squadron, No 138, would also be at Wittering, and as it also housed the co-located Valiant Trials Flight, No 1321, the station was at the very heart of the RAF'S V-Force build-up.

Development work on the first atomic bombs progressed well, and for some time it was unclear if the bomb would be completed before the first Valiants or vice versa. Doctor Penney commented:

> 'My philosophy is that the RAF has handled aircraft for a long time and can fly Valiants as soon as they come off the production line. But the Royal Air Force has not yet handled atomic weapons, therefore we must get some bombs to the RAF at the earliest possible moment, so that the handling and servicing can be practised and fully worked out.'

He added that these first weapons would be essentially the same as later developed versions, but that they might require modifications to the In-Flight Insertion (IFI) cartridge and, if these first bombs had to be used, the cartridge might have to be loaded just before take-off, rather than in flight.

During 1953 plans had been laid for the storage of atomic weapons at Wittering, Marham, Honington and Waddington, and for a bomb depot to be completed at Barnham in Norfolk. The first atomic bombs were finally delivered to the RAF (the Bomber Command Armament School) on the nights of 7 and 14 November 1953, when convoys arrived at Wittering. Trials with the Valiant were to begin during 1954, but the first aircraft could not be delivered to Wittering until the following year, and it was not until 15 June that Valiant WP201 arrived. The MoS had formed No 1321 Flight during 1954, specifically for the dropping trials, and the unit completed some bomb delivery and handling training with Vickers at Wisley. It was decided that bombs would only be dropped by operational units, so, following the initial drop trials, No 1321 Flight became C Flight of No 138 Squadron in February 1956, changing to No 49 Squadron on 1 May 1956.

With the arrival of Valiant WP201 the MoS flight trials began, the first ballistic store being dropped from 12,000 feet at 330 knots over the Orfordness range on 6 July. Progress was satisfactory, but a setback occurred on 29 July when No 138 Squadron's WP222 crashed shortly after taking off on a cross-country training flight. The Valiant entered a left-hand turn that continued through 300 degrees until the aircraft impacted at about 300 knots just three minutes after take-off, killing the entire crew. Subsequent investigations revealed that a runaway actuator had fixed an aileron tab in the 'up' position, causing the uncontrollable roll. Fixing the potentially fatal flaw was simple, but the crash raised more concerns about the ability of aircrew to abandon all three V-Bomber types in emergencies.

In fact, this controversial subject was never properly addressed, for a variety of reasons. Most importantly it was accepted that the original concept of fitting each bomber with a jettisonable crew compartment would have created a design and engineering problem that would have inevitably delayed the aircraft's entry into service. Likewise, such a system would have been prohibitively expensive, and, as service experience on all three V-Bombers began to grow, there was an understandable belief that the V-Bombers were proving to be remarkably safe aircraft, which rendered expensive and exotic escape systems unnecessary. Not surprisingly, following the Heathrow tragedy, the whole question of escape systems for the V-Bombers was raised once more, and on the day after the accident the Air Minister asked the Air Staff to outline its policy on ejection seats. The DCAS replied on 15 October, stating that when the Valiant, Vulcan and Victor were first conceived they were to have been fitted with jettisonable cabins that would separate from the aircraft and make a parachute-retarded descent:

> 'As design and development proceeded it became clear that this facility could not be provided, and agreement not to have a jettisonable cabin was reached in the case of the Valiant in June 1948, the Vulcan in May 1949 and the Victor in October 1952. In all three bombers the layout of the cabin, which was operationally very satisfactory, made it impossible, for structural reasons, to produce ejection facilities for aircrew other than the pilots. It was, however, agreed to provide ejector seats for the pilots so that they could remain with the aircraft for longer and help other crew members to escape. Facilities for the other crew members were provided by means of side doors in the Valiant and Victor and through an underneath hatch in the Vulcan, which has virtually no fuselage. The result of this is that all three bombers and the developments of them will, according to present planning, have ejector seats for the pilots and escape by door or hatch for the three other crew members. A trained crew takes approximately 20 sec from the time the order to jump is given until the last man leaves the aircraft, but it is important to remember that it is unlikely that the three non-pilot crew members would escape in conditions where high "g" forces are being applied through battle damage or loss of control, when the aircraft is at low level. On the other hand, when the first Valiant had a fire in the air all live members got out of the aircraft successfully at high altitude – unfortunately the second pilot was killed by striking the fin. I have discussed a possible modification plan for the V-Bombers with Mr James Martin [managing director and chief designer of the Martin-Baker ejection seat company] and with the Ministry of Supply, and am of the opinion that it is certainly not impossible to incorporate ejection facilities for the three non-pilot members of the crew of the V-Bombers, but the implementation of such a policy would naturally raise very grave issues. The first issue is whether or not we would be right to go in for such a policy, and the second is whether we could afford to do so, both in terms of money and effort as well as the delay of the V-Force build-up. A retrospective modification programme would naturally be an immense undertaking but it is not technically impossible, and if we do go in for it we must realise what may be involved. My own view is that we should not attempt to adopt such a policy.'

Consequently it was always accepted that in most potentially catastrophic situations the rear crews would probably have sufficient time to abandon the aircraft by more conventional means – something that proved to be only partially correct – and the notion of the pilot and co-pilot being afforded the luxury of ejection seats while the remaining crew had only parachutes was an issue that was never properly settled throughout the history of the V-Force.

As the Blue Danube bomb-dropping trials continued, it became clear that the RAE designers had produced a near-perfect ballistic casing for the atomic warhead. When the instrumented test specimens were released they often 'flew' beneath the Valiant's rear fuselage before falling clear; the drops were tracked visually by theodolite. To counter this potentially dangerous tendency the Valiant was equipped with strakes forward of the bomb bay that created an airflow disturbance to push the bomb downwards, away from the fuselage, but later experience with the Vulcan and Victor confirmed a cleaner release from both types.

Equipped with six Valiant B.1s, No 49 Squadron was quickly assigned to Operation 'Buffalo', which would culminate in the first air drop of a live British nuclear weapon. While other Valiant squadrons became involved with Operation 'Musketeer' (the Suez saga with which neither the Vulcan nor Victor was involved), No 49 Squadron's specialised tasks continued at Wittering until two Valiant B.1s, W3366 and WZ367, departed for Australia on 5 August. Further training was to have been conducted from Wittering, but poor weather and difficulty in obtaining sufficient weapons range access meant that the crews could operate more effectively by using the Maralinga and Woomera ranges. The work-up towards Operation 'Buffalo' did not proceed smoothly, as a report by Gp Capt Menaul, the Commander of the Air Task Group, indicates:

> 'The Valiants arrived at Edinburgh having completed part of a bombing training programme in the UK. It was planned to complete their training in Australia using the range facilities at Maralinga or Woomera as required. The main reason for the non-completion of training in the UK was the late delivery of aircraft and the lack of flight clearance for certain items of equipment, notably the bombing system, the automatic pilot and the radar altimeter. Unsuitable weather and difficulties in obtaining bombing ranges also added to the delays. During the training which followed, ten practice 10,000lb bombs and 60 x 100lb bombs were dropped in the UK by the two Valiant aircraft. The 10,000lb bombs were primarily to prove the weapon and aircraft systems and the 100-pounders to prove the accuracy of the aircraft's bombsight, particularly in the hands of inexperienced crews. On completion of this training programme in the UK, the results of which were not entirely satisfactory, it was decided that the standard obtained, considering the time available, was adequate and the aircraft and crews were prepared to fly out to Australia. Technical defects discovered during the UK training phase were corrected, and modifications to the bombsight sighting head and the Green Satin output improved the system and gave considerably better bombing results at a later date. The whole of the training programme in the UK could have been considerably improved if more emphasis had been placed on overseas operations.'

The report by the Operation 'Buffalo' Air Task Commander also includes a description of the historic sortie during which the first air drop took place:

> 'Valiant WZ366 took off from Maralinga airfield with the live nuclear weapon on board. The crew consisted of Sqn Ldr Flavell (captain), Group Capt Menaul, Flt Lts Ledger and Stacey, Flg Off Spencer and Pilot Off Ford. The aircraft climbed to 38,000 feet in a wide arc, avoiding the range area until it reached the emergency holding area. The bombsight was levelled, contact was established with the air controller on the ground by VHF and HF and the aircraft then descended to 30,000 feet ready to begin the fly-over sequences, using precisely the same drills and procedures as in the concrete and HE drops. At 14:25 the first fly-over, Type A, was successfully completed, with all equipment, both in the air and on the ground, working satisfactorily. Types B and C fly-overs were then completed in turn, and by 15:00 all was in readiness for the final Type D fly-over and the release of the nuclear weapon. The final D Type fly-over was completed according to plan with all equipment functioning perfectly, and the weapon was released at 15:27. Immediately after release a steep turn to starboard on to a heading of 240 degrees true was executed in order to position the aircraft correctly for the thermal measuring equipment to function. During this turn 1.9g was applied. The weapon exploded correctly and the aircraft, after observing the formation of the mushroom cloud, set course for base, where it landed at 15:35. The operation had gone smoothly and exactly according to the plans drawn up during training. The bombing error was afterwards assessed at 100yd overshoot and 40ft right.'

The Blue Danube's explosive yield had been fixed at 40 kilotons, but for this drop the figure had been reduced to 3 kilotons in order to avoid the risks of extensive radioactive fallout, which would have been created if the bomb's barometric fusing had failed and the airburst had become a ground burst. This was mentioned in a report sent to the Secretary of State for Air by Air Marshal Tuttle, which said:

> 'The weapon, a Blue Danube round with modified fusing, in-flight loading and with the yield reduced to 3-4 Kilotons, was dropped from the Valiant aircraft at 30,000 feet. The weapon was set to burst at 500 feet and telemetry confirmed that the burst occurred between 500 and 600 feet. The bomb was aimed visually after a radar-controlled run-up.'

This first 'live' air drop demonstrated to the world that Britain not only possessed nuclear know-how, but that the RAF now had a practical means of delivering nuclear weapons to their targets. The British nuclear deterrent had been created.

Of course, the story of Britain's nuclear ambitions did not end with the Blue Danube. While the RAF was concentrating on the entry into service of its first atomic weapon, the Government was already turning its attention to the next generation of destructive power, and by this stage Britain was, essentially, a generation behind both the USA and the Soviet Union in terms of nuclear weapon development. The development of the thermonuclear (hydrogen) bomb represented yet another quantum leap in technology, which was almost as great as the initial development of atomic weaponry.

Although the concept was essentially a development of atomic fission (and had been established as a theoretical possibility for almost as long as the atom bomb), the physics are rather different. In simple terms, hydrogen bombs use the energy of a fission bomb in order to compress and heat fusion fuel. In the Teller-Ulam design, which accounts for all multi-megaton-yield hydrogen bombs, this is accomplished by initially placing a fission bomb and fusion fuel (tritium, deuterium or lithium deuteride) in close proximity within a special radiation-reflecting container. When the fission bomb is detonated, gamma and X-rays emitted at the speed of light first compress the fusion fuel, then heat it to thermonuclear temperatures. The ensuing fusion reaction creates enormous numbers of high-speed neutrons, which can then induce fission in materials that are normally not prone to it, such as depleted uranium. Each of these components is known as a 'stage', with the fission bomb as the 'primary' and the fusion capsule as the 'secondary'. In large hydrogen bombs, about half of the explosive yield comes from the final fissioning of depleted uranium. By chaining together numerous stages with increasing amounts of fusion fuel, thermonuclear weapons can be made to an almost arbitrary yield.

Just six weeks after the first British atomic test at Monte Bello, the USA successfully detonated the first two thermonuclear (hydrogen bomb) devices at Eniwetok, the Soviets following with their own test on 12 August 1953. In 1954 British Prime Minister Anthony Eden said that the hydrogen bomb had 'fundamentally altered the entire problem of defence', and in February 1955 the government produced a paper that included the following:

> 'In the Statement on Defence 1954, HM Government set out their views on the effect of atomic weapons on UK policy and on the nature of war. Shortly afterwards the US Government released information on the experimental explosion at Eniwetok in November 1952, of a thermonuclear weapon many hundred times more powerful than the atomic bombs which were used at Nagasaki and Hiroshima in 1945. On 1 March 1954 an even more powerful thermonuclear weapon was exploded in the Marshall Islands. There are no technical or scientific limitations on the production of nuclear weapons still more devastating. The US Government have announced that they are proceeding with full-scale production of thermonuclear weapons. The Soviet Government are clearly following the same policy, though we cannot tell when they will have thermonuclear weapons available for operational use. The United Kingdom has the ability to produce such weapons. After fully considering all the implications of this step the Government have thought it their duty to proceed with their development and production. The power of these weapons is such that accuracy of aim assumes less importance; thus attacks can be delivered by aircraft flying at great speed and at great heights. This greatly increases the difficulty of defence. Moreover, other means of delivery can be foreseen which will, in time, present even greater problems.'

The precise background to Britain's thermonuclear weapon development remains unclear, but there is no doubt that British scientists returned from the Manhattan Project with more than a little generalised knowledge of how such a weapon might be produced. Theoretical work on the hydrogen bomb appears to have begun in 1951, but early results were evidently disappointing, as during 1952 Lord Cherwell, the Prime Minister's chief technical advisor, told Churchill that hydrogen weapons were 'beyond our means'.

Despite this pessimism, Britain's thermonuclear weapon programme was remarkably successful. The decision to manufacture hydrogen bombs was made by a Cabinet Defence Committee on 16 June 1954, and was discussed by the full Cabinet during the following month. The final Cabinet decision was taken on 26 July, when approval was given to the proposal that 'the current programme for the manufacture of atomic weapons should be so adjusted as to allow for the production of thermonuclear bombs'. Churchill stated, 'We could not expect to maintain our influence as a world power unless we possessed the most up-to-date nuclear weapons.' Likewise, the Cabinet agreed that thermonuclear bombs would be more economical than their atomic equivalents, and that in moral terms the decision to manufacture these new weapons would be no worse than accepting what would amount to the (rather unreliable) protection of America's hydrogen bombs.

Towards the end of 1955 the Defence Research Policy Committee (DRPC) commented, 'The earliest possible achievement of a megaton explosion is necessary to demonstrate our ability to make such weapons, as part of the strategic deterrent against war.' In fact, the main reason for the almost frantic development of hydrogen bombs was not directly due to any perceived deterrent policy or any immediate threat – it was simply in response to growing pressure for an international ban on atmospheric nuclear tests that would effectively prohibit Britain from testing (and therefore creating) a hydrogen bomb. Britain maintained what was perceived to be an even-handed diplomatic position by backing calls for a test ban while maintaining an active developmental programme.

Aubrey Jones, the Minister responsible for the nuclear tests, said:

> 'In the absence of international agreement on methods of regulating and limiting nuclear test explosions – and Her Majesty's Government will not cease to pursue every opportunity of seeking such an agreement – the tests which are to take place shortly in the Pacific are, in the opinion of the government, essential to the defence of the country and the prevention of global war.'

Britain was caught up in a race to develop and detonate a hydrogen bomb before being forced to accept a ban on nuclear testing. The DRPC said, 'It is essential that this first series should be planned in such a way as to safeguard the future by obtaining the greatest possible amount of scientific knowledge and weapon design experience as the foundation of our megaton weapon development programme.' The committee identified four basic requirements: a megaton warhead for a free-falling bomb; a similar megaton warhead for a powered guided bomb (which eventually became Blue Steel); a smaller and lighter warhead for a medium-range ballistic missile (which was to have become Blue Streak); and a multi-megaton warhead intended to demonstrate that Britain could match the devices that had been tested by America and the Soviets.

Because Britain's scientific knowledge of thermonuclear weaponry was restricted to theoretical analysis, the DRPC report stated that the most certain way of achieving a megaton explosion in 1957 would be to use a large, pure-fission assembly in a Mk 1 case, surrounded by sufficient fissile material to ensure a megaton yield (a so-called 'boosted' device). The bomb would consequently be big, heavy and extravagant in fissile material, but it would work, although it could only be developed into a free-fall bomb because of its size and weight. The Government agreed to proceed with a series of test explosions, and on 5 June 1956 the Cabinet agreed that the Prime Minister should make a Parliamentary statement about the tests. Two days later Eden announced to the House of Commons that Britain was to conduct a limited number of test explosions in the megaton range.

No 49 Squadron at Wittering began training for Operation 'Grapple' on 1 September 1956, using standard Valiant B.1s until the first specially modified 'Grapple' aircraft, XD818, arrived during mid-November. The 'Grapple' Valiants featured a number of changes to equip them for their live-weapon trials. Most notably they were sprayed with an anti-flash white paint capable of withstanding 72 calories of heat energy per square centimetre. The control surfaces were strengthened to withstand the bomb's pressure wave, the flight deck and bomb-aiming positions were fitted with metal anti-flash screens, and a number of sensors and cameras were installed. Valiant XD818 departed for Christmas Island on 2 March 1957, via Aldergrove, Goose Bay, Namao, Travis and Honolulu, as it had not been fitted with underwing fuel tanks at this stage. It was followed by XD822, XD823 and XD824 at one-day intervals.

Flown by No 49 Squadron's CO, Wing Commander K. G. Hubbard, XD818 arrived over Christmas Island on 12 March, descending from a 1,500-foot familiarisation circuit to a rather more spectacular 50 feet for an ultra-low-level fly-by over the airfield dispersal area before landing. After a settling-in period, the four Valiant crews began a series of training sorties, establishing a precise bomb-aiming capability and the bomb-drop manoeuvres that would be used for each trial. The training was completed by 5 April, despite the unexpected torrential rain, which made life in Christmas Island's 'tent city' rather unpleasant for some time. Two aircraft, XD818 and XD823 were prepared for the first live drop, which was designed to test the 'Green Granite Small' warhead. This was a two-stage device weighing 4,200lb, and contained in a lead bismuth casing with a total weight of 10,000lb. It was a true thermonuclear fusion weapon, the spherical fission primary having a composite uranium U-235 and plutonium Pu-239 core with a spherical secondary comprising U-235 and a U-238 tamper, all packed in lithium deuterate. Valiant XD818, flown by Wg Cdr Hubbard, was to be the drop aircraft for the first test, with XD824 acting as observer aircraft, flying a second crew to give them experience of flash and blast from a thermonuclear weapon. The historic 2hr 20min sortie took place on 15 May 1957, as reported by Hubbard:

> 'The aircraft became airborne at 09:00 V-time and all anti-flash screens were in position prior to the aircraft commencing its first run over the target. After one initial run to check telemetry, the Task Force Commander gave clearance for the live run. The bombing run was made at 45,000 feet true, and as Green Satin drift was fluctuating badly, the set was put to Memory on average drift. The bombing run was steady on a course of 203 degrees and the weapon was released at 10:36 W-time. Immediately after release the aircraft was rolled into the escape manoeuvre which averaged a turn of 60 degrees bank, excess g 1.8 to 1.9, airspeed Mach 0.76, rolling out on a heading of 073 degrees. The time taken for this turn was 38 sec and at the time of air burst of the weapon, the slant range between aircraft and burst was 8.65 n/ms.'

Neither crew nor aircraft felt any effect of flash, and the air blast reached the aircraft two and a half minutes after release; the effect of the blast was to produce a period of five seconds during which turbulence alike to slight clear air turbulence was experienced. Six minutes after release all shutters in the aircraft were removed, and after one orbit to see the mushroom cloud effect, the aircraft returned to base and made a normal landing.

Hubbard's professionalism disguises the excitement of the event, although he later commented, 'It really was a sight of such majesty and grotesque beauty that it defies adequate description.' Nick Wilson, an able seaman on board HMS Warrior, witnessed the explosion from a distance of 30 miles:

> 'I felt my back warming up and experienced the flash, though I had my hands over my face and dark goggles on. Five seconds after the flash we turned round and faced the flash, but it was still bright, so I replaced them.
>
> There in the sky was a brightly glowing seething ball of fire. This rapidly increased and became more cloudy. Soon it was looking like a very dark ripe apple with a snow-white sauce being poured over it. On the horizon at sea level a cloud appeared that must have been dust and spray from the island. The whole sight was most beautiful and I was completely filled with emotions.'

Impressive though the sight may have been, 'Green Granite Small' was not a perfect success. Yielding 0.3 megatons from a predicted yield of up to 1.0 megaton, the lithium deuterate only partly ignited, but the test did at least demonstrate the bomb's potential (and an explosive force 100 times that of the 'Buffalo' air drop), and it also provided the opportunity for another airborne test of a live Blue Danube casing. Despite the mixed results, Britain had undoubtedly entered the thermonuclear age.

The next drop, on 31 May, was 'Grapple 2', or 'Orange Herald', which tested Dr Penney's fall-back high-yield fission bomb; this was again contained in a Blue Danube casing. The warhead, which was developed into an operational physics package known as 'Green Grass', was, as on the first drop, detonated by barometric means at 8,000 feet, and yielded 0.72 megatons. Squadron Leader Roberts, captain of the release aircraft, XD822, was the most experienced 'Grapple' pilot, but the 'Orange Herald' flight nearly ended in disaster, as his flight report calmly records:

'My crew was detailed to take off at 09:00 on 31 May in Valiant XD822 to drop Orange Herald on the target area south of Malden Island. The forecast weather for the target area was one- to two-eighths of cumulus, and wind velocity 090 degrees/20 knots at 45,000 feet; conditions at base line. In view of this, fuel load was reduced to 5,000 gallons in order to give an all-up weight of 99,000lb immediately after release of the bomb. The crew reached the aircraft at 07:40 and completed cockpit checks by 08:10. AWRE [Atomic Weapons Research Establishment] had connected the bomb batteries by the time the crew entered the aircraft at 08:40, but then the take-off time was delayed on orders from JOC [Joint Operations Centre]. At 09:00 permission was given to start engines and we were airborne at 09:07. The flight to the RV [radar vector] took 50 minutes and was uneventful. Good contact on HF and VHF was established and maintained with the appropriate authorities throughout the flight. The first run over the target was navigation-type and the weather was found to be as forecast.

After the first run the remaining black-out shutters were fitted and we went straight round on the initial run. Shortly after completing this, permission was given to carry on with the live run. The run-up was steady, and the bomb was released at 10:44, heading 202 degrees, IAS [indicated airspeed] 216, IMN [Indicated Mach Number] 0.75. After a slight pause I initiated a steep turn to port at 60 degrees bank. At this stage the second pilot should have started to call readings on the sensitive accelerometer, but on this occasion he was silent for a few seconds. I looked up and saw that the instrument indicated unity. Experience told me to believe the instrument, so disregarding my senses I increased the backward pressure on the control column. At that instant the second pilot and I realised that the instrument had failed at the time of release; simultaneously, the aircraft stalled, and the bomb-aimer, who was making for his seat, returned to the bomb-aimer's well with some force. After regaining control, the manoeuvre was completed in 43 seconds, using the mechanical accelerometer. This instrument might have been referred to earlier had it not been so far from our normal instrument scan. At 53 seconds by the navigator's countdown a bright white flash was seen through chinks in the blackout screens, and the coloured glass in the first pilot's panel was lit up. At 2min 55sec after release the blast waves were felt, first a moderate thump followed a second later by a smaller one. I waited a further two minutes before turning to port to allow the crew to see what had happened. The cloud top at this time appeared to be some 10,000 feet above our flight level, and it is a sight which will not easily be forgotten. The symmetry and the colours were most impressive, especially against the dark blue background provided by the sky at that height; as we watched, the upper stem and mushroom head started to glow with a deep peach colour. We then set course for base and landed at 12:47.'

The third test, code-named 'Purple Granite', took place on 19 June. This time the air drop involved the 'Green Granite Large' warhead, an enlarged version of the 'Green Granite Small' with a total weight of 6,000lb, excluding the HE.

Yellow Sun Mk 1 pictured during loading trails at Farnborough.

XH532 was eventually operated by the Waddington Wing, although the aircraft first took part in flight trials associated with the Yellow Sun Mk 1 bomb. *Robin A. Walker*

Surprisingly, it was the least successful of the three drops, yielding just 0.2 megatons after being dropped from XD823. Clearly the tests had been useful, but they were rather disappointing in terms of results.

Consequently the situation was summed up at a Progress Meeting on 16 July, when the development of 'Yellow Sun' (the code-name for what would become the standard free-fall megaton weapon for the V-Force) was discussed. While 'Grapple' had been successful in providing data on the performance of two different types of megaton warhead, it had not provided sufficient data to enable a firm decision to be made regarding the warhead to be chosen for Yellow Sun. On the evidence of the trials, a 'Green Bamboo'-type warhead had been chosen by the Air Staff for use in what would be an interim megaton weapon, pending deliveries of Yellow Sun. Further trials, 'Grapple X', were scheduled for November 1957, and No 49 Squadron was again tasked with the provision of aircraft and crews.

This time the bomb drop at ground zero would be just 20 miles from Christmas Island (instead of nearly 400), to avoid the expense and delays involved in setting up a naval task force to monitor the tests. Valiant XD825, captained by Sqn Ldr B. T. Millett, made the fourth drop on 8 November, the weapon being fitted with another 'Green Granite Small' warhead. This time, however, the scientists were pleasantly surprised at the result when the bomb delivered a yield of 1.8 megatons, the first true British megaton explosion. Sapper Arthur Thomas witnessed the test from Christmas Island:

> 'Then it happened, the blast, a lightning speed of wind and whistle of trees – a bang – it hit us all unexpectedly, lifting us off our feet and depositing us three to four yards away landing on top of each other in a pile of bodies. We were not told to expect anything of this nature.'

The apparent success of this drop explains why just one test was made when the Task Force Grapple Air Plan called for the 'air drop of two thermonuclear weapons with minimum risk to all concerned'. The records of No 49 Squadron state, 'The results were entirely satisfactory, precluding the necessity for any further tests in this particular phase of Operation Grapple.'

However, the test programme was far from complete, and 'Grapple Y' took place during April 1958. Valiant XD824, captained by Sqn Ldr R. M. Bates, made the next (fifth) drop on 28 April. Evidence suggests that the warhead was a 'Green Granite Large' device, and the resulting explosion delivered the biggest yield of any British nuclear tests, a tremendous 3.0 megatons, which demonstrated that Dr Penney and his scientists could, if necessary, produce weapons that easily matched the destructive power of either the American or Soviet Union bombs. Further tests ('Grapple Z') took place during the summer of 1958, on a faster time-scale, as Britain slowly moved towards an agreement to end nuclear testing. No 49 Squadron's records state: 'Due to the decision to accelerate the entire dropping programme for political considerations, the intensity of the high-explosive drops in preparation for the nuclear drops has been increased.'

These (final) nuclear detonations took place in September, as the squadron's records indicate:

> 'The month of September brought to fruition all the training for Operation Grapple Z with the dropping of two more nuclear weapons by the squadron. On the 2nd Sqn Ldr G. M. Bailey and crew in Valiant XD822 dropped the first device of the series. This weapon was the first to be dropped by ground-controlled radar. A grandstand aircraft on this occasion, Valiant XD818, was flown by Flt Lt S. O'Connor and crew. On 11 September Flt Lt S. O'Connor and crew, in Valiant XD827, dropped a second nuclear device. This weapon was released on a visual attack. Sqn Ldr H. A. Caillard and crew in Valiant XD824 flew as grandstand. Immediately after the second air drop the aircraft were prepared for the return trip to Wittering.'

With the Grapple Valiants back at Wittering, No 49 Squadron continued training for an anticipated further series of test drops, but by the end of November 1958 Britain had effectively decided to abandon nuclear testing, and the government later stated that the United Kingdom would no longer carry out further nuclear tests, whether in or above the atmosphere, underwater or underground. Christmas Island was gradually reduced to a minimum holding state and HQ Task Force Grapple was disbanded on 3 June 1960. With effect from 1 December 1958, No 49 Squadron reverted to a standard bomber role and the 'Grapple' Valiants were de-modified and refitted with the standard radar navigational bombing system.

Although the Cabinet decision to manufacture thermonuclear bombs had been taken more than five months previously, Churchill made no reference to this historic date in British history when he said:

> 'The advance of the hydrogen bomb has fundamentally altered the entire problem of defence, and considerations founded even upon the atom bomb have become obsolescent, almost old-fashioned. Immense changes are taking place in military facts and in military thoughts. We have for some time past adopted the principles that safety and even survival must be sought in deterrence rather than defence and this, I believe, is the policy which also guides the United States.'

Of course, Operation 'Grapple' demonstrated quite clearly to the USA that Britain was more than capable of producing thermonuclear weapons that were, if anything, more efficient than its foreign equivalents. It is hardly surprising, therefore, that relations between America and Britain gradually improved after the first nuclear tests (Operation 'Buffalo') in 1956, and even survived through the days of the McCarthy period. The Atomic Energy Act, effectively a revision of the infamous McMahon Act, was signed in 1954, permitting the transfer of data concerning the external characteristics of nuclear weapons: size, shape, weight, yield and effects. The USA and Britain agreed to co-operate within the terms of this Act in a bilateral agreement signed on 15 June 1955.

However, the real breakthrough came, rather ironically, after the huge transatlantic rift that developed during the Suez Crisis. Prime Minister Harold Macmillan and President Eisenhower met twice in 1957. During their first meeting, in Bermuda during March, agreements were made for the deployment of Thor missiles to the UK, which would be under 'dual key' control. Eisenhower commented that this was 'by far the most successful international conference' he had attended since the war. When they met for a second time, in Washington during October, revisions were made to the Atomic Energy Act to allow scientific co-operation between 'Great Britain, the United States and other friendly powers'. It can hardly be a coincidence that the President's enthusiasm for restoring the Special Relationship came in the same year that Britain embarked upon the first 'Grapple' tests, just months after the first operational hydrogen bombs had entered US service. Britain's thermonuclear advances had been remarkable, and the USA identified potential advantages for a new co-operative arrangement.

Another important consideration was the launching of the Soviet Union's Sputnik satellite, which threw into question many assumptions of American technical superiority, and underlined the Soviet's growing ability to deliver nuclear weapons, to Europe and the United States, by missile. The Suez Crisis had also served to illustrate just how far apart Britain and America were, and possibly served as a catalyst in repairing the faltering Special Relationship. Finally, in 1958 the Agreement for Co-operation on the Uses of Atomic Energy for Mutual Defence Purposes was signed, enabling both countries to exchange virtually all types of nuclear information. Britain quickly learned a great deal from American expertise in engineering and weapons assembly techniques. Conversely, American officials were amazed at the scientific and technical knowledge that Britain possessed, which in many respects was ahead of the Americans. The intimate and totally reciprocal collaboration continued into the 1960s and beyond, long after America could possibly have hoped to derive any further benefits from the co-operation.

A British official illustrated the position in the 1970s when he stated:

> 'The United States has two laboratories and we have one; they spend five times as much as we do on these establishments; they have conducted some 870 tests – how many of which were really necessary I wouldn't say – and we have conducted 30. That gives, I think, a fair indication of the "hardware balance", although in the idea end of the business, the relationship is rather more equal.'

The first Air Staff Requirement for a thermonuclear bomb, ASR OR 1136, was issued on 6 June 1955. It called for a bomb that was not to exceed 50 inches in diameter (and would be made smaller if possible), was not to exceed 7,000 lb in weight, and would be capable of carriage internally by Valiants, Vulcans and Victors, to be in service in 1959.

When the Cabinet had made its historical decision to proceed with development of hydrogen bombs during June 1954, the meeting agreed that work on the programme should be performed 'as unobtrusively as possible', and that it was 'desirable that costs be concealed as much as possible'. Before them was a memorandum from the Chiefs of Staff that was based on a report by the Working Party on the Operational Use of Atomic Weapons, whose members included Sir William Penney, the Deputy Chiefs of Staffs and scientific advisors. The report concluded that hydrogen bombs would 'go a long way towards overcoming the difficult problems of terminal accuracy simply by delivering a huge explosive force, and that only a relatively limited number of bombs would be required, beyond which any increase of stocks would not offer any corresponding military advantage.'

During a meeting on 6 April 1955 the Chiefs of Staff decided to develop, as a first priority, a thermonuclear bomb with a yield of approximately 1 megaton, and ASR OR 1136 was accepted by the Ministry of Supply on 28 July, leading to the start of development work on the weapon, called 'Yellow Sun'. During March 1956 members of the Operational Requirements staff visited Farnborough, where drawings and a wooden mock-up of Yellow Sun were prepared. The bomb carcass was 240 inches long and 48 inches in diameter, and

A rare photograph of the Red Beard tactical bomb that was occasionally carried by Vulcans. Although the Yellow Sun and Blue Steel were the Vulcan's primary weapons (prior to the introduction of the WE177), Red Beard appears to have been regarded as a second-strike weapon for possible use after the stocks of Yellow Suns and Blue Steels had been exhausted.

had small cruciform stabilising fins similar to the Second World War 'Grand Slam' design (larger ones would have had to fold out to enable the weapon to fit into the Valiant's bomb bay). Unusually, however, the bomb had a flat nose, which was intended to slow the bomb during free-fall to increase the weapon's stability and to simplify the requirements of the internal barometric detonation device. The warhead weighed approximately 3,500lb, and the completed weapon had a weight of around 6,500lb.

While Yellow Sun was being developed, a second weapon, Red Beard, was also produced, primarily as a smaller tactical bomb but also partly as a replacement for the first-generation Blue Danube, which, as a crude fission device, delivered a relatively low explosive yield of around 20 kilotons, despite weighing 10,250lb. However, it was quickly accepted that, if the development of these two new weapons proceeded smoothly (and much depended on the outcome of Operation 'Grapple'), the warheads would be ready before the bomb bodies being designed to carry them. Consequently it was suggested that an interim weapon should be manufactured so that the RAF could acquire a thermonuclear capability as soon as possible. Three warheads, Green Granite, Green Bamboo and Orange Herald, could be made available, and could be incorporated within a standard Blue Danube bomb casing. A report prepared by the RAE in April 1957 described progress with the RAF's hydrogen bomb:

> 'The Yellow Sun weapon to meet Air Staff Requirement OR 1136 will provide the first British bomb having a yield in the megaton range; as such it is the

An interesting post-Operation 'Corporate' publicity photograph showing a 'Black Buck' Vulcan at Waddington complete with a representative selection of weapons that the Vulcan could carry. Items missing include Yellow Sun, Red Beard, Blue Danube/Violet Club and WE177.

keystone of the offensive deterrent policy. It is intended for carriage in the V-class bombers and will have a diameter of 48 inches and a length of approximately 20 feet. The weight will be about 7,000lb. The weapon is being designed around the Green Bamboo warhead under development at the Atomic Weapons Research Establishment. The means of making a warhead in this range wholly safe in storage and transport has not been finalised, but all schemes of providing in-flight insertion of some part of the fissile material have been abandoned.

Consideration has also been given to the alternative warheads Green Granite and Short Granite, which are being tested at Operation Grapple. Both are fission-fusion-fission types and differ only in that Short Granite is smaller and lighter. Neither warhead requires ENI [External Neutron Initiations]. Nuclear safety is ensured by some form of in-flight insertion.'

After referring to other considerations, the report continued:

'Preliminary investigation indicates that if after Grapple Short Granite becomes the preferred warhead, no serious delay should occur, but the Green Granite would require a larger and heavier weapon, so that much of the ballistic and fuzing work already well advanced would have to be repeated, and the in-service date would have to be set back at least nine months.'

A second Progress Statement, issued later in 1957, included the following:

'The Operational Requirement OR 1136 Issue 2 calls for the development of a megaton bomb, the type of warhead to be carried not being specified except that it shall be capable of use in both Yellow Sun and Blue Steel. The original requirement was for incorporation of the Green Bamboo warhead and much of the work done has been on the assumption that this warhead will be used. It has, however, been evident for some time that as a result of the Grapple trials another warhead might be preferred, and preliminary investigations were made at an early stage into the problems which would arise if one of the Granite type of warheads were chosen. These have been followed by further work, especially in connection with the possible use of Short Granite. The general position now is that an early decision as to the type of warhead, and information on associated matters such as nuclear safety systems, is essential if development of the weapon is not to be held up.'

The revised Operational Requirement stated that Yellow Sun would be carried by the Vulcan and Victor, the earlier requirement for carriage by the Valiant having been cancelled. Despite this decision, many of the Yellow Sun trials were conducted by Valiants until Vulcans and Victors could be made available. The report continued:

'Yellow Sun is being developed as a fully engineered weapon to meet the requirements of the OR. The provision of an interim megaton weapon only partially meeting these requirements is planned for introduction into service considerably earlier than Yellow Sun.'

A paper issued by the MOS in August outlined the indecision that had surrounded the choice of warhead for Yellow Sun:

'The object was to give the RAF a megaton capability at the earliest possible moment. It was proposed to base the interim weapon on one of the bombs to be dropped in Operation Grapple and it was stated that the date of introduction to the Service of Yellow Sun would not be affected by the interim missile. The results of Operation Grapple were such that none of the rounds dropped was immediately applicable to the interim weapon, but AWRE were satisfied that the principles had been cleared sufficiently for them to offer various alternative warheads to the Air Staff for consideration. The Air Staff, largely on the basis of numbers which could be provided, chose a warhead similar in outside shape to Green Bamboo but having a yield of half a megaton ... known as Green Grass.'

The MoS also commented:

'Throughout the discussions on this interim megaton weapon the general approach has been that, in the interests of providing a megaton capability to the RAF at the earliest possible moment, the Service is prepared to sacrifice rigorous testing, proofing, and clearance of the weapon and to introduce special maintenance procedures in association with AWRE. Furthermore, the Air Staff were willing for the same reason to discount many of the provisions of OR 1136.'

With reference to Yellow Sun, the report said:

'A re-assessment of the Yellow Sun programme has recently been made primarily to examine the possibility of offering an earlier capability to the RAF in view of the successful progress of the development, but also taking into account the desirability of switching over to the Short Granite type of warhead if and when this is cleared by AWRE. This later type of warhead is very desirable because of the smaller amounts of fissile material needed.'

After much deliberation it was agreed that an interim weapon ('Violet Club') would be brought into service until the first deliveries of a limited-approval version of Yellow Sun could be made. The Deputy Chief of the Air Staff commented:

'We are anxious to get megaton weapons into Service as soon as we can. We most certainly think it worthwhile to have even as few as five by the time the Yellow Suns come along. At the same time we are anxious to get Yellow Sun as soon as we can because it does not have the serious operational limitations of Violet Club.'

On 24 February 1958 the Assistant Chief of the Air Staff wrote to the AOC in C Bomber Command as follows:

'I am directed to inform you that the first Violet Club which is now being assembled at Wittering is expected to be completed by the end of the month.

A total of five of these weapons will be assembled on Bomber Command stations by July of this year, when deliveries of Yellow Sun should commence. Violet Club is still in some degree experimental and it will be subject to a number of serious handling restrictions. The extent of these and their effect on operational readiness are still under discussion with the Ministry of Supply. Until Violet Club has been formally cleared by the Ministry of Supply and it is possible to issue specific instructions on storage, handling, and transport, the weapon is to remain exclusively in the custody and under the control of the Atomic Weapons Research Establishment. It is possible that some such arrangement will continue throughout the life of these weapons as it is intended to replace them with Yellow Sun as early as possible. This will be done as soon as sufficient aircraft are modified to carry Yellow Sun. I am to say that the operational limitations of Violet Club, particularly those affecting readiness, are serious. Nevertheless, it provides a megaton deterrent capability several months earlier than would otherwise have been possible.'

Externally similar to its kiloton predecessor, Violet Club had the same ballistics as Blue Danube and could therefore be used with the same bombing equipment, suitably adjusted for the required burst height. The Bomber Command Armament School (BCAS) Operations Records state that:

'...on 28 February a convoy from AWRE was stuck in a snowdrift at Wansford Hill at 15:00. An officer from this unit was sent to investigate. At 17:00 the vehicle was still unable to be moved. Rations and bedding were sent to the convoy and an officer and a team of airmen were detailed to stand by throughout the night to give help if required. It was not until the following day that vehicles began using the A1. The convoy arrived at the main Guard Room at 12:00. Personnel were sent immediately for a meal. Unloading was commenced at 14:00, when the convoy arrived at BCAS.'

This wintry scene marked the arrival of the first Violet Club, and with it the very beginnings of the RAF's thermonuclear capability. By July a total of five bombs had been completed. Bombs 1 to 4 remained at Wittering, where personnel were trained by AWRE staff to install the Green Grass Warhead at the relevant site. Bombs 5, 6, 7, 8 and 11 were despatched to Finningley, and 9, 10 and 12 went to Scampton, production of all twelve weapons being completed by the end of 1958. Violet Club was, as DCAS commented, 'rather delicate' and could only be assembled at the base from where it would be used. Likewise, road transport was limited to relatively short trips from the assembly point to the storage building. Further details of Violet Club's use were contained in a document issued in July:

'In view of the very small number of Violet Clubs being made available to the Service, and the difficulties of clearing the Victor for the carriage of this weapon, it has now been decided to limit its carriage to the Vulcan. Vulcan aircraft modified to carry Yellow Sun are now being returned to service and it is desirable to transfer the warheads from all Violet Clubs to Yellow Suns as soon as possible, and at the same time to redeploy the weapons to Scampton and Finningley.'

In November 1958 a further report stated:

'Bomber Command wish to get rid of Violet Clubs as soon as possible, but they are anxious to have a number of Yellow Suns in store before they do this. As the obvious time to change over is at the six-monthly inspections, it is suggested that Violet Clubs should start to phase out at the rate of about one a month from April or May, providing the Yellow Suns are not late.'

Consequently, by the end of 1958 the RAF possessed a force of fifty-four Valiants, ten Victors and eighteen Vulcans, the last including suitably modified aircraft capable of carrying the 12-megaton-range Violet Clubs. The remaining V-Force aircraft were, of course, equipped to carry kiloton-range Blue Danubes.

But even as the V-Force was being created, the concept of guided nuclear missiles had already been established, and by 1954 the Air Staff had issued OR 1132, calling for a propelled air-to-surface missile for the V-Class bombers. Capable of being launched at up to 100 nautical miles from its target, the missile would rely on the parent aircraft's NBS system for aiming, and would use Green Satin Doppler equipment to determine ground speed and drift, and to provide an accurate heading reference. Responsibility for the weapon, named Blue Steel, was divided between the MoS and Avro's Weapons Research Division, which had completed a design study for a stand-off bomb that resulted in a development contract being awarded in March 1956. Tests of a two-fifths-scale model of the guided bomb were conducted in 1957-58, a series of drops being made from Valiant WP204 over the Aberporth range off the Welsh coast.

Development of Blue Steel was far from easy, not least because virtually the entire programme represented a move into completely unknown territory as far as technical knowledge was concerned. In essence, Blue Steel was an aeroplane, and Avro treated the missile as such. Some 35 feet long, with small delta foreplanes, rear-mounted 13-foot-span wings, and vertical fins, it was powered by a hydrogen peroxide and kerosene Armstrong Siddeley Stentor rocket motor. Guided by inertial navigation, and with automatic flight control and trajectory decision-making, the missile manoeuvred at supersonic speeds before delivering a 1-megaton ('Red Snow') warhead to its target. The all-up weight of the bomb was about 17,000lb, which included 400 gallons of High Test Peroxide (HTP) fuel and 80 gallons of kerosene. After release at 40,000 feet the bomb would free-fall to 32,000 feet, at which stage the motor would ignite and the missile would climb to 59,000 feet, where the speed would increase to Mach 2.3. The missile would then cruise-climb to 70,500 feet, where the engine would burn out and a steady dive towards the target would begin. Bearing this very complicated system in mind, and the fact that Avro had no previous experience of designing and manufacturing guided missiles, it is hardly surprising that the development programme time-scale began to slip, especially when the MoS was unable even to supply a Valiant test-bed aircraft on time. The Air Staff's hope

The complex nature of Blue Steel (and the long fuelling and preparation process) made the weapon less than practical for short-notice dispersal or 'scramble' launches. Likewise, there was never any guarantee that the Stentor motor would successfully fire when launched, and this prompted the RAF to anticipate using the weapon as a free-fall bomb if such action had been necessary.

that Blue Steel would be operational by 1960 was quickly discounted, despite the fact that it was already looking for a missile that could fly up to ten times further.

In May 1958 Air Staff Requirement OR 1159 called for an extended-range air-to-surface guided missile, stating:

> 'By 1963 it is expected that the Russian SAGW [surface-to-air guided weapons] and the fighter defences will be so improved and expanded that the V-Bombers, even with Blue Steel and RCM, will find it increasingly difficult to penetrate to many of their objectives. In order to maintain an effective deterrent during the period commencing with the decline in effectiveness of Blue Steel and continuing during the build-up of the RAF ballistic missile force, it will be necessary to introduce a replacement for Blue Steel having a range for attacking targets from launching points outside the enemy defence perimeter. It is envisaged that V-Bombers equipped with this missile should be able to supplement the ballistic missile deterrent for several years.'

The statement continued that 'a missile range of 600 nautical miles will be acceptable as an initial operational capability, but a range of 1,000 nautical miles is desirable.' Duncan Sandys, the Minister of Aviation, recognised that it would be foolish to distract Avro's efforts to get Blue Steel into service by adding the complication of a long-range Blue Steel development, so the subsequent cancellation of both Blue Streak and Blue Steel Mk 2 enabled Avro, the RAE and the Royal Radar Establishment to concentrate on getting Blue Steel Mk 1 into service at the earliest opportunity. On 28 April 1961 Minister of Aviation Peter Thorneycroft reported that:

> 'Blue Steel was accepted as a requirement in January 1956. It was then thought that the first delivery of missiles would be made to the RAF in 1961/2. By the end of 1960, however, it had become apparent (owing to delays in the development programme) that the number of trial firings that could be expected to have been made by early 1962 would not be sufficient to enable the fast deliveries of missiles to the RAF to be approved for normal operational use. It is, however, expected that by mid-1962 the functioning and safety of the weapon (including its warhead) will have been sufficiently proved to enable the missile to be used in an emergency, if required, thus providing a deterrent capability. Further trials will continue during the succeeding months to enable approval to be given for normal operational carriage and use of the missile.'

A typical scene at Scampton as a Blue Steel missile is towed away from the Vulcan carrier. Loading and unloading the cumbersome weapon was never easy. Thanks to the 'hand-made' craftsmanship that was prevalent in the aerospace industry in the 1950s, no two Vulcans were precisely the same, and some missiles tended to fit the aircraft (or vice versa) rather more easily than others.

The Minister added that the delays in the programme were due to 'detailed engineering faults and problems which are a normal part of the development process and may be expected to continue'.

An order was placed for fifty-seven missiles, made up of a Unit Establishment of forty-eight operational rounds, plus four backing rounds and four proof rounds. Additionally, there would be sixteen training rounds, ten of which would be manufactured from light alloy, and six with steel carcasses. During 1959 Vulcan B.1 XA903 joined the Blue Steel programme, and a variety of full-sized missile test bodies, mostly powered by de Havilland Double Spectre engines, were dropped from both the Valiant and Vulcan. Most of the later test flights were made over the Woomera range in Australia, where two Vulcan B.2s, XH538 and XH539, were employed.

The development programme was dogged by a series of relatively minor problems that led to increasing frustration on behalf of the Air Staff; Avro was accused of poor management and of wasting time on projected plans for a long-range Blue Steel when it should have devoted all of its efforts to the completion of Blue Steel Mk 1. In reality, Avro simply suffered from a combination of initial over-optimism shared by the Air Staff and a whole range of new technical problems which, as an aircraft manufacturer, the company had never before encountered. Considering that the typical development period for a conventional military aircraft could be anything up to ten years, it was perhaps unrealistic to expect a system as complicated as Blue Steel to be completed in a significantly shorter time-scale.

The Minister of Aviation also said in 1961 that:

> 'Blue Steel was a fully navigated cruise-type missile with a range of 100 nautical miles designed for launching from Mk 2 Victors and Vulcans. It was intended to provide the main deterrent weapon between the time when bombers equipped with free-falling bombs were likely to become less effective against enemy defences, and the introduction of Skybolt. Its cost was currently estimated at £60 million for R&D and £21 million for production; of this total, some £44 million had been spent or committed. Trial firings had proved disappointing in some respects, but it appeared that the difficulties were caused by teething troubles rather than by any basic fault which might invalidate the concept of the weapon. Further trials were proceeding at Woomera and it should be possible to make a comprehensive review towards the end of the year.'

When progress was reviewed at the beginning of 1962, the Cabinet Defence Committee learned that the firing of W.100A rounds, closely representative of the final production version, was about to start, and that by August or September enough preliminary information would be available from launchings to enable the Air Ministry to assume an emergency capability. Finally, in July 1962 a production-model Blue Steel was successfully fired after being air-launched from a Vulcan over Woomera, and on 25 July the Minister of Aviation, now Julian Amery, wrote to the Air Minister, stating:

> 'I am glad to be able to inform you that Sir George Gardner [Controller of Aircraft, Ministry of Aviation] has today forwarded to DCAS a CA Release for Blue Steel to be carried on Vulcan Aircraft, complete with its operational warhead, in a national emergency. The clearance does not specifically authorise the launching of the missile because the required trials to prove the safety of the systems are not yet complete. However, no difficulties have been experienced up to date which affect the safety after launch and we are confident that

The Blue Steel's HTP (High Test Peroxide) fuel was a nightmare for the groundcrews who were obliged to deal with it. Protective clothing was vital, and the volatile nature of the fuel meant that it required very careful handling. Water pits were positioned next to servicing areas (and even next to some ORPs) to enable missiles (and personnel if necessary) to be rapidly immersed, should combustion of the fuel occur.

> further trials will provide the necessary proof. We expect to issue the operational launch clearance in December 1962. We have issued the Present Clearance on the understanding that, should a national crisis occur which warrants the carriage of the operational Blue Steel with its warhead, limitations as to its use could be overridden. In effect this means that you could declare an operational capability with Blue Steel as soon as you consider that you are in a position to do so.'

The first unit to be declared operational on Blue Steel was No 617 Squadron at Scampton. A great deal of time was spent discussing when the RAF should officially declare that the squadron had an operational capability, as Bomber Command wanted to arrange a press facility to show the Blue Steel system to the public. It was feared that a premature display of Blue Steel might lead to embarrassing questions as to the true extent of the RAF's capability at that time, so the date of the press day was continually delayed until 14 February 1963, by which time the squadron was fully operational with at least six missiles available. Unfortunately, by this time the Skybolt programme had been cancelled, and the press was more concerned with plans for the V-Force's future than with the event being celebrated at Scampton. However, Nos 17 and 83

XM576, complete with Blue Steel missile, was damaged significantly during a landing at Scampton in May 1965. The aircraft's wing struck the ground, causing the aircraft to swing off the runway. The aircraft was subsequently struck off charge and didn't fly again.

Squadrons subsequently re-equipped with Blue Steels at Scampton, followed by No 139 Squadron on Victors at Wittering, their conversion beginning in October 1963, followed by No 100 Squadron. On 21 August an Air Ministry Nuclear Weapon Clearance for the use of Blue Steel on Quick Reaction Alert (QRA) standby was issued, although the weapons were to be unarmed and unshelled except in an emergency. Clearance for a fully armed and fuelled Blue Steel on QRA was finally issued on 16 April 1964.

The delivery of the last Victor B.2s in 1962 represented the culmination of Britain's long-established plans to equip the RAF with a fully effective nuclear deterrent. By 1961, with a combined force of 144 Valiants, Victor B.1s and B.2s and Vulcan B.1s and B.2s, Bomber Command possessed awesome striking power beyond anything imaginable during the Second World War.

The first-generation 20-kiloton Blue Danube fission weapon was gradually withdrawn during 1960, Red Beard replacing it as the RAF's main fission weapon. Red Beard was essentially a tactical bomb, however, developed into Mk 1 (15-kiloton) and Mk 2 (25-kiloton) versions to be carried by RAF and Fleet Air Arm tactical strike aircraft (the Canberra, Scimitar, Sea Vixen and Buccaneer). Valiants, Vulcans and Victors were also capable of carrying the weapon, generally as a potential 'second strike' bomb that could have been used for repeat attacks, following a first strike with megaton-class weapons (assuming there was any need for a second strike after the first exchange).

The Yellow Sun Mk 1, with a yield of 500 kilotons, was superseded by Yellow Sun Mk 2 beginning in February 1962, initial stocks going to Waddington. The Red Snow warhead, which was also fitted to Blue Steel, was manufactured to deliver one of three different yields, and it is likely that both the Blue Steel and Yellow Sun warheads were largely of the 1-megaton variety, the remainder being 500-kiloton. For much of the 1960s the Yellow Sun Mk 2 became the 'standard' free-fall nuclear weapon for the Vulcan squadrons, while the Scampton Wing concentrated on Blue Steel operations.

XA903 gets airborne on another test flight. Despite being the very last active Vulcan B.1, the aircraft was scrapped at Farnborough following retirement in 1979. As ever, while Second World War aircraft continue to be restored and saved for future generations, unique and historical aircraft like this Vulcan are simply ignored. *Robin A. Walker*

Above: 617 Sqn B.2 XL360 flies low over Finningley in September 1968, flaunting her Blue Steel capability. *Fred Martin*

Left: XL361 on public display at Farnborough, wearing 617 Squadron's low-visibility (pink) markings on the tail. *Ken Elliott*

Below: Gleaming in what appears to be a relatively fresh coat of paint, XL321 sits on Finningley's ORP during a Battle of Britain 'At Home' Day. The fluorescent orange 617 Squadron tail markings were the second of three distinct designs used by the squadron while operating the Vulcan.

Above: 617 Squadron's XL446, pictured after returning from a 'Giant Voice' exercise in the USA during which a SAC badge has been applied to the aircraft's nose. *Michael J. Freer*

Right: XM574 proudly wearing the markings of No 617 Squadron, pictured on short finals to home base at Scampton. *Terry Senior*

Below: A photo-montage of Scampton's southern taxiway showing a line-up of Vulcans from each current squadron, in celebration of the aircraft's 25th year. *Bruce Woodruff*

An interesting image showing a Blue Steel missile awaiting delivery to a 617 Squadron Vulcan at Scampton, while in the foreground is the grave of Guy Gibson's dog 'Nigger', killed in a road accident on the night of 617 Squadron's famous Second World War attack mission.

Blue Steel missiles inside their storage hangar at Scampton.

XJ823 taxiing out for take-off at Scampton. The aircraft is one of the few Vulcans to have survived and remains on display at Carlisle Airport. *Terry Senior*

XH537 with 'everything down' on final approach to Scampton. In addition to the extended undercarriage and airbrakes, the underwing air sampling pods make an otherwise aerodynamically clean aircraft look distinctly 'dirty'. *Terry Senior*

Below: 617 Sqn B.2 XL427 with Blue Steel at Finningley in September 1968. *Fred Martin*

Left: 617 Sqn Vulcan B.2 XM595 was displayed with Blue Steel at Farnborough in 1964. Squadron motifs of the Scampton Wing are painted on the entrance hatch. *Martin Derry*

Below: 617 Sqn Vulcan B.2 XM595 was displayed with Blue Steel at Farnborough in 1964. *Martin Derry*

The government's unease at the prospect of relying on fixed-site ballistic missiles, which could take at least 15 minutes to prepare for launching, prompted a search for a more suitable nuclear delivery system to replace Blue Steel towards the end of the 1960s. The difficulties and costs associated with the production of Blue Steel Mk 1 effectively killed off the Mk 2 project before it started, and with the Thor missile system as a stopgap, Britain seemed destined to rely upon the developing Blue Streak missile as the country's main nuclear deterrent for the late 1960s, though the system, which differed very little from Thor, was likely to be obsolete even before it entered service.

It was at this stage that Britain again looked towards America, where the Douglas GAM-87A Skybolt missile was being developed. This air-launched nuclear-tipped guided bomb was to be capable of delivering a sizeable warhead from a range in excess of 1,000 miles, far beyond the capabilities of Blue Steel. In essence, it was everything that the Air Staff had wanted from Blue Steel Mk 2. Talks with the US President at Camp David suggested that there would be no American objection to the British purchase of Skybolts if necessary, and it was also indicated that the sea-launched Polaris might also be made available, should Britain be interested. The ever-increasing cost of the Blue Streak programme and the pressing fears that the weapon would be very vulnerable to Soviet attack, led to the conclusion that an updated V-Force, equipped with Skybolts, would be a much cheaper and much more credible option than relying on another non-mobile IRBM, which would be an easy target for Soviet attack. Consequently, Blue Streak was abandoned.

Skybolt was originally designed in response to a USAF requirement for an air-launched strategic-range nuclear missile to be carried by SAC Boeing B-52s and Convair B-58s. It was quickly established that development of the weapon would be completed to enable the first Skybolts to enter RAF service in 1966, although a new carrier for the weapons would probably be required by 1970 because the Vulcans and Victors would reach the end of their useful life at that time. Macmillan was keen to place a provisional order for 100 Skybolts, but the Chancellor was rather more cautious, not least because the missile was only just beginning development. However, following the meeting between Macmillan and Eisenhower in March 1960, a minute was issued by the US Government that stated:

> 'In a desire to be of assistance in improving and extending the effective life of the V-Bomber force, the United States, subject only to United States priorities, is prepared to provide Skybolt missiles minus warheads to the United Kingdom on a reimbursable basis in 1965 or thereafter. Since Skybolt is still in the early stages of development, this offer is necessarily dependent on the successful and timely completion of its development programme.'

The minute also made mention of Polaris, stating:

> 'As the United Kingdom is aware, the United States is offering at the current NATO Defence Ministers meeting to make mobile Polaris missiles minus warheads available from US production to NATO countries in order to meet Saceur's requirements for MRBMs [medium-range ballistic missiles]. The United States is also offering to assist joint European production of Polaris if our preference for United States production proves unacceptable.'

Given that Britain and France were the only NATO countries outside the USA with a nuclear capability (and therefore, the only countries able to operate the missiles), this offer was of particular significance to Britain, which effectively had the luxury of a choice between two weapons systems. However, Britain continued to pursue the Skybolt project, even though American concerns over the possible success of the development programme were being expressed as early as 1961.

With more than 150,000 component parts, some 60,000 of which had to function perfectly in order to launch the missile, Skybolt was a very complex piece of hardware. The first live launches in 1962 ended in failure, and the US Secretary of Defense advised his British counterparts that, although the USA did not believe Skybolt was a technical failure, it did believe that continuing the programme would be a waste of money in view of the emergence of other delivery systems at that time, such as Hound Dog and Minuteman. It was pointed out that the Skybolt missile was essentially a research programme at the time when Britain first requested it, rather than being a production weapon. Consequently, the path would be open for Britain to continue development of Skybolt in association with a scaled-down American effort, or to develop Skybolt in isolation, albeit by using American technology.

Another option would be to adopt the Hound Dog missile, or to participate in a multilateral force of sea-launched (Polaris) missiles, under the terms being offered in March 1960. The Hound Dog option was quickly discounted because the weapon could not be carried by V-Bombers, and the Ministry of Defence concluded that the real options for Britain were to acquire Polaris, to complete Skybolt in America, to complete Skybolt in the UK, or to produce a ballistic weapon in co-operation with France.

It had already been agreed that the V-Force dispersal concept, whereby the bombers would leave their home bases in a wartime emergency and relocate in groups of four or two to thirty-six dispersal airfields scattered around the UK, would not be a credible option for very long without improvement, as the increasing sophistication of Soviet attack forces meant that bombers on the ground in any location would be at risk from attack. The best alternative to a dispersal policy would be to mount a continuous airborne alert, with Skybolt-equipped bombers maintaining a round-the-clock airborne presence, immune from attack. It was an expensive and technically difficult concept, but one that America was pursuing at the time. If Skybolt was abandoned, Britain would have to choose one of two expensive alternatives: either adopt an Anglo-French ballistic missile system, which would mean accepting all the disadvantages of Blue Streak again, or buy the Polaris missiles and a submarine fleet from which to launch them.

When Prime Minister Macmillan met President Kennedy at Nassau in the Bahamas during December 1962, Macmillan pressed Kennedy to continue with the Skybolt programme, but as a post-talk report stated, Kennedy, who was fairly ambivalent towards British interests, wanted to abandon the project:

Above: The Vulcan's sturdy undercarriage gave it excellent ground clearance, which was ideal for carriage of the Skybolt. Unfortunately the Victor's low-set wings were less than suitable for the missile, and had the project continued it is likely that only the Vulcan would have been fitted with the weapon.

Above left: XH537 was one of two aircraft that operated from Woodford on Skybolt trials, prior to the cancellation of the project.

Left: The huge Skybolt missiles required significant re-working of the Vulcan's airframe, and although the project was ultimately abandoned, many Vulcan B.2s were suitably modified with wing hard point locations and associated ducting, which subsequently proved invaluable for the 'Black Buck' missions in 1982.

'The President and the Prime Minister reviewed the development programme for the Skybolt missile. The President explained that it was no longer expected that this very complex weapon system would be completed within the cost estimate or the timescale which were projected when the programme was begun. The President informed the Prime Minister that for this reason, and because of the availability to the United States of alternative weapon systems, he had decided to cancel plans for the production of Skybolt by the United States. Nevertheless, recognising the importance of the Skybolt programme for the United Kingdom, and recalling that the purpose of the offer of Skybolt in 1960 had been to assist in improving and extending the effective life of the British V-Bombers, the President expressed his readiness to continue with the development of the missile as a joint enterprise between the United States and the United Kingdom, with each country bearing equal shares of the future cost of completing development, after which the United Kingdom would be able to place a production order to meet its requirements.'

This was a very generous offer to Britain, bearing in mind that Kennedy had little practical reason to provide any further funds for the Skybolt programme, and was not even a great believer in the Special Relationship. Macmillan recognised the value of Kennedy's offer, but the continuing doubts over technical difficulties, rising costs, and delays in the delivery timescale prompted him to decline the opportunity to divert more financial responsibility towards Britain. Likewise, he could not accept Hound Dog because of the time and expense that would be involved in modifying the Vulcan and Victor to carry it. Instead he opted for Polaris, and Kennedy agreed that America would provide Polaris missiles, minus warheads, for British submarines. Consequently, Skybolt was officially cancelled by America on 31 December 1962, just a few days after a test round had made a perfect launch from a B-52 over the Eglin weapons range. As far as the RAF was concerned the programme officially ended on 3 January 1963, and, having abandoned Skybolt in favour of Polaris, the future of the RAF's nuclear force had been sealed; the British airborne nuclear deterrent was to be transformed into a seaborne system.

However, until Polaris could be brought into service the V-Force would have to remain viable, and the Air Staff turned their attention to ways in which Bomber Command could maintain a credible nuclear deterrent until the end of the 1960s. They concluded that, beyond the mid-1960s, improvements in Russian air defences would mean that it would be practically impossible to penetrate Soviet airspace at high level – the V-Force would have to attack at low level if it was to survive. Soviet radar systems had been geared towards the detection of high-flying bombers, and although the range and sensitivity of their equipment was continuing to improve, radar detection was not effective at heights of around 3,000 feet or lower, where aircraft radar returns began to merge with other forms of ground-generated clutter. However, if the V-Force was prepared to attack at low altitude (500 feet or less), the Soviet radar systems would probably be unable to detect them at all.

Fears were naturally expressed that the V-Bombers would not be capable of performing a low-level penetration role,

Former B.2(MRR) XJ823 takes to the air at Finningley in September 1982, her refuelling probe having been requisitioned for use in the Falklands conflict. The aircraft is now preserved at the Solway Aviation Museum, Carlisle. *Fred Martin*

bearing in mind that they were designed for high-altitude operations, but the Bomber Command Development Unit (BCDU) at Finningley pioneered a series of test flights to prove that low-level sorties could be conducted safely. A Vulcan and Victor were sent to Libya to perform a series of ultra-low flights over the desert to demonstrate the capability still further. In any case, the Valiants that had been assigned to Saceur had already begun low-level operations without experiencing any significant problems (although serious problems were to be encountered later). The Vulcan B.1A squadrons at Waddington (Nos 44, 50 and 101), and the Victor B.1A squadrons at Honington (Nos 55 and 57) and Cottesmore (Nos 10 and 15), were assigned to low-level operations in March 1963. Training flights were normally made at 1,000 feet, a conservative figure aimed at preserving aircraft fatigue life, but every third sortie was flown at 500 feet.

The Mk 2 Victors and Vulcans assumed a low-level role as of 1 May 1964. Blue Steel also had to be re-roled as outlined in a revised version of the original Operational Requirement:

> 'The Air Staff requires the further development of the Blue Steel missile to enable it to be launched from Mk 2 V-Bombers flying at the lowest possible level in the height band 250 feet-1,000 feet.'

After a further series of trials in the UK and at Woomera, the Blue Steel missile was found to be readily capable of being launched at low level, and after release its motor was ignited and Blue Steel comfortably zoom-climbed to 17,000 feet before beginning a terminal descent to its target. Having been designed for high-altitude launch, then having had new operational profiles developed around it, Blue Steel entered RAF service primarily as a low-level weapon, requiring a completely different training programme from the one that had first been envisaged.

The V-Force's main nuclear weapon, Yellow Sun Mk 2, was, however, incapable of low-level delivery, and bombers assigned to Yellow Sun delivery were trained to make low-level penetrations followed by brief 'pop-up' ascents to medium altitude, from where the bomb could be released. This was a far from ideal situation, and the Air Staff accepted that a completely new weapon would be required to suit the low-level delivery environment. Joint Naval/Air Staff Requirement NASR.1177 gave the appropriate details:

> 'Because of envisaged enemy counter measures and the need to change aircraft approach and delivery tactics, the existing British nuclear bombs Yellow Sun, Blue Steel and Red Beard will be unsuitable as primary weapons beyond 1975. Moreover, with the cancellation of Skybolt as the planned replacement for Yellow Sun and the introduction of Polaris unlikely to become effective before 1970, an urgent need exists for a new bomb to maintain the United Kingdom independent deterrent

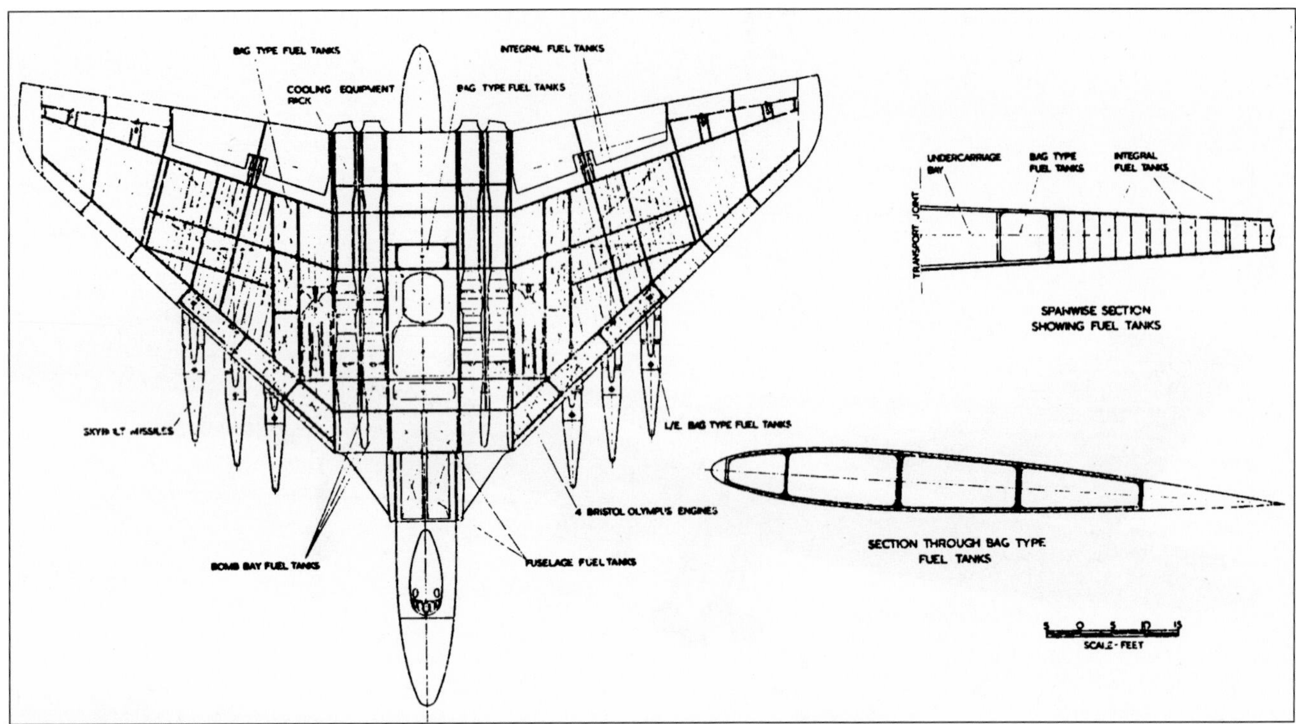

Had the Skybolt programme continued, Avro may well have proceeded with the Vulcan B.6, capable of carrying six Skybolts and maintaining a 24-hour airborne deterrent capability.

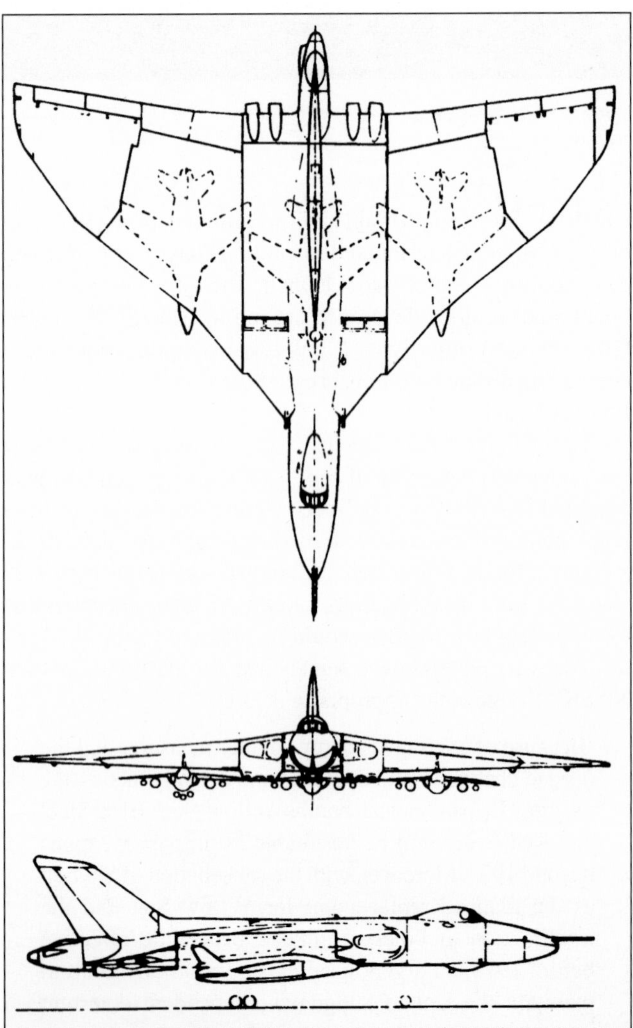

during the interim period and as a supplementary capability thereafter. By 1966, the manned bomber aircraft may survive enemy defences in the European theatre and deliver a successful strike only by flying at high speed at very low level. Yellow Sun and Blue Steel are designed for release at medium/high altitude where the delivery aircraft and/or bomb is vulnerable to interception, whilst Red Beard cannot withstand the low-level flight environment, is limited in method of fuzing and delivery, and possesses some undesirable safety restrictions when held at readiness in an operational state. Early replacement is essential. The replacement bomb must be multi-purpose by design. It must satisfy joint Naval and Air Staff requirements for carriage and delivery in current medium-bomber aircraft and planned high-performance aircraft, to exploit fully their low-level strike capability against strategic and tactical, hard and soft targets, with corresponding different warhead yields. Research and development studies show clearly that such a bomb can be produced fully within the timescale. However, to maintain an effective United Kingdom nuclear deterrent during development of the Polaris weapon system, priority is given to production of the high-yield version for the RAF medium-bomber force.'

The design, which became the WE177 bomb, was to be used as a laydown weapon, either by ballistic/loft mode or by

A well-known but fascinating study was conducted by Avro into the possibility of utilising Gnat fighters, each carrying an atomic bomb, as a substitute for the aborted Skybolt programme. Needless to say, the Gnat pilots would have been faced with a one-way mission and the idea was not pursued.

retarded (parachute) delivery. It was also to be as small and light as possible, the Type B version not to exceed 1,000lb (in fact, it weighed 950lb, while WE177A weighed 600lb). The A and B designations referred to the different yields of 200 kilotons and 400 kilotons; they were 144 inches long, had a carcass diameter of 161/2 inches and a tail-fin span of 24 inches. Deliveries of the WE177A began in September 1966, and trials were conducted with a Vulcan B.2 at Cottesmore. The WE177C was a 10-kiloton naval derivative of the same weapon, carried by Buccaneers and Sea Harriers until the weapon was withdrawn in 1992.

The aircraft in the re-roled low-level Vulcan and Victor force were modified with an updated ECM fit, sidescan radar, roller maps, ground position indicator equipment, and terrain-following radar. The all-white anti-flash paint was replaced on their upper surfaces by a disruptive grey/green camouflage, the first aircraft so finished being Vulcan XH505, which emerged from Hawker Siddeley Aviation's Bitteswell plant on 24 March 1964. Thus the V-Force was assigned to high-low-high delivery profiles, with a low-level phase of up to 1,000 nautical miles in the extreme case (for training sorties this would normally be 350-500 nautical miles). All-weather operations were practiced, and the height of the low-level phase of an operational sortie would be left to the discretion of the aircraft captain – in poor visibility conditions this might be 1,000 feet, but in good weather it could be as low as an incredible 50 feet. For delivery of Yellow Sun, and possibly of Red Beard if required for a second strike, the captain would fly a pop-up manoeuvre to 12,000 feet. Consequently, the unmistakable shape and sound of Valiants, Vulcans and Victors became a more regular part of rural life in the UK as the V-Force crews descended from the heavens and began to thunder around hills and valleys at 500 feet, or even lower in specific low-flying areas, much to the surprise (and eventual displeasure) of those who were unfortunate to live underneath the new low-level air corridor that ran around the UK.

No 44 Squadron's XL319 trailing the familiar braking parachute at Waddington.

CHAPTER SEVEN

Down Under

AS the Vulcan settled into service, the RAF's Chiefs of Staff were keen to demonstrate the aircraft's global capabilities. Likewise, the British Government was happy to emphasise to every country that the United Kingdom was an independent nuclear power, and had the means to project that power to any part of the world. Long-range exercises became a regular part of life for Vulcan crews and, as Australian writer Denis O'Brien describes, the Vulcan was by no means a stranger to the most distant parts of the globe:

'On 27 June 1956 it was announced by the Minister for Air that an RAF Vulcan B Mk 1 aircraft would visit Australia for Air Force Week in September. This heralded the beginning of what would prove to be a long association between the RAF Vulcans and the RAAF. The ensuing years witnessed the involvement of Vulcan aircraft in weapons testing at Woomera, joint Air Defence Exercises (ADEX) in Darwin and at Butterworth, and, of course, participation at numerous air shows in the capital cities.

On 9 September 1958 Vulcan B Mk 1 XA897 departed Boscombe Down in Wiltshire on the first leg of a journey to Australia and New Zealand. Staging via Aden and Singapore, XA897 arrived overhead at RAAF Avalon about 1700 hours on the 11th, having logged a flying time of 22 hours 30 minutes. On board the Vulcan was the Commander-in-Chief of RAF Bomber Command, Air Marshall Sir Harry Broadhurst. Sir Harry was taking turns in the "right-hand seat" with regular Vulcan pilot Squadron Leader James Stroud. The aircraft flew the 3,730-mile United Kingdom to Aden leg in 7 hours 20 minutes, the 4,105-mile sector from Aden to Singapore in 8 hours 20 minutes and the final sector to Avalon in 6 hours 50 minutes.

Pictured during an exercise in New Zealand, the Vulcan's distinctive sooty exhaust trail is clearly evident when compared to the accompanying RNZAF Skyhawks. Various trials and modifications were made to the engine cans and burners, particularly on 200 series engines, but the generation of smoke (which made the aircraft conspicuous even at long distance) was a problem that was never properly solved. *RNZAF*

Soon after arriving over Avalon, the Vulcan circled the field, then flew on to Melbourne, returning via Port Phillip Bay to Avalon, only to return again to Melbourne before finally landing at Avalon RAAF near Geelong. Among the VIPs waiting on the tarmac to greet Sir Harry were Air Marshall Sir John McCauley, Chief of Air Staff RAAF, and Air-Vice Marshall W. L. Heley. The following day, Wednesday 12 September, Sir Harry Broadhurst attended the Parade of the Queen's Colour at Point Cook RAAF Base. The Vulcan remained at Avalon while Sir Harry flew to Canberra for talks with Air Force staff and politicians. On his arrival at RAAF Fairbairn, he was greeted by the Minister for Air, Mr Townley, the Commanding Officer of Fairbairn, Group Captain A. D. Garrisson, and Mr J. R. Fraser, the representative of the High Commissioner for the United Kingdom. The following morning, at 0930 hours, Sir Harry laid a wreath at the War Memorial before attending a sitting of the House of Representatives and the Senate. He was a guest at a Cabinet luncheon before returning to Melbourne to rejoin the Vulcan for the trip to Sydney.

Flying low over Canberra en route, the Vulcan arrived at Sydney's Kingsford-Smith Airport, where it taxied to a heavily guarded enclosure near the old ANA terminal. The aircraft was on public display the following day, albeit from a secure distance. While in Melbourne, Sir Harry Broadhurst created quite a controversy over statements he had made concerning the Avon Sabre. He was quoted in the press as saying that "the Sabre was well suited to its task when first built, but has been completely outdated by recent developments". This statement produced a quick response from the Minister for Defence, Sir Philip McBride. The Sydney press was quick to revive the issue when it interviewed Sir Harry. In an attempt to quell the debate, Sir Harry denied saying that the Avon Sabre could not stop an atomic attack by a bomber like the Vulcan: "What I said was, the Vulcan flies at night, and there was not a night fighter in the world today which could stop her. By day, however, the Vulcan is vulnerable and can be caught by a fighter like the Avon Sabre."

On Saturday 15 September the Avro Vulcan had to share the limelight when a BOAC Bristol Britannia, on her inaugural visit to Australia, landed at Mascot. The Britannia was, at the time, the world's largest turboprop commercial transport. While a 24-hour guard was maintained on the Vulcan, which was, of course one of Britain's top secret aircraft, Sir Harry attended a Battle of Britain Commemorative Service at the Cenotaph. Air Force Week concluded on the Sunday with an open day at RAAF Richmond. Having completed the ceremonial duties, Sir Harry boarded as RAAF Douglas Dakota at Richmond to fly to Mascot to rejoin the Vulcan. After take-off from Mascot, XA897 flew to Richmond where it gave a five-minute display before flying to Adelaide, where it would undergo a routine service before the flight to New Zealand. The crowds that had flocked to RAAF Mallala for the Air Force Week air display were not disappointed when, as expected, the Vulcan arrived to give a display before landing at Edinburgh.

Servicing complete, the Vulcan departed from Edinburgh on the 18th for the trip to Christchurch, New Zealand. The route to New Zealand was via Launceston and Hobart. Landing at Christchurch's Harewood airport saw the trip successfully completed with an elapsed time of 4 hours 35 minutes. Avro claimed a point-to-point record for the 1,200-mile sector, Hobart to Christchurch, of 635mph. An Avro Shackleton aircraft landed at Amberley, Queensland, on 19 September, in preparation for the return of the Vulcan to Australia. This was one of three aircraft accompanying the Vulcan and carrying spares and maintenance personnel.

Anticipation was running high in Brisbane as the time for the Vulcan's visit approached, this enthusiasm having been generated by the local press coverage of the aircraft's visit to the southern capitals. Variously described as the "flying triangle" and a "giant bat" by the newspaper journalists, XA897 arrived over Brisbane from Ohakea Air Base in New Zealand on the morning of 22 September. A traffic jam was created as an estimated 20,000 people motored to Amberley Air Force Base to inspect the Vulcan.

The following day, after a farewell circuit of Brisbane, the aircraft set course for Darwin, the last Australian port of call on the homeward leg. The departure of XA897 from Darwin completed a celebrated and notable visit to Australia and New Zealand, but as already mentioned the trip ended in tragedy when XA897 crashed at Heathrow at the very end of its journey.

On 13 October 1959 four Vulcans of 617 Squadron departed from Scampton for the RNZAF base at Ohakea, New Zealand. These aircraft staged via Darwin, arriving at their destination on the 18th. They departed the following day with their support aircraft, a Bristol Britannia. The visit to New Zealand was marred by a near disaster involving XH498. On the third attempt to land at Wellington's relatively short runway after two previous "go rounds", the aircraft's main undercarriage struck the undershoot area. The undercarriage damaged the wing and pierced a fuel tank, but the Vulcan pilot managed to retain control of the aircraft and climb away. Returning to Ohakea the aircraft landed safely. XH498 was destined to remain in New Zealand for a further eight months before repairs were completed, not returning to the United Kingdom via Edinburgh RAAF Base until June 1960; departing from Edinburgh, XH498 overflew Sydney on 24 June en route to Darwin.

The Royal Aircraft Establishment and Avro began studies on a "stand off" bomb in 1954. In November 1960 Vulcan Mk B.1 XA903 ferried to Australia one of the first full-scale Blue Steel missiles. This aircraft, piloted by Avro's chief test pilot in Australia, Mr J. D. Baker, arrived at Edinburgh on 17 November, having been delayed in Darwin due to a minor technical

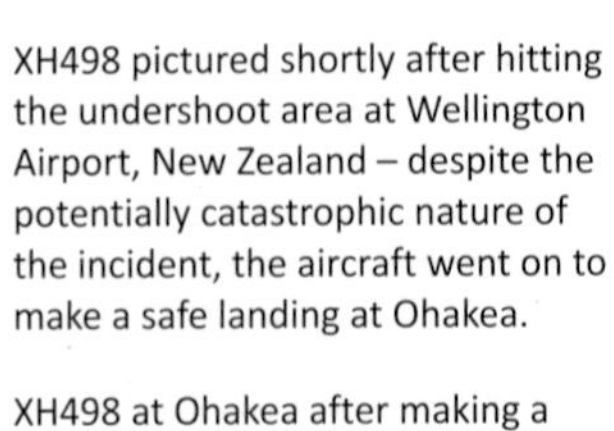

XH498 pictured shortly after hitting the undershoot area at Wellington Airport, New Zealand – despite the potentially catastrophic nature of the incident, the aircraft went on to make a safe landing at Ohakea.

XH498 at Ohakea after making a successful landing in October 1959. The port main undercarriage had been severely damaged during the undershoot approach at Wellington. *RNZAF*

problem. All trials connected with weapons research were conducted under a joint UK-Australian project agreement. Two other Vulcans, both early production B Mk 2s, XH538 and XH539, were also sent to Australia for deployment in the Blue Steel trials at the Woomera range. The range, in red desert country consisting of sand hills, mulga scrub, claypans and clumps of spinifex, was ideally suited to the tests because of the sparseness of population and the long hours of sunshine and clear skies. The first test firing took place from XH539 in early 1961, and the programme continued until October 1964. The history of Woomera, the aboriginal name for a spear-throwing device, is quite fascinating, and a full account of the area has been produced under the auspices of the Defence Research Centre.

The non-stop flight to Australia by a Vulcan B Mk 1A from No 101 Squadron was indeed an historic occasion. While it did not eclipse the round-the-world non-stop flight by three USAF Boeing B-52 bombers in 1957 or the supersonic flight from New York to Paris by a USAF Convair B-58 Hustler, averaging 960 knots, in May 1961, it was certainly significant in terms of RAF long-range deployment. The England to Australia flight was the longest non-stop flight undertaken by the RAF and demonstrated the speed and long-range strike capability of the aircraft, and it still stands as a record for the Vulcan, although the "Black Buck" operations flown against the Argentinian forces in the Falkland Islands in 1982 were not much shorter; the Ascension to Ascension round trip in the "Black Buck" missions was just some 3,000 nautical miles shorter than the 10,000-nautical-mile England to Australia trip.

Vulcan XH481 departed from its base at 11.36am on 20 June 1961, arriving over the control tower at RAAF Richmond at 4.39pm on the 21st. The elapsed time of 20 hours 3 minutes and 17 seconds for the flight gave an average speed of 500 knots. Favourable winds on one sector east of Alice Springs gave a ground speed of 600 knots. En route refuelling from Vickers Valiant tankers occurred over Nicosia, Cyprus, near Karachi, near Singapore, and a final "top-up" about 500 miles out of Singapore. Each refuelling took about 12 minutes, the Vulcan taking on about 5,000 gallons (23,000 litres) during that period. To complete the transfer the Vulcan reduced speed to 350 knots. The aircraft went on display at Richmond on 22 June, and while in Sydney its crew – Squadron Leader M. Beavis (Captain), Flight Lieutenants D. Bromley, R. Taylor and G. Jukes, Flying Officer J. Knight, and Chief Technician W. Alpine – were guests of the De Havilland Aircraft Company on a harbour cruise.

The following day the aircraft departed for RAAF Edinburgh, although not before a spectacular flying display at 1,500 feet down Sydney Harbour to Vaucluse and Watsons Bay, returning to the Harbour Bridge and thence to Bankstown. While stationed at Edinburgh, XH481, in company with one of the Victors, overflew Hobart, Launceston and Melbourne on 27 June. Departing from Edinburgh on the 30th, the aircraft flew to RAAF Pearce, where it went on display to the people of Perth on 2 July before returning to the United Kingdom on the 3rd.

The British Commonwealth and Empire Games held in Perth in November and December 1962 afforded RAF Bomber Command another opportunity to display its Vulcans. The three Vulcan squadrons sharing RAF Scampton in Lincolnshire participated in the opening and closing ceremonies of the Games, together with nine CAC Sabres of No 75 Squadron RAAF, Williamtown. The Vulcan B.2s – XH556 of No 27 Squadron, XL319 of No 617 Squadron and XL392 of No 83 Squadron – arrived at RAAF Pearce on 19 November, and the aircraft's servicing crew arrived in a Bristol Britannia of RAF Transport Command. The following day a fourth Vulcan arrived carrying Air Vice-Marshall P. Dunn, OC of No 1 Group RAF. It was anticipated that one of the Vulcans would go on public display at Perth Airport, but after hurried consultations with the Department of Civil Aviation it was considered that the Vulcan's tyre pressure of 190psi was too high and the proposed landing at Perth was banned; while the all-up weight of the Vulcan B.2 was less than that of the Boeing 707s using the airport, its tyre pressure was 50psi higher.

In the interval between the flypasts for the opening and closing ceremonies on 23 November and 1 December, the aircraft flew to the eastern states, departing from Pearce on 26 November and returning on the 28th. Vulcan XL392 was involved in a spectacular flying exhibition at the Games on 29 November. Resuming their round-the-world flight, the Vulcans departed from Pearce on 2 December for RAAF Edinburgh, then flew on to New Zealand.

A major air-to-air refuelling exercise involving three Vulcans from the Waddington Wing once again saw the aircraft "down under". In July 1963 the trio, XH481 of No 101 Squadron, XH482 of No 50 Squadron, and XH503 of No 44 Squadron, arrived in Perth having completed a non-stop flight from RAF Waddington. The first two aircraft arrived on 9 July, the third aircraft arriving the following day. The first Vulcan to arrive completed the 8,600-nautical-mile trip in 18 hours 10 minutes. The commander of this aircraft, Wing Commander A. Griffiths, commented that they were not trying to establish any speed records. Unfavourable jet streams were encountered on the leg to Aden, producing headwinds of 100 knots. On the 3500-nautical-mile final sector from Gan, Maldive Islands, more favourable jet streams produced ground speeds of 625 knots. The co-pilot, David Bromley, had been the co-pilot of the 1961 non-stop UK to Sydney mission in XH481. Refuelled by Vickers Valiant tankers over Libya, Aden and the Maldive Islands, they flew at 40,000 feet for most of the journey. Flight Lieutenant J. R. Ward, the Captain of the second Vulcan to land, said the only problem encountered was the need to descend from their cruising

Vulcan XL392 was involved in a spectacular flying exhibition at the British Commonwealth and Empire Games on 29 November 1962. Seen here wearing 617 Squadron markings, she finally served with No 35 Squadron before being retired to Valley for crash rescue duties in March 1982. *Shaun Connor*

altitude to 25,000 feet to enable the fuel transfer, over the Gan position, to proceed clear of cloud. Before returning to the United Kingdom the aircraft visited RAAF Richmond and RAAF Edinburgh. They arrived at Richmond after a 2-hour 50-minute dash from Perth, 2,000 nautical miles away, ground speeds of 670 knots being recorded en route.

The confrontation between Indonesia and Malaysia in the early 1960s focused government attention on the northern approaches to Australia. Not only was the efficacy of the defence complex in question, but the lack of a modern strike aircraft capable of swift retaliation meant a serious lack of a credible deterrence. The now obsolete Canberra was, indeed, no match for Indonesia's Russian-built Tupolev T-16 Badger with its range of 4,000 miles and 6000lb weapons load. Britain's treaty obligations to Malaysia had resulted in the maintenance of a detachment of the V-Force in Tengah, Singapore, or Butterworth, Malaysia, during the period of the confrontation, in addition to its existing air force detachments. These "Sunflower" detachments consisted of eight aircraft travelling in two waves, either by the west route, via Canada, USA, Wake Island and Guam, or the Middle East route. Against this background the decision was taken to test out northern defences by simulated attacks by elements of the V-Force.

The Vulcans operated initially out of RAF Tengah, Singapore, flying a series of war training exercises in and around Malaysia and later flying on to Darwin, flying day and night sorties to test the Australian defences. These visits to Darwin became a highlight of the "Sunflower" exercises, providing realistic war training in tropical conditions and uncrowded skies. From the British Ministry of Defence's point of view these exercises and overseas flights had the following broad objectives: a) to demonstrate that the V-Bombers could be deployed quickly around the world, and to practice such deployments; b) to maintain the deterrence and its credibility; c) to allow Allied and friendly nations to manoeuvre their own air forces, with and against large jet bombers; d) if required, the force could be dispersed in times of international tension, and such dispersals needed to be practiced; and e) to give crews practice in long-distance flying.

On 17 December 1965 Vulcans under the command of Sqn Ldr D. S. Harris, and Canberra bombers deployed from "enemy bases" at Alice Springs, Townsville and Amberley, raided Darwin. The ADEX "High Rigel" had moved into its major phase. RAAF Sabres and Mirages equipped with Matra and Sidewinder missiles opposed the attacking bombers. The defending fighters were directed to their targets by the No 2 Control & Reporting unit, the RAAF's most modern radar unit, which first became operational in 1961. This major exercise involved not only the RAAF and a Vulcan detachment but also some forty National Servicemen of the 121st Anti-Aircraft Battery, Civil Defence and Emergency Services, the Fire Brigade and the St John's Brigade. It was generally conceded in Air Force circles that the exercise would herald the end of the Bloodhound missiles as an effective weapon. Indeed, it was considered that the Vulcans were unlikely to be threatened by the Sabres, SAMs or the Bofors of the AA Battery, although it was not until 1968 that the Bloodhound SAM Squadron was ultimately phased out of service.

The realism of the exercise was significantly enhanced by the weather pattern around Darwin. Tropical storms extending up to 50,000 feet ensured that the attacking Vulcans did not have a clear run to their targets. Darwin itself was subjected to intense thunder activity and heavy tropical rain showers. Following the exercise, the OIC, Air Vice Marshall C. T. Hannah, commented that the capability of the Mirage as a day and night fighter had been fully tested. The aircraft involved had flown more than 800 sorties, and some 700 personnel had been deployed.

The Vulcans visited Darwin again in March 1966, this time to participate in ADEX "Short Spica". Four Vulcans combined with Canberras from No 1 Squadron Amberley to test the defences of Darwin and its protective mantle of Mirage and Sabre fighters. Attacking bombers were deployed from bases at Amberley, Townsville and the civilian airport at Mt Isa. The "hostilities" commenced at 0900Z (GMT) on 17 March, the exercise continuing under the direction of Group Captain W. N. Lampe until 1000Z on the 23rd. Following one of the missions, Flying Officer David Lee experienced a hydraulic failure in his Vulcan, necessitating the use of an emergency system to lower the landing gear. Several low passes over the control tower enabled engineers to visually inspect the gear and to conclude that it appeared locked. Finally, after flying several extended circuits to use up fuel, the Vulcan came in for an uneventful landing with fire tenders and rescue helicopter on standby.

XH562, complete with her unofficial RNZAF roundels.

ADEX "High Castor", held in August 1966, was the next major exercise in Darwin. Following along the lines of the previous exercises, all of the RAAF operational elements were involved including Lockheed Neptune maritime reconnaissance aircraft from No 10 Squadron, Townsville, three RAF Vulcans and two RAF Canberras.

November 1967 was the date selected for a major joint exercise involving the Canberra bombers of No 14 Squadron, RNZAF, based at Ohakea, in addition to four RAF Vulcans from No 9 Squadron and an RAF photo-reconnaissance Canberra. The Canberras of No 14 Squadron, RNZAF, were based together with Canberras of No 3 Squadron, RAAF Amberley, at the "enemy" base of Tindal, situated some 200 miles south of Darwin near Katherine. Named after the first Australian officer killed on the mainland in the Darwin air raids of 1942, Tindal cost $7 million to construct. Built largely by personnel of the No 5 Airfield Construction Squadron, it consisted of a 9,000-foot runway with two 1,000-foot over-runs, providing more than adequate runway length for aircraft of the Vulcan's size and weight. Some 1,000 personnel, including eighty men of the Royal Australian Army, participated.

Exercise "High Jupiter", conducted in June 1968, brought together in the north the largest peacetime deployment of men and equipment. More than 2,000 personnel and fifty aircraft were involved in this exercise under the command of Air Vice Marshal K. S. Hennock. The newly formed RAF Strike Command was represented by Air Officer Commanding Strike Command, Australian-born Sir Wallace Kyle, assuming the role of an observer. This exercise also saw the Mirage in its new guise as a low-level strike aircraft. Several Mirages were heavily camouflaged and specially equipped for the lower-level attack role. The RAF contributed four Vulcans, under the command of Wing Commander D. J. Mountford, and two Canberra bombers from the Far East Airforce HQ in Singapore. The heaviest raid on Darwin occurred on 23 June: two Vulcans came in at high level and a third, evading the defence, flew in at low level.

The exercise in December 1968, "Rum Keg", was more in the nature of a training exercise than an operational exercise. The OIC for the exercise, Group Captain Mick Mather, stated that there would be no "open go" flying or "dog fights" over the city, as had been the case in "High Jupiter". A detachment of four Vulcans of No 44 Squadron arrived from Singapore on 5 December, returning to their Singapore base on the 11th. This exercise would see the end of the Bristol Bloodhound SAM Squadron, a part of Darwin's defence since 1961. It had been announced in the House of Representatives in August by the Minister of Air, Mr Gordon Freeth, that these missiles would be phased out of service, their role being taken initially by the Mirage but ultimately by the General Dynamics F-111.

The Vulcans of No 50 Squadron were accompanied by RAF Lightnings of No 74 Squadron when they visited Darwin in June 1969 to participate in "Town House". This exercise commenced officially at 1500 hours CST on 18 June and ended on the 25th. The initial Vulcan raid occurred soon after 1000 hours on the 19th. As the Vulcans tracked to Darwin they left a prominent "footprint" in the form of contrails; the cloud streamers formed in the wake of jet aircraft were of considerable important militarily, indicating not only the presence of the Vulcans but also giving their approximate heading and information on the winds aloft. Contrails were visible long before a visual contact could be established and often before onboard radar could detect the intruder. In its European theatre of activity, the high-level operations were conducted at altitudes that put the Vulcan above the tropopause (the interface between the troposphere and stratosphere). In this environment contrails are less likely to be produced because of the low ambient moisture above the troposphere. However, in the low latitudes of Darwin the tropical tropopause is much higher, extending up to 18km. The raids on Darwin consisted of high-level attacks by the Vulcans and low-level attacks by the Canberras, the latter aircraft being based at Tindal. The raids were preceded by live bombing runs on the Quail Island and Leanyer Swamp ranges. Live bombs were not carried over the populated areas of Darwin, and the Sidewinder and Matra missiles carried by the Mirages were unarmed.

Air Vice Marshal Bill Townsend was the OIC of the next major ADEX, "Castor Oil", held in February 1970. While many air-to-air exercises were conducted in the south-eastern states, Darwin, as the northern approach to Australia, was favoured for the ground-to-air defensive exercise. The logistics of moving large numbers of men together with equipment was a vital part of these exercises, simulating a potential scenario; thus the Darwin exercises involved all elements of Operational Command. Following the pattern of previous exercises, Darwin was to be subjected to attack by "enemy" aircraft from Tindal. In this respect the exercise was rather one-sided, as no counter-offensive was planned against Tindal. Four Vulcans from RAF Strike Command arrived to support the "enemy", which consisted of six Canberras, six ground-attack Mirages and six Sabres. The defence of Darwin again rested with the Mirages of No 76 Squadron. Paratroops, dropped from RAAF Caribou in a pre-dawn raid, signalled the commencement of the exercise, and at its conclusion the defenders were claiming four Vulcans, four Canberras, three Mirages and two Sabres as "kills". This was not without heavy damage to the RAAF base, the city, fuel depots and the power station. The senior medical officer at the RAAF Base Hospital stated that some 65 casualties were treated, three being considered to have received fatal wounds.

In November 1970 four Vulcans of No 27 Squadron departed from their base at Scampton and flew to Singapore and Australia via the Westabout Route, staging through Gander, Offutt AFB, San Francisco, Honolulu, Wake Island and Guam. Their arrival in Darwin was

Vulcan B.2, XH570, No 27 Squadron, Scampton, 1973. Medium Sea Grey/Dark Green upper surfaces with White undersides; 'Tactical' national markings. Black serial and radome; unit badge on fin. *Copyright © Richard J Caruana*

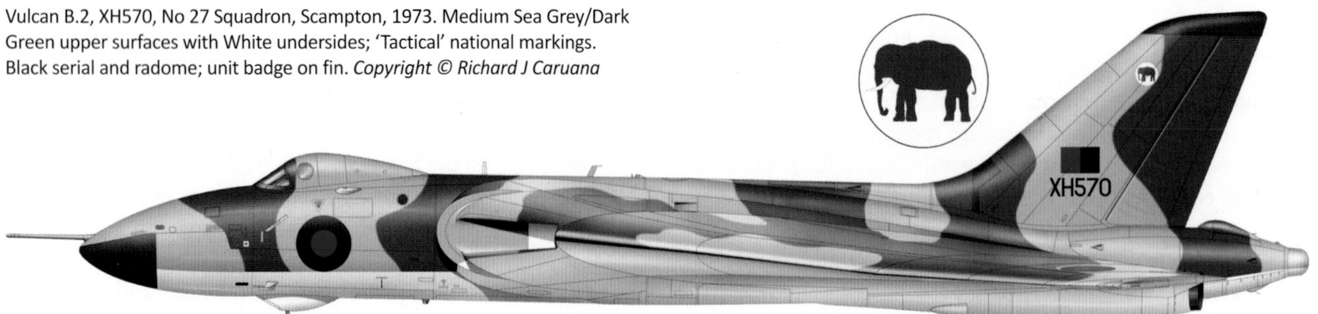

timed so that they could participate in "Opal Digger", an ADEX consisting of high-altitude exercises in conjunction with eight Mirages of No 76 Squadron and the Darwin Base Radar Unit, 2CRU. The Mirages, half of the No 76 Squadron complement from Williamtown, staged through Townsville, flying the Townsville-Darwin sector in 2 hours 15 minutes. Compared with "Castor Oil" this was a minor exercise. The OIC was Wing Commander Stewart Creswell, with Air Commodore G. H. Steege from HQ Operational Command, Glenbrook, acting as an observer. As a salute to Darwin, one Vulcan and four Mirages overflew the city at the completion of the exercise. Two Vulcans returned to the UK, the other two first visiting RAAF Williamtown.

Vulcans from No 44 Squadron visited Darwin in February 1972 to participate in "Whisky Sour". Two Vulcans arrived on 4 February, the other two arriving two days later. Eight Mirages under the command of Squadron Leader "Joe" Owens from No 76 Squadron had already arrived on 1 February, after having staged via Townsville. It was almost thirty years to the day since Kittyhawk aircraft of No 76 Squadron had been involved in the defence of Darwin against the Japanese. This ADEX involved day and night exercises with the Mirages defending Darwin from the attacking Vulcans. The Vulcans would depart from Darwin and later return from any direction at high or low level to attack the RAAF Base.

A common spoof employed by the Vulcans was to send one aircraft in towards the target to ensure radar detection. When the Mirages were scrambled the Vulcan would hastily turn away. This ruse was designed to get all the Mirages into the air so that they would be low on fuel when the real strike came. The Vulcans could carry 90,000lb of fuel and their fuel burn of some 10,000lb per hour gave them sufficient time in the air to permit some element of surprise in their attack. It was not too difficult for the Mirage pilots to establish visual contact with the Vulcans; they were large aircraft, and even without contrails they left a tell-tale trail of black exhaust. However, establishing a "kill" was not so easy; at low level, coming in under radar, they were often able to "release" their weapons before being intercepted. At high altitudes they posed another series of problems, their ECM equipment and chaff being effective in deflecting radar-guided missiles, while heat-seeking missiles required the Mirages to be at a similar altitude to their target. Success of the intercept and rocket launches was assessed by on-board equipment and ground radar assessment. A "probability of kill" factor was established using statistical information that considered such extraneous factors as post-launch failure. During night exercises a safety parameter required aircraft to use their navigation lights. On a lighter note, the Mirage crews endeavoured to see that all the Vulcans were suitably decorated with red kangaroos before their departure.

The joint exercise "Top Limit" in May 1972 involved aircraft from the RNZAF as well as the RAF and RAAF. This exercise involved thirteen Phantoms from No 1 Squadron, Amberley, led by Wing Commander Lyall Klakker. The RNZAF contributed eight Skyhawks, commanded by Wing Commander F. M. Kinvig, from its base at Ohakea in the North Island of New Zealand. The RAF was represented by four Vulcans from No 9 Squadron, including XM600 and XM656. The exercise commenced on 9 May and ended on the 15th, and was followed by a debriefing the next day. Tindal was brought out of mothballs to serve as a base for the "enemy", while RAAF Darwin was again defended by Mirages.

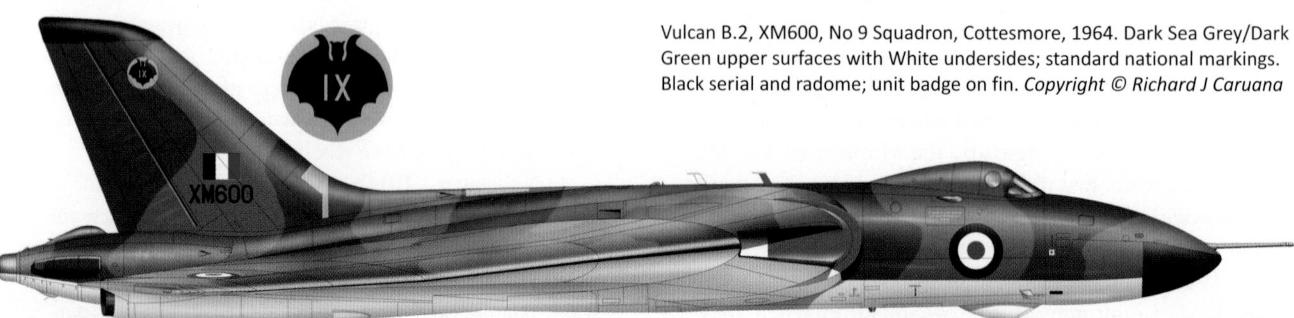

Vulcan B.2, XM600, No 9 Squadron, Cottesmore, 1964. Dark Sea Grey/Dark Green upper surfaces with White undersides; standard national markings. Black serial and radome; unit badge on fin. *Copyright © Richard J Caruana*

Pictured during an exercise in New Zealand, the Vulcan's distinctive sooty exhaust trail is clearly evident when compared to the accompanying RNZAF Skyhawks. Various trials and modifications were made to the engine cans and burners, particularly on 200 series engines, but the generation of smoke (which made the aircraft conspicuous even at long distance) was a problem that was never properly solved. *RNZAF*

Pictured while visiting Malta, the entrance door to this Vulcan B.2 reveals Akrotiri Strike Wing markings comprising the station badge and those for the component squadrons, Nos IX and 35. *Richard Caruana*

A third exercise in 1972 involving the Vulcans commenced on 17 July when the Vulcans arrived to participate in "Dry Martini". This exercise was marred by the grounding of the fourteen Mirages from No 76 Squadron; all the RAAF's Mirages were grounded following an engine failure at Butterworth in Malaysia. However, the Vulcans continued to fly bombing missions at the Quail Island range. While engineers conducted checks on the Mirages, the pilots and other groundcrew were kept busy with survival exercises, sport and recreational activities. The checks completed, the Mirages recommenced flying operations on 29 July. On the same day two of the Vulcans flew to RAAF Pearce, returning to Darwin on the 31st. All four Vulcans departed from Darwin for Britain on 3 August.

The participation of the Vulcans in Australian air shows always proved to be a great crowd-pleaser. In the mid 1960s the Vulcans were often a feature at the various RAAF bases during Air Force Week celebrations, and in 1971 a Vulcan participated in the Golden Jubilee celebrations of the RAAF and appeared in the skies accompanying aircraft of the RAAF. The final visit to Australia was by XM571, which was in Adelaide for the 617 Squadron reunion in April 1980.

The Vulcan was indeed a highly successful aircraft design and one that became familiar in skies over Australia. In summary, the Darwin Air Defence Exercises were as follows: "High Rigel", December 1965; "Short Spica", March 1966; "High Castor", August 1966; "High Mars", November 1967; "High Jupiter", June 1968; "Rum Keg", December 1968; "Town House", June 1969; "Castor Oil", February 1970; "Opal Digger", November 1970; "Whisky Sour", February 1972; "Top Limit", May 1972; "Dry Martini", July 1972.'

Bill Pearsey was one of many RAF crew members who regularly participated in long-range Vulcan exercises, and recalls his first trip to New Zealand:

'The ranger was planned as a navigation exercise by the OC IX Squadron, RAF Akrotiri, Wing Commander Ron Dick, who "wanted to get away from the northern climes' winter for a couple of weeks in the sun" – as good a reason as any! The ranger would also involve representing the RAF at New Zealand's annual air show, and also included in the itinerary was a trip to Christchurch where the widow of a previous IX Squadron CO, Mrs R. Warner, was to present the squadron with a large model of a bat – IX Squadron's emblem. The model had belonged to Captain Warner since 1930, and Mrs Warner was anxious that it would have a good home. The crew comprised of the Boss, Flight Lieutenants Adrian Sumner (co-pilot), Bob Sinclair (nav-radar), Nobby Clarke and myself (crew chiefs), Flight Lieutenants John Clarke (nav-plotter)

XM571 of the Waddington Wing acquired a union flag for the 1979 'Giant Voice' exercise. XM571 was the last Vulcan to visit Australia, when in April 1980 she attended the 617 Sqn reunion in Adelaide. *Fred Martin*

and Maurice Stocks (AEO) kneeling. Take off time was at 0130 on Wednesday 23 February 1972; this was to enable us to get two eastbound legs done in daylight.

Arriving at Masirah in time for breakfast and a quick turn-round, we also had a word with Bill Kid, who was doing a Masirah ranger. The next leg to Gan and the night stop were uneventful, as was the next day's flight to Tengah in Singapore. After refuelling and an after-flight check, it was on to the transport to take us to the Equatorial Hotel on the Bukit Timah road. Singapore had changed beyond recognition since I was stationed there in 1965. The night stop at Singapore was a quick shower, change and a dash around the shops – nothing very exciting – as we had another double leg to go the following day, 25 February.

We landed at Darwin in a tropical rainstorm and Nobby and myself were busy refuelling, checking the aircraft and trying to get a bite to eat, all the while being pestered by the Aussie Customs man who wanted to search all our kit and check our passports. By the time we reached Richmond RAAF base outside Sydney it was quite late (locally), and the Darwin Customs gentleman had phoned ahead to Richmond to tell the guy there that we hadn't done a full Customs clearance, so we had to perform again. The Aussies had kept the mess bar open for us, but there was no food available.

The following morning (Saturday) we had a phone call from the crew; they had decamped down to the Rotunda Hotel at King's Cross, Sydney, and asked us to follow them there. We managed to cadge a lift in a posh RAAF staff car, which apparently was picking up an AOG from Sydney International airport (very handy!). The time in Sydney was mainly spent sight-seeing, with a trip across the harbour to see the aquarium at Manley, returning in very fast time by the hydrofoil.

Off to New Zealand on the Monday, where we were met and escorted by an RNZAF A4 Skyhawk, which formated on our starboard wingtip for a while. It then disappeared, only to re-appear a short while later in a steep dive just ahead of us. We were not allowed to put a foot on New Zealand ground until we had had a swig of a welcoming can of Double Diamond while we were still on the ladder. The rest of the welcome ritual was a get-together in 75 Squadron RNZAF crew room. Apparently the road around Ohakea had been jammed for some time by people eager to see the aircraft landing; there was also a big front page spread, with photos in the Palmerston North evening paper.

The following day we were flown in a VIP Dakota on a sight-seeing tour of the North Island, more or less picking our own route from the charts that we had been given, and we ended up at Rotorua with a minibus and a local guide, a guy called Judd, who drove us around all the local tourist spots, which included the Hot Springs and bubbling hot mud pools. We paid a visit to a Maori craft and culture centre and several other local points of interest. We arrived back at Ohakea in the early evening, and as we were taxiing in No 75 Squadron members were just finishing a paint job on the biggest Zap I have ever seen, and our Vulcan now sported a large kiwi roundel on both sides. They must have spent all day doing it, as it was a full professional paint job.

I did mention that on our return trip to Cyprus we would do a low-level run under Sydney Harbour Bridge. That night the 75 Squadron RNZAF A4s all broke out in a rash of green bats.

Wednesday we were off to see the South Island, doing low-level fly pasts and a display at what seemed every hole in a hedge until we arrived at the RNZAF base at Wigram on the outskirts of Christchurch, were the Boss went through the whole display routine. After the display we carried on to land at Christchurch International Airport, where we were greeted by the whole razzmatazz of a VIP welcome by the local press and TV. This was to record and cover the presentation of the large model bat to the squadron by Mrs Warner. A quick service and refuel of our Kiwi Vulcan followed the ceremonies, after which we were transported to Wigram where we were to spend the night. The Sergeants Mess were holding a function that night and the members were very reluctant to admit Nobby and myself initially as we did not have the appropriate dress; however, once we explained that we were members of the Vulcan crew who had given them a big display earlier, the hospitality was terrific. Late in the night we were driven to a point high above Christchurch to see a fantastic panorama of the city lights.

Back at Christchurch the following morning, our first technical problem reared its ugly head: the Rover refused to start and we had no GPU. Eventually we were offered a GPU by a member of the USAF who was attached to Christchurch. The lead from the power unit was so short that it had to be placed immediately behind the No 2 ECU jet pipe to reach the socket, so we had to start No 3 engine first, remove the GPU and do a cross-feed start. Eventually we were on our way, going south at low level following the coast road; at times we could see the cars on the road and where the number plates were, but we couldn't quite read the numbers. Again the Boss was giving the locals a show at various locations.

We turned north at the southern end of the South Island, and it was during this leg of the trip that we spotted "The Kingston Flyer", a narrow gauge railway that years later contributed to a story involving Tony Regan, Roy Heaton and Ron Dick, which needs a full article of its own. After catching the "Flyer" on the F95, we continued north over a sheep-rearing area. You could see the sheep as agitated white spots in the distance, and as we approached they seemed to be drawn together by a piece of string, then as we flew overhead they scattered to the four corners of the globe – they could still be running! The rear crew took turns to stand on the steps between the pilots to see as much as possible of the scenery. When my turn to look out came, all I could see in front of us was a sheer wall of rock and snow; as we cleared the ridge, one could see that it was very sharp, not eroded and rounded as older mountains. Once over the crest it was down again into the next valley. At one stage I was sort of dozing on the spare seat and looked up to the radar's porthole to see sheep – I was looking up at sheep! The trip ended with a sedate recovery into Ohakea.

That evening the Boss received a telegram from Russell Glendenning, the driver of "The Kingston Flyer" saying: "To the Vulcan crew from the driver of the Kingston Flyer. You look very beautiful – but you have a dirty exhaust and a fly spot on the left wing." (There are other stories relating to the "Kingston Flyer" that are better related by the people concerned.)

Friday was spent preparing for the Hamilton display on the Saturday. Nobby Clarke and I were "adopted" by a Sgt Fireman called 'Rangi Lewis' who looked after us like a Dutch uncle. One night in the bar was spent playing 7, 14 & 21 with poker dice. The evening started out quite sedately, but eventually became slightly more boisterous.

The weather at Hamilton deteriorated on the Saturday, which meant that the air display was postponed until the following day. The rest of Sunday, after our Kiwi Vulcan had finished the display, was spent preparing for the return trip to Akrotiri, re-fitting the pannier, etc, followed by a little farewell party in the mess.

While we were waiting for the rest of the crew to arrive on the Monday morning Nobby took my photo, which later appeared in two of Robert Jackson's books. I wasn't "really surprised" as the Kiwi had been painted on nearly a week earlier. During the "crew in" Rangi Lewis came to say cheerio (again!) and brought Nobby and myself the prize that we had apparently won in a mess draw, although we had not bought any tickets. The prize was a "chook" (chicken) and a bottle of local wine each. The chooks had been cooked that morning and were still hot and in insulated bags. When eventually we opened our prizes the cabin was filled with the smell of roast chicken. The rest of the crew were soon looking round to see where the aroma was coming from, as they had to be contented with the normal Vulcan buttie in flight rations!

The direct flight to Darwin was covered at high altitude and took more than six hours. When we landed at Darwin we were subjected to the normal Aussie Customs welcome, a guy climbing up the ladder and spraying everything and everybody in sight before we even realised that he was on board.

It was during the next day's leg to Tengah that the aircraft started to unravel – a crack started on the outer skin of the co-pilot's windscreen. Being highly efficient, I marked it with a "chinagraph" and asked him to keep an eye on it. No sooner had I settled back in my corner than I was called up again as another crack had started on the opposite edge of the screen – this was also marked. A short while later, the same thing happened to the centre windscreen – two cracks progressing down each side. Eventually the outer glass of both windscreens broke off with a sharp crack. We landed at Singapore with a very sick budgie – two u/s windscreens, no NBS radar, a surging No 3 donkey and a few other niggling bits. We stayed at the Equatorial Hotel again, travelling back and forth daily to Tengah to check on the progress of the spares, which turned out to be no progress at all.

It was decided that the Boss and his crew would return to Akrotiri; this also included Nobby Clarke, the other crew chief. As soon as the rest of the crew left for Cyprus, I had to move into the camp at Tengah. I was joined at Tengah by Bob O'Dowd and Ivor Maggs, who came to assist me nail all the broken bits together. The windscreen replacements had to be carefully timed so that we didn't have an empty hole during the afternoon downpour. Singapore was a new experience to Bob and Ivor, so it was my pleasant duty to provide a conducted tour of Change Alley, Bougis Street and various macan stalls.

Eventually we had 562 serviceable again and Jon Tye's crew arrived to fly us back to base in Cyprus. As Nobby Clarke had returned to Cyprus with the Boss's crew, Bob flew 7th seat with us. The trip back was uneventful and consisted of a refuelling and meal stop at Gan and a night stop at Masirah, arriving back at Akrotiri on 18 March. 562 was almost immediately taken into the hangar for the "Kiwi Roundel" to be replaced. Pity, really!

We will all remember the detachments we took part in during our time on the Vulcan Force. Some were not so good perhaps, but most of the ones I did were pretty enjoyable. I thought I would mention a few of the better ones I was on while I was on No 101 Squadron.

One of the first ones I did was a detachment to Singapore in the early '70s. It was the first really long dicing trip I had ever done and was a bit of an adventure as far as I was concerned. I had never done a co-pilot's tour on the aircraft so it was a bit new to me. Bit of a change from instructing on Chipmunks on London University Air Squadron on my previous tour! I still remember during the climb on my first solo when the cabin pressurisation valve went with that very loud rattle and I nearly hit the roof. We all set on to Goose Bay initially and after that to Offutt, followed by McClellan AFB. We taxied out from there to take off for Hawaii, but the brakes failed in a tricky S-bend, which was downhill of course, and it was a bit of a game to stop the aircraft before we hit a hangar! Having been towed back, the brake motors were bled and we set off again. I had been feeling tired and about an hour out over the Pacific my back began to hurt badly. Eventually I handed over to the co-pilot and fell asleep! He woke me up after a while and said we were approaching Hawaii. I landed the aircraft, quite well I thought, except that the crew told me later it was truly awful. When we got to the hotel I was delirious with Hong Kong flu and had to go to bed for three days. My crew thought it was Christmas! One source of amusement was that the co-pilot cut his hand on some coral while trying surfing and spent all one evening looking at it secretively in case the coral was growing inside his hand! Must have been something we said…

We eventually continued to Guam via Wake Island. My Crew Chief slipped on some spilled oil and hurt his neck while refuelling the aircraft. He and I went to Sick Quarters where this enormous chap twisted his head – straight ahead again! He decided to continue the trip immediately. The rest of the detachment was really outstanding. The aircraft we took there stood on the tarmac for six weeks acting as a "Christmas tree" to keep the others going. The tail number was 606 and I discovered later it was nicknamed "sick-o-sick" because it tended to go u/s whenever it landed somewhere overseas. I bet a few will remember that one. I managed to get a trip in a Lightning and a Hunter in exchange for trips in the Vulcan for the fighter pilots. I think they thought it would be a slow old aircraft at first, but they soon changed their minds, and by the time we came back to base each time and did a few roller landings it was all I could do to get them to let go of the controls!

We saw in the Chinese New Year while we were there – very noisy business that was, with all the crackers going off. This was the old Singapore before all the skyscrapers, etc, and the sights on Buds Street were best remembered but not discussed. One thing I remember was that I was in the habit of running in for a low-level break and landing, ending up slightly inverted on the downwind leg. My Flight Commander asked me pointedly why I kept doing this and I said because I wanted to do air displays. It worked a month later, I am delighted to say!'

Having become a regular visitor to Australia and New Zealand, it is interesting to note that the Vulcan almost established a permanent presence in the southern hemisphere. Historian Denis O'Brien describes the fascinating story of Australia's interest in the Vulcan:

'Analysis of military equipment purchases, particularly of equipment as sophisticated as aircraft, can provide valuable corroborative evidence of contemporary military and political thinking. To concentrate attention on those aircraft taken on charge by the RAAF provides but part of the story of military thinking and planning. Often analysis of the evaluation process of those aircraft considered and perhaps even recommended for purchase yields a broader view of military and political reasoning of a particular epoch. The consideration of the Avro Vulcan is a case in point.

The mid and late 1950s, when the Vulcan came under RAAF scrutiny, was to be a transitional period where the whole question of aircraft requirements would be closely evaluated against a background of rapid advances in technology and changing defence requirements. This evaluation encompassed not only the mission requirements but also weapon systems, including the nuclear option. In 1954, with less than a quarter of the GAF Canberra aircraft delivered into squadron service, an Air Mission departed from Australia charged with the responsibility of selecting aircraft to meet the RAAF's needs into the 1970s. The Air Attaché in Washington and the Air Officer Commanding the RAAF Headquarters in London were instructed to obtain preliminary information relating to the new generation of aircraft being developed. The relevant aircraft and engine manufacturers were contacted and requested to prepare documents detailing the information that the Air Mission would require.

A. V. Roe & Co Ltd had prepared a paper 'A Comparison of Bomber Performance', and this was made available to the RAAF; it compared the V-Bombers, Canberra and Boeing B-47B Stratojet. Armed with RAAF Operational Requirements for bomber aircraft (No OR/AIR36), fighter aircraft, medium-range transport aircraft and an Applied Jet Training aircraft, the Mission left Australia for the United Kingdom on 17 October 1954. Apart from the more precise requirements stipulated in OR/AIR36, the Mission was to consider bomber aircraft on the basis of the selected aircraft entering squadron service from mid-1959, replacing the Canberra. Without precluding the local manufacture of some components and spares, it was felt that, "Because of the high cost, the complexity, and the relatively small numbers required of this type, these aircraft may be purchased overseas."

The brief given to the Mission in relation to the bomber and OR/AIR36 originated in 1954, when the Air Staff were directed to submit a recommendation on the most suitable aircraft to fulfil the current bomber requirements of the RAAF. The Directorate of Operations produced a document in May 1954 detailing the bomber requirements to meet Australia's defence commitments "of the foreseeable future". These defence commitments were established by the lessons of the Korean War, the United States' desire to rebuild Japan and contain Communism to the Asian mainland, and the changing relationships with Britain. The Korean War was to have political consequences extending far beyond the contemporary events. It highlighted the political instability within South East Asia and emphasised the need to improve the living standards of the peoples of the region as a bulwark against Communism. On a strategic level, Australia's prompt response to North Korea's invasion of the South influenced American opinion and aided the signing of the ANZUS Treaty in 1951.

The Korean conflict again demonstrated the lack of preparedness of the RAAF in terms of aircraft. CAC Mustang aircraft of No 77 Squadron deployed from Japan were totally outclassed by the MiG-15. The Gloster Meteors hurriedly acquired from Britain were also no match for the MiG-15, although both these aircraft were successfully employed in a ground attack role. The RAAF was once again in the position of having to function with second-best equipment. Against this background Australia's strategic defence policy towards the mid-1950s, in broad terms, involved a balance between:

a) Meeting her commitments as a member of the United Nations.

b) Meeting her commitments as a member of the British Commonwealth of Nations.

c) Meeting her commitments in Regional Defence Schemes.

d) Safeguarding her own security.

While the defence of Australia held prominence, it was considered that such defence planning "should assist in the preparation for a global war". Concerns expressed about the security of Malaya and Australia if "Indo China falls under Communistic domination" were realised when, later that year, the French were defeated in Indo-China.

The Geneva Conference failed to resolve the problems of Korea and Vietnam, heightening the tensions and reinforcing the prevailing belief of Chinese Communist expansionism. This was the geopolitical landscape that predicated defence planning and equipment requirements in the early 1950s. The bomber requirement was for an aircraft that could mount an offensive strike deep into enemy territory to destroy the enemy air force. Together with this objective was a requirement for a major interdiction task of hitting the main concentrations of fuel and equipment, the communications centres and airfields. The operational range for the defence of Malaya was based on the assumption that the majority of suitable targets were in the Hanoi-Haiphong area. The requirements, with significant reference to the British MoS Specification No.B35/46 document were more precisely defined in OR/AIR36. The operational aspects can be summarised:

Summary of Selected Operational Requirements for a Bomber Aircraft

* Operating height above 50,000 feet

* Speed 500 knots at operating height

* Range not less than 4,000 nm in still air

* Take-off to 50 feet in 2,400 yards at max AUW

* Bomb load: maximum load up to 40,000lb including one "special bomb"

If the politicians were undecided about a nuclear capacity, the RAAF Air Staff were certainly anxious to obtain a nuclear-capable aircraft should this option become available. While nuclear arms control was a major objective of the United Nations in the early 1950s very little progress had been made in that direction. When the Treaty of the United Nations came into force on 24 October 1945 only the United States had a nuclear capability. The first Resolution of the United Nations, adopted unanimously on 24 January 1946, proposed the establishment of an Atomic Energy Commission (AEC). Superpower mistrust and suspicion was intensified as the US postured to preclude the Soviet Union from developing nuclear weapons. The discussions continued to be frustrated by arguments relating to the precedence of inspection and verification of weapons versus weapon destruction until the Soviets detonated their first atomic bomb on 23 September 1949. A few months later the Soviet Union withdrew from the AEC.

In 1952 the General Assembly of the United Nations established the Disarmament Commission. While members of this Commission investigated solutions to nuclear arms control, the superpowers and others continued the "arms race", with Britain exploding its first atomic bomb and the US detonating its first hydrogen bomb on 1 November 1952. In 1953 a more conciliatory and business-like attitude evolved in relation to nuclear arms control. The events at the United Nations in relation to nuclear arms control were quite peripheral for an RAAF aiming to be prepared to meet any current threat. On a political level the objectives of nuclear arms control were openly supported. Australia was to be indirectly involved in nuclear weapon development when it was announced in June 1953 that the UK was developing nuclear weapons and that these weapons would be tested in Australia.

This involvement in the UK's nuclear programme certainly maintained the RAAF's support for the nuclear option. When the "Murdoch Mission", in their report dated March 1955, unanimously recommended "the purchase of the Avro Vulcan or Handley Page Victor", the desire to be nuclear capable was confirmed. Both these aircraft could accommodate a 10,000lb "special bomb". In reference to the Vulcan it was stated:

> "Although a heavy and varied bomb load may be carried for a tactical mission, a strategic mission bomb load would consist of approximately 21 x 1,000lb or LC bombs or 1 x 10,000lb special bomb, depending on the operational conditions and requirements."

As it was not planned to order a medium bomber until the 1956/57 financial year, a decision between the Vulcan and the Victor was to be delayed so that evaluation of the aircraft could continue as their development and flight testing programme progressed. If, however, a decision had to be made at the time of writing the report, the Mission would recommend the Vulcan. As if to give assent to this recommendation, the Minister for Air, William McMahon, in a letter to the Minister for Defence, Sir Philip McBride, proposed that if additional funds became available for the defence vote, they should be "used to acquire British V-Bombers". More direct evidence of the RAAF's desire to be nuclear capable is contained in a letter to Air Commodore N. Ford, Overseas HQ, London, from Air Marshal Sir John P. J. McCauley, dated 5 July 1956:

"For your personal information only, I am taking the initial steps in an endeavour to have a supply of tactical atomic weapons made available from the United States for use from our Canberras and Sabres. Much will depend on the outcome of these negotiations."

In his reply, dated 5 October 1956, Air Commodore N. Ford advised:

"The only nuclear bomb at present available to the RAF of UK origin is the 10,000lb otherwise known as the Blue Danube. This bomb has only just been cleared for Valiants. Vulcan trials are still proceeding. A smaller nuclear bomb – 2,000lb – is being developed for the Canberra force."

The assignment of the Air Mission to investigate the available aircraft, compile a report and make recommendations was the easy task. To convince the government that the massive expenditure was warranted was a far more difficult assignment. Any plans to re-equip with Vulcans would have to go before Cabinet. Following release of the Murdoch Report a series of papers were produced by the RAAF with that objective in mind. In June Group Captain G. C. Hartnell produced a paper reviewing the role of the bomber and assessing the relative costs of buying and operating Canberra and Vulcan aircraft. Validating the economic concerns relating to such a purchase, the introduction of this paper states that "some doubt has been expressed as to the ability of the RAAF to operate such aircraft (Vulcans) within the limits of the national economy". The aircraft requirements for the defence of Malaya, as assessed by the ANZAM planners in Singapore in December 1954, was for thirty-two medium bombers and 193 light bombers. The tasks to be fulfilled to meet the defence of South East Asia were considered to be:

a) A favourable air situation must be firmly established over Malaya and the air situation should be made as difficult as possible for the enemy over the rest of the Kra Peninsula and Southern Indo China.

b) A major offensive must be developed against Chinese military, industrial and economic targets.

c) A thorough denial and interdiction programme. This would be confined to enemy communications from Bangkok southwards.

It was considered that the RAAF's role would encompass a) and c) above, the wider task of developing the Offensive to South East Asia as a whole, and, more particularly, that China proper would be considered to be a task more applicable to a major force such as the United States Strategic Air Command. Having validated the need for an offensive bomber force, the paper then continued to assess the relative "hitting power" of the Canberra and Vulcan, concluding that one Vulcan had the same "hitting power" as four Canberras. What they were trying to justify was the high capital cost, the higher maintenance costs and the significant increase in personnel required to maintain operational Vulcan squadrons. In essence, the Vulcan would cost as much as two Canberra aircraft, maintenance for each Vulcan would be nearly three times that of a Canberra, and twice the personnel would be necessary to maintain a Vulcan in an operational condition than for one Canberra. Balancing this equation was the increased "hitting power" over the operational range of the Canberra or the same "hitting power" over an increased offensive radius of action.

In response to this paper the Directorate of Operations, in a minute paper dated 24 June 1955, advised that the perceived "hitting power" should be increased as the Vulcan had superior bombing accuracy with an all-weather capability – a convenient argument that ignored the fact that the loss of one aircraft would be more significant than the loss of one Canberra and did not allow for a future upgrade of the Canberra's weapons delivery system. Overlooking the relative effect of the loss of an aircraft was perhaps justified as the nuclear capability of the Vulcan was not factored into the assessment of the aircraft's destructive power. This paper also considered that the area of operations for both the establishment of air superiority and the denial and interdiction programme should extend north of the Kra Peninsula to the South China border.

Continuing to promote the introduction of V-Bombers, the Directorate of Operations presented a paper in September 1955 examining the suitability of airfields, hangarage and fuel dumps. While none of the existing RAAF bases fulfilled the RAF Bomber Command Class 1 airfield criteria, it was considered that limited AUW operations (150,000lb) would be possible at Amberley, Darwin, Williamtown and Pearce. The main deficiencies were the restricted AUW, runway width (200 feet recommended), and the size and strength of the hardstanding areas. The more immediate work required was a widening of the taxiways at any base selected for full-time training operations. To completely comply with the RAF standards would have required significant additional expenditure. It could be argued, and indeed it was pointed out, that the UK requirements embraced operational and meteorological conditions quite different from those in Australia. Peacetime flying training requirement were estimated at 480 hours per month or thirty hours per aircraft per month. Fuel requirements were for 60,000 gallons per month.

If the recommendation to acquire V-Bombers was based upon the less tangible and more hypothetical concepts of strategic planning, then its subsequent rejection resulted from far more pragmatic considerations. These considerations, based upon an interaction of economic, political, operational and technical factors, evolved with economic factors playing the dominant role. The RAAF planners were not the only ones examining the Murdoch Report: Mr M. B. Woodfull, General Manager of the Government Aircraft Factories, was also requested to comment. In a detailed report dated 23 May 1955 he concentrated on the ability to manufacture and/or maintain and modify the selected aircraft. Assessment of the selected aircraft he viewed as a matter for the RAAF because "only the RAAF personnel are really acquainted with their own operational requirements". In conclusion, he commented:

> "It is my opinion that both the bomber and fighter, including the engine, have now become so costly and complex, that Australia has neither the technical nor productive manpower and money to build them and I would therefore recommend that all our first-line aircraft be purchased from the USA and that our available money be spent over a much broader field than hitherto, to the end that we shall, at least, develop the facilities necessary for the maintenance of these aircraft in the event of War."

So concerned was Woodfull about the reliability of the UK to provide spares, particularly in time of war, that he suggested that if the Vulcan was selected then a re-design of the aircraft installations "to take a US engine and Appendix 'A' equipment should be performed".

However, time and events would overtake the Vulcan recommendation. At an estimated cost of £538,000 per aircraft ex-works, it was always going to be difficult to justify the purchase. The rapid development of airframes, engines and avionics in the 1950s was more than matched by the rapid escalation in cost, a fact that disturbed both defence planners and politicians alike. By August 1956 the cost of an aircraft had increased to £760,000, and £900,000 with the Olympus 6 engine. This figure did not include costings for engine spares, ground handling equipment, bomb gear, vocabularies and technical publications or flight simulators.

The cost of infrastructure was also a significant factor. Concerned that a V-Bomber purchase could seriously compromise defence spending for many years, the RAAF looked critically at the number of aircraft required and the strategic requirements. The initial estimate, based on the current Order of Battle, was for thirty-nine bomber aircraft with an expected service life of sixteen years. In the 1955 Air Force Programme the number was down to twenty-eight, with an expectation of approval for the 1957/58 financial year. Intrinsic to the economics was the political reality of satisfying the other Defence Chiefs as well as keeping the Defence vote in a proper relationship to the whole budget strategy. After a close election in 1954 the Coalition had a comfortable electoral victory in 1956. Furthermore, the split in the Australian Labor Party that occurred in October 1954 rendered the Government less susceptible to opposition pressure. The amounts allocated to the Air Force in the Defence Programme 1955/56 and 1956/57 were substantially less than the programmes submitted by the Department. In a scathing attack, Aircraft magazine announced "RAAF in Jeopardy" and that "The Government has reneged on the pledge it made a few years ago to build up the RAAF as the chosen vessel of defence".

In June 1956 the Minister for Defence, P. A. McBride, wrote to the Prime Minister echoing the concerns of the Minister for Air, Athol Townley, relating to the "tremendous cost of re-equipping the Royal Australian Air Force". Townley had been advised by Lockheed that it could provide sixty-six F-104A fighters for £A44 million, with delivery in 1957/58, as compared to an estimated cost of £A55 million for local production with deliveries over the period 1960-65. Against this proposal was the effect it would have on the local aircraft industry. Townley had also pointed out that "although fighters may be within our financial capacity, modern medium bombers appear to be well beyond it". McBride's letter continues:

> "The Americans might be induced to assist our defence effort by storing some of their reserve aircraft in Australia. He (Townley) mentions that the US Air Force is re-equipping with the B-52 Bomber which is too heavy for our requirements, but that the B-47 which it replaces would be suitable; he suggests that 30 of the latter could be stored here, and we would pay a hiring fee for those used."

Just as economic and political factors were intertwined, so too were the operational and technical considerations. The rapid advances in airframes and engines was certainly appreciated by the Murdoch Mission. Details of experimental aircraft such as the Avro 720, Bristol 188 and Chance Vought XF8U-1 were included in its report. While not considered by the Mission, the Convair B-58 Hustler was known to be under development. A letter from the Office of the Air Attaché, Australian Embassy in Washington, to the secretary of the Air Board, dated 30 November 1954, gave details of this development and quoted an attack speed of 2.0 Mach. Had a firm order for the Vulcan been placed by mid-1956 delivery would not have been possible before mid-1959 to late 1960. Would it be outdated by then, possibly replaced by a supersonic equivalent? By late 1956 this question had been answered in the affirmative.

A secret meeting attended by the Prime Minister, the Ministers for External Affairs, Defence, Army, Navy and Air, and the Chiefs of Staff was held on 10 October 1956. At that meeting Air Marshall Sir John McCauley advised that "the Air Force was looking for a supersonic light bomber". By 1957 another major political factor affecting defence thinking appeared with the release in April 1957 of the White Paper by the UK Defence

Minister, Duncan Sandys. This paper drastically slashed Britain's defence spending by £128 million. Fighter development was cut, with many highly developed experimental aircraft designs abandoned overnight. The plans for a supersonic bomber were dropped. The highest priority was to be given to the development of nuclear weapons suitable for delivery by existing manned bombers and ballistic rockets.

The Commonwealth Government had received prior warning of this policy change when Lord Home and the Chief of Air Staff, Sir Dermot Boyle, visited Australia for the ANZAM Defence Committee Meeting in March 1957. When Duncan Sandys, Lord Carrington and Sir William Dickson, Chairman of the Chiefs of Staff, visited Australia in August 1957 the question of Defence Policy in South East Asia was fully examined. Sir William advised that Tengah airfield in Singapore was to be rebuilt to cater for the three squadrons of V-Bombers for permanent garrison there. During these meetings it was further advised that the United Kingdom intended "to stockpile nuclear weapons in Singapore. Atomic bombs for Canberras will be available in 12-18 months."

Public confirmation that the Vulcan proposal had been abandoned came with the release of the 1957 Australian Defence Review. This Review, by the Minister for Air, F. M. Osborne, was drafted in response to two papers prepared by the Defence Committee: "The Strategic Basis of Australian Defence Policy" and "The Composition of Australian Defence Forces". Osborne stated:

> "Though it is of the opinion that modern bomber aircraft are strongly desirable, the Air Staff has adopted the realistic view that they are at present beyond our economic capacity, and indeed no supersonic bomber is yet in service."

Australia's defence planners, in the preparation of the 1957 Defence Review, had moved to a position where it was considered desirable to operate equipment "standard or compatible as far as possible with that used by the United States Forces with whom they are likely to be associated in war". The Defence Programme 1957/58 to 1959/60 gave the Air Force's principal new Proposals as:

a) The re-arming of one fighter squadron with US Lockheed F.104 aircraft.

b) The re-arming of one Dakota squadron with Lockheed C-130 (Hercules) medium-range transport aircraft towards mobility requirements of all Services.

c) The introduction of the first RAAF surface-to-air guided weapons unit.

d) The procurement of additional light aircraft for the Army's expanded requirement, and the formation of two additional Control and Reporting Units.

The Minister for Air, quoting a speech by Sir Winston Churchill given at Boston in 1949, was critical of the allocation of the defence vote given to the Air Force: "None of these proposed alterations does more than make minor adjustments to the pattern of Australian post-war defence." While a strategic bomber was no longer an option, the recognition that the F-104 was capable of carrying conventional guided weapons and nuclear weapons was highlighted in both the Defence Review and the paper on "The Composition of the Australian Defence Forces". Thus the nuclear option remained a viable defence objective.

At Cabinet level the nuclear option was canvassed in a Top Secret atmosphere: "the recent changes in the arrangements between Britain and the United States raise the expectation that we may have early success in obtaining tactical nuclear weapons". The issue was more openly canvassed by the Victorian Liberal, Mr W. D. Bostock, a former Air Vice-Marshal of the RAAF. He was quoted as saying that "Australia should arm herself with atomic bombs and long-range bombers".

During a meeting with the British Prime Minister at Parliament House, Canberra, on 29 January 1958, Mr Menzies raised the question of nuclear weapons. While referring to "internal pressures" for Australia to develop a capacity to produce such weapons, Menzies expressed personal doubts "about the wisdom of any such action". Macmillan advised that the United States Government had a strong desire that no further powers should develop a nuclear capability. Macmillan expressed the view that Australia should look closely at its infrastructure, runways and technical equipment, so that nuclear weapons provided either by the UK or USA could be used if the need arose. This may have been an appropriate objective but it failed to appreciate the problem facing the Service Chiefs – the denial of technical information on the nuclear weapon.

Both Mr R. G. Casey and Sir Philip McBride highlighted this point at a further meeting at Parliament House on 11 February 1958. There was insufficient practical knowledge in Australia concerning nuclear weapons. The Australian Services needed to know "what organisational and other changes the use of nuclear weapons would involve". Sandys promised to make such arrangements upon his return to the United Kingdom.

The Vulcan chapter was closed, or almost. Perhaps – somewhat ironically – there was a proposal in 1961 that Vulcan aircraft be made available to Australia as an interim measure pending the delivery of the British TSR.2, should Australia choose that aircraft as a Canberra replacement. Although it is unlikely to have influenced their decision, that plan was wrecked by the British Air Staff. Instead of a no-strings deal, the Australians were told that they might have the Vulcans provided the operational control remained with the RAF. The Australian reaction was natural and typically caustic.'

CHAPTER EIGHT

From the Cockpit

NO account of the Vulcan's history would be complete without a proper description of the aircraft from some of the people who had the privilege of being part of the V-Force. Squadron Leader John Reeve (with input from his colleagues) provides a fascinating account of how the aircraft would have been used, had the aircraft been required to perform its designated wartime role:

'The V-Force posture was obviously based on deterrence, the famous Mutually Assured Destruction (acronym "MAD") and also on the four-minute warning that the radar at Fylingdales was expected to provide of any incoming ballistic missiles. That said, it was reasonable to assume that the war would not come out of nowhere, but would reflect a steady build-up of tension over a period of time – the Cuban missile crisis being the obvious example. The Force therefore had a series of measures to progressively increase its readiness in response to the international situation, and these were regularly practiced on generation exercises. These generations were either ordered by Command under the code-names Exercise "Mickey Finn" or Exercise "Mick", or by the Ministry of Defence under the Taceval (TACtical EVALuation) programme. Aircrew and ground crew would be recalled to base and Engineering Wing would generate as many aircraft as possible. This was taken seriously enough for aircraft in deep servicing to be re-assembled and placed on the flight line. Crews would meanwhile be changing into flying kit and, as a particular aircraft became available, they would be detailed to check it by running the normal checklists up to and including engine start. Once declared as fully serviceable, the aircraft was handed back to the engineers, and particularly the armourers, to load the nuclear weapon.

While this was being done, the crew would report to the Operations block where, behind double locked

A typical Vulcan crew, with the captain, Squadron Leader Joe L'Estrange, on the extreme left, pictured after delivering XM603 back to its home at Woodford, upon retirement from RAF service.

doors in "The Vault", the station had begun allocating crews and aircraft to targets. This was complicated by the priority given to certain targets that had to be covered first, and also because some targets were so far away that they required bomb bay fuel tanks to be fitted. There not being enough tanks to equip all aircraft, obviously only certain aircraft could be allocated to those longer-range targets. Moreover, crews were not allocated targets at random; each crew had previously studied only a very small selection of the targets and so were familiar with, and able to fly, only a very small proportion of the overall plan. Matching crews, targets and aircraft was the job of the station weapons team, who were usually very experienced navigators on a ground tour, and it required close liaison between the engineers, squadron planners and individual crews to get it right first time.

While this was being done, the armourers would have loaded the nuclear weapons, which, for the Vulcan of the 1970s, was the WE177B, a 400-kiloton device that replaced the Yellow Sun, Blue Steel and Red Beard weapons that had previously provided the Vulcan's nuclear capability during the 1960s. That these bombs are now on display in museums (and no longer in RAF service) must be good news. On most exercise generations, the WE177B was simulated by loading a "shape" – a dummy weapon that gave all the right electronic indications as the relevant switches were made. These switches were the preserve of the navigators, with the Nav Plotter reading the special checklist, which was subsequently actioned by the Nav Radar. With a "shape" this usually took about ten seconds. However, you occasionally found a genuine nuclear weapon had been loaded, and with this it was very noticeable how much more steadily and slowly the checks were done and also – as you cannot be too careful with a nuke – that the Nav Radar would lean back very slightly when he moved a switch, and especially the arming keyed, just to be that bit further away in the event of a nuclear detonation!

The crew would then place the target material in the aircraft, stow the in-flight rations where they could be easily reached, lock the aircraft entrance door, and retire. Armed RAF Police would then mount guard with an access list of names and photographs – the list being specific to that particular aircraft – and instructions that nobody, not even the Station Commander, was ever allowed into the restricted area unless accompanied by another authorised person – the "No Lone Zone" policy. What the Americans call Use of Deadly Force would be invoked to enforce this rule. The weapon system – the aircraft, crew and weapon – could now be declared to Command as At Readiness State 15. If this sounds like a highly organised and totally secure system, the illusion is somewhat spoiled by the fact that there was actually only one key for all Vulcan aircraft, reference number FA501, which could be, and often was, bought at Halfords in Lincoln High Street to replace the keys that got lost. WE177B arming keys were also not as unique as we had hoped, a similar design being used in one-armed bandits and fruit machines to change the payout percentage.

Readiness State (RS) 15, as you might imagine, required the crew to be airborne within fifteen minutes of the order. This was not too onerous to achieve; the crew had to stay together, usually at their squadron or the mess, and be within a few minutes of their allocated transport. It was perfectly possible to sleep and live a "normal" life at RS15, so it could be held for 28 days, and in reality it would probably have been longer. RS15 also required good communications. The line from Command utilised what were called the Bomber Boxes, comprising small public address system speakers that beeped every ten seconds to prove that they were alive, and these would be used by Command to raise the Readiness States. It was fitted in the Operations Room, and from there the Operations Staff could use phones or the station Tannoy to pass the message on.

At this stage on an exercise there would be a Command decision, and, on an actual alert, a political decision too as to whether to disperse the V-Force.

On an exercise, any genuine nuclear weapon would be downloaded and returned to the bomb dump before the aircraft flew. Groups of up to four aircraft would then disperse to a whole range of UK airfields. You can still recognise some of them today by the four concrete

Right and overleaf: rare images of a WE177B on a loading trolley at Cottesmore, ready for insertion into the Vulcan's bomb bay.

fingers close to, or at the end of, the runway – the Operational Readiness Platforms (ORPs) – where the aircraft would be parked. Dispersal on a genuine alert would be a very public action, which might raise tension as it was a very high-level decision, but obviously once dispersed the force as a whole was far less vulnerable. On arrival at the dispersal the aircraft would be refuelled and declared back at RS15 as soon as possible.

The next stage of a generation was RS05 – obviously a capability of being airborne in just five minutes. This would be ordered by the Bomber Controller over the Bomber Box in a standard format:

> "Attention, this is Bomber Controller for Bomblist Charlie. Readiness State Zero Five for Bomblist Charlie. Readiness State Zero Five for Bomblist Charlie. Readiness State Zero Five. Bomber Controller Out."

This message was not authenticated on the assumption that there was no advantage in any enemy sending the V-Force to RS05. Crews would run or drive to their particular aircraft, be admitted by the RAF Police and board the aircraft in a set order. The Captain would go first to get ready to start the checks, the AEO would be next to start the APU, the co-pilot would follow him, then the Nav Plotter and Nav Radar. The door would be closed and checks completed to get the aircraft to RS05, which was everything up to, but not including, actual engine start. With no fuel being burned – apart from the APU – RS05 could be held for up to five hours, the limit being essentially crew fatigue. Again I expect that it would have been a flexible limit if necessary.

The next stage was RS02, which was ordered in a similar format to the RS05 message, and involved starting all four engines, probably using the rapid start, which could spin an Olympus up to self-sustaining speed in about five seconds – a modern big fan engine today will take up to a minute to achieve the same steady state. The message could be passed over the radio or over a dedicated external intercom plugged into the Vulcan between the engine jet pipes – the telescramble. If the aircraft was not on an ORP it would be taxied to the runway and lined up for take-off. As the aircraft was now burning fuel, RS02 could only be held for a short time. The requirement to get airborne in two minutes was obviously linked to the four-minute warning that Fylingdales could provide. Crews would now be waiting for the scramble message. This would be authenticated, as a false launch of the V-Force would involve them in recovering and being off readiness for some time. The Navs would have a sealed envelope containing the authentication codes and would be ready to confirm the order, which would be in the format:

> "Attention, Attention. This is Bomber Controller for Bomblist Charlie. Scramble. Authentication Echo Two Bravo, Echo Hour One Nine Four Fife Zulu for Bomblist Charlie. Scramble. Authentication Echo Two Bravo, Echo Hour One Nine Four Fife Zulu for Bomblist Charlie. Scramble. Authentication Echo Two Bravo, Echo Hour One Nine Four Fife Zulu. Bomber Controller Out."

This would be repeated endlessly by the Bomber Controller and relayed by ATC on any and all airfield frequencies. The message also implied take-off clearance. ATC might announce that you were clear to take off, which was nice of them, but you were going anyway. The "E Hour" was a time datum, and if you were scheduled to bomb at, say, E plus 3 hours 10 minutes, then your time on target was 2255 Zulu. Achieving this should de-conflict you from other "friendly" nuclear weapons. As the accompanying map shows, once airborne the aircraft would be turned towards the north-east and climbed to high level. The force had been ordered airborne, but it had not yet been given clearance to proceed on its mission. Therefore a Positive Control Line was established at 8 East and we would, if necessary, orbit at this position until

Most of the UK's major military airfields were modified to include an Operational Readiness Platform (ORP) for V-Bomber dispersal. This aerial picture of Valley reveals a two-aircraft ORP attached to the eastern end of the main runway.

A rare aerial photograph of Finningley's ORP with four Vulcans on dispersal detachment. Unlike the similar ORPs at other V-Bomber bases, Finningley's ORP fed into the runway from the right-hand side.

our fuel dictated a return to base.

Obviously the Russians would be aware of our scramble as their satellites would have picked up the mass of radio traffic ordering the take-off. This was part of the increasing diplomatic and military pressure that would hopefully generate some common sense at political levels. Equally the scramble might have been to get the aircraft airborne in response to incoming weapons, but without a political decision to retaliate. We thus headed north-east in hope of a recall, but, to avoid giving away any additional information, all aircraft transmissions were minimised. There would be no radio transmissions and, once in the holding positions at 8 East, the radar would only be used when facing west – away from Soviet listening devices – to update our navigation kit. We would, however, be listening out on all relevant frequencies, which would include base RT

Right: Vulcans on the ORP at Waddington as an Avro product from an earlier era flies by in salute.

Below: Although taken some years after the withdrawal of the Vulcan fleet, this aerial photo of RAF Leeming clearly illustrates that, as at many other airfields across the country, the former Vulcan ORP is still present, providing a home for ground vehicle shelters. Still just visible are traces of the taxiway that directly linked the ORP to the runway.

Scrutiny of the crew door reveals markings for the Cottesmore Wing comprising the station badge and those for the component squadrons, Nos IX, 12 and 35. *Joe L'Estrange*

frequencies, the long-range Bomber Command High Frequency radio net, and even the old Light Programme and Home Service, which used the powerful transmitters at Daventry. The Positive Release Message would be in the format "… Is Now In Force".

The Nav Radar would open another envelope to confirm that the codeword was correct, and once this was received it was irrevocable, and Armageddon was inevitable. As you can see from accompanying maps, the crews would now turn east and, with a few exceptions, head towards Norway and Sweden; the exceptions were aircraft going to the Kola Peninsula. Strict emission control (EMCON) would hopefully hide us from the Soviet air defence system until we penetrated their long-range early warning (EW) line, which was based on the Tall King radar. This was a large and immobile radar whose locations we knew. It had a range at altitude of some 250 nautical miles, so well before this range we would start our descent to low level so as to stay out of its coverage. The assumed EW lines shown included radars based in Finland, as we calculated that by this time Finland might have been over-run and its radars linked back to the Soviets via the Finnish TV system, which was compatible with Soviet radar data. Norway is, of course, in NATO, and knew our routeing, but our route would also have taken us over neutral Sweden, which had – and

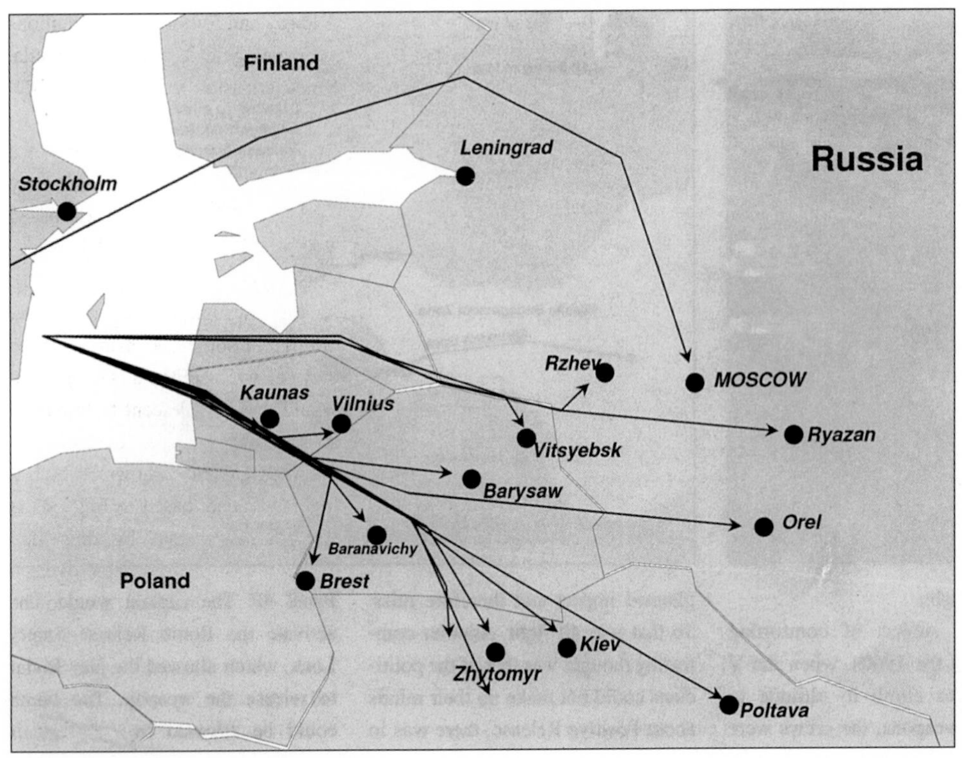

Although each Vulcan crew had their own pre-planned attack route towards a designated target, the basic routes into Soviet territory were designed to take into account projected radar and anti-aircraft defences. *Courtesy of Vulcan Restoration Trust*

Its low-level flying ability was a Vulcan speciality. Here Squadron Leader Joe L'Estrange displays the Vulcan in typically flamboyant fashion, making a (very) low fly-by at a Passing-Out ceremony at Swinderby. *Joe L'Estrange collection*

still has – a first-class air defence system. Hopefully we would make our transit over Swedish territory without incident and settle down over the Baltic, flying as low as common sense and captaincy would allow. There would be a final visual or radar position update on Oland or Gotland Islands, scanning backwards to the west with the radar to conceal its emissions from the Soviets. The only radio emission would be from our radar altimeter, which pointed straight down, and its transmissions should not have been detectable at great ranges – allegedly.

The forward-looking Terrain Following Radar would definitely be off over the sea and would only go on as we approached the southern Baltic coast. We would also switch off cabin ventilation and go to 100 per cent oxygen to avoid breathing any contaminated air. Also at this point the pilots might put on the famous eye patches. The cockpit blinds would be up, and if a pilot had to look out he wore a patch over one eye so that any nearby nuclear detonation would only cost him one eye. Having done his bit, he then flew the aircraft, and the remaining two-eyed pilot did any necessary looking out … or at least that was the theory. Traditionally, Soviet defences were shown on our maps in red, and when I joined the V-Force in the 1960s the defences were a solid red line down the Baltic, but at least it was a fairly thin red line. When I last saw them in the 1980s it was as if somebody had suffered a nosebleed over the East German and Baltic states coastlines. Whatever else the V-Force was, it was a marvellous economic weapon against the Soviet State, since all of these defences were largely there because of the Force; bombers from continental USA would come over the Pole, not via the Baltic.

The first line of defence would be elements of the Soviet fleet thrown forward in the Baltic to provide an early warning line. They would have to operate their radars to be of any real use and we hoped to detect their emissions and aim between the gaps. In principle, we would not jam these radars as we might give away more by active jamming than we might gain by silence. Of course, by day there was nothing we could do to prevent the visual detection of a large triangular aircraft that trailed a long line of black smoke, so we hoped to fly by night. Indeed, on IX Squadron that was our official motto – "Per Noctem Volamus" – and we hoped it was an omen.

XM603 making another low flypast over Greenham Common during another memorable display given by Squadron Leader Joe L'Estrange.

The main line of defences would be at the coast. The first problem would have been the Barlock Ground Control Radars and the Thinskin height finders that controlled the Soviet fighters. Barlock has a low-level range of about thirty miles. These radars we would have to accept and, on the same logic as with the Soviet fleet, we would not jam them as they did not in themselves constitute a hazard to our aircraft. We would, however, jam any Soviet fighter radar that locked on to us, deploy Type 22 chaff and flares, evade violently, and hopefully fade away into the darkness. Another ploy was that most aircraft would penetrate just south of Riga, giving the defences in that area too many targets to engage at once – the old Bomber Command stream principle. However, we had to expect to be illuminated by the Firecan radars that controlled the Anti Aircraft Artillery, and the Fansong radars associated with the SA2 missile system.

Despite the earlier "nosebleed" comment, we took comfort from the fact that the Baltic has a long coastline and the defences could not physically be everywhere at once. Also, the guns had a fairly short engagement zone although they could shoot at very low-level targets; however, the SA2 missile, with a much longer range, had little capacity against low-level targets. The SA3 was more capable, but also more thinly spread. In principle we would jam any radar that directly controlled a threat system while manoeuvring to stay outside the engagement zone of that weapon. There was a formalised evasion plan for skirting missile sites called the Veronica Manoeuvre – an impressive name for basically turning away and flying around them.

Once through the coastal defences, the sheer size of Russia meant that even the Soviet military could not provide low-level radar coverage except round high-value military or civilian sites. Our routes would be planned to avoid such areas except, of course, that our target would, by definition, be a high-value site. The routes were also designed to avoid our being hit by other "friendly" nuclear weapon detonations, so major deviations from your planned route were considered to invalidate your Lloyds flying risk insurance policy. The original Vulcan weapons,

Opposite top: A diagram illustrating the projected routes for Vulcans launching a nuclear strike into the heart of the Soviet Union. Note that many of the missions were initiated from dispersal sites rather than designated V-Bomber bases. *Courtesy VRT*

Bottom: Post-strike routes for the V-Force varied depending on the location of each aircraft's target. Although most aircraft anticipated returning to the UK (although there was obviously no guarantee that any British airfields would remain intact by that stage), some aircraft would have routed south towards locations such as Cyprus and Malta. *Courtesy of Vulcan Restoration Trust*

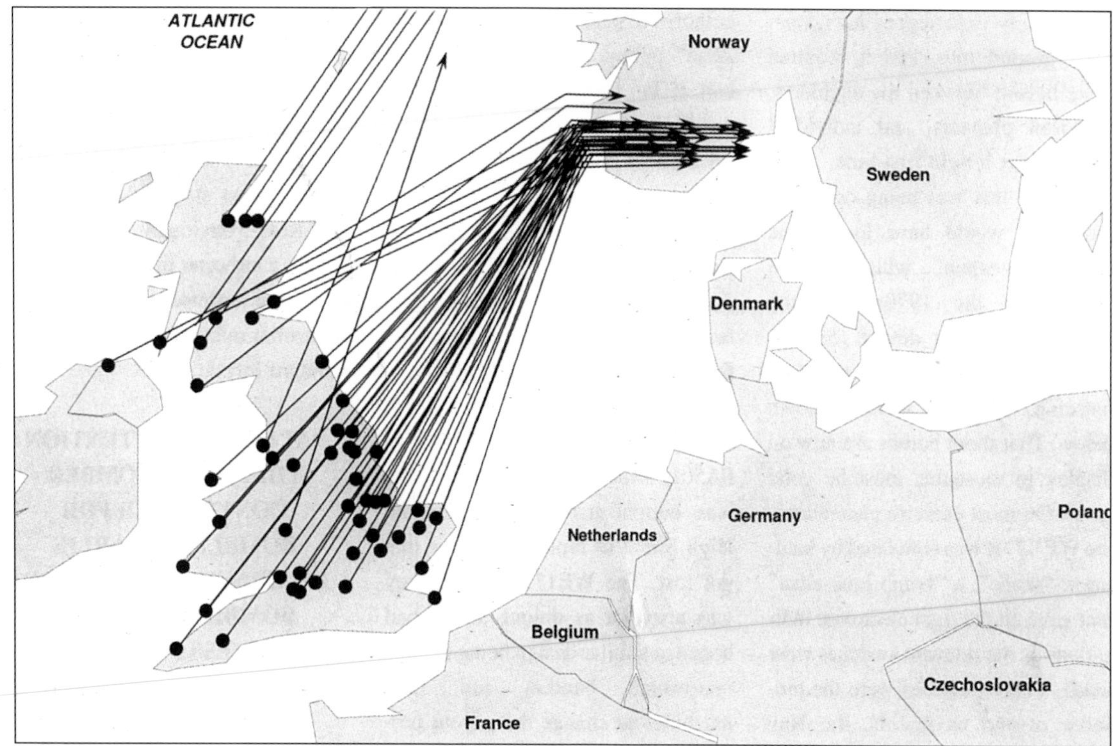
ATLANTIC OCEAN
Norway
Sweden
Denmark
Germany
Poland
Netherlands
Belgium
Czechoslovakia
France

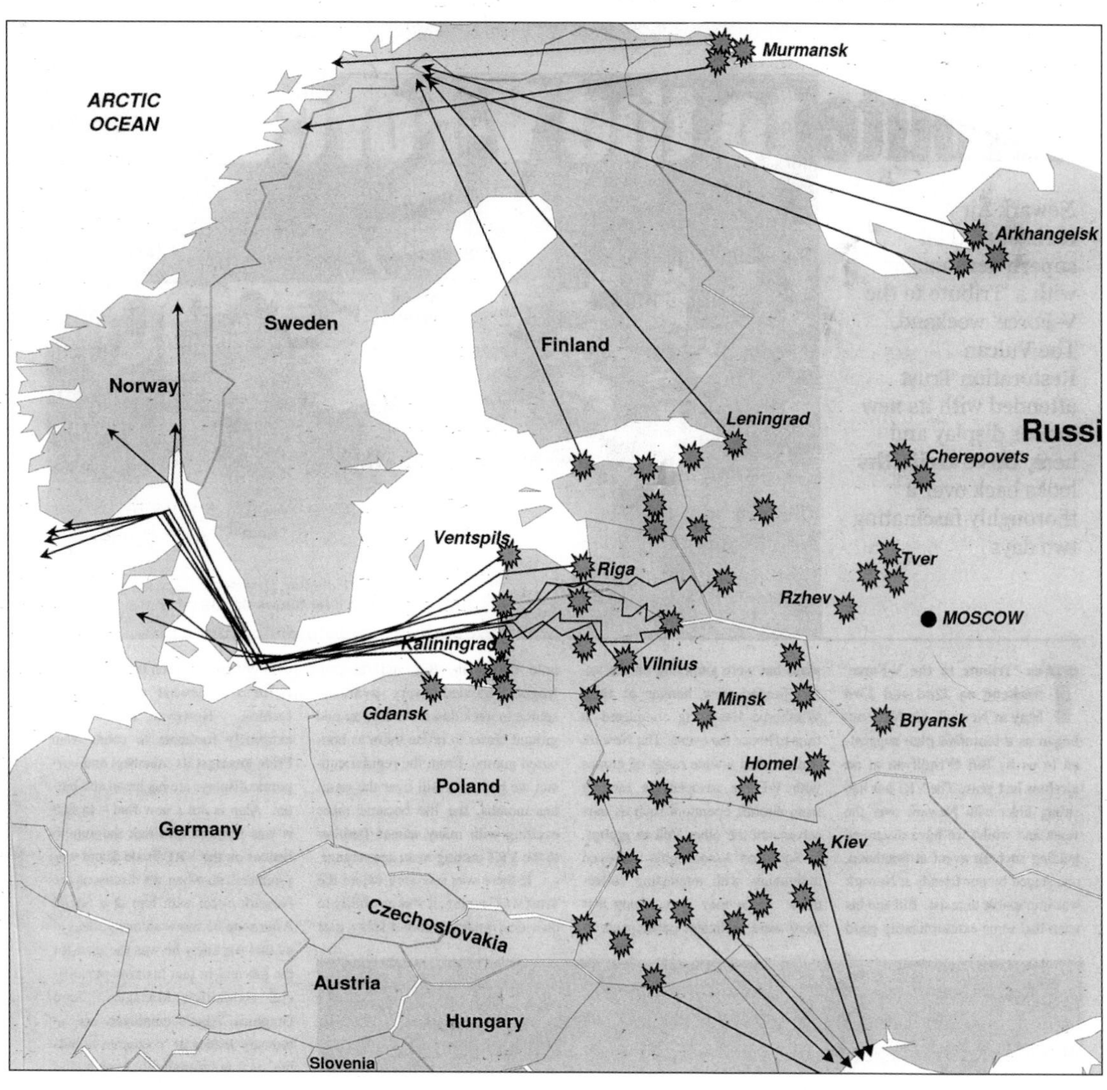

Murmansk
ARCTIC OCEAN
Arkhangelsk
Sweden
Finland
Norway
Leningrad
Russi
Cherepovets
Ventspils
Riga
Tver
Rzhev
MOSCOW
Kaliningrad
Vilnius
Gdansk
Minsk
Bryansk
Homel
Poland
Germany
Kiev
Czechoslovakia
Austria
Hungary
Slovenia

Blue Danube, Red Beard and Yellow Sun, had to be dropped from altitude, but WE177B could be dropped from 50 feet. Therefore our WE177B plan was to go straight in and straight out of the target area with little manoeuvring and as fast as possible, partly to counter Soviet defences and partly to avoid the effects of our own weapon. There was a problem here in that the Vulcan was designed for high-level flight where the airspeed is quite low, even when the Mach number or groundspeed is quite high. The aircraft structure was therefore not designed for very high airspeeds. Our general wartime low-level speed would have been only 375 knots with a once-only acceleration to 415 knots. This speed of 415 knots was not based on scientific calculation, but rather on a rough guess of how fast the Vulcan was going when it broke up in mid-air over RAF Syerston on an air display in the 1960s – not a comforting thought.

On the subject of comforting thoughts, in the 1960s, when the V-Force had to climb to altitude to release its weapons, the crews were somewhat concerned that SA2s and 3s would shoot them down in the climb. Not so, they were told, because the Soviets needed 112 seconds to detect and engage a target. The good news was that your bomb would detonate 103 seconds after the Soviets first detected you, which would burn out the ground radars, and thus the missile would loose guidance nine seconds before the planned impact and therefore miss. So that was all right! Another comforting thought was that if the politicians could not make up their minds about Positive Release, there was in certain cases an option to hold until you were down to a planned 1,000lb of fuel – say, five minutes' flying – over your target. In that case the book said that the recovery plan would be at the "captain's discretion".

On the run to the target, the Nav Radar would arm the nuclear weapon. There were two fusing systems, and if they did not work we had to check (all together now, ex-V Force crew!) Fuse 616 in Panel 3P and Fuse 1167 in Panel 4P. The captain would also activate the Bomb Release Safety Lock, which allowed the Nav Radar to release the weapon. The bomb could be released by radar or, if visual, we used the TLAR system, lining up the AAR probe with the target (TLAR stands for That Looks About Right). On release, the bomb would be retarded by a parachute, and, for Vulcan operations, would be detonated by a thirty-two-second timer that allowed us to reach a relatively safe 51/2-mile distance. "Safe" meant not only that we would not be blown out of the sky, but also that the shockwave would not rush up the engine jet pipes and flame the engines out.

Assuming we survived, we would then send a message estimating our aiming accuracy and weapon yield. The message was in the format "Call Sign … Alpha Alpha" – the first letter being accuracy and the second yield. Command would then decipher this as a Category 1 accuracy and Category 1 yield, so we had a code for a code. I never found out why we just couldn't say "Call Sign One One" instead – after all, the Russians would know all about the yield and accuracy by then. I also never found out how we were meant to actually estimate a weapon yield, never having experienced one before. Depending on which dispersal you started from, your TOT would vary because of the different distances involved. More distant dispersals were often given targets that were not time-critical – perhaps cities with a military function rather than bomber bases. For example, a typical profile flown from Ballykelly in Northern Ireland against Kiev would require bomb bay tanks and a high-level phase after bomb release. It had the small advantage that the Soviet defences should, by the time it got there, have been severely degraded.

The bomber's route after bomb release at Kiev would include turns that would be put in every ninety seconds to break lock-on SA2 missile radars. The targets covered by the V-Force extended across much of western Russia. The map also shows the recovery routes, some of which were back to the UK but others ended up in Cyprus and elsewhere. What we did after landing was somewhat unclear, and obviously could not be planned. The reaction would be to go home, but would we have a home to go to? Would it have worked? I don't know, and nor does anybody else, which is perhaps the whole point of deterrence. Less semantically, by day I doubt if many aircraft would have got through. Any half-decent Soviet fighter pilot who saw a Vulcan would have rammed it if all else failed. By night or in bad weather, however, I think we would have stood a good chance, especially if the Soviet defences had been softened up by earlier strikes. I can only close by saying that the one Soviet fighter pilot I have ever had any contact with was, when I met him, a First Officer on the A300 Airbus that I used to fly. I first met him over a drink in Athens in 1999, and I am glad that the world has so moved on that we did not first meet twenty years earlier, somewhere south of Riga.'

Of course, the Vulcan relied upon the skills of the rear crew as well as the pilot and co-pilot. Alan Steel provides a fascinating account of the Navigator Radar's role:

'The best way to begin describing the role of the Nav Radar is to describe the radar itself. For those of you who have not seen it in operation, we viewed the radar through something called the Indicator 301. In typical Air Force terminology, everything becomes abbreviated, so this was always called the 301. It was in fact a 301C, and I have no idea what the A and B model looked like, but they never got any further than the 301C. It was a True North presentation, which means simply that north was always at the top of the picture and the on-board compass system kept that so. The reason that it was designed like this was a belief that it made it easier to interpret the radar. Other radars are heading-orientated, like the radars looking out of the front of the aeroplane, scanning left and right. There would be no point in having a True North picture there because you might be heading east, in which case you wouldn't see very much. Some of my colleagues would have found that a help on some occasions! The

The distinctive smoky trail of a Vulcan made visual detection rather too easy, and became a familiar sight to air show audiences.

radar normally scanned 360 degrees and by convention we had it scanning clockwise; the engineers told me (out of sheer devilment, after I had flown for five hours with it scanning anti-clockwise) that the reason it should go clockwise is because that's the only way you get any lubrication on the thing, and the scanner was now completely ruined. Still, you live and learn! The aircraft position was normally the centre of the screen – it did not have to be, but it normally was. We called that the "origin", and we had the line, known as the time-base, sweeping around the screen, and you would pick up an almost map-like presentation of the ground underneath, ahead and to the sides of you. Now, those of you who have looked carefully at the Vulcan will be thinking that you can't see behind you because somebody's put a great big aeroplane there. Very true, but at high level you could tilt the antenna down towards the ground and by doing so you missed out maybe as little as 5 or 10 degrees directly behind you and effectively had a full 360-degree sweep. The screen was 9 inches across. So there you have it – that was the 301.

I said that normally the origin would be in the centre of the radar displays, but sometimes you could move it off, and you did that by moving the little joystick in front of the indicator. When you moved the origin away, the system produced a range as a curved marker and a bearing marker, and when you wanted to define a point on the ground that you were going to use to fix your position, you put that point underneath the intersection of the two markers. Effectively you drove the marker intersection around using the joystick in front of the radar display. The range marker was compensated for altitude such that it produced a plan range to the system, in a wonderful piece of very adult Meccano.

Before she was famous; B.2 XH558 in early gloss camouflage when with the Waddington Wing in 1968. *Martin Derry*

XH558 at Scampton in February 1982 when she was a B.2(MRR), seen with an air sampling pod beneath the wing. *Fred Martin*

The whole navigation and bombing system was actually called NBS, the Navigation and Bombing System, but those of us that knew it and loved it called it Navigation by Bits of String! The reason for this was that inside the equipment were all these wonderful electromechanical devices, the best of which, I always thought, was something called the triangle solver. The triangle solver did what it said on the box – it solved triangles. The short side was your altitude, and had a little runner that tracked your altitude. One of the other sides was slant range – the distance from where you were to the object on the ground; but that's not what you want to know when you're going to drop a bomb – you want know your actual plan range. Now, when you're at low level the plan range and the slant range are exactly the same, but when you're 7 miles above the earth, that starts to change, and that is why the marker was height-compensated.

Literally when you open up the Calculator 5 (I think – my instructor would be turning in his grave if he knew I'd forgotten that!) there is this triangle solver and it really is like adult Meccano. Instead of using bits of string, it used the equivalent of metal measuring tape. It may seem as if I am taking the mickey out of the technology, and although it was almost a case of having to wind the thing up before flight, it did work incredibly well.

Now we come to Offset Aiming Technique. In our war role, where we were meant to go across into the Soviet Union and bring some heat and light into their lives, you might have been required to drop a bomb, for example on the middle of Red Square in Moscow. Unfortunately, on the radar it is impossible to distinguish such a target. You could just go for the middle of the town, and from that came the expression Good Enough for Government Werk! If you were going to nuke somebody then you did not really need to be terribly accurate. However, the Royal Air Force insisted that our aim needed to be refined, and that was done through a technique called Offset Aiming.

To do this, you use an easily recognisable point on the ground that is close to your target and you aim at that, but you effectively tell the bombing computer that you don't want to take out this place, you want you to bomb 2,000 yards to the south and 15,000 yards to the east. And that is exactly what we did, and there were two sets of offsets, one in the table to the right of the radar display with the little joystick – which allowed offsets of up to 40,000 yards in any direction – and the other inside the control unit in front of that table, but this only allowed offsets of 20,000 yards. The idea was that the Nav Radars would go off and would be studying their targets well before flight, and would work out (guess) what would show on the radar that they could easily identify. You were looking for something nice and big to get you into the target area. A power station springs to mind – something that big. So you could start your aiming on the power station and as you got closer to the target, you were looking for something a lot smaller.

Choosing the offsets was very difficult, and one way you could do it would be to steal somebody else's ideas. There were some very generous Nav Radars around, particularly experienced ones, who used to leave the maps lying around on which they used to pick offsets. So as new boys we used to sneak in and copy them! However, later on, when we got a bit more experienced, we used to pick up any gash old map, stick a few random pinholes in it, then leave it lying around for just this very same purpose. All's fair in love and war!

Having calculated the values, you loaded them into the offsets; the internal offsets were 20,000 yards or 10 miles north, south, east and west, and the external offsets were 40,000 yards, which is around 20 miles. On the bombing run you selected the appropriate offset aiming point into the system and started aiming on that. Now, this was not a foolproof method, and the unwary among us had been known to bomb the offset! That normally resulted in you wearing your best blue uniform with your heels together and your hat on in front of the Squadron Commander!

The Nav Plotter (the guy who sits in the middle of the aeroplane) was tasked with checking the actual values that the Nav Radar had set in. Now, the next time you are in a Vulcan please look at the external offset box – the one hidden in the desk – and try and put in something like 18,153 yards. There's only a tiny little scale on the thing and it is very easy to enter the offsets incorrectly.

The Nav Plotter also had the task of making sure that I was not going to drop a bomb on the offset. For instance, if I was saying that the range to the target was 15 miles and his calculations gave it as 18 miles, it was a clue that something had gone wrong. I used to blame him, anyway! When it came to choosing the offsets, operationally you were very limited as to how you would approach the target. For instance, if you were doing a round-robin of the Soviet Union, you tended to come from the west and you wanted to go back home to the east, so you were reluctant to vary your track to come at it from the south or the north. There might well have been one or indeed a whole cluster of Surface to Air Missile (SAM) sites that you wanted to route around as far away as possible – always considered a good idea! The upshot was that we maybe only had a 10 or 15 degree flexibility in how we could change our attack track. For training missions we chose whatever was going to give us the best result.

A large external offset like a power station was ideal. You would not be terribly accurate aiming on the middle of a power station, but hopefully that would get you a lot closer to your release offset, which ideally was nice and small. To give an example, we used to practise bombing Lincoln using a target that was the middle of a little bridge, now in the middle of the shopping precinct. You could not see the target directly on the radar, so we had to use offset aiming points. Somebody had built a nice big power station to the west and to the east, and somebody had built a sugar beet factory at the little village of Bardney, so if you came in from the north-west, tracking to the south-east, you would easily see the power station and, at around 6 to 10 miles, you would start to pick up the return from the Bardney sugar beet factory. Having defined the target very precisely to the radar system on the aeroplane, we needed to tell the pilots up front what we wanted them to do to take us over the target. The same marker system that produced the aiming markers also produced a steering signal, compensated for the wind effect. All the pilots needed to do was fly exactly down this steering signal and we would go right over the target.

The plan range that my dear old friend the triangle solver had worked out would be fed into the bombing computer, which would, in turn, calculate how far away from the target you would have to release the bomb for it to hit. You cannot release the bomb when you are directly over the target because, when you release anything from an aeroplane, at the instant of release it still has the same speed as the aeroplane. Once released it will start to slow down at a rate determined by its shape. The distance between release and impact was known as "forward throw". Before talking about this further, let me briefly return to the steering problem.

The steering signal did not just tell the pilots to go left or right – it actually indicated a bank angle to zero the steering signal. It was really a "feed the monkey" system for the pilots! I could control how many degrees of deflection the steering signal would give the pilots on a little control down by my left knee, but normally that was set to something like 2.4 degrees. Normally at high level we would prefer to use the autopilot, because then you did not even have to wake up the Captain! When the pilots selected bomb function on the autopilot, that then linked my little joystick to the autopilot. The maximum angle of bank it would ever give you was 45 degrees, I think, but the pilots used to chicken out well before it got anywhere near there. It would only control the aircraft in azimuth – you could make it go left or right but not up or down. So the Nav Radar could actually be flying the aeroplane in azimuth using his little joystick, and, when we were totally serviceable, the Captain would select the bomb door opening switch to automatic, and at a pre-determined time interval before release the bomb doors would open up and we would get an automatic weapon release. While this system guaranteed maximum accuracy, there was also some psychology going on, as it saved you personally having to press the release button to nuke Moscow (or the little church that is just off to the north).

Returning to forward throw, there was a computer under the co-pilot's chair called the Calculator (or Calc) 3, and that used to work out how far away from the target you would have to drop the bomb for it to hit. At low level, to stop the Vulcan being blown up by its own bombs, someone had the clever idea of fitting a parachute to the bomb, effectively dragging it behind you quite a way before it went bang on the ground. Very safe, but incredibly boring, because you don't hear the bangs – well, hopefully you don't! I was assured by the armourers that if the parachute did not open, the bomb would not be live. That's a bit like the assurance from the guy that gives you the parachute and tells you to bring it back if it doesn't work.

Each weapon type was loaded into the ballistic computer using a 35mm film strip, and that computer sat on top of the Calculator 3 underneath the co-pilot's seat. To give us great reassurance that it had worked out how far away to drop the bomb, the system displayed the forward throw to us. A smooth 1,000lb bomb would typically fly for about 6 miles if dropped at 40,000 feet at a cruising speed of Mach 0.84 – and that's a fair old way. For low level, where we tended to overfly the target, drop the bomb, big parachute and on our way, a 1,000lb bomb with a parachute has a forward throw of 800 yards – about half a mile – when dropped from 300 feet at an attack speed of around 350 knots. As you can appreciate, there is a fair difference in distance there.

Once you knew the distances, height, speed and forward throw, you could introduce a set time interval at which to open the bomb doors before release, and that was done as late as possible to keep the bomb nice and warm, before you threw it into the minus-56-degree air temperature; you did not want your nuke to freeze up until it absolutely had to. It was normally set at eight seconds on the Vulcan, which was fine operationally, but there was a snag. If you had unknowingly taken some damage from, say, an air-to-air missile, and this had affected the hydraulic system, and therefore the bomb door operation, the first you would know about it would be when you tried to open the bomb doors – and they didn't. You would then have to start up the emergency hydraulic power pack, but that took something like thirty seconds to get the bomb doors open. So, here's the situation: you've got Moscow in your sights – eight seconds to release – and the pilot says that the bomb doors aren't working. Tell you what, why don't we use the hydraulic power pack? He does that, and thirty seconds later the doors open – twenty-two seconds after you wanted them open. Unfortunately you only get one shot at the release signal, so you have to grab the button and release the bombs manually. But you are twenty-two seconds further down the road – not a good idea. In practice, most crews would open the bomb doors manually, probably about 4 or 5 miles from the release point to make sure the damn things were open before it came to the crunch.

Another little problem we had when dropping a stick of conventional bombs – maybe a stick of twenty-one 1,000lb bombs – was that you would always be aiming the middle bomb at the target. This meant that you had to release the first one extremely early, the middle one at exactly the right time and the last one very late. You could actually programme the stick length into the aeroplane, and this length could be maybe a mile. To get your stick length, you could drop the bombs at a different time interval between each bomb, and that was

B.2 XL426 at Waddington in March 1982. In 1984 she was allocated to the Station Flight for the Vulcan Display Flight and is now preserved in taxiing condition by the Vulcan Restoration Trust at London Southend Airport. *Fred Martin*

the Nav Radar's job to work out. This was done on the Weapons Control Panel on the left-hand side. It was called a Ninety Way because there were ninety ways of dropping the bombs. Depending on the bomb type you could set the release interval from 0.24 seconds up to around five seconds, and the time interval gave you the distance between individual bombs. However, during peacetime dropping bombs is normally to be avoided. In fact, the rules said that if you had practice bombs on board you were not even to open the bomb doors over populated territory – unlike the crew I witnessed doing a flypast at a display in Manchester in the early 1970s; they flew past on a wingtip, showing off the lovely delta shape, then someone decided to open the bomb doors to show all the nice people all the practice bombs on board! There was a wonderful photograph in the Manchester Evening News of sixteen practice bombs, all rattling and itching to get out!

The Air Force decided that it would have to introduce an accurate system of no-drop bomb scoring. This resulted in something called the Radar Bomb Scoring Unit (RBSU). Effectively, a very accurate radar picked up the Vulcan coming in on its attack track, followed it all the way through and, at about ten seconds to release, the Nav Radar would switch on a tone on the radio; at the instant of release, the computer switched the tone off. The Nav Radar would then lie through his teeth, telling the operatives on the ground about the Vulcan's airspeed, altitude and other details, and he would also tell them, in a highly secret code, what bomb type was being used. Depending on how the bomb run went, I used to like to change that at the last minute – sometimes it was advantageous but sometimes you'd get it the wrong way round and create an even bigger error. The ground operatives would calculate your track over the ground, your precise distance at the instant of release and your flight parameters, and work out your bomb score. They would then pass the bomb score back to you in code. I always liked the wag on the ground who passed you "M…I…S…S…E…D"! The RAF code was actually eight letters long, and when we used to go out to America, doing the low-level competitions, their code was something like twelve letters long. I think included your inside leg measurements! Unfortunately the score went on the telephone and the fax machine back to your base, so if you missed by quite a bit people knew about it before you got back on the ground. There were RBSUs at RAF Lindholme near Doncaster (now a prison), RAF Coningsby and down in Devon.

For some of the UK bombing competition work, they used to take the RBSU on the road and take it to somewhere where you'd never bombed in your life, nor indeed had any of the experienced guys, so you really had to pick good offsets. I reckon that legions of V-Force Nav Radars paid people like Wimpey to build power stations where they did – we needed those power stations! If the RBSU was not available, we could still score bombs by assessing the photographs that the system took. Right above the Indicator 301 screen was the R88 Radar Camera, which had a wonderful headrest – or was it a lens cover? – on it! Very conducive to checking the inside of your eyelids! Unfortunately, when you went into a bombing run, this thing took a photograph every seven seconds and a little lamp illuminated at the instant of release. From this your bombing run could be reconstructed on the ground and your score calculated.

For low-level attacks we used to use a conventional vertically mounted camera in the visual bomb-aimer's position. In fact, I have to thank one of those for giving me my worst ever bomb score. It was something like twelve miles – we weren't even in the right country. We were supposed to be bombing England, the target was on the border and I let it go around twelve miles to the north! It was heels together on the Squadron Commander's carpet for that one!

Bombing accuracy was always very contentious amongst V-Force people. At high level (40,000 feet and above) the overall accuracy was in the order of 400 yards – not very good. However, when you bear in mind that we were doing it with Meccano and navigation with bits of string and the primary weapon was a big one, it was perfectly acceptable. A lot of system errors contributed to this inaccuracy: incorrect airspeed, incorrect ground speed, and the forward throw calculation could be out by plus or minus 0.2 miles (400 yards). But the main problem really was caused by the Nav Radar not aiming at the right response, which might seem like a damning thing to say but it wasn't that easy, especially if you were a new boy. At low level the accuracy was the same, because we were still using the same system, but you could cheat a little bit by asking the pilots to resolve the track problem. To do this you told them not to follow the bomb steer totally, but to compensate for the wind, putting the target either right or left of the nose, which would allow the bombs to drift on to the target.

The other good thing the pilots could do, after the nose probe was fitted to the Vulcan, was to use it for range marking. A point exactly halfway down the nose probe, as seen from the average pilot's sitting position, would be 800 yards in front of the aircraft. That worked between 300 and 500 feet and speeds of around 350 knots. In fact, the Near East Air Force Wing out in Cyprus from the mid to the late 1970s used to put fluorescent tape round the nose probes. "Just going to calibrate the nose probes," we used to say! It was a remarkably accurate, if unorthodox, system.

As for system problems in the air, an experienced crew never had a system problem in the air because they would always spot it on the ground and throw the aircraft back at the groundcrews, asking for another aircraft. Generally speaking, when we got airborne we were committed to carrying out all the attacks we had planned, and all those attacks would count towards your crew rating at the end of the year. The major prize was Command Rating, and the big advantage of that, other

A unique image of a Vulcan releasing a full load of twenty-one 1,000lb HE free-fall bombs. Although the Vulcan was designed as a nuclear bomber, the aircraft always retained a conventional bombing capability for which crews practiced regularly.

The view from the bomb-aimer's panel, in this case an unusual vertical view of Buckingham Palace during a flypast to celebrate Her Majesty's Birthday. *Joe L'Estrange*

than getting a lovely bright badge for your flying suit and the undying respect of your brother aircrew, was that next year you had to drop fewer bombs to maintain your rating. To achieve something like Command Rating, you had to drop six bombs within 400 yards of the target, which is tricky when you are working with an overall system accuracy of 400 yards! For every bomb dropped outside 400 yards, we would have to drop two inside to bring our average back down – quite a challenge.'

The Vulcan was, of course, a crowd-stopper whenever the aircraft appeared at shows, both in the UK and around the world. Squadron Leader David Thomas, in addition to his regular RAF duties as a Central Flying School Instructor, was one of two pilots qualified to fly XH558, prior to her retirement from RAF service. The following is his own description of display-flying the mighty Vulcan:

'The Ministry of Defence Participation Committee issued a list of venues that XH558 would be released to appear at. The situation was slightly different back in 1992, as this was acknowledged as being the Vulcan's last year, so the Vulcan Display Flight was given some choice as to the show selections. Amazingly, although we obviously wanted to fit in as much as we could that year, we only managed to add another two or three displays than normal as we just ran out of crews. People were just not available away from their normal flying tasks.

So a great deal of time was spent putting together a show programme for the season, making sure the aircraft was fully serviceable and that the crews were available at the right times, then we would liaise with each individual show venue, finding out exactly what they wanted, and what time they wanted us to appear and so on. At Biggin Hill in 1992, for example, we flew out there on the Friday, and they used us as a static exhibit. Then on the Saturday we flew out to do a display at Locking, a flypast at Lyneham, then back to Biggin to do a display and land. That was quite a difficult one, as I had to fly the aircraft in a heavy configuration at Locking, and of course the aim is always to have the aircraft as light as possible during displays, not only because it is then more manoeuvrable and therefore more impressive, but also because she also uses less fatigue … and it also causes less fatigue for the pilots! A weight of 10,000lb makes an incredible performance difference, and we try to display with around 20,000lb of fuel, but at Locking I had 31,000lb of fuel. On the Sunday I launched early to do a short display and land, then, after a quick turn-round, we took off again to fly to Cosford for another display, then on home to Waddington. In that way we flew two sorties, which means more groundwork, but the actual flying element is much easier. It sounds fairly easy, but sometimes it was very difficult to tie three different venues together on time. Some organisers are better than others at doing it. We always had freedom in deciding where we would stop and mount our displays from, but there would always be naturally imposed limitations such as flying hours and fatigue, which was always kept to a minimum.

Vulcan B.2 XH558 in its final form as the last flying example of its type, reconverted to B.2 standard from K.2. Dark Sea Grey/Dark Green wrap-round camouflage overall with 'Tactical' national markings; No 1 Group Panther's Head on fuselage sides, Lincoln Coat of Arms on fin. Black serial; Union Jack, outlined in white, on fin. Note fin RWR housing is Dark Green not black, as usual. *Copyright © Richard J Caruana*

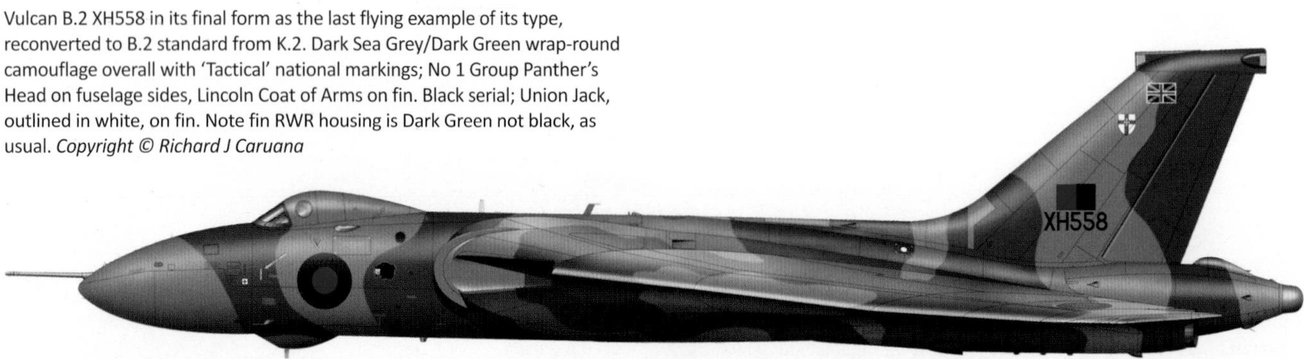

Hastings aircraft were operated by a Flight within 230 OCU at Scampton, tasked with the training of Vulcan rear crews, a role formerly undertaken by the Bomber Command Bombing School at Lindholme near Doncaster. Here XL320 poses for a photographic opportunity in formation with a Hastings T5.

About to touch down at Waddington, XJ824 wears the markings of No 44 Squadron. After retirement early in 1982 the aircraft was flown to the Imperial War Museum at Duxford, where she remains on public display. *Shaun Connor*

XJ824 at the point of touchdown, illustrating the huge inboard elevons and the extended upper wing airbrakes. *Shaun Connor*

Once we got to the venue, we always tried to get maximum exposure, so whenever possible we tried to park close to the crowd, in order to let everyone see the aircraft. However, both Paul Millikin and I always preferred to be removed from any outside interference. People are very well-meaning and interested in the aircraft, asking lots of questions and so on, but about an hour before take-off I liked to get myself under control, calming everything down, thinking about the wind, working out how I'm going to fly the manoeuvres and things like that. So we always aimed to keep the spectators away from the aircraft at least an hour before take-off. That allowed me to get into the right frame of mind, so that I could fly the aircraft to the limit of its performance, but still have spare capacity to see what was going on around me. Obviously the last thing I wanted was to climb into the aircraft all hot and bothered, irritated and angry at someone maybe, or perhaps just distracted by something that someone might have said. It didn't always work like that, though, and it's amazing how things could disrupt the flow. One of the Vulcan's problems was that it didn't have air-conditioning, and sometimes you could be operating in cockpit temperatures of up to 130 degrees, so you got pretty hot. You always took cans of drink along to keep replacing lost fluid, but it was always a problem.

The amount of information we received from the display venue tended to vary quite considerably too. For example, taking Biggin Hill again, they were always very good, providing the material well in advance, creating no hassle. Other show organisers wouldn't come clean until the last minute, and obviously you need to know the time you're wanted on display, how long you've got to display, where the crowd and display

XM605 takes a rest between display appearances at Greenham Common.

lines are, and where the holding points are. Otherwise I expected the show organisers to already know what the display would be like, but things were kept flexible and, for example, where we usually included a "touch-and-go" we would convert this to a low flypast when the runway was either unsuitable or didn't exist. But organisers knew our display sequence pretty well.

However, although we kept ourselves flexible, once we'd agreed a timing I expected them to stick to it. If they came to me and said that we'd be ten minutes late, that would then create a knock-on effect right down the list of displays for that day, making things very difficult. So we insisted they stuck to their word, although in all honesty we always included some timing flexibility, but we kept that as an emergency back-up, rather than allowing show organisers to go crazy. We would receive an instruction "package" from each venue, with things like a drawing of the airfield layout, support services that would be available, who else would be displaying, and a time schedule. We always flew to the clock, even on pre-show arrival days, and we planned to arrive at each venue within a couple of seconds of our scheduled time, purely as a matter of professionalism. We didn't include any diversions, "touch-and-goes", anything like that in our transit flights – the aim was always to go straight to the venue, fly a circuit to look at the airfield traffic pattern, then land. The only exceptions were things such as the occasional flypast, not least at nearby Cranwell during their graduation ceremonies, and we would time our departure to include things like that.

For pilot continuation training we would fly specific dedicated sorties, often going over to Marham to run through a practice display. We had a fourteen-day currency requirement, and if we hadn't displayed within the preceding fourteen days we would have to fly a practice display before flying before a show crowd. We always flew our display practices over Marham, simply because XH558 was actually a part of No 55 Squadron, and we had to fly there in order to be seen by our supervisors. It may seem rather odd that the aircraft was part of a Marham squadron but hangared at Waddington; however, one should bear in mind that Waddington was an established Vulcan base, where lots of technical support remained, and where all the spares were kept. Hangar space was also available.

The aircraft was always flown clean, without any equipment, although our golden rule was never to separate ourselves from our personal baggage, so we carried that with us. Otherwise we transited as we displayed, to the clock, and with a light fuel load, to

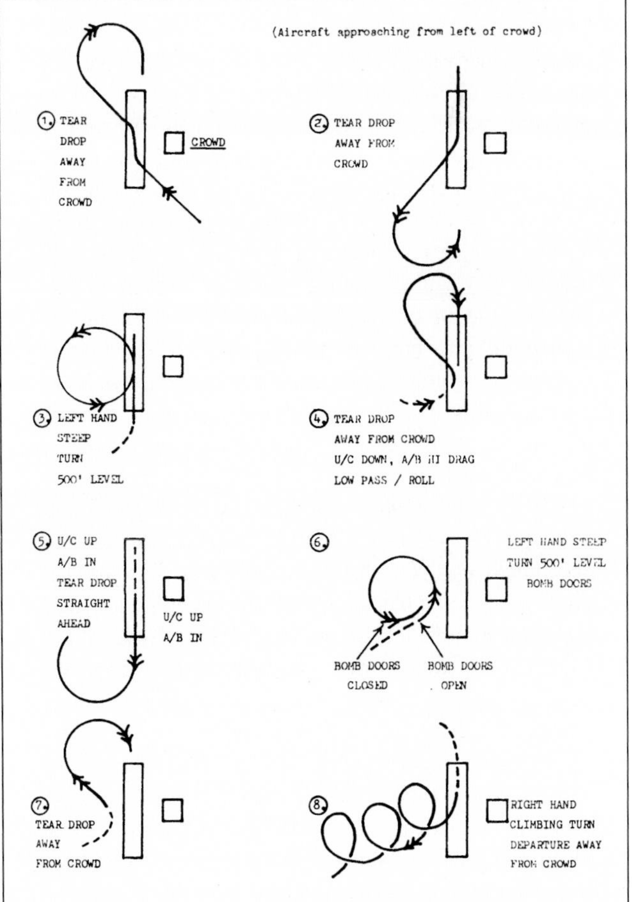

An air display sequence chart as drawn up by a Vulcan display pilot.

minimise flying time and fatigue. We usually carried two members of the groundcrew with us so that there would be someone with us as soon as we arrived at a show site. We would often send a ground party ahead of us by road, but we carried people with us, just in case their transport broke down. We could then drop them off on a taxiway and let them guide us into our parking position. Most people these days are not familiar with Vulcans, and it's not a very manoeuvrable aircraft on the ground. One thing we had to watch was the organiser who stuck you in a slot that you couldn't taxi out of. If you didn't have a towing arm with you, then you had a problem. It happened maybe once a year, and if there was any doubt we would take a towing arm with us by road.

I tended to become a little hardened to the whimpering of some show organisers who expected us to move heaven and earth to accommodate their wishes, but after so many years of display flying we knew where the line was to be drawn and we'd do so much and then no more, simply saying that if we were messed around any more we'd go home. And that had a very sobering effect on people! There were some shows where we came very close to just going home, but in the end the organisers saw sense. It's not a case of us being bloody-minded, it's just that when you have two or three shows to do in one day, you figure that if the organiser can't give you a take-off time within five minutes of a desired time, there must be something wrong with their ability, and you can't disrupt maybe three displays because of a delayed take-off at just one place. Over the years we tended to display at the same venues, so there wasn't much that surprised us. However, you did see some incredible contrasts. For example, when we flew at Mildenhall we had 9,000 feet of concrete and an enormous airfield. But after displaying there we went to Barton in Manchester, where there's about 3,000 feet of grass, and the visual perspective is completely different. Instead of performing well within the boundaries of the airfield, you get a great urge to fly a really tight display, in order to stay within the boundaries of a much smaller airfield, and of course it just doesn't work, and it's something you have to watch carefully. In a more conventional aircraft you could inadvertently stall the aircraft by trying to fly in a tightly confined space, whereas the Vulcan's very good as it just doesn't stall no matter how tight it gets.

All the military shows we flew at were provided free, although the organiser had to provide the cost of accommodation, food, bringing equipment and so on. The civil shows had a blanket charge of something less than £2,000, a nominal fee that obviously didn't pay for much.

It's interesting to note that towards the end of the Vulcan's display career, one show organiser simply stated that they would pay whatever the necessary asking price was, because they just had to have the Vulcan at their show. It's an incredible display aircraft, as it's a unique design, and XH558 was the only one of its kind. It flies very slowly, it's very manoeuvrable, it's big and it's noisy – quite unmistakable.

A great deal of the pre-flight preparation was entrusted to the groundcrew, but we still made a walk-round check, really just looking for the more obvious things. Personally I didn't regard the Vulcan as a lump of metal, but as something closer to an animate object, and each aircraft had its own characteristics. I tended to touch the aircraft, talk to her even, just to get myself together, to become at one with the aircraft. Sure, you look at the tyres and hydraulic lines, but it's more to do with becoming part of the aeroplane. Sounds strange with a 120-ton aeroplane, maybe, but that's how it was. Once inside the aircraft we ran through the whole pre-flight checks right from square one in complete detail. You could rush things, but that's no way to go into a display, and we'd always allow plenty of time. If everything went smoothly you could expect to go from climbing in to getting airborne in maybe thirty-five minutes, but you don't want to erode your margins, and we'd aim to work with about fifty minutes to spare, then we could sit and hold at engine start until the required time. It's an electrical jet, so if the aircraft had been standing out in the damp for any length of time, we'd get silly little electrical failures, especially in the navigation system, but luckily that wasn't really necessary for most displays.

Since the Vulcan was dedicated purely to flying displays, we lost just two, one because of engine intake cracks and one because the undercarriage failed to retract, and with only a minimal amount of fuel it would have been impossible to fly the display with the gear down, so I recovered back to Waddington. Problems were mainly of a minor nature. The vital things such as the flying controls and engines never caused us any problems. It's only when the weather is bad that things like the navigation kit and the bombing system become important. It's amazing that back in the days of regular squadron service you'd have to drag the rear crew into the aircraft kicking and screaming, whereas now you couldn't keep 'em out! Certainly, if the rules allowed, you could have flown displays without the navigators on board, but you had to have the Air Electronics Operator on board as certain switch selections had to be made in the event of an electrical malfunction, in order to safeguard the integrity of the aircraft.

Basically there are two systems to start the engines. One is to use a ground air supply, to pump air into the engine. The other is to use stored air in high-pressure bottles contained inside the aircraft. We generally insisted on having a Palouste air starter at each display site so that we could use the internal rapid-start system as a back-up, rather than our primary means of starting the engines; another instance of not eroding your margins. Additionally you can use one engine to start the others by cross-bleeding air to turn the other three in turn, the normal procedure being to start each engine directly from the Palouste unit. The internal rapid-start system has sufficient pressure to make six individual engine starts, so there was always a great deal of redundant capacity in terms of engine start capability. The starting procedure was indicative of 1960s technology, and unusual by modern standards in that you would deliver air to the engine, let it wind up, then manually control the fuel flow into the engine before igniting it. So you started with a small fire, helping the engine to accelerate, then you'd add more and more fuel until it was self-sustaining. Of course you could start all four engines at the same time with the rapid-start system, and you could prepare the aircraft so that just one button needed to be switched to fire up everything automatically, but that was obviously not something we needed for display flying.

There was more than sufficient thrust to get a good taxi speed with the engines at idle speed, but we would use a bit of thrust to gain a little inertia, then throttle back and use the brakes to keep speed in check. You can use differential braking to steer, but the Vulcan also has a nose-wheel steering system using hydraulics in relation to the position of the rudder pedals. You can barely see the wingtips from the cockpit, and they're a long way behind the main wheels, so when you make a tight turn you get what is called swept wing growth, and the wings seem to swing out much wider than you expect, but we were aware of the situation. At display venues you had to be very conscious of the problem as there was often great pressure to put the aircraft into tight spots.

For take-off we'd run up to 80 per cent power and hold on the brakes, check that the engines were functioning correctly and go to full power as the brakes were released. You checked that the engine acceleration stabilised at a predetermined rpm, which was dependent on air temperature. On a hot day we'd be looking for a minimum of 98.5 per cent rpm. OK, the power isn't critical with so much in reserve, but if I didn't get that figure I'd obviously want to know why, as I might have an engine failure on my hands. In the take-off phase, if you lose an engine you're likely to lose the adjacent one as well, which can be serious if you're flying slowly, or flying a display. Operationally we would practice three-engine take-off procedures, as the aircraft has more than adequate performance with just two engines, and it has a climb-away capability even at a heavy weight on only one engine, but for display flying we always required all four to be functioning perfectly. XH558 has the Olympus 200 series engines, whereas many Vulcans had the more powerful Olympus 300s, and these aircraft featured a combat cruise selector restricting maximum power to the same figure as that attained by the less powerful aircraft like XH558. After some years in service, the aircraft with 300 series engines had the cruise selector permanently wired up in order to conserve engine life, which was degraded at full power, due to resonance inside the engine.

Normally we aimed to take off into a holding position, to give us time to settle down, but we often went straight into the display sequence. The rotation speed was usually 135 knots, and that remained the same until the aircraft weight rose quite significantly. With no wind the take-off run was about 2,000 feet. The display datum speed was around 155 knots, and even the tight turns were entered at that slow speed. You were able to manoeuvre fully at slow speed, and the aircraft always appeared to be close to the crowd because of its size, so the display was always impressive. Because the aircraft was essentially built to fly long straight lines, it has a very heavy feel when you're flying a display sequence, and we tended to use full control deflections. For example, in a steep turn we'd have full up elevator, and that gets very tiring. An eight-minute display is like twenty-four hours of normal work. It's physically demanding to fly a display in the Vulcan, and despite its fighter-type control stick, we tended to use both hands in a steep turn. Another point to consider is the need to balance the adverse aileron yaw that is created by the Vulcan. If you roll quickly, one set of ailerons goes up, the others go down, and you start to yaw. So before you move the ailerons you must move the rudder to counter the yaw.

There's no finesse involved in Vulcan display flying. The only reason we used so much bank at the top of the turns was to bring the nose down more easily. The climb angle was very steep, and trying to bring the nose down with wings level would create a significant bunt, and the Vulcan isn't designed to fly negative-g manoeuvres, so we rolled the Vulcan onto its side and allowed the nose to fall through.

The Vulcan's negative-g limit was zero, the positive limit being just two, so the secret of how we made the display look so impressive was to fly slowly. Modern aircraft can't perform well when they're close to their stalling speed whereas it was no problem for the Vulcan. At the top of a tight turn the speed was even lower than the datum display speed. All the manoeuvring was done in a heavy buffet, which in a conventional aircraft would mean a deep stall, real trouble, but in the Vulcan we didn't have that problem.

Basically you cannot stall the Vulcan's wing. If you look at the stalling angle, the point where lift starts to decrease, it's 48.5 degrees angle of attack, but by that time the aircraft would have so much drag that it would be difficult to recover. What you actually get is a great deal of airflow separation over the top of the wing, so when we're manoeuvring we're sitting in very heavy buffeting but at a low speed, flying a big, flat plate through the air. The only stall warning as such would be a low speed indication and a very high angle of attack, but with a unique aircraft like the Vulcan you could get out of that situation easily by just unloading the wing. However, there is so much thrust that you could quickly get into a situation where you could overstress the airframe, so it literally requires just a couple of seconds at most to just unload the wing, get the speed back, then heave back into the deep buffet, because we control the aircraft speed by flying in the heavy drag area. That's why the aircraft display was so noisy, as we were just sat there with the engines roaring away, maintaining a minimum radius turn, using drag to control speed.

Our minimum manoeuvre height was 500 feet, the level flypast 300, and for a "touch-and-go" we'd hope it was zero feet! The top of the turns was around 1,600 feet, sometimes a little lower. As you can imagine, there was quite a level of interest from the guys in the back, and the AEO called out heights, while one of the navigators called out speeds, especially if it went significantly high or low. It was very much a team effort, as when things start to go wrong they go wrong pretty quickly, so a nudge from the rear crew about our speed or height was always very useful. As for other things going wrong during the display, we had very few problems. Once at a display over Barton we had an aircraft call short finals with an engine failure while I was flying a display, but I was able to reposition and continue the display without creating a hazard to either his safety or mine. It was always our policy to have no other aircraft flying over the airfield while we were displaying. You'd think that was pretty sensible, but many display organisers like to maximise their earning potential by having pleasure flights taking place during displays –you'd see them flying as you were running in, which is very dangerous.

Touchdown speed was around 130 knots, and with the high nose angle you couldn't actually see the touchdown point, so you were looking some way ahead of you down the runway and you used your vision either side to judge your position. Close to the runway the Vulcan sat on a cushion of air, so with the nose high it was very difficult to have a bad landing. You allowed the aircraft to simply sink gently. If it was slightly fast it would balloon quite significantly. With the whole trailing edge acting as a flap, pulling the stick back would reverse the flap, causing the aircraft to smack onto the ground, so to stop an impact you pushed the nose down, although that would provide only a very temporary respite, because the aircraft pitches quite rapidly. So the easy way out of a difficult landing was to hold a high-nose attitude, put on power and go around again. Going down the runway with the nose held high, the aircraft remains quite responsive, but you reach a speed where you no longer have any control. It will sit there quite happily, but we tended to lower the nose at around 70 knots. With short runways where we used the brake parachute, the nosewheel had to be on the ground in order to steer the aircraft.

The final retirement of XH558 really did mark the passing of an era. It was tremendously sad for me and the rest of the Vulcan's crew. In the military you have to get used to losing loved ones sometimes, and the Vulcan was yet another loved one that we simply had to say goodbye to.'

A Vulcan captain's view as the aircraft streaks over the A15 on to a wintry runway at Waddington. *Joe L'Estrange*

A Vulcan captain's view of another Vulcan during an AAR training sortie. *Joe L'Estrange*

CHAPTER NINE

Into Battle and Beyond

EARLY in April 1982, when the prospect of war with Argentina seemed likely, nobody imagined that the Vulcan would play a key part in Operation 'Corporate', the recapture of the Falkland Islands. Only three Vulcan squadrons remained operational, and the type's complete retirement from RAF service in June seemed to be certain. However, after looking at a variety of military options involving the use of air power, the concept of using Vulcans did not seem quite such a ludicrous idea. There was every likelihood that RAF Harriers (operating from aircraft carriers) could support Navy Sea Harrier strikes on the islands, but in terms of offensive capability it seemed to be the most that the RAF could offer. The huge distances from the nearest operating base (Ascension Island) meant only carrier-borne aircraft stood any chance of reaching the Falklands; both the Jaguar and Buccaneer would have been unable to remain airborne long enough, even with the aid of air-to-air refuelling (AAR). The new Tornado GR1 (which was just entering service with the RAF at the time) was certainly better suited to an ultra-long-range mission, but as the aircraft was still very much a 'new toy' there was no way that it could be pressed into combat action at such an early stage in its career. Only a bomber with 'long legs' could conduct a land-based strike against the Falklands, and with the aid of AAR it looked as if the Vulcan could just do the job. Ironically, the aircraft was in the process of being withdrawn from RAF service at the very moment when it was needed. Scampton had already lost its aircraft and the wind-down at Waddington was already under way, but there was still a sizeable Vulcan fleet at Waddington from which a pool of suitable aircraft (and crews) could be drawn.

On 5 April groundcrews at Waddington were instructed to begin restoration of air-to-air refuelling capability in ten selected Vulcans, to enable aircrew to begin training in refuelling procedures as soon as possible. Although the Vulcans had

A Vulcan B.2 (MRR), complete with underwing air sampling pods, leads a formation of Tornado GR.1s across Essex towards Buckingham Palace to celebrate Her Majesty's Birthday in 1982, the year in which the Tornado finally began to assume the Vulcan's strike/attack role in RAF service. *Both Joe L'Estrange collection*

retained their bolt-on refuelling probes since they were first fitted during the early 1960's, the fuel system had not been used since the aircraft had switched from strategic strike to tactical bombing operations, and the internal fuel transfer system had either been de-activated or, in the case of many aircraft, the seals and valves had deteriorated to such an extent that the system could no longer be used. To add to the mechanical engineering team's problems, most refuelling probes had recently been removed for attachment to Nimrods and Hercules (which were already assigned to Falklands duties) and personnel had to search far and wide (including the USA and Canada) to locate and obtain sufficient numbers of usable items. A fortuitous discovery was also made at RAF Stafford, where a batch of 'new' refuelling non-return valves was found still in storage.

From the ten aircraft that had their refuelling systems restored, five machines were selected for conversion to full conventional bombing operational standard: XL391, XM597, XM598, XM607 and XM612. They were the only Vulcans remaining in service that had full fore and aft Skybolt attachments, and the associated refrigeration ducting that would be required for the new cabling necessary for the carriage of external stores. Under the code-name Operation 'Black Buck', these Vulcans would be prepared for attack missions over the Falkland Islands, carrying twenty-one 1,000lb HE bombs over the staggering distance of 3,900 miles from the nearest operating base at Ascension Island.

XL391, one of the aircraft that was modified for 'Black Buck' operations, was retired to Blackpool Airport in February 1983. Unfortunately the aircraft was largely abandoned on public display outside the airport and a combination of weather conditions and vandalism eventually reduced it to little more than scrap. Ultimately it was cut up and removed in 2006. *Shaun Connor*

XM597, another of the Falklands aircraft, touches down with elevons extended in the 'up' position. The 'Black Buck' mission markings are just visible on the nose section.

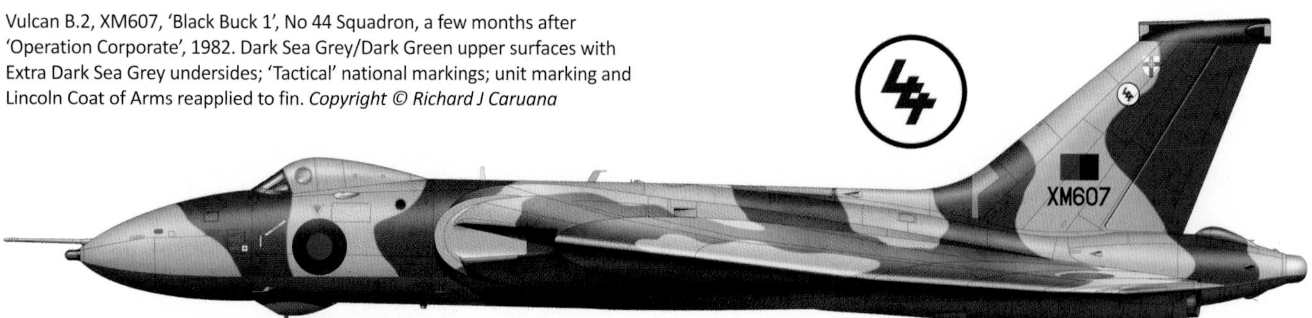

Vulcan B.2, XM607, 'Black Buck 1', No 44 Squadron, a few months after 'Operation Corporate', 1982. Dark Sea Grey/Dark Green upper surfaces with Extra Dark Sea Grey undersides; 'Tactical' national markings; unit marking and Lincoln Coat of Arms reapplied to fin. *Copyright © Richard J Caruana*

Preparing the aircraft for a conventional bomb load mission was a fairly simple task, for although the Vulcans had been operating primarily in the tactical nuclear role for more than ten years, they retained a conventional capability that crews practised quite regularly, and only a change in bomb carriers and cockpit control panels was required. A Delco inertial navigation system was also installed. This was purchased 'off the shelf' from British Airways, which used the same INS in its Boeing 747s, and also taken from former British Airways Super VC10s, which were in storage at RAF Abingdon, awaiting conversion to tanker standard.

The first trial INS fit was made at Marham, where the resident Victor fleet had already been equipped with Carousel and Omega, but although the idea of fitting Omega in the Vulcans was considered, the modification was not made. Five crews were selected (although only four were eventually fully trained) for 'Black Buck' duties, two from No 50 Squadron, one from No 44 Squadron, one from No 101 Squadron, and another from the recently disbanded No 9 Squadron. Some of the crews had recent Red Flag exercise experience, which it was felt would be of value when faced with a real combat situation.

Aerial refuelling training sorties began on 14 April – catching up on thirteen years of inexperience in thirteen days. It quickly became apparent that the Vulcan' s refuelling probe was prone to fuel spillage and leakage, and various short-term fixes were improvised to prevent fuel flowing back over the windscreen, almost totally obscuring the pilot's forward view. No perfect solution was found, and two rows of flat plates were eventually attached to the nose behind the refuelling probe in an effort to direct the stream of fuel away from the canopy. On the evenings of 16 and 17 April the crews flew night refuelling sorties with Victor tankers, and even the Victor crews had to familiarise themselves with the skills necessary to refuel from other Victors at night, as until then there had been no requirement for such operations for many years.

Left: Although a rare event (indeed, the conventional role had been effectively dropped by 1980), the Vulcan was capable of delivering a very respectable load of twenty-one 1,000lb HE bombs, and of course this capability was ably demonstrated during Operation 'Corporate'.

Below: The 'Black Buck' mission markings applied to XM607. Following a post-retirement repaint the markings were re-applied but not in their original form.

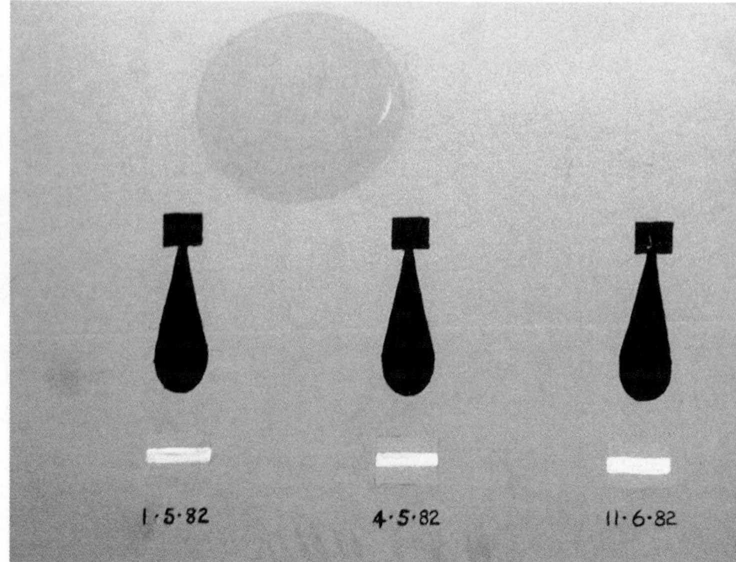

A magnificent plan view of XM607 performing at Greenham Common in the experienced hands of Squadron Leader Joe L'Estrange. *Michael J. Freer*

The Vulcans selected for 'Black Buck' were all equipped with Olympus 301 engines, and in an effort to extend their service life the throttle mechanisms had been fitted with a restriction device during RAF service, which had effectively given the engines a maximum thrust of 18,000lb and saved on fuel burn and engine fatigue caused by intake resonance. These engine inhibitors were removed and the engines were restored to their original full power status, which – as far as the regular Vulcan crews were concerned – effectively gave the engines 103 per cent power.

Because the Argentine Air Force was known to be operating a Boeing 707 in the airborne early warning role over the South Atlantic, and because ground-to-air radar had been positioned in the Falklands, it was also decided to equip the Vulcans with additional ECM protection, and the only readily available equipment was the Westinghouse AN/ALQ-101 pod normally carried by Buccaneers. Fortunately Avro's Skybolt modifications now proved to be invaluable, as the wing hard points provided a suitable location to attach the ECM pod. More good fortune revealed the presence of some mild steel girders at Waddington, which had been mistakenly ordered some time previously, and when suitable sections were welded together a near-perfect weapons pylon was produced. The

The rather bleak view of RAF Waddington as seen from the Vulcan's cockpit as a long approach is made from the north over the A15. During AAR operations fuel flowing back over the windscreen almost totally obscured the pilot's forward view. *Joe L'Estrange*

Few photographs exist showing the various modifications that were made to the noses of the 'Black Buck' Vulcans in order to solve the problem of fuel spillage obscuring the pilot's vision during AAR. This picture (believed to be of XM597) shows two rows of deflector plates forward of the windscreen

Below: 'Black Buck One' refuelling plan diagram.

VULCAN REFUELLING PLAN

WAVE 3 (RECOVERY)
WHITE FORMATION
White 3 White 4 White 2 White 1
Airborne Reserve

WAVE 1 (OUTBOUND)
RED FORMATION
Red 5 Red 6 Red 2 Red 1 Red 4 Red 3
Primary Vulcan

WAVE 2 (OUTBOUND)
BLUE FORMATION
Blue 2 Blue 1 Blue 4 Blue 3 Blue 6 Blue 5 Blue 7
Airborne Reserve

Key
Victor Vulcan
From Stanley
To Stanley

REFUELLING PLAN FOR BLACK BUCK 1

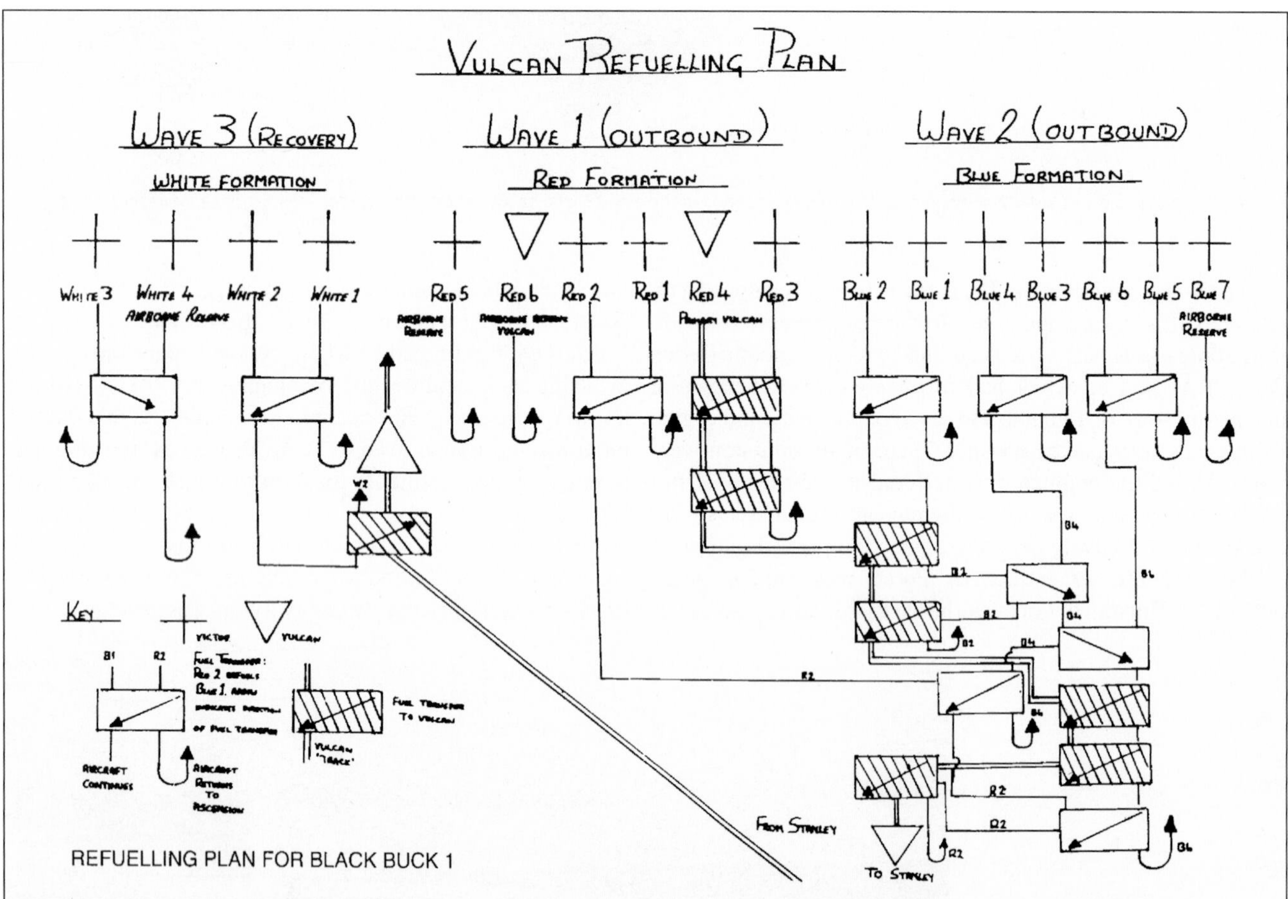

electric cables necessary for operating the pod were run through the Skybolt's coolant pipes and connected to a control panel fitted in the Air Electronics Officer's crew position. XM645 had also originally been selected for 'Black Buck' modifications, but because it was a late production machine, manufactured after Skybolt had been abandoned, it was consequently devoid of the appropriate hard points and ducting, so no further modifications were made to this aircraft. The 'Black Buck' Vulcans retained their normal service camouflage, although all squadron markings were removed and the undersides were hastily painted Dark Sea Grey. With no sign of a likely political solution to the Falklands crisis, the crews continued intensive training, flying night missions down to 200 feet AGL – much lower than ever before.

The first two Vulcans (XM607 and XM598) left Waddington for Ascension Island at 0900Z on 29 April, with XM597 taking off as a reserve but later returning to base. Supported by Victor tankers, the two Vulcans made their 4,100-mile journey to Wideawake Airfield on Ascension, arriving at 1800Z.

The first mission ('Black Buck One') was to be an attack on the airfield at Port Stanley, which the Argentine Air Force was expected to use as a forward base for Mirage, Skyhawk and Super Etendard fighters. Only one Vulcan would be used, as the tanker support required to support just one Vulcan on such a lengthy mission was phenomenal and the RAF simply didn't have sufficient Victors or crews to support anything more than a solo mission; in any event, Wideawake Airfield's ramp was not large enough to handle any more aircraft and was already filled to capacity.

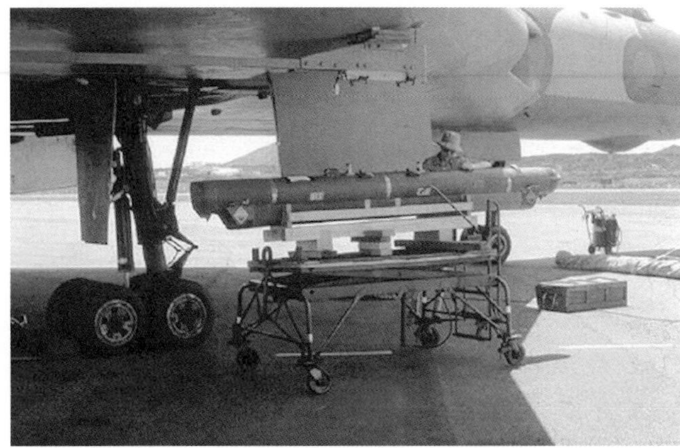

An AN/ALQ-101 pod being attached to XM607 on Ascension Island. As can be seen, the attachment trolley was suitably improvised.

Four 'Black Buck' Vulcans making a 'farewell tour' over locations associated with the aircraft, shortly before the disbandment of No 44 Squadron. Note that the Dark Grey paint applied to the undersides of the aircraft has also been applied around the wing leading edge on to the upper surfaces of XM612.

Groundcrew busy sorting supplies at Wideawake Airfield as XM612 departs on a sortie.

Shortly before midnight on 30 April, the two Vulcans (XM598 with XM607 as a spare) taxied to the threshold of Wideawake's runway accompanied by four Victors (including one spare). At one-minute intervals, in radio silence and with all navigation lights switched off, the fearsome and magnificent gaggle of V-Bombers then thundered into the air. Each Vulcan carried a bomb load of twenty-one 1,000lb bombs (with an overload take-off weight of 210,000lb compared with the normal maximum of 204,000lb), while the Victors were heavily laden with transfer fuel required both for each Victor and the Vulcan. Shortly after their departure, a second wave of seven Victors (one a reserve) also rumbled skywards, and the thirteen-aircraft formation began to head south, climbing to 27,000 feet.

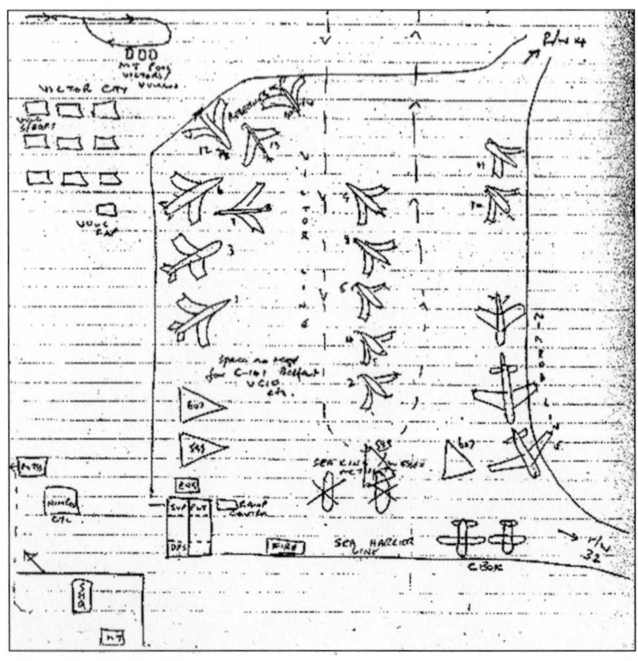

The crew of the primary Vulcan (XM598) encountered continual problems with the aircraft's port direct-vision window, which refused to seal properly, thus preventing the crew cabin from pressurising the cabin. Consequently XM607, captained by Flight Lieutenant Martin Withers, became the primary aircraft for the mission, and as XM598 returned to Wideawake the formation climbed at 260 knots to 30,000 feet, a compromise altitude between the optimum cruising height for each aircraft type. The plan was for the Vulcan to make five refuelling contacts ('prods'), but six were actually made as the Vulcan's unusually high operating weight, together with the additional drag of the ECM pod, required even more fuel than envisaged.

The long haul south was maintained at heights varying between 27,000 and 32,000 feet, severe thunderstorms and turbulence being encountered along the way, making the task of formating and refuelling even more difficult and hazardous than it already was. Most of the supporting Victors gradually

A fascinating sketch drawn up by RAF personnel on Ascension Island during 'Black Buck' operations, illustrating the difficulty in fitting a huge number of aircraft into a very small hardstanding area.

peeled off and headed north, eventually leaving just two Victors and XM607 to continue south to the Falklands. Although the Vulcan crew were obviously unaware of it at the time, the Victor tasked with the Vulcan's final pre-attack refuelling 'prod' flew into severe turbulence while refuelling from the second Victor (XL189), causing contact to be broken. This made it necessary for the receiver to head back to base, leaving XL189 to refuel XM607. Squadron Leader Bob Tuxford, captain of the Victor, was forced to wave off the Vulcan (using signal lights) before XM607's tanks were completely topped up, otherwise there would have been no way in which the Victor could have returned to Ascension safely.

At 300 miles from the target, the stage of the attack began and XM607 descended to approximately 250 feet for an under-the radar approach to the target, the co-pilot re-taking his cockpit seat from an AARI (Air-to Air Refuelling Instructor), who had sat there temporarily while the refuelling 'prods' were being made. At forty miles from the target, Flight Lieutenant Withers hauled the Vulcan back up to 10,000 feet (the necessary 'pop-up' manoeuvre) and turned onto a heading of 235 degrees, directly towards Port Stanley's airfield. Some initial difficulty was encountered in establishing a radar picture, as the H2S had been switched off for most of the flight (the nose probe fuel pipe runs through the radar bay and the mixture of electrical activity and fuel was not judged to be a good combination), but a good picture of the Falklands was eventually secured.

At least one Argentine radar illuminated the Vulcan en route to the target, but it fell silent when the AN/ALQ-101 pod was activated. The Vulcan crossed the airfield runway diagonally, ensuring that at least one bomb stood a good chance of hitting the tarmac runway, with a long 'stick' of bombs straddling the target. In just five seconds (although the crew later said that it seemed like an eternity) the twenty-one bombs dropped from the bomb bay, and a few seconds later the airfield erupted into a brief blaze of light. A few of the bombs were fitted with thirty- and sixty-second delay fuses, adding to the confusion of the occupying Argentine forces, who had been enjoying a peaceful night until XM607's appearance. At 0746Z the codeword 'Superfuse' was relayed to base, signifying the success of the attack (a 1982 equivalent of the Dambusters' famous 'Nigger' codeword). This finally enabled Squadron Leader Tuxford in Victor XL189 to radio for another Victor to be scrambled from Ascension, so that his aircraft would not run out of fuel some 400 miles from Wideawake.

Turning back northwards, the plan had called for an initial outbound height of 300 feet, but to conserve fuel XM607 began to climb towards the first refuelling rendezvous, which was eagerly awaited, the Vulcan already being some 8,000lb short on fuel. As the time for the Victor rendezvous came and went, the Vulcan crew began to worry, but thirty minutes later contact was made, thanks to the assistance of a Nimrod crew in the vicinity who directed the two aircraft to each other. With roughly one hour's-worth of fuel remaining, the refuelling was complicated by a considerable amount of leakage, making forward vision difficult. This prompted the Vulcan's Nav Radar officer to stand halfway up the pilot's access steps, to peer through a small portion of the lower windscreen that remained clear, in order to co-ordinate the task.

XM607 being armed with a full load of twenty-one 1,000lb HE bombs on Ascension Island.

The rest of the flight back to Ascension was long and tedious but uneventful, and XM607 touched down at Wideawake at 1452Z to complete what (at the time) was the longest bombing mission ever flown by any aircraft, and still remains as the longest bombing mission in the history of British military aviation. The attack was successful; the oblique approach to the airfield had virtually ensured that one bomb would fall on the runway, and indeed one produced a 115-foot crater, 84 feet deep, on the runway, almost mid-way along its length. Additional damage was also caused to other parts of the airfield, including the destruction of a hangar – pretty impressive for a crew that had been accustomed to the task of delivering nuclear bombs on to rather larger targets.

The Argentinians were denied the unrestricted use of the airfield and only aircraft types such as the Hercules were able to use the strip for the rest of the conflict (and one of these flights almost ended in disaster because of the damage caused to the runway). Although the huge crater was repaired, the Argentinian forces failed to prevent the area from regularly subsiding, and the attack had permanently halved the useable length of runway available. It had also produced an obvious deterrent effect, which demonstrated to Argentina that even her mainland bases were not invulnerable and prompted the allocation of fighter resources away from the Falklands, in

An aerial reconnaissance photo of Port Stanley's runway shortly after the first 'Black Buck' attack. The image clearly shows the almost perfectly positioned crater that effectively ended any possibility of the runway being used by Argentina's fighters and fighter-bombers.

defence of these airfields, thus providing a much greater degree of local air superiority for British Harriers and Sea Harriers. Of course, the practicalities of actually attacking Argentina directly were never actively considered by the RAF as the task would have been too ambitious to even contemplate. But the seeds of doubt had been sown, and the Vulcan raid achieved everything that had been required of it.

A second mission ('Black Buck Two') was flown on 3 May with the same aircraft, effectively making a repeat performance of the first sortie. However, the attack altitude was raised to 16,000 feet, and sadly no bombs hit the runway, even though significant damage was caused to surrounding parts of the airfield. A third mission ('Black Buck Three') was scheduled for 16 May, but forecast winds would have reduced fuel reserves beyond an acceptable limit and the mission was cancelled before take-off. In the meantime, XM607 had returned briefly to Waddington, where the port pylon was removed (the crews thought, erroneously, that it would cause excessive drag), and XM598 had been replaced by XM612.

Perhaps the final word on these bombing missions should go to Flight Lieutenant Withers:

> 'Thanks to the massive team effort involving twenty aircraft and about 200,000 gallons of fuel, we managed to put a bomb on the runway, which was the aim of the exercise. We thus denied its use to high-performance aircraft and showed the Argentinians that we had the capability to attack their mainland – a threat that they certainly took seriously, because many of their Mirages which had been deployed south were recalled to defend their own bases. This must have considerably helped the Harriers to attain air superiority.'

Following the attacks on Port Stanley's airfield, attention now turned to Argentine radars, in particular a Westinghouse AN/TPS-43F and a Cardion TPS-44. Skyguard and Super Fledermaus were also in use, and back at Waddington plans were being made to equip the Vulcan for anti-radar attacks. The AS.37 Martel anti-radar ASM was selected as a suitable weapon, as the RAF already had these missiles in good supply for the Buccaneer squadrons. With roughly eight times as many wires to attach than were needed for the ECM pod, the engineering crews at Waddington worked long and hard until, on 4 May, XM597 carried the missile into the air for the first time under the port wing hard point, with an ECM pod attached to the starboard pylon. A live firing was made over the Aberporth range the next day, after a cold-soak at altitude. Although the test firing was successful, there was some concern over the reliability of the missile after a long flight south at high altitude. However, the US Government stepped in and offered to supply the AGM-45A Shrike weapon after the Secretary of State (Alexander Haig) had ended his famous 'even-handed' mediation between Britain and Argentina. The Vulcan's pylons were reconfigured to accommodate one Shrike under each wing and more trial flights began; the pylons were subsequently adapted to carry twin-missile launchers on each.

An oblique view of the damaged runway. Apart from rendering the runway useless for fast jet operations, the long-range nature of the bombing missions encouraged Argentina to believe that its mainland could also be under threat of attack, and this led to the withdrawal of hostile fighter assets from action over the Falklands.

XM612, a Falklands veteran and now preserved, spent most of her service life with the Waddington Wing. The City of Lincoln coat of arms was applied to the tail of virtually every Vulcan that operated from the base, although similar markings were never applied to Scampton's aircraft. *Michael J. Freer*

On 26 May XM596 flew to Ascension, followed a day later by XM597, which was fitted with missiles at Wideawake. The first operational mission ('Black Buck Five') was launched on the night of 28 May, but the attack was aborted five hours into the mission when one of the supporting Victor's HDUs (Hose Drum Units) failed. 'Black Buck Five' began again the following evening and everything went well, with XM597 descending to 300 feet when approximately 200 miles from the Falklands. At twenty miles the aircraft climbed to 16,000 feet and quickly picked up signals from the primary target, the AN/TPS-43F radar, although the unit was quickly switched off. It was later learned that the Argentine operators were attempting to make the radar's signal strength weaker than normal in order to fool the Vulcan crew into thinking that the radar was further away than it really was, thus luring the Vulcan within the range of anti-aircraft guns. However, forty minutes later the radar was re-acquired and two Shrikes were launched at 0845Z on 31 May. One of the missiles caused damage to the radar, but it was quickly repaired and a revetment was constructed to protect it from further damage.

'Black Buck Six' was launched on the evening of 2 June, this time with four Shrikes, and the same radar unit was again identified during the run-in to the Falklands. After spending forty minutes in the area waiting for that, or any other radar, to illuminate the Vulcan, a dummy run was made towards Port Stanley's airfield and a Skyguard radar was quickly identified. Two Shrikes were launched and made a direct hit; four Argentine soldiers were killed. As the Vulcan's fuel state deteriorated, XM597 began the long journey north, making a rendezvous with a Victor with assistance from a Nimrod crew.

The Vulcan's refuelling was not without incident, and during the 'prod' the aircraft's probe broke off at the tip, leaving XM597 with insufficient fuel to reach Ascension. The only alternative for the crew was to divert to the nearest airfield, which was Rio de Janeiro in Brazil. Even so, there was little chance of reaching the Brazilian coast if the Vulcan remained at the refuelling height of 20,000 feet, so a climb to 40,000 was initiated to conserve the remaining fuel. The two remaining Shrikes were fired off, but one 'hung up'; there was no other way to jettison the missile and it remained attached to its pylon. Various sensitive documents were collected and loaded into a hold-all bag, before being dumped through the crew entrance. In order to open the door at 40,000 feet, the cockpit was first de-pressurised while the crew breathed oxygen. At a distance of around 200 miles, a 'Mayday' call was made to the airport at Rio. Contact with Brazilian ATC was difficult at first because XM597's crew attempted to keep their identity concealed for as long as possible, while diplomatic

XM596 and a gaggle of No 44 Squadron machines pictured during 1982. The dispersal area is now the home to Waddington's E-3Ds.

50 Sqn B.2 XM597 on dispersal at Finningley in September 1976. *Fred Martin*

staff contacted embassy officials in Rio. However, an initial descent to 20,000 feet was eventually cleared and, with the runway just six miles away by this stage, a straight-in approach was initiated, involving a dramatic steep spiral descent with airbrakes out and throttles closed. Thanks to some careful flying by the captain, Squadron Leader Neil McDougall, XM597 straightened out at 800 feet, just one and a half miles from touchdown (at twice the normal speed) and then made a gentle touchdown with only 2,000lb of fuel remaining – less than that necessary for just one airfield circuit.

After landing, the remaining Shrike was made safe before it was impounded by Brazilian authorities. The Vulcan crew were well treated and, although they were soon allowed to leave, they elected to stay with XM597 until she too was cleared to leave, minus the Shrike. On 10 June the aircraft returned to Ascension, and on the 13th flew home to Waddington, complete with a new refuelling probe.

'Black Buck Seven' took off from Ascension on the evening of 11 June, with Vulcan XM607 flying a bombing mission against airfield facilities at Port Stanley; as the British forces anticipated capturing the airfield, there was no desire to cause further damage to the runway. Apart from an engine flame-out that required three relight attempts, the mission was completely successful, and a full load of 1,000lb HE and anti-personnel air-burst bombs were delivered over the airfield. Four days later the Argentine forces on the Falkland Islands surrendered, and no further Vulcan missions were undertaken.

While the 'Black Buck' sorties were being made, other weapons options for the Vulcan were considered, including the use of Sidewinder AAMs and laser-guided bombs; indeed, LGB trial flights were made in the UK. However, none of these options was used operationally. After the Vulcans had returned to Waddington, conventional bombing training flights continued, together with some anti-radar defence suppression sorties – XM597 and XM607, among others, were noted carrying Shrike training rounds – until No 44 Squadron finally disbanded on 21 December 1982, finally marking the very end of Vulcan bomber operations.

The Martel anti-radar missile was held in relatively large stocks by the RAF (for the Buccaneer squadrons) and was considered a suitable weapon for use against Argentine radar sites on the Falklands. However, the missile's reliability was in some doubt and an offer from the US to supply Shrikes prompted the RAF to choose the American option.

Above: A Martel missile installation pictured during flight trials at Waddington in May 1982.

Left: The twin-pylon attachment for two Shrike missiles as fitted to XM597. Following the end of Operation 'Corporate', some former 'Black Buck' aircraft (including XM597 and XM607) regularly flew training missions from Waddington carrying the missile. *Mike Jenvey*

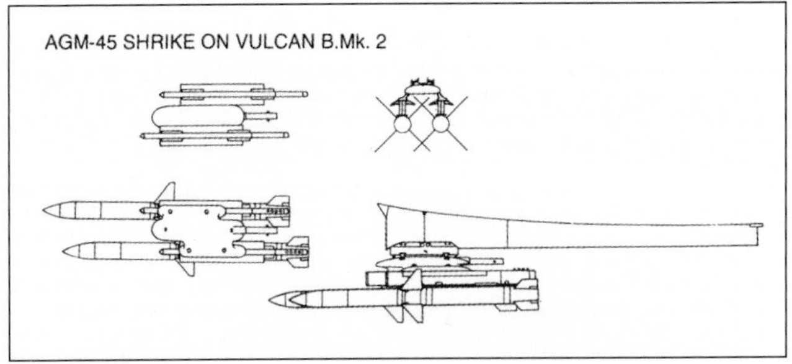

Right: A Shrike missile prepared for an operational launch from XM597 during Operation 'Corporate'. As can be seen, operational missiles appear to have been hastily camouflaged with spray paint.

Below: 9 Sqn B.2 XM597 touches down at Waddington in June 1979. *Fred Martin*

But even this event was not to be the end of the RAF's long association with the Vulcan, as No 50 Squadron remained in business, operating Vulcans as AAR tankers. As a direct result of the Falklands conflict, the RAF suddenly had a huge requirement for air-to-air refuelling tankers, to support the seemingly endless supply flights to and from the Falklands (and Ascension Island), while still maintaining day-to-day activities back in the UK, supporting fighter and offensive support squadrons.

The Victors were heavily committed to the AAR task, and although the conversion of VC10s into tankers was now under way, there was still a short-term requirement for even more tankers. While the USAF supported RAF operations in the UK with Boeing KC-135 tankers, it was decided to convert a number of Hercules and Vulcan aircraft into single-point refuelling tankers. The initial proposal, sent to British Aerospace at Woodford, was to install a Hose Drum Unit (HDU) in the aft section of the Vulcan's bomb bay, the Vulcan then being designated B(K) 2. But this idea was dropped, mainly because of the resulting proximity of the receiver to the tanker aircraft. It was felt that, for safety, the HDU should be placed as far aft as possible, to provide adequate separation between the tanker and receiver. The ECM bay was identified as being a suitable location (the internal equipment being unnecessary for tanker operations), and this would also allow an additional fuel tank to be installed in the bomb bay. XM603, which had been delivered to Woodford after retirement for static display, was used as a mock-up platform and, as J. J. Sherratt, BAe's Assistant Chief Designer for Victor tanker systems, explains, the task of fitting the HDU was far from simple:

> 'Sunday morning saw a group of us standing around a crated HDU, thinking that if it was anything like the size of the crate, we wouldn't stand much of a chance of fitting it. Even with the crate removed it looked big, but by this time we had resolved to get it into the Vulcan even if it meant restyling the back of the aircraft. There was no way of straight-lifting the 51/2-foot-wide HDU through the existing ECM opening of 4 feet, but we noticed that the top part of the HDU could be separated from the bottom, and we might be able to get the top half through the opening, leaving the bottom half to be straight-lifted in. A piece of wood the same size as the top section was called for to investigate the possibility. The verdict was that there was plenty of room if we had a good shoehorn, and if necessary we could put the odd blister here and there to cover any awkward bits.
>
> After a design team meeting in the afternoon, we agreed to tell the MoD that we could do the job, and in the general euphoria a target of three weeks to first flight was set. On Monday two representatives from Flight Refuelling arrived to advise us on splitting the HDU, and at around midday we received authority from MoD to proceed with the conversion of six aircraft. The first aircraft, XH561, arrived on Tuesday, by which time a whole army of workers had been mobilised to work all the hours it needed to do the job. Seven weeks later to the day, on Friday 18 June, the first converted aircraft made its first flight at 12.32pm. An interim CA release was granted on 23 June, and the first aircraft was delivered to the RAF on the very same day.'

Sadly, the Vulcan K.2 did not enjoy a particularly long career with the RAF, as although the Vulcan proved to be an excellent tanker aircraft, the HDUs had been out of production for some time, and the Mk 17 HDUs fitted to the Vulcans had already been allocated to the VC10s, which were being converted to tankers. As the VC10s were completed, the HDUs were removed from the Vulcans, starting with XJ825 on 4 May 1983.

Three Waddington Wing Vulcans en route to London to take part in the official 1982 Falklands Victory celebrations. *Joe L'Estrange*

Opposite: An unusual view of the trio flying low over London during the celebration ceremony. *Joe L'Estrange collection*

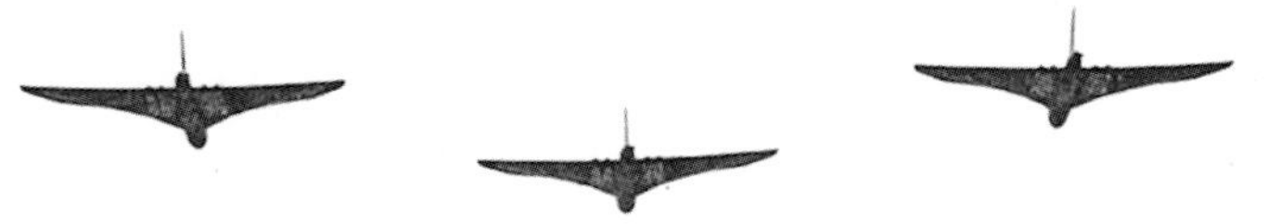

B.2 XM597 of 101 Sqn about to touch down at Waddington, shortly before being despatched to Ascension Island in April 1982. *Fred Martin*

On 31 March 1984 No 50 Squadron, the last operational Vulcan unit, disbanded at Waddington, leaving its fleet of six K.2s and three B.2s to be delivered to museums and fire dumps.

XM597, which had been used for anti-radar strikes during the Falklands conflict, was flown to East Fortune, where it remains today with the Museum of Flight. XM652 was dismantled at Waddington and moved by road to Sheffield, where a private purchaser planned to house it in a leisure complex. Sadly the new owner failed to establish the precise proportions of the Vulcan's airframe and decided that it was simply too big. The aircraft was therefore offered for re-sale and only the nose section survived with a private collector. A happier fate was in store for XL426, as the MoD decided to retain the aircraft on a temporary basis for air display purposes.

Of the Vulcan tankers, XM571 was flown to Gibraltar, where it was placed on display next to the runway and became a popular attraction for thousands of passing tourists, but – astonishingly – the aircraft was scrapped during 1989. XL445 was flown to Lyneham, where it was used for crash rescue training until 1991, when it too was scrapped. XJ825 remained at Waddington in use as a crash rescue and battle damage repair airframe until 1992, when the scrap merchant arrived to dismantle the aircraft. Likewise XH561 was destroyed at Catterick after serving for a few years as a crash rescue trainer. XH558 was flown to RAF Marham on 17 September 1984 on what was to have been the aircraft's last flight, before suffering the same fate as other Vulcan tankers as a crash rescue training airframe.

However, XH560 managed to survive, being designated as a replacement for XL426 (which was running out of hours) on the air display circuit. But as a crew from Waddington began to remove all usable items from XH558, a study of XH560's paperwork revealed that only 160 flying hours were left before a major service was due. After examining the logs of the other recently retired aircraft, it was discovered that XH558 had 600 hours left until its next major service was due. Therefore it made good sense to perform a swap, and XH560 was flown to Marham (where it was eventually scrapped during 1992) while XH558, looking rather battle-worn, was restored to a basic flying condition and ferried back to Waddington in November.

The next major task was to convert XH558 back to standard bomber configuration. A draft SEM (Service Embodied Modification) was compiled by Chief Technicians Brian Webb, Al Hutchinson and Bob Leese, and was submitted to both the MoD and British Aerospace at Woodford for approval, then in February 1985 the conversion began. The HDU housing, associated fuel pipes, air pipes and huge quantities of wiring were removed and the original electrical system was restored. After two months (and roughly 1,200 man-hours) XH558 was once again a B.2, albeit in a slightly modified state, as the ECM equipment was not replaced in the tail fairing because of structural modifications made by BAe during tanker conversion. The associated cooling duct on the starboard side of the tail cone was not replaced either, and was the only external difference between XH558 and a standard B.2. Internally, however, some tanker systems were retained, and one of the three bomb bay fuel tanks was also left in place to counterbalance the missing ECM equipment.

Following minor servicing, the aircraft departed for RAF Kinloss where it was stripped and resprayed in a rather odd wrap-around camouflage paint scheme that was presumably intended to represent the tactical scheme worn by many Vulcans towards the end of their careers. However, XH558 had never carried such colours, and the tactical camouflage was given a high-gloss polyurethane finish, helping to protect the airframe from the harsh British climate, but this finish was also distinctly non-standard when compared to the matt finish applied to the bomber fleet. Quite why more authentic camouflage, or even overall gloss white, was not chosen has never been established.

Above: As part of No 44 Squadron's disbandment celebrations in 1982, XL426 was displayed in spirited fashion, complete with a suitable farewell message inside the bomb doors.

Right: Among a variety of weapon options examined during 1982 was a fit of three laser-guided bombs. Although flight trials were conducted, the weapon system was not used operationally or for training profiles.

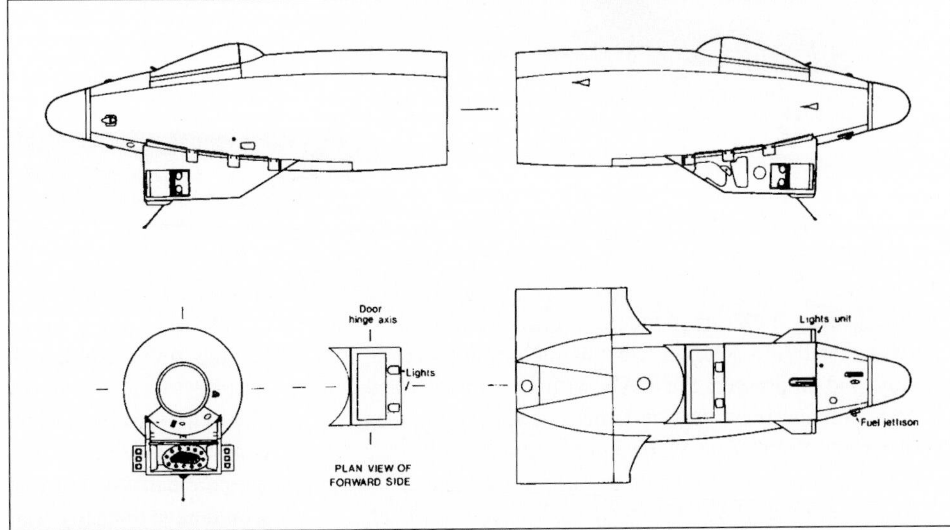

The Vulcan K.2 HDU.

Emerging from the paint shop at Kinloss in November, XH558 made its public debut the following year. Meanwhile XL426 was retired and sold, and as Dave Griffith, from the Vulcan Restoration Trust (VRT) describes, the story of its post-retirement existence at Southend provides a fascinating insight into the complicated and expensive nature of preserving a Vulcan in running order:

'The smiling faces on the press photographs taken in December 1986, when XL426 flew into Southend, held promise of a flying future for XL426, although even by this time the bold plan to return her to the skies as a civilian aircraft was encountering its first problems. XL426 was the penultimate Vulcan airframe disposed of by the RAF, having survived two years beyond the retirement of the Vulcan fleet as an operational aircraft. She had been sitting at Scampton throughout the autumn awaiting disposal. Two years previously had seen the mass exodus of Vulcan airframes to museums, one of which was XM655, the third from last Vulcan ever built. This aircraft was flown to Wellesbourne Mountford by Sqn Ldr Joe L'Estrange, where its owner, Roy Jacobsen, had intended to return it to flight, much as he was intending to do in 1986 with XL426.

XH560 is seen after conversion to tanker standard, complete with 50 Squadron tail markings. Still wearing glossy camouflage indicative of her former role as a maritime radar reconnaissance Vulcan, the aircraft continued to fly occasional sorties with air sampling pods attached. *Michael J. Freer*

Following the resumption of Vulcan AAR training in 1982, and the subsequent introduction of the Vulcan K.2 tanker, this picture shows a slight twist on 'normal' RAF operations, with a Victor K.2 tanker taking on fuel from the Vulcan.

Reading the correspondence from the time, it is clear that flying XM655 was not such a pipe-dream. Roy had impressed the CAA with his professional approach and they were certainly not against the idea of a ferry flight from Wellesbourne to a maintenance base for a routine inspection prior to a limited display flying season in 1985. British Aerospace was, however, unwilling to provide product and design support (other than answering the odd query), and CAA regulations on take-off distances effectively stranded XM655 at Wellesbourne. Despite a cogent argument from Joe L'Estrange describing accepted Air Force procedures for operating from shorter runways, the CAA would not be swayed on the matter of the length of Wellesbourne's runway, and for a Permit-to-Fly aircraft it was simply too short in the CAA's opinion. Although XM655's dream of flying seemed to have stalled, one contact Roy made during this period was to shape the future of XL426, the then current RAF Vulcan Display Team aircraft.

In early 1985 Roy wrote to Heavylift Aircraft Engineering at Southend to see whether they would be willing to carry out XM655's restoration to flight. Heavylift confirmed that they would be happy to consider carrying out the work on a commercial basis, which clearly meant that significant amounts of cash would have to change hands. With XM655 indefinitely grounded at Wellesbourne, looked after by the embryonic enthusiasts' group whose successors operate her in magnificent form today, Roy realised that the Vulcan Display Flight's current display airframe XL426 would be up for grabs within the not too distant future. While harbouring the hope that one day XM655 would fly, when the tender documents for XL426 became available in 1986, Roy, buoyed up by the positive noises from Heavylift at Southend, contacted the airport and obtained their cautious agreement that he could operate the Vulcan from there. Roy's bid was successful and his correspondence with Heavylift intensified.

Vulcans back on Woodford's factory floor during 1982 when the K.2 tanker conversion programme was completed.

XH560 receiving her HDU equipment during tanker conversion at Woodford in 1982. As can be seen, the aircraft is still carrying underwing air sampling pods.

Three weeks before the aircraft was due to be delivered, Heavylift wrote a long and detailed letter to Roy confirming their capability to carry out the airworthiness procedures but pointing out that any further work would have to be subject to a contract entered into between themselves and Roy. Once the aircraft had landed at Southend, Roy discovered that there was significant fuel contamination in the tanks. As part of the disposal agreement, the fuel tanks were largely drained by the RAF and we have confirmation from one of the de-fuellers there that day that the fuel leaving XL426 was "almost milky" in consistency. Understandably annoyed, Roy fired off a letter to the Ministry of Defence and a subsequent survey confirmed Roy's worst fears – significant traces of Claudisporium Resinae, a fungal problem that the VRT is fighting even today, many years later.

In the spring of 1987 the MoD offered Roy some new seals and filters as a goodwill gesture, provided he settle the outstanding payments due in respect of XL426's delivery flight. Heavylift's insistence that everything be carried out in accordance with CAA regulations, and the fact that Roy was not in a position to engage their services on a commercial basis, delayed the restoration to flight, and Roy's ambitious plan of attack was to attempt to persuade the CAA to ease off on their requirements. Things started to deteriorate during 1988. Clearly, Heavylift's warnings that the project would have to be undertaken on a commercial basis did not impress Roy, and when one reads Roy's subsequent letters, which

Former 'Black Buck' B.2 XM597 with dark grey undersides, post-Falklands at Finningley in September 1982. She is now preserved at East Fortune with mission markings reapplied. *Fred Martin*

B.2 XM612 of the Waddington Wing was modified for Falklands operations and is carrying a Martel missile and an AN/ALQ-101 jamming pod, in June 1982 after the Argentinian surrender. *Martin Derry*

become increasingly irate, one can sense that the project was steadily imploding. Heavylift eventually threw in the towel, forcing Roy to explore new engineering avenues. He made overtures to British Air Ferries Engineering on the other side of the airport, and while his relationship with them remained cordial, the lack of major financial backing was still hampering progress.

By now the airport had not received any parking fees for the Vulcan parked on the apron in front of the terminal for two years, and decided that it needed the space. The ground was fairly hard at the time, so they pushed the Vulcan on to the adjacent grass, prompting Roy to launch a press campaign with the hope of generating sufficient interest to reverse the decline of the Southend Vulcan. Press coverage ignited the interest of local enthusiasts, who made contact and Roy prepared a circular that was issued in January 1990, calling all local people interested in the Vulcan to a meeting.

There is no doubt that Roy's short fuse and tendency to put pen to paper whenever he felt under attack may have made him unpopular with quite a few key figures in the aviation community, but there was a great deal more to Roy than his fiery temperament. His venomous blasts directed at British Aerospace, the CAA and the Ministry of Defence during his Vulcan days certainly point towards him being a bit of a loose cannon, but Roy was an enthusiast – a dreamer of wild dreams. Portrayed in the press at the time as "Surrey businessman", one only had to visit him at home to instantly warm to this fabulous British eccentric. He lived in a state of controlled chaos, but was witty, eloquent, well-educated and genuinely charming.

XM603, shortly after retirement, is pictured in front of the Woodford factory from which she had emerged some twenty years previously. *Joe L'Estrange*

Right: An atmospheric image of XM603 in her final in-service paint scheme. After retirement, the Avro Heritage team repainted the aircraft in overall gloss white paint.
Shaun Connor

Below: XH561, a B.2 of 50 Sqn seen at Waddington in April 1982.
Fred Martin

Back in March of 1990, Kevin Packard, the current Chairman of the VRT, was getting suspicious about how keen British Air Ferries Engineering (BAF) was in getting involved and the extent to which the fuel contamination was a problem ('severe', according to a magazine article, but dismissed as 'light' by Roy). Roy had apparently offered an undisclosed four-figure sum to BAF in order to get them to progress work on XL426's airworthiness, but had not had a reply; a similar approach to Heavylift had already failed.

Later that month, during a committee meeting of the "698 Club" (which became the VRT), Roy expressed an interest in purchasing an Avro Shackleton, a suggestion that was quickly dismissed as impractical, but one wonders whether Roy, jaded by years of fighting over the Vulcan, saw the Shackleton as an easier option! At that same meeting the committee was of the view that offering BAF £1,000 may not be adequate, but Roy was confident that it would cover preliminary investigations. Nevertheless, the committee decided that a proper quotation was necessary and tasked Roy with the job. The Club had by this point paid £1,000 of Roy's £16,000 parking debt to the airport, and the name of the organisation had been extended to the wonderfully verbose Vulcan Memorial Flight Supporters Club!

XL445 on 50 Squadron's dispersal area at Waddington in 1983. Unusually, the HDU housing appears to have been removed for servicing access, although the actual HDU equipment is still installed. *Shaun Connor*

Roy had been making enquiries with another company, Inflite, based at Stansted, and while it was not adverse to giving involved, its cautious interest hardly merited Roy's triumphant press release entitled "Vulcan Breakthrough", which surprised even the committee. The tone was extremely upbeat and stated that an agreement had been entered into with Inflite and that the Club would raise £100,000 through a man-hour sponsorship scheme, allowing the Vulcan to fly "within a matter of weeks". But the story became more and more vague until October, when Roy's contact at the company left and Inflite withdrew from the project.

Further enquiries were made with Dan Air to survey the airframe, but the quoted costs were prohibitive. While the club continued to build as a supporters' organisation, with Norman Skinner taking over as Chairman, the aircraft engineering aspect took a more positive turn when BAF made contact and quoted

XL448 of 35 Sqn at Binbrook in July 1975. *Fred Martin*

XM575 flies over its home base. *Avro Heritage*

B.2 XM652 of the Waddington Wing at Finningley in September 1975. *Fred Martin*

£5,000 to carry out a full airframe survey, allowing them to put together a cost estimate for returning the aircraft to flight. In hindsight, it is difficult to see how the depth of a survey that could be carried out for such a modest sum could glean anything particularly useful, but it was a step in the right direction. The survey was arranged for December 1990 and the job of moving XL426 from her grassy home in front of the terminal began. It took two air tugs to pull her back on to the tarmac, but she seemed none the worse for her agricultural ordeal and was towed gently down to the Charlie threshold, then across the main runway towards BAF's facility. It was then that an awful realisation dawned – she was too tall to fit into the hangar. Only a touch too tall, but too tall nonetheless. In a move that the VRT regretted bitterly in years to come, BAF's engineers drained the main shock absorbers on the undercarriage in order to squeeze the Vulcan's tail fin under the door jamb of their hangar. It was considered so innocuous at the time that no mention is made of it in meeting minutes, but it was to prove an almighty headache when the VRT wanted to begin taxiing XL.426 at high speed five years later.

B.2 XM652 at Finningley 1980. *Fred Martin*

The engines were run, albeit with a few spectacular flame-outs, and half the aircraft was surveyed, on the basis that the other half would be in much the same condition! Furnished with the survey results, BAF quoted between £450,000 and £500,000 to restore XL426 to flight subject to CAA requirements, a caveat with such far-reaching implications that it rather brings into doubt the reliability of the quotation. However, the sight of XL426 in the hangar prompted the committee to approach BAF to see if they would allow XL426 to return to the hangar for a public open weekend that Easter. BAF agreed and one of the key events in the organisation's history was staged over the Easter Weekend of 1991. finally putting the Southend Vulcan on the map.

After the open weekend, the Vulcan was returned to the far end of the cross-runway where she had been moved after the survey. She was away from public gaze but, most importantly, given the spiralling parking fees (£15,000 at that time), she was parked in an area where the airport felt able to grant free parking. The only engineering carried out at this stage was the regular engine inhibiting, which involved spraying a protective oil (still used on XL-426 to this day) known as DWX-24 'Rustillo', provided free of charge by Castrol UK Ltd. This foul operation involved donning protective overalls and breathing apparatus, and spraying this noxious oil into every accessible part of the engines. The undersides were fairly easy other than gravity providing one with a constant shower of the stuff, but the intakes were treacherous as the only method of accessing them at the time was using a domestic ladder and sliding inelegantly down toward the first compressor stage of each engine. However, this discomfort paled into insignificance when it was time to tackle the jet pipes, as that involved sending a poor soul up each jet pipe to the back of the engine in order to apply the Rustillo. Kevin manufactured a wheeled scaffold pole system, which removed the need for a body to be shoved up the jet pipe, but we reverted to the old Victorian chimney-sweep method as it was quicker and resulted in a more thorough job. Unscientific though this treatment was, it is certain that, without it, the engines would have long since seized.

The big news on the Vulcan front at this time was the sudden demise of the Vulcan Association (supporting XH558) and speculation over the implications for the Vulcan Display Team. Meanwhile, XM655 had been suffering from vandalism, and in July 1992 the Southend enthusiasts learned (from the Observer newspaper) that Roy had been forced to sell XM655 to Radar Moor, the owners of Wellesbourne Mountford airfield, for a nominal sum of £1 in lieu of outstanding parking charges. The VMFSC was powerless, and its only significant action in response was to remove XM655 from the club's logo, but it did serve to step up the campaign to wrest ownership of XL426 from Roy, which was not a pleasant prospect. However, to continue fund-raising for an aircraft over which they had no control was pointless.

A familiar view for users of the A15 as Vulcan XM594 lumbers across the traffic-light-control zone at the end of Waddington's runway. Typical of many Vulcans during 1982-3, the refuelling probe has been removed for use on other RAF aircraft types assigned to Falklands duties. *Robin A. Walker*

The Aircraft Working Party, as the engineering team was then known, received a vital boost with the arrival of Derek Boron, an ex-Vulcan electrician and avionics man. Derek was employed by British Airways on the Concorde fleet, but injected some badly needed expertise into the team. Work to date had consisted of removing access panels, inhibiting behind them, then putting them back. Finally, the campaign to persuade the Ministry of Defence to continue flying XH558 hit the buffers and, while the new owners (C. Walton Ltd) maintained a long-term goal of flying XH558 again, it was clear that the VMFSC was in the forefront, and we were not entirely sure how sustainable the project was when there were no Vulcans left "flying the flag" for such projects. How long could the interest in a retired type be maintained? One thing was certain: XL426 would have to be restored to taxiing condition, but nobody had a clue how to bring this about.

In the spring of 1993 Southend Airport was being transferred from municipal to private ownership, and outstanding matters were being settled. Having contacted Roy requesting the outstanding parking charges (which they erroneously calculated to be £66,000, overlooking the free parking deal negotiated two years previously), they had received no reply and had contacted Hanningfield Metals with a view to scrapping the Vulcan. Not being the owners of the aircraft, the VMFSC knew nothing of this, but after negotiations with Roy Jacobsen, Don Carrick, a member of the group and a solicitor, offered his services and was immediately pressed into drawing up documents to transfer ownership of XL426 from Roy to the club. The issue that dogged the process was the £15,000 outstanding parking charges owed by Roy to Southend Airport. Simply transferring the aircraft ownership without reference to the debt would achieve nothing; the Vulcan had been a second-class citizen at the airport for far too long, largely thanks to its status as a debtor. However, if the club was to buy the aircraft from Roy for however much, he would have no obligation and little inclination to settle the debt, so VMFSC had to devise a better solution.

Don suggested drawing up a Deed of Gift in which Roy gifted the aircraft to the club in return for which his debt to Southend would be no more. £15,000 was way beyond the club's finances, so Norman Skinner contacted the airport to broker a deal. Norman argued that the airport had no realistic chance of getting any money from Roy, but that the VMFSC was in a position to make a payment in order to remove the debt. The airport agreed to accept £8, and in return for wiping out his debt Roy would give the VMFSC the Vulcan. Persuading Roy was not easy, but it was for the good of the aircraft, and the change in airport ownership certainly heralded a new era for XL426.

As the Vulcan's new owners, we changed our name from the morbidly obese Vulcan Memorial Flight Supporters Club to the low-fat Vulcan Restoration Trust, and within a year we were a registered charity.

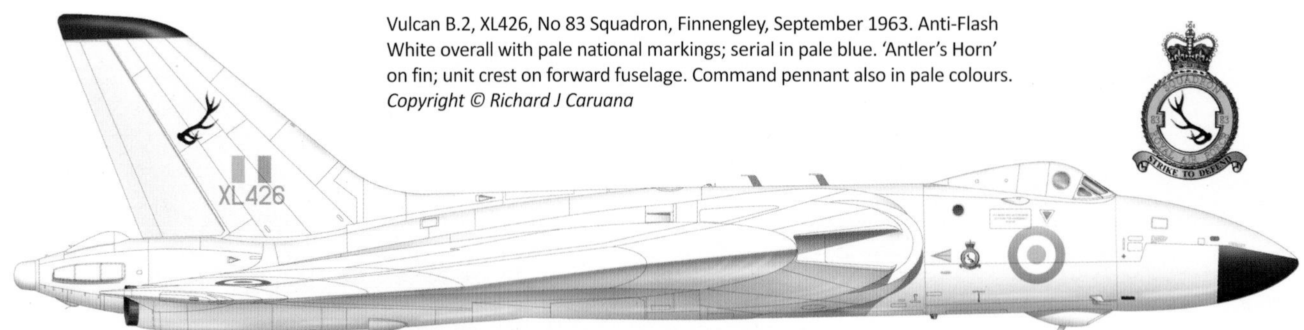

Vulcan B.2, XL426, No 83 Squadron, Finnengley, September 1963. Anti-Flash White overall with pale national markings; serial in pale blue. 'Antler's Horn' on fin; unit crest on forward fuselage. Command pennant also in pale colours. *Copyright © Richard J Caruana*

Left: XM651 awaits the attention of the scrap merchant at Waddington. Having been retired in August 1981, the aircraft's service life extended to only seventeen years, a fairly short life span when compared to more modern aircraft, but a typical figure for much of the Vulcan B.2 fleet. *Joe L'Estrange*

Below: Following withdrawal XM571 was flown to Gibraltar to become a tourist attraction. In earlier years she is seen trailing her single refuelling hose and displaying the final definitive underside paint markings applied to the tankers. The 101 Squadron markings have yet to be removed.

With the ownership transferred, and with some trepidation, I contacted Heavylift on some engineering matter or other and found the reaction most pleasant. Happy to let bygones be bygones, Southend's largest aircraft maintenance company became a valuable and generous supporter of the VRT from that day forward, and our relationship with the airport took an equally encouraging turn for the better. A Ground Power Unit was purchased from the erstwhile Vulcan scrappers Hanningfield Metals, and at last we could connect a 200V electrical supply on the aircraft, making electrical engineer Derek Potton's life so much easier as well as revealing a whole wealth of new electrical snags that required his attention. At this point all our tools and consumables were stored in XL426's huge bomb bay, in the absence of any other storage at the airport, but the availability of a power supply also gave our store a whole new and unwanted feature: a floor that fell away at the touch of a button in the cockpit! Norman Skinner was sent back to the airport to enquire about a 40-foot container that was minding its own business on the other side of the runway, and a week or so later the very same container found itself behind the Vulcan. For the first time ever we had a proper workshop and store, albeit a small one; it gave the whole project its first taste of permanence, which had been perilously lacking until then.

In early 1994 the VRT changed for ever thanks to the arrival on the scene of one Trevor Bailey. Now an airframe and engineer technician on the Vulcan To The Sky project at Bruntingthorpe, Trevor at this point was a young, tall, madly enthusiastic lunatic, working at BAF with exhausting energy reserves and a technical intelligence that left us breathless. He knew nothing about Vulcans, but we had manuals and he intended to find out. The cautious days of the VMFSC were about to end as we began to realise that Trevor was the key to the future. Derek Potton eyed Trevor with quite understandable suspicion; doing things by the book was not Trevor's strength in those days, but he could problem-solve like no one else we ever met.

In February we hired the least serviceable van in Essex and made our way over to Roy Jacobsen's Selsdon home to pick up all of his Vulcan spares, which he had offered to the Trust. The spares were everywhere: his garage, the garden, the outside loo, you name it. Some were Buccaneer spares, but there was a great deal of good Vulcan stuff emerging from this suburban semi-detached. The day was vintage Roy Jacobsen and a very happy memory for all of us present.

Milestones came thick and fast with Trevor Bailey at the helm. One day in particular, in March 1994, is etched on my memory. Trevor borrowed an air start unit from Heavylift in order to dry-cycle all four engines, none of which had been run since the end of 1990. Our inhibiting proved its worth as he confirmed good rpm and oil pressure readings after filling the nearly dry oil tanks on each engine. Having finished the dry cycles, Trevor called us over to the aircraft to ask us whether he could start engine No 2. He retired to the cockpit and a few minutes later a loud rumble and an unmistakable heat haze from the back of the aircraft confirmed that XL426 was irrefutably alive; writing this, the lump in my throat has returned!

B.2 XL444 of 66 Squadron at Leuchars in February 1976. *Fred Martin*

Above: XL445 was used for crash rescue training at Lyneham until 1991, when it was scrapped. This grainy but fascinating image shows two iconic strategic bombers together as a Soviet Bear makes a visit to the edges of the UK's Air Defence Identification Zone (ADIZ), and XL445 remains on station to provide fuel for the RAF's interceptors.

Left: Part of Waddington's 'Vulcan Graveyard' in 1982 as XH557 and XM654 (notable as the trials aircraft that carried both 201 and 301 engines simultaneously) awaits the scrap merchant. *Joe L'Estrange*

The Aviators Flying Club contacted the Trust with regard to a small concrete pan to the north of their flying club, which they thought might be suitable for the Vulcan. We paid them a visit and were shown a diminutive blister hangar base, way too small for a Vulcan but with potential. There were various paperwork issues, but the principle was good and it was an opportunity to relocate the Vulcan somewhere where she could be seen by the public. I vividly recall a man visiting our stand on Southend's seafront during the 1992 airshow, accusing us of taking money under false pretences as he personally knew that the Vulcan had been scrapped years ago! In April 1994 the members were invited over to the airport to watch an engine run. Trevor had been working furiously to make this happen and had arranged to once again borrow the air start from Heavylift. Cotton wool was handed out to be shoved in any orifice that the owner chose, but I decided for the first (and last) time that I would stand in front of XL426 and suffer the unattenuated racket of the Vulcan on full power – not something I would recommend, as I could not hear clearly for the next two days! The faces of those present were a picture and the significance of such progress was not lost on them. Ladies and gentlemen, we have a live Vulcan!

Our early attempts to de-corrode the Vulcan were stepped up when Trevor joined the team; Alochrom and Deoxidine were purchased and the repaint began in earnest. Ernie Lawrence arrived on the scene with his son Matt (now our Chief Engineer), and they were part of a general influx of people crucial to the project in years to come, including Steve Williams, who single-handedly pulled the Trust out of several engineering problems that had even thwarted the ubiquitous Bailey. The ambitious job of repainting XL426 was under way in 1994 using paint left over from East Midland Aeropark's recent repaint of XM575. The grey was a trifle on the brown side and the green a bit close to British Racing Green, but it was better than the sad brown and off-white camouflage that had been steadily deteriorating since she arrived at the airport. Once the technique had been established, painting continued at a furious pace, sometimes with only a couple of people present, one of whom was invariably Ernie. He eschewed spraying equipment, preferring a 3-inch paintbrush, but even then this did not prevent Ernie applying equal quantities of paint to himself and the aircraft; in fact, he spent most of 1994 and 1995 sporting a very dashing wrap-around camouflage.

On 12 January 1994 a team of us set off for a rainy RAF Marham with one thing on our minds: wheels. Once admitted to the base we drove around to the Marham fire dump where we stumbled into a scene of utter horror: the Bird Group had been on site for a day or two and had been steadily eating their way through the last few aircraft of Britain's mighty V-Force. The Victor had retired from service during 1993 and these few airframes were being reduced to mounds of unrecognisable scrap, but not before we had liberated a huge stock of main wheels, which are identical to those on the Vulcan. The weather was foul, and the sound of the aircraft being torn apart sounded like someone breaking plastic cups (on a huge scale), but the mood brightened when the sheer quantity of the spares haul we would be returning to Southend with became apparent. I will never forget the late Dennis Marriott drenched to the skin with the biggest grin stretched across his grubby face, bowling Victor wheels he had liberated from various disembodied undercarriage legs across the tarmac towards our van.

Progress on drawing up a lease on the new pan was painfully slow, and the cost was certainly going to stretch the Trust. Fund-raising continued through 1994, and by October of that year Aviators Flying Club had changed hands, but eventually work started on the new pan in December and, flying by the seat of our pants, the Trustees decided to hold another Easter Open Weekend on the new pan, which was still far from complete. As the work neared completion, a date for moving XL426 to her new home was settled: it would be 4 March 1995 or die! The big day dawned bright and sunny and it was the first time many of us had seen XL426 move at Southend, and that sight was moving in itself. A large crowd had gathered on the other side of the airport, and the vision of XL426's huge wing emerging from the other side of the shimmering heat haze on the runway as it approached the threshold was, by all accounts, a tear-jerking moment.

Preservation candidate XH558 rolls on Bruntingthorpe's 10,000-foot runway, demonstrating the size of the Vulcan's brake parachute. Although rarely required, the brake 'chute gave the Vulcan a respectably short landing run that would have been vital for dispersal to smaller airfields.

Above: XH560 was later converted to a K.2 with 50 Sqn and is seen here at Wyton in June 1983. Her nose survives in Essex. *Fred Martin*

Left: XM648 was one of only fifteen aircraft to receive the tactical low-level wrap-around camouflage towards the end of her career. *Robin A. Walker*

Below: During its service days XM655 shivers in the cold in this wintry scene, awaiting de-icing from the Waddington groundcrews. *Joe L'Estrange*

XM595 ended her service career with No 35 Squadron at Scampton during March 1982 and was (as illustrated) cut up for scrap on site later the same year.

Above: XL426, wearing the markings of No 83 Squadron, is pictured on Finningley's ORP. The aircraft was eventually retired to Southend where it is maintained in excellent condition (and occasionally taxied) by a team of dedicated volunteers.

Left: Squadron Leader Joe L'Estrange gives one of his notoriously spirited displays, in this case over RAF Akrotiri. *Joe L'Estrange collection*

That open weekend was a great success, although various technical problems dogged the aircraft at this point: No 2 engine was refusing to accelerate beyond idle, the nosewheel steering, which had been broken by BAF when they manoeuvred XL426 into the hangar in 1990, had not been repaired, and the aircraft was sitting very low at the back thanks to the empty shock absorbers drained by BAF on that same day. In fact, one of the shock absorbers was damaged beyond repair, so a replacement was vital. This period also marked Roy Jacobsen's final and rather disappointing contact with the Trust when he began proceedings to take back ownership of XL426. This sad episode dragged on for several months before it eventually petered out, but not before we received a fair amount of bitter correspondence from a very disenchanted Roy.

On the engineering side, the Trust purchased a tug from a scrap yard in Warwickshire in May 1995, and through a complete twist of fate a replacement shock absorber had also arrived at Southend Airport, courtesy of an enthusiast who had been given the unit some years previously by Andy Ward, one of the informal team working on XM655. Trevor traced the problem with No 2 engine to fuel contamination blocking microfilters within the engine's chassis-mounted fuel system, which he diagnosed by dismantling the offending unit on his kitchen table, much to the delight of his wife Zoe.

On 7 October 1995 XL426, with no steering other than differential braking and with a saggy behind, moved for the first time under her own power since 1986. Joe L'Estrange and Roger Frampton took the controls on a gentle trundle from the far end of the disused cross-runway to the fence on the live side of the airport. The distance was a little over 500 yards, but it was momentous. I had the honour of marshalling the Vulcan and it was more than a little terrifying to stand in front of this huge noisy beast as it bore down on me! A week later we towed XL426 down to the apron in front of Heavylift's facility. It was a calm, foggy morning, foul weather for flying but ideal conditions for what we were about to do. Borrowing four jacks from within the hangar, a huge gang of us steadily jacked XL426 clear of the ground and the replacement shock absorber was fitted. The fluid, OX-16, was pumped into both shock absorbers, and on Trevor's signal we began lowering the aircraft back on to the concrete. Towing XL426 back to the pan, those of us walking behind the aircraft were astounded at how different she looked, although it did make fitting the jet pipe blanking boards more of an athletic task. We still had the troublesome steering problem to solve, but we were not far off thinking about high-speed taxiing.

In early 1996 the first Vulcan to be scrapped for years went under the axe – XM569 at Cardiff. The museum collection was being broken up and a team from Southend arrived on site and spent two freezing snowy days removing a valuable hoard of spares from this doomed airframe. Back at Southend, slow-speed taxiing continued despite the lack of nosewheel steering, but the demise of the Cardiff Vulcan did bring with it the solution to our problems – a nose undercarriage leg. The leg was badly damaged during its removal, but the steering eyebolt, broken on XL426, was intact. The pin holding XL426's steering motor to the leg had also been bent when BAF moved the aircraft, which was a quite remarkable feat. It was hard to see how a pin, designed to fit snugly within the accurate casting of a nose leg, could possibly have become so twisted, but that was by the by – Steve Williams's next job was to make a new one. Steve worked at Thurrock College at the time and stayed behind at work to use the college's equipment to make a new pin. He finished this precision engineering task late one evening and, rather than go home, he hot-footed it to the airport just to know whether it fitted! It did, and Bill Burnett subsequently took XL426 for a slow-speed run to show that she really did go round corners rather elegantly.

In March 1997 another major milestone was achieved. Meridian TV was filming a series called Big Day Out and was keen to focus one programme on the Vulcan and our Chairman, Norman. The nosewheel steering was fixed, the aircraft was at the proper ride height and it seemed an ideal opportunity to take her out for a proper blast. With the film crew in place, Bill Burnett negotiated the difficult turn at the Charlie 06 threshold and lined the Vulcan up for her first high-speed taxi run. It was a memorable day in so many ways, and those who were there will never forget it. The sight of XL426 pouring on the power and charging along the runway was the realisation of a dream for many of us. A week later Roger Campbell, the Airport Director, and I were chatting in his office. Do you fancy taxiing her at our next airport open weekends, he asked. It was an offer we leapt at, and a tradition was established that we have continued on and off to this day.

In the years since a shabby XL426 whistled over the airport perimeter to begin her retirement, the Vulcan Restoration Trust has achieved the extraordinary, and XL426 is a fully live airframe once again. We wondered how we would ever have the time and money to restore the Vulcan to flight, and ultimately we relinquished that goal. We had many difficult decisions to make over the years, but that was the hardest. In doing so we reluctantly accepted that Roy's, and indeed our, dream to fly the Southend Vulcan was over. The lessons learned from valuable and crucial involvement with the XH558 project proved to the VRT that, while flying XL426 would not be impossible, it just wouldn't be possible for the VRT. But at least a year doesn't go by without XL426's throaty roar rattling the windows in Southend and beyond.

So many people have been involved with this project over the years, and for every one named one will inevitably omit someone else. However, I will raise a proverbial glass to Roy Jacobsen – his original dream to fly XL426 may be no more, but thanks to his enthusiasm and determination in the early days, and the Vulcan Restoration Trust's tireless work since then, his baby will be part of Britain's aviation heritage for many years to come.'

With XL426 gone, the RAF's Vulcan fleet now comprised just one aircraft – XH558 – which remained active as a hugely popular air show attraction. The future was not exactly secure, however, as the continuation of the Vulcan Display Flight at Waddington (XH558 actually belonged to No 55 Squadron at Marham for administrative purposes) was regularly reviewed by the MoD. While XH558's countless fans watched with delight as Squadron Leader David Thomas and Flight Lieutenant Paul Millikin put her through her paces every summer weekend, there was continual doubt as to how long the aircraft could actually remain in RAF service. Sufficient airframe hours remained to allow XH558 to fly a full 1992 display season, and possibly a more restricted number of venues during 1993, but no official decision was given until shortly before the aircraft was offered for sale.

It was this final act of disposal by the MoD that prompted an endless number of campaigns and petitions to keep XH558 flying. The Vulcan Association was formed in 1987, dedicated to publicising the uncertain future of XH558 in the hope that continual public support would encourage the MoD to keep the Vulcan flying. Sadly, internal 'political' problems befell the group and, despite its many thousands of members, the VA effectively disappeared long before the aircraft it was supposed to be supporting. But even this severe blow did not silence the Vulcan fans. The Vulcan Display Flight continued to receive huge quantities of fan mail, all of which said basically the same thing: what can I do to help keep XH558 flying? Unfortunately the answer seemed to be not much, unless you happened to have around a million pounds to spare.

Prior to becoming a display jet XH558 was modified as a K.2 tanker with 50 Sqn, seen here at Coningsby in May 1983. *Fred Martin*

The future for XH558 continued to be just as uncertain as ever. There was no official statement from the Secretary of State for Defence, just a bewildering range of varying statements from MoD departments. The cost of maintaining XH558 was doubtless a major factor in the decision to ultimately ground the Vulcan, but the biggest factor appears to have been the cost of refurbishing the aircraft for another five to ten years of display flying, combined with the projected costs of continued operation and support. Cost estimates from the MoD varied wildly from a quarter of a million pounds to more than two million, depending upon who was asked. Whatever the true figure, it was clearly a ridiculously small amount when compared with the annual operating costs of the Red Arrows, for example, and no one could claim that the Vulcan was any less effective as an RAF publicity and recruiting medium.

The petitions continued to be signed (estimated at over 100,000 signatures) and MPs continued to be questioned, but the final disposal of XH558 remained in the balance. In November 1992 the MoD announced that XH558 was, at least for the time being, no longer for sale. In addition, the RAF confirmed that the aircrew had been instructed to remain current, and that the Vulcan's groundcrew had not received postings to new units. Consequently, there was great optimism within the many enthusiast communities that XH558 would be retained for further display flying, and the possibility of civilian sponsorship began to look like a workable proposal. Most notably, the Save the Vulcan 558 Campaign made great efforts to convince the MoD that XH558 could be operated by the RAF on a no-cost basis with civilian finance, and the future began to look a little brighter.

Sadly, however, the Ministry of Defence had evidently decided to dispose of the Vulcan almost regardless of any sponsorship proposals that might have been organised given sufficient time. The final blow was the announcement that the RAF would not be able to provide suitably qualified manpower

B.2 XM603 of the Waddington Wing at Finningley. She is currently displayed in an anti-flash white scheme at the Avro Heritage Museum, Woodford. *Fred Martin*

to service the Vulcan beyond the end of 1993. The MoD based this statement upon the fact that the remaining Victor tankers would be retired at that time and consequently the RAF would no longer retain any technicians with V-Bomber experience. Whether this really was a valid point is open to question, as the RAF continues to train crews to service relatively aged heavies such as the VC10 and Nimrod, and the predictable reaction of the Vulcan's supporters was that the MoD was simply washing its hands of the whole affair, and was looking for a suitable excuse.

But whatever the reasons, the MoD stood by its view that XH558 was a public expense that simply could no longer be afforded, and early in 1993 it was announced that XH558 was once again for sale. The announcement was perhaps inevitable, but the debate as to whether it was the right decision will doubtless continue for many, many years. It was estimated by some commentators that the Vulcan had become the RAF's most popular air show attraction, and yet the MoD had decided to end XH558's career, merely authorising monthly continuation flights to keep her crew current, until a final delivery flight could be made to her permanent resting place.

For some time the final fate of XH558 was unclear, and while the aircraft continued to make occasional crew continuation sorties from Waddington, there appeared to be some hope that she might be kept airworthy once the RAF had finally disposed of her. However, romantic notions of seeing a Vulcan operate under civilian ownership did seem to be a little unrealistic when the Civil Aviation Authority had made its position quite clear. As far as the CAA was concerned the Vulcan was a 'complex' aircraft that could not be permitted to fly on the civil register, at least not without the direct support of its manufacturer (British Aerospace, which was equally unenthusiastic). There were various bids made to the Ministry of Defence for the aircraft, including one bizarre plan to turn the aircraft into a restaurant, but among the interested parties was David Walton, head of a profitable family business based at Bruntingthorpe in Leicestershire.

XH561 when modified as a K.2, on approach to Waddington in April 1984; the wrap-around camouflage compromised by addition of gloss white undersides for tanker duties. *Fred Martin*

Among Walton's business interests was a huge airfield facility at Bruntingthorpe, once used by the USAF as a B-66 and B-47 base, and now used primarily as a car storage and test facility. A number of preserved aircraft had already arrived on the site, and Walton was persuaded to put in a bid for XH558. His offer was accepted, and on 23 March 1993 the aircraft was prepared to make the relatively short ferry flight from Waddington to Bruntingthorpe. Not surprisingly, the event had reached the attention of countless aviation enthusiasts, and a large number of spectators assembled around Waddington's perimeter to see the Vulcan's very last flight. Although there was plenty of optimism surrounding the aircraft's future, almost everyone believed that this would indeed be the last time that a Vulcan would fly.

Shortly after 11am XH558's four engines were started, and the assembled crowds (which by now resembled a small air show audience) prepared to witness the historic event. Although there was a great deal of excitement, there was a also a tangible feeling of sadness, and more than a few tears were wiped away as XH558 made her way to the runway threshold for the very last time. Finally, the Vulcan's Olympus engines gave out their ear-splitting roar, and XH558 thundered skywards, heading west to begin a short tour of some familiar Vulcan sites, including Finningley, Coningsby and, of course, Woodford. Returning via Scampton, the Vulcan was steered above Lincoln Cathedral for the last time before heading directly over Waddington.

Bomb doors open (with the legend 'Farewell' painted inside) for a final salute, a gentle wing rock signified the very end of the Vulcan's long and illustrious RAF career. Just a few minutes later, XH558 arrived over Bruntingthorpe before making her final touchdown on the airfield's 10,000-foot runway. This would have been the end of the story, but David Walton persevered, and continually maintained contact with both the CAA and British Aerospace, carefully exploring the many and varied concerns that the authorities expressed as being good reasons to keep the Vulcan grounded. Months of inactivity began to run into years, and although XH558 was occasionally treated to a full-power take-off run, her wheels remained on the runway, seemingly doomed to remain flightless for ever.

K.2 tanker XJ825 of 50 Sqn at Waddington in May 1984. *Fred Martin*

An excellent view of B.2 XM602 of 101 Sqn at the Greenham Common Air Tattoo in July 1981. Her nose survives at the Avro Heritage Museum, Woodford. *Fred Martin*

However, Walton's dedication and persistence were rewarded, and finally the CAA agreed in principle that a Permit-to-Fly could be given to the Vulcan (now registered as a civilian aircraft, G-VLCN) if a whole series of very detailed but vital steps were taken. First, a feasibility study was undertaken, revealing that the project was indeed a viable one, and this led to a technical survey, specifying the actual maintenance and restoration that would be required to make the aircraft fit to fly again. Both British Aerospace (now BAE Systems) and Marshall Aerospace were heavily involved in this task and, following a very thorough investigation, it was concluded that XH558 was in remarkably good condition, and that restoration could begin.

And so XH558 disappeared into her hangar at Bruntingthorpe before being slowly disassembled into her main component parts. Many of these items were outsourced to companies appointed to undertake specific restoration tasks, before each component was gradually returned to Bruntingthorpe for re-assembly, ready to begin flight testing. However, the engineering task was only part of the restoration project. Most importantly, the necessary funds had to be found with which to finance the programme, and despite a continuing trickle of money from donations and fund-raising, the cost of restoring XH558 continued to climb steadily. Thankfully, the Heritage Lottery Fund eventually accepted an application for a grant, which effectively made the project remain viable, but even this wasn't enough to get the aircraft to flight status.

The entire project, which had been running for some seven years and had cost some £6 million, reached a crisis point late in 2006. Literally at the last minute it was saved by Sir Jack Hayward, who donated a further half a million pounds to provide sufficient cash to continue working on the Vulcan to the flight test stage. Sir Jack commented, 'It should never have been allowed to stop flying. It's a lovely aircraft that will give a real thrill to the British public.' Thankfully his generous donation was enough to complete the project, enabling XH558 to finally fly once again after having remained earthbound for more than fourteen years.

Of course, getting the aircraft back into the air is a magnificent achievement, but finding further funds with which to operate the aircraft as an air show performer requires yet more money. In the short term it is quite likely that major show organisers will be happy to pay for a star exhibit like the Vulcan, but there are no guarantees that commercial interest in the aircraft will continue for long. If this proves to be the case, XH558's return to the skies may be a short-lived affair, but after so much money and effort has been poured into the project, everyone must hope that the sight and sound of the mighty Vulcan will be around for years to come.

Fifty-five years on, the Vulcan story has turned full circle with another 'first flight', performed rather ironically by XH558, the RAF's first Vulcan B.2 and the RAF's last. The very fact that so much money and effort has been put into returning XH558 to the skies illustrates the fascination and affection for the Vulcan that still exists to this day, even though the dark days of the Cold War (for which the Vulcan was created) are long gone. It is strange to witness the very evident public affection for a machine that was designed exclusively for the business of mass murder, but then the Vulcan is a one-off, which, like the legendary Spitfire, has the ability to impress just by its very presence. By any standards the Vulcan was an excellent aeroplane that more than fulfilled the promise of its manufacturer.

XM652 of 44 Sqn at Coningsby in June 1979. *Fred Martin*

Capable of delivering a megaton-range bomb into the heart of the Soviet Union, the Vulcan was at the very forefront of the West's striking power and undoubtedly ensured the credibility of Britain's (and therefore the West's) deterrent posture for many years. The aircraft's adaptability, flexibility and sheer strength enabled the RAF to use it in a variety of roles for which it was never designed, but for which it was equally well suited. Certainly it would be impossible to find a former member of the V-Force who would utter so much as one criticism of the mighty Vulcan – it was, and still is, a true icon of British aerospace technology.

K.2 XL445 taxiing for take-off at Coningsby in April 1983. *Fred Martin*

Vulcan B.1s XA892 (7746M) & XA898 (7856M) at Halton in July 1966.
Martin Derry

Appendix One

Vulcan Production List

Publisher note: *Conversion dates are listed below where known. Placement in text is not an indication of chronology.*

Prototypes

Contract 6/Air/1942/CB.6(a), dated 6.7.48.

VX770 Delivered August 1952. Avon/Sapphire/Conway engines. A&AEE and manufacturer trials, suffered mid-air explosion due to structural failure at Syerston 20.9.58, and was destroyed.

VX777 Delivered September 1953. Olympus 100 engines. Trials aircraft, converted to prototype B Mk 2, making first flight in this configuration 31.8.57. Further trials in new configuration before being used for non-flying runway trials with RAE. Last flight 27.4.60. Broken up at Farnborough 7.63.

Vulcan B Mk 1

Contract 6/Air/B442/CB.6(a) for twenty-five aircraft, 14.8.52.

1/XA889 Delivered 4.2.55. Olympus 104 engines. A&AEE trials. Bristol Siddeley trials at Patchway. Withdrawn and scrapped at Boscombe Down 1971.

2/XA890 Delivered 1955. Olympus 104 engines. A&AEE trials. RAE Farnborough and Thurleigh trials. Manufacturer trials. Radio and radar trials, blind landing trials, and ballistics research. Withdrawn and scrapped at Bedford 1971.

3/XA891 Delivered 1955. Olympus 104 engines. A&AEE trials. Bristol Siddeley trials at Patchway (Olympus 200 series). RAE trials at Farnborough. Manufacturer trials. Crashed on test flight 24.7.59 near Hull due to electrical failure.

4/XA892 Delivered 1955. Olympus 104 engines. Manufacturer trials and A&AEE armament trials. Delivered to Halton for ground instruction (became 7746M). Scrapped 1972.

Vulcan B.1 XA896 of 230 OCU at the Finningley air show in September 1961. In May 1964, Rolls-Royce at Hucknall began to convert her to be a test bed for the BS.100 vectored thrust engine, intended to power the Hawker-Siddeley P.1154 'supersonic Harrier'. Conversion was stopped when both projects were cancelled in 1965 and the Vulcan was scrapped on site the following year. *Fred Martin*

Vulcan B.1 XA898 (7856M) was allocated to No1 School of Technical Training (SoTT) at RAF Halton in July 1964, this photograph being taken two years later with the aircraft carrying the marking '30'. *Martin Derry*

5/XA893 Delivered 1956. Olympus 104 engines. A&AEE electrical trials, connected with B Mk 2 variant. Broken up at Boscombe Down 1962, XA893 nose is now in store at RAF Museum Cosford.

6/XA894 Delivered 1957. Olympus 104 engines. A&AEE trials, engine development trials. Operated by Bristol Siddeley at Patchway, used as engine test bed for Olympus 22R as part of TSR2 trials programme. Destroyed during ground fire while ground running at Patchway 3.12.62.

7/XA895 Delivered 16.8.56. Olympus 104 engines. Converted to B Mk 1A. 230 OCU, Bomber Command Development Unit, A&AEE. Scrapped by Bradbury Ltd 19.9.68.

8/XA896 Delivered 7.3.57. Olympus 104 engines. 230 OCU, 83 Squadron, 44 Squadron. Bristol Siddeley test bed for BS100 vectored-thrust engine, intended for Hawker P.1154 the 'supersonic Harrier'. Partially converted for this role until fighter development was abandoned. Withdrawn during 1966 and scrapped at Patchway.

9/XA897 Delivered 20.7.56. Olympus 104 engines. 230 OCU, A&AEE trials. Crashed during approach to Heathrow Airport 1.10.56, and destroyed.

10/XA898 Delivered 3.1.57. Olympus 104 engines. 230 OCU. Used exclusively by OCU before being delivered to Halton 25.8.64, for use as an instructional airframe (7856M). Scrapped 1971.

11/XA899 Delivered 28.2.57. Olympus 104 engines. A&AEE trials, RAE trials at Thurleigh, blind landing experiments. Auto-pilot development. Delivered to Cosford as instructional airframe (7812M). Scrapped 1973.

XA899 never saw squadron service but flew with the A&AEE, RAE and Avro on a variety of test and development duties before being withdrawn for use as a ground instructional airframe at Cosford.

12/XA900 Delivered 25.3.57. Olympus 104 engines. 230 OCU, 101 Squadron. Delivered to Cosford as instructional airframe (7896M) 28.2.66. Withdrawn from use and transferred to Aerospace Museum, Cosford. Scrapped 1986, the last intact Vulcan B Mk.1.

13/XA901 Delivered 4.4.57. Olympus 104 engines. 230 OCU, 44 Squadron, 83 Squadron. Delivered to Cranwell as instructional airframe (7897M) 1965. Scrapped 1972.

14/XA902 Delivered 10.5.57. Olympus/Conway/Spey engines. 230 OCU. Damaged in landing accident 28.2.58. Engine trials (Conway and Spey) with Rolls Royce. Scrapped 1963.

15/XA903 Delivered 31.5.57. Olympus 101 engines. A&AEE, RAE Farnborough. Blue Steel trials aircraft. Delivered to Rolls Royce as test bed for Concorde Olympus and Tornado RB 199 engines. Experimental 27mm cannon fit at A&AEE. Last flight by B.1 at Farnborough 22.2.79. Scrapped 1980. Nose with private owner, Stranraer, Scotland.

16/XA904 Delivered 16.7.57. Converted to B.1A standard 1960. Olympus 104 engines. 83 Squadron, 44 Squadron. Damaged in crash landing at Waddington 1.3.61. Disposed as instructional airframe (7738M). Scrapped 1974.

Although destined to be destroyed after being shamelessly abandoned by the RAF Museum at Cosford, XA900 is pictured in happier times during the mid-1960s while serving with 230 OCU at Finningley. *Ray Deacon*

Vulcan B.1 XA901 (7897M) at RAFC Cranwell. The aircraft first flew in March 1957 and joined 230 OCU the following month, also serving with 44 Sqn prior to returning to the OCU. *Martin Derry*

17/XA905 Delivered 11.7.57. Converted to B.1A standard 1960. Olympus 104 engines. 83 Squadron, 44 Squadron, 230 OCU, Waddington Wing. Delivered to Newton as instructional airframe (7857M). Scrapped 21.9.74.

18/XA906 Delivered 12.8.57. Converted to B.1A standard 1962. Olympus 104 engines. 83 Squadron, 44 Squadron, Waddington Wing. Stored at St Athan 10.3.67. Sold as scrap to Bradbury & Co 6.11.68.

19/XA907 Delivered 29.5.57. Converted to B.1A standard 1961. Olympus 104 engines. 83 Squadron, 44 Squadron, Waddington Wing. BCDU. Withdrawn from use 3.11.66. Sold as scrap 20.5.68.

20/XA908 Delivered 18.9.57. Olympus 104 engines. 83 Squadron. Crashed near Detroit, Michigan, USA, 24.10.58.

21/XA909 Delivered 1.10.57. Converted to B.1A standard 1962. Olympus 104 engines. 101 Squadron, 50 Squadron, Waddington Wing. Crashed in Anglesey 16.7.64 following engine explosion.

22/XA910 Delivered 31.10.57. Converted to B.1A standard 1962. Olympus 104 engines. 101 Squadron, 230 OCU, 50 Squadron, 44 Squadron. Became instructional airframe (7995M) at Cottesmore. Scrapped 27.5.70.

23/XA911 Delivered 1.11.57. Converted to B.1A standard 1962. Olympus 104 engines. 83 Squadron, 230 OCU, Waddington Wing. Delivered to St Athan 2.2.67, sold as scrap 8.11.68.

24/XA912 Delivered 2.12.57. Converted to B.1A standard 1960. Olympus 104 engines. 101 Squadron, Waddington Wing. Sold for scrap on 20.5.68.

25/XA913 Delivered 19.12.57. Converted to B.1A standard 1961. Olympus 104 engines. 101 Squadron, Waddington Wing. Stored St Athan 21.12.66, sold as scrap 20.5.68.

Following a loss of power controls during a landing at Waddington on 1 March 1961, XA904 veered off the runway and suffered an undercarriage collapse. The aircraft was scrapped, although the nose section was transferred to Finningley.

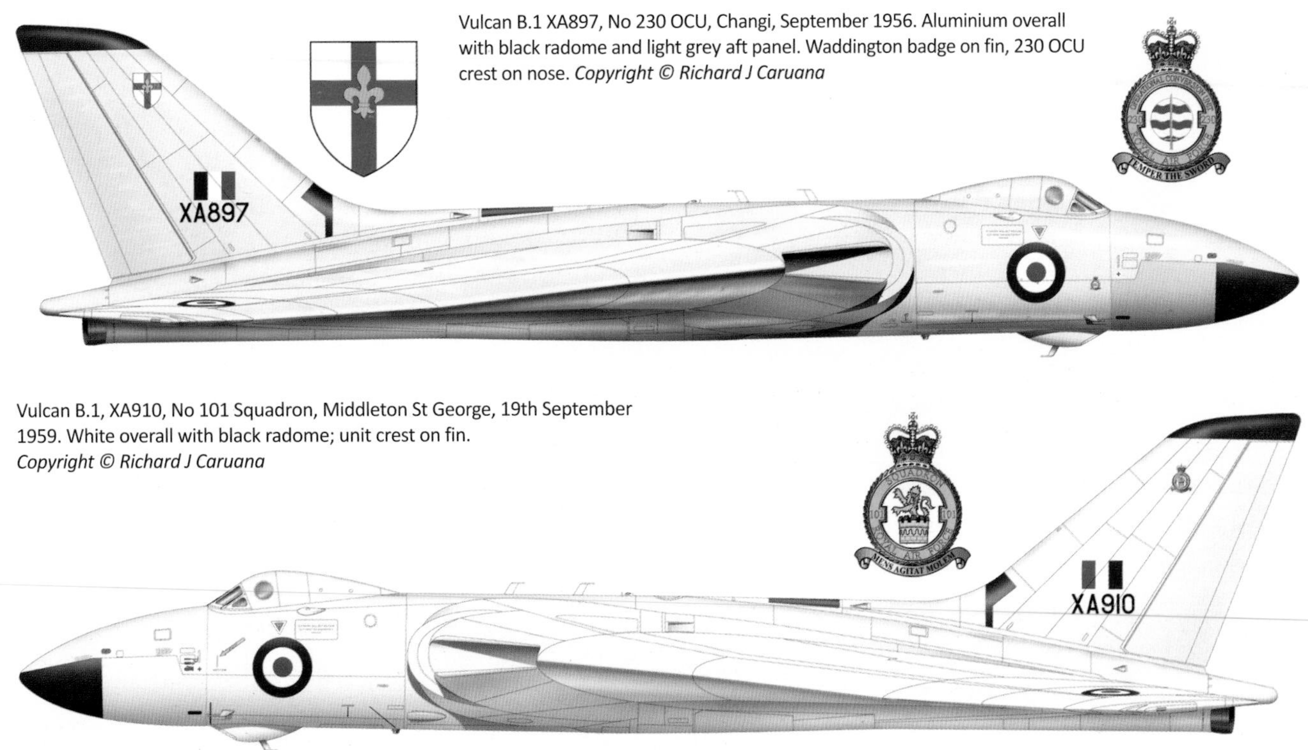

Vulcan B.1 XA897, No 230 OCU, Changi, September 1956. Aluminium overall with black radome and light grey aft panel. Waddington badge on fin, 230 OCU crest on nose. *Copyright © Richard J Caruana*

Vulcan B.1, XA910, No 101 Squadron, Middleton St George, 19th September 1959. White overall with black radome; unit crest on fin. *Copyright © Richard J Caruana*

Vulcan B Mk 1

Contract 6/Air/11301/CB.6(a) for thirty-seven aircraft, 30.9.54.

26/XH475 Delivered 11.2.58. Converted to B.1A standard 1962. Olympus 104 engines. 101 Squadron, Waddington Wing. Became instructional airframe (7996M) 20.11.67. Scrapped 7.6.69.

27/XH476 Delivered 4.2.58. Converted to B.1A standard 1962. Olympus 104 engines. 101 Squadron, 44 Squadron, Waddington Wing. Withdrawn from use 4.5.67, sold as scrap 21.1.69.

28/XH477 Delivered 17.2.58. Converted to B.1A standard 1960. Olympus 104 engines. 83 Squadron, 44 Squadron, 50 Squadron. Crashed near Aboyne, Scotland, 16.6.63.

29/XH478 Delivered 31.3.58. Converted to B.1A standard 1962. Olympus 104 engines. Ministry of Aviation (in-flight refuelling trials), Waddington Wing. Delivered to Akrotiri as instructional airframe (MC8047M) 3.69. Later scrapped.

30/XH479 Delivered 28.3.58. Converted to B.1A standard 1961. Olympus 104 engines. Waddington Wing. Delivered to Halton as instructional airframe (7974M), scrapped 1973.

31/XH480 Delivered 22.4.58. Converted to B.1A standard 1962. Olympus 104 engines. 83 Squadron, 44 Squadron, Waddington Wing. Delivered to St Athan 10.11.66, sold as scrap 30.9.68.

32/XH481 Delivered 30.4.58. Converted to B.1A standard 1960. Olympus 104 engines. 101 Squadron, Waddington Wing. Delivered to Cottesmore fire dump 11.1.68, scrapped 1977.

Vulcan B.1A, XH475, Waddington Wing, 1966. Medium Sea Grey/Dark Green upper surfaces with White undersides; standard national markings. Black serial and radome. *Copyright © Richard J Caruana*

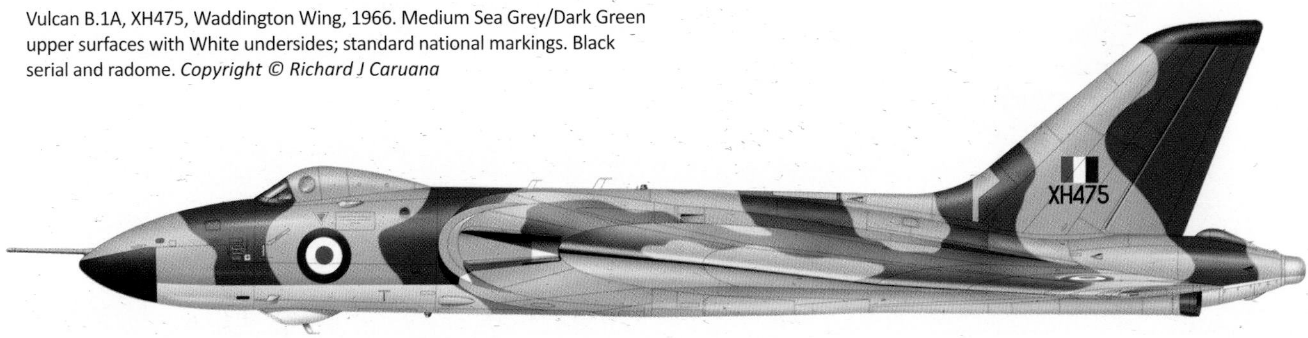

B.1A XH478 ended her days at RAF Akrotiri, as a crash rescue training airframe. The 56 Squadron markings have been applied by a visiting Lightning detachment.

B.1A XH500 active at Waddington in September 1963. After her flying days were over she was employed for crew drill and fire-fighting training. *Martin Derry*

Vulcan XH501 touches down at Waddington in September 1964, a replacement rudder being apparent. She was scrapped four years later, together with many other B.1As. *Martin Derry*

XH502 was delivered directly to No 617 Squadron after completion, but eventually transferred to the Waddington Wing, ultimately being used as a crash rescue trainer on the airfield.

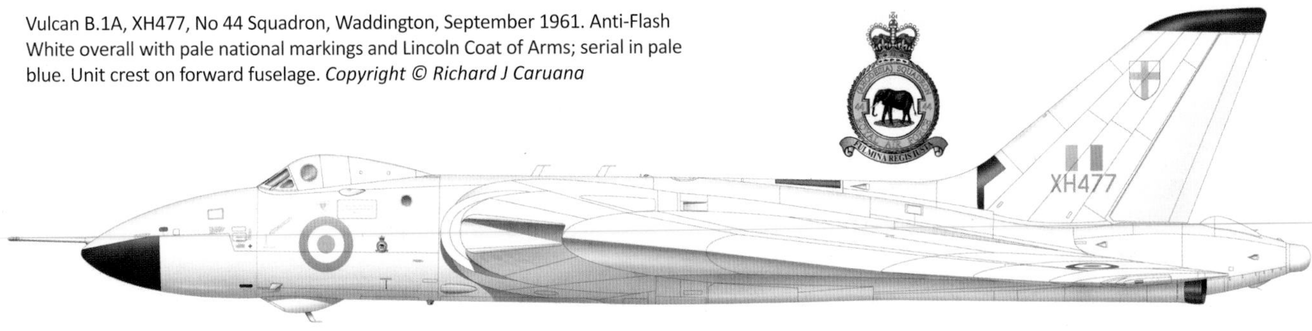

Vulcan B.1A, XH477, No 44 Squadron, Waddington, September 1961. Anti-Flash White overall with pale national markings and Lincoln Coat of Arms; serial in pale blue. Unit crest on forward fuselage. *Copyright © Richard J Caruana*

33/XH482 Delivered 5.5.58. Converted to B.1A standard 1962. Olympus 104 engines. 617 Squadron, 50 Squadron, 101 Squadron, Waddington Wing. Delivered to St Athan 13.10.66, scrapped 19.9.68.

34/XH483 Delivered 20.5.58. Converted to B.1A standard 1961. Olympus 104 engines. 617 Squadron, 50 Squadron, Waddington Wing. To Manston fire dump 3.8.67, scrapped 1977.

35/XH497 Delivered 29.5.58. Converted to B.1A standard 1962. Olympus 104 engines. 617 Squadron, 50 Squadron, Waddington Wing. Withdrawn from use 17.5.66. Sold as scrap 21.1.69.

36/X11498 Delivered 30.6.58. Converted to B.1A standard 1962. Olympus 104 engines. 617 Squadron, 50 Squadron, Waddington Wing. Became instructional airframe (7993M). Scrapped 2.70.

37/XH499 Delivered 17.7.58. Converted to B.1A standard 1962. Olympus 104 engines. 617 Squadron, 50 Squadron, 44 Squadron, A&AEE. Withdrawn from use 11.65. Scrapped at Bitteswell 12.65.

38/XH500 Delivered 15.8.58. Converted to B.1A standard 1959. Olympus 104 engines. 617 Squadron, BCDU, 50 Squadron, Waddington Wing. Became instructional airframe (7994M), to Waddington fire dump, scrapped 1977.

39/XH501 Delivered 3.9.58. Converted to B.1A standard 1961. Olympus 104 engines. 617 Squadron, 44 Squadron, 44/50 Squadron. To St Athan 3.11.66, sold as scrap 8.11.68.

40/XH502 Delivered 10.11.58. Converted to B.1A standard 1962. Olympus 104 engines. 617 Squadron, 50 Squadron, Waddington Wing. To Scampton fire dump 1.68, nose section to Waddington for instructional duties.

41/XH503 Delivered 30.12.58. Converted to B.1A standard 1963. Olympus 104 engines. 83 Squadron, 44 Squadron, Waddington Wing. To St Athan 6.12.66, sold as scrap 8.11.68.

42/XH504 Delivered 30.12.58. Converted to B.1A standard 1961. Olympus 104 engines. 230 OCU, Waddington Wing. Delivered to Cottesmore fire dump 4.1.68, later scrapped.

43/XH505 Delivered 13.3.59. Converted to B.1A standard 1960. Olympus 104 engines. 230 OCU, 617 Squadron, 50 Squadron, Waddington Wing. Delivered to Finningley fire dump 9.1.68, later scrapped.

44/XH506 Delivered 17.4.59. Converted to B.1A standard 1960. Olympus 104 engines. 101 Squadron, 617 Squadron, 50 Squadron, Waddington Wing. Withdrawn from use 10.1.68, sold as scrap 8.11.68.

45/XH532 Delivered 31.3.59. Converted to B.1A standard 1962. Olympus 104 engines. Last production B.1. 230 OCU, 101 Squadron, Waddington Wing. To St Athan 17.5.66, sold as scrap 8.11.68.

Vulcan B Mk 2

Contract 6/Air/11301/CB.6(a) for seventeen aircraft, 30.9.54.

1/XH533 First flight 19.8.58. Olympus 200 engines. A&AEE 26.3.59. St Athan Engineering Squadron (8048M). Sold to Bradbury & Co as scrap 15.10.70.

2/XH534 Completed 17.7.59. Olympus 201 engines. Manufacturer trials 4.3.60, A&AEE, 230 OCU, 6.12.66. Manufacturer for storage 7.4.72. Converted to B.2 (MRR) 8.73. Air sampling pod modifications. 27 Squadron 14.8.74, St Athan 7.4.81. Sold to Harold John & Co as scrap 16.2.82.

3/XH535 Completed 27.5.60. Olympus 201 engines. A&AEE 27.5.60. Crashed near Andover 11.5.64.

4/XH536 Completed 17.7.59. Olympus 201 engines. A&AEE 31.5.60. Waddington Wing 24.11.65. Crashed in Brecon Beacons 11.2.66 during TFR trials.

5/XH537 Completed 27.8.59. Olympus 201 engines. Manufacturer trials 31.8.60. 230 OCU 31.5.65. St Athan 14.2.78. Conversion to B.2 (MRR), air sampling pod modifications. 27 Squadron 8.5.78. Abingdon 24.3.82, instructional/exhibition airframe (8749M). Scrapped 10.90, nose to Bournemouth Aviation Museum 29.10.91.

6/XH538 Completed 23.9.59. Olympus 201 engines. Manufacturer/A&AEE trials 30.1.61. Scampton Wing 14.5.69, Waddington Wing 29.4.70, 230 OCU 21.4.71, 27 Squadron 3.12.73, 230 OCU 15.1.75, 35 Squadron 28.7.77, Waddington Wing 16.8.78, 35 Squadron 23.11.79, St Athan 11.3.81, sold to W. Harold & Co as scrap 31.8.81.

7/XH539 Completed 30.9.59. Olympus 201 engines. Manufacturer/A&AEE trials 25.5.61 . Blue Steel modifications. Withdrawn from use 12.71, to Waddington fire dump 7.3.72, later scrapped.

XH535 was destroyed near Boscombe Down after having entered a flat spin. The wreckage reveals the initial impact point, from which the airframe appears to have shunted forward before breaking up.

XH537 following a touch and go at Leconfield in August 1976. *Fred Martin*

B.2 (MRR) XH537 of 27 Sqn on a misty day at Scampton in February 1982. *Fred Martin*

Following her early career as a trials aircraft (fitted with both 201 and 301 engines) XH557 settled into regular squadron service, her final days being with No 50 Squadron, minus a refuelling probe removed during Operation 'Corporate'. The aircraft was scrapped at Waddington late in 1982. *Joe L'Estrange*

Vulcan B.2(MRR) XH558 enjoyed a long service career including many years with 27 Squadron in the maritime reconnaissance role which included regular visual monitoring of offshore rigs.

8/XH554 Completed 29.10.59. Olympus 201 engines. 83 Squadron 10.4.61, 230 OCU 1.11.62, Firefighting School Catterick 9.6.81 (8694M), later scrapped.

9/XH555 Completed 6.61. Olympus 201 engines. 27 Squadron 14.7.61, 230 OCU. Manufacturer for fatigue tests, St Athan for structural integrity tests. Scrapped 1971.

10/XH556 Completed 9.61. Olympus 201 engines. 27 Squadron 29.9.61, 230 OCU. Struck off charge following undercarriage collapse 19.4.66, to Finningley fire dump, later scrapped.

11/XH557 Completed 13.5.60. Olympus 201/301 engines. To Bristol Siddeley for engine trials 21.6.60. Fitted with Mk 301 engines in outer nacelles (one followed by pair), later fitted with four 301s. First B.2 with enlarged intakes. Cottesmore Wing 6.12.65, Waddington Wing 8.2.66, Akrotiri Wing 19.4.74, Waddington Wing 15.1.75, 50 Squadron 3.81. Sold to Bird Group as scrap 8.12.82.

12/XH558 Completed 30.6.60. Olympus 201 engines. 230 OCU 1.7.60, Waddington Wing 26.2.68. Conversion to B.2 (MRR) 17.8.73, air sampling pod modifications. 230 OCU 18.10.76, 27 Squadron 29.11.76, Waddington Wing 31.3.82. Conversion to K.2 30.6.82. 50 Squadron 12.10.82, Waddington Station Flight 1.4.84, allocation cancelled Waddington 14.11.84. Conversion to B.2 4.85. 55 Squadron 9.85. Last Vulcan in RAF service. Sold to C. Walton, and delivered to Bruntingthorpe 23.3.93. Struck off charge 23.3.93. To Vulcan to the Sky Trust, registered as G-VLCN 6.2.95. First flight after restoration 18.10.07. Last flight 20.10.15, displayed at Doncaster Sheffield Airport (Finningley).

13/XH559 Completed 30.7.60. Olympus 201 engines. 230 OCU 24.8.60, St Athan 27.5.81. Sold to Harold John & Co as scrap 29.1.82.

A dramatic shot of B.2 XH555 putting on the power and going around at Finningley in September 1965. *Martin Derry*

14/XH560 Completed 30.9.60. Olympus 201 engines. 230 OCU 3.10.60. Manufacturer 28.11.60. 12 Squadron 26.9.62, 230 OCU 29.11.63, Cottesmore Wing 23.8.65, Waddington Wing 10.4.67, Cottesmore Wing 2.2.68, Akrotiri Wing 15.1.69. Manufacturer for storage 20.10.71. Conversion to B.2 (MRR) 1.2.73, air sampling pod modifications. 27 Squadron 15.3.74, Waddington Wing 25.3.82. Conversion to K.2 15.6.82. 50 Squadron 23.8.82, Waddington Station Flight 4.84. To Marham dump 29.11.84, scrapped 1.85. Nose preserved by The Cockpit Collection, Essex.

15/XH561 Completed 31.10.60. Olympus 201 engines. 230 OCU 4.1 1.60, Waddington Wing 7.8.67, Cottesmore Wing 8.5.68, Akrotiri Wing 19.3.69, 35 Squadron 6.3.75, 50 Squadron 4.9.81. Conversion to K.2 4.5.82. 50 Squadron 18.6.82, Waddington Station Flight 1.4.84. Allocated 8809M 22.3.84. Delivered to Firefighting School Catterick 14.6.84, later scrapped.

16/XH562 Completed 30.11.60. Olympus 201 engines. 35 Squadron 1.3.63, 230 OCU 11.3.63, 35 Squadron 30.4.63, 230 OCU 19.9.63, Cottesmore Wing 1.8.65, 50 Squadron 8.3.66, Waddington Wing 10.2.67, Cottesmore Wing 24.4.68, Akrotiri Wing 15.1.69, Waddington Wing 9.5.75, 230 OCU 27.9.77, 35 Squadron 16.12.80, 9 Squadron 7.81, 101 Squadron 6.82, Firefighting School Catterick 19.8.82 (875851). Scrapped 1984.

17/XH563 Completed 22.12.60. Olympus 201 engines. 83 Squadron 28.12.60, 12 Squadron 26.11.62, 230 OCU 5.3.65, Waddington Wing 6.8.68, 230 OCU 18.3.69, Scampton Wing 3.5.71 , 230 OCU 7.5.71. Conversion to B.2 (MRR) 9.2.73, air sampling pod modifications. 27 Squadron 17.12.73. Allocated 8744M for preservation at Scampton 31.3.82. Scrapped 11.86. Nose privately owned, arrived at Durham Tees Valley Airport 25.4.17.

XH563 completed her service career with 27 Squadron before being retired and placed on display at Scampton, carrying the markings of the four Scampton Vulcan units on the tail. Sadly, the aircraft did not survive further changes in command and the airframe was scrapped in 1986 with only the nose section surviving.

Vulcan B Mk 2

Contract 6/Air/l1 830/CB.6(a) for eight aircraft, 31.3.55.

18/XJ780 Completed 10.1.61. Olympus 201 engines. 83 Squadron 16.1.61, 12 Squadron 26.11.62, 230 OCU 16.8.63, Waddington Wing 10.10.67, Cottesmore Wing 6.12.68, Waddington Wing 18.4.69, Akrotiri Wing 12.1.70, Waddington Wing 17.1.75. Modification to B.2 (MRR) 31.3.76. 27 Squadron 23.11.76. Allocated for spares recovery 31.3.82. Sold to Bird Group as scrap 11.82.

19/XJ781 Completed 20.2.61. Olympus 201 engines. 83 Squadron 23.2.61, 12 Squadron 29.10.62, 230 OCU 4.2.64, Waddington Wing 10.2.66, Cottesmore Wing 22.4.68, Akrotiri Wing 18.4.69. Damaged during landing at Shiraz, Iran, 23.5.73, struck off charge 27.5.73.

20/XJ782 Completed 2.61. Olympus 201 engines. 83 Squadron 2.3.61, 12 Squadron 23.10.62, 230 OCU 20.12.63, Waddington Wing 25.3.66, Cottesmore Wing 9.4.68, Akrotiri Wing 19.3.69, Waddington Wing 8.1.75. Modification to B.2 (MRR) 1.77. 27 Squadron 15.2.77. Flew last Vulcan sortie at Scampton 31.3.82. Allocated to Scampton dump 31.3.82. Re-allocated to 101 Squadron 22.5.82. To Finningley for preservation 4.9.82 (8766M). Later transferred to dump and scrapped 5.88.

21/XJ783 Completed 6.3.61. Olympus 201 engines. 83 Squadron 13.3.61, 9 Squadron 7.11.62, 230 OCU 28.2.64, Waddington Wing 3.1.66, Cottesmore Wing 22.3.68, Akrotiri Wing 15.1.69, 35 Squadron 16.1.75, 230 OCU 11.8.76, 35 Squadron 23.8.76, 617 Squadron 23.11.78, 35 Squadron 3.4.81. Spares recovery 1.3.82. Sold to Bird Group as scrap 11.82.

B.2(MRR) XJ780 of 27 Sqn at Scampton in February 1982. *Fred Martin*

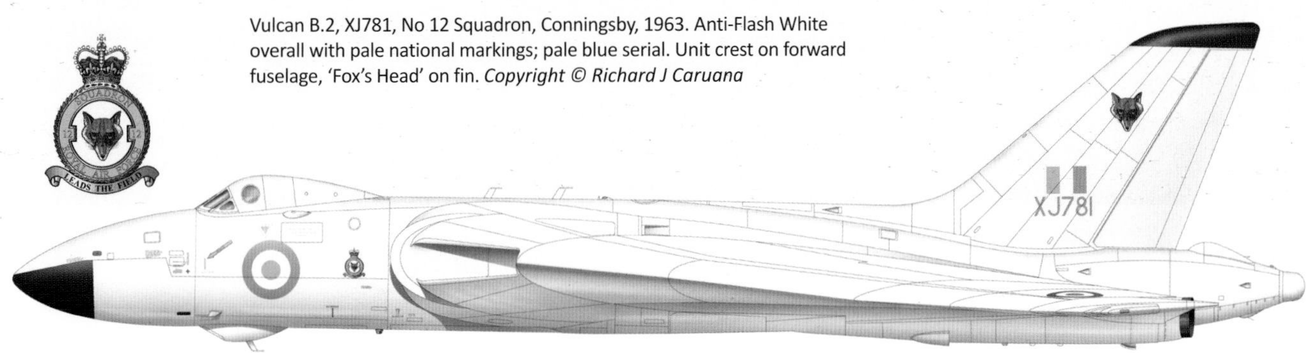

Vulcan B.2, XJ781, No 12 Squadron, Conningsby, 1963. Anti-Flash White overall with pale national markings; pale blue serial. Unit crest on forward fuselage, 'Fox's Head' on fin. *Copyright © Richard J Caruana*

27 Sqn B.2 XJ782 at Leconfield in August 1976. *Fred Martin*

35 Sqn B.2 XJ783 at Scampton in February 1982. *Fred Martin*

XJ823, a former 27 Squadron B.2 (MRR), spent her final months in RAF service with No 50 Squadron, the nose-mounted refuelling probe having been 'stolen' for use during Operation 'Corporate'. The aircraft survives at Carlisle Airport.
Michael J. Freer

Vulcan B.2, XJ824, No 35 Squadron, Scampton, 1975. Medium Sea Grey/Dark Green upper surfaces (with areas repainted in Dark Sea Grey); Light Aircraft Grey undersides. 'Tactical' national markings; unit badge on fin.
Copyright © Richard J Caruana

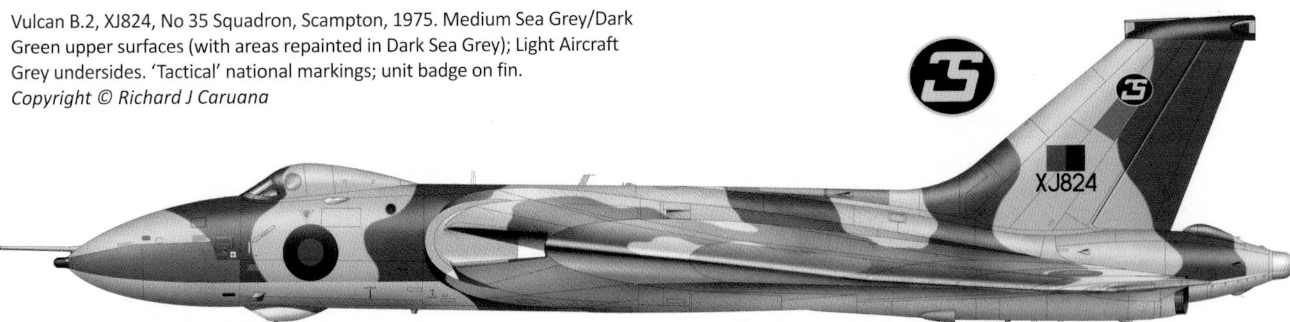

22/XJ784 Completed 30.3.61. Olympus 201 engines. Refitted with 301 engines. A&AEE 29.3.61, 230 OCU 22.12.66, Akrotiri Wing 21.7.70, Waddington Wing 15.1.75, 9 Squadron 2.75, 44 Squadron 6.79, 101 Squadron 6.80. Spares recovery 10.9.82. Sold to Bird Group as scrap 8.12.82.

23/XJ823 Completed 20.4.61. Olympus 201 engines. 27 Squadron 21.4.61, 35 Squadron 2.1.63, Cottesmore Wing 3.64, 230 OCU 11.5.64, Waddington Wing 1.11.66, Cottesmore Wing 29.4.68, Akrotiri Wing 5.2.69, Waddington Wing 17.1.75, 9 Squadron 2.75. Modified to B.2 (MRR) 3.77. 27 Squadron 27.4.77, 35 Squadron 2.4.81. Waddington Wing 1.3.82, 9 Squadron 3.82, 50 Squadron 4.82, Station Flight Waddington 4.1.83. Sold to T. Stoddart 21.1.83, delivered to Solway Aviation Society, Carlisle, 24.1.83. After 83 add: Displayed at the Carlisle Aviation Museum.

24/XJ824 Completed 11.5.61. Olympus 201 engines. 27 Squadron 16.5.61, 9 Squadron 25.2.63, 230 OCU 2.12.63, Cottesmore Wing 4.7.66, Waddington Wing 4.10.66, Cottesmore Wing 19.6.68, Akrotiri Wing 5.2.69, 35 Squadron 24.1.75, 230 OCU 14.2.77, 44 Squadron 10.79, 101 Squadron 7.82. Last Vulcan to leave Bitteswell after manufacturer's modifications 8.6.81. Delivered to Imperial War Museum, Duxford, 13.3.82.

25/XJ825 Completed 27.7.61. Olympus 201 engines. 27 Squadron 28.7.61, 35 Squadron 4.2.63, 230 OCU 30.4.64, Cottesmore Wing 3.9.65, Waddington Wing 11.4.67, Cottesmore Wing 19.2.68, Akrotiri Wing 26.2.69, 35 Squadron 16.1.75. Modification to B.2 (MRR) 13.1.76. 27 Squadron 15.12.76, 35 Squadron 6.4.81, 101 Squadron 1.3.82. Conversion to K.2 11.5.82. 50 Squadron 25.6.82. Allocated 8810M for battle damage repair duties 22.3.84. Struck off charge 5.4.84, scrapped 1.92.

XL318 inside one of Scampton's hangars being dismantled prior to transportation to Hendon. *Bruce Woodruff*

Vulcan B Mk 2

Contract 6/Air/13145/CB.6(a) for twenty-four aircraft, 25.2.56.

26/XL317 Completed 14.7.61. Olympus 201 engines. Blue Steel modifications. A&AEE 13.7.61, 617 Squadron 7.6.62, 230 OCU 24.4.74, 617 Squadron 1.5.74. To Akrotiri as 8725M for crash rescue training, delivered 1.12.81, scrapped 12.86.

27/XL318 Completed 30.8.61. Olympus 201 engines. Blue Steel modifications. 617 Squadron 4.9.61, 230 OCU 22.5.72, 27 Squadron 31.1.74, 230 OCU 1.2.74, Waddington Wing 18.6.75, 230 OCU 5.8.75, Waddington Wing 7.1 1.79, 230 OCU 21.2.80, 617 Squadron 1.7.81. Last sortie by 617 Squadron Vulcan 11.12.81. Assigned to RAF Museum 4.1.82 as 8733M. Transported to Hendon 12.2.82.

28/XL319 Completed 19.10.61. Olympus 201 engines. Blue Steel modifications. 617 Squadron 23.10.61, 230 OCU 14.5.70, Scampton Wing 12.11.70, 617 Squadron 22.4.71, 230 OCU 19.9.72, 35 Squadron 16.10.78, 44 Squadron 1.3.82. Sold to North East Aircraft Museum 20.1.83. Delivered to Sunderland 21.1.83 and displayed at the North East Land, Sea and Air Museum.

29/XL320 Completed 30.11.61. Olympus 201 engines. Blue Steel modifications. 617 Squadron 4.12.61, 83 Squadron 6.71, 27 Squadron 9.71, 230 OCU 29.3.72. Flew 500,000th Vulcan hour 18.12.81. To St Athan 2.6.81, sold to W. Harold & Co as scrap 31.8.81.

30/XL321 Completed 10.1.62. Olympus 201 engines. Blue Steel modifications. 617 Squadron 11.1.62, 27 Squadron 1.71, 230 OCU 29.3.72, 617 Squadron 15.9.72, 230 OCU 11.10.72, 617 Squadron 13.4.73, 44 Squadron 8.6.76, 230 OCU 8.11.76, 35 Squadron 1.7.81, 617 Squadron 14.9.81, 35 Squadron 6.10.81, 50 Squadron 21.1.82. Delivered to Firefighting School Catterick 19.8.82 (8759M). Highest individual Vulcan operational flying hours (6,952.35). Later scrapped.

31/XL359 Completed 31.1.61. Olympus 201 engines. Blue Steel modifications. 617 Squadron 1.2.62, 27 Squadron 3.71, 230 OCU 21.10.71, 35 Squadron 1.7.81. Allocated as gate guard at Scampton 1.3.81, but later dumped. Sold to Bird Group as scrap 11.82.

Vulcan B.2, XL321, No 617 Squadron, Scampton, 1964. Anti-Flash White overall with pale national markings; serial in pale blue. Unit crest and command pennant on forward fuselage. *Copyright © Richard J Caruana*

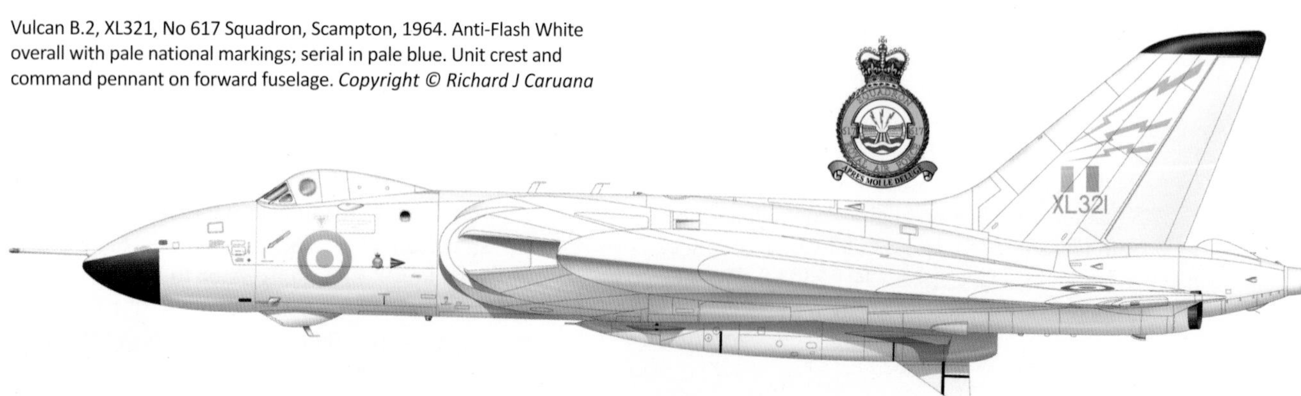

XL318 inside one of Scampton's hangars being dismantled prior to transportation to Hendon. *Bruce Woodruff*

A very clean-looking B.2, XL318 of 617 Sqn, at Waddington in August 1981. Five months later the aircraft was allocated to the RAF Museum, Hendon, for preservation. *Martin Derry*

Above: XL319, seen at Scampton in February 1982, is now preserved at the North East Aircraft Museum at Usworth, Sunderland. *Fred Martin*

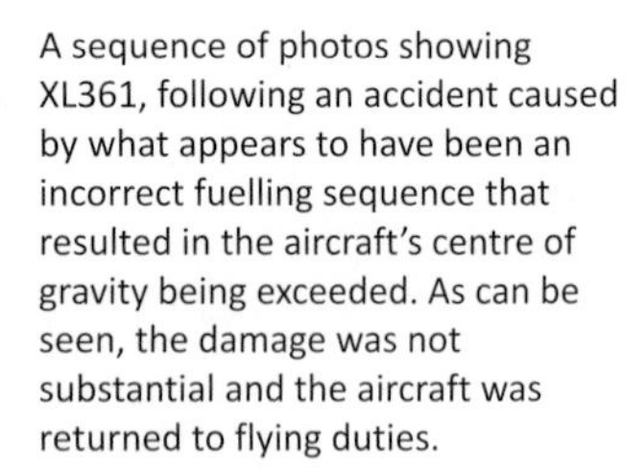

A sequence of photos showing XL361, following an accident caused by what appears to have been an incorrect fuelling sequence that resulted in the aircraft's centre of gravity being exceeded. As can be seen, the damage was not substantial and the aircraft was returned to flying duties.

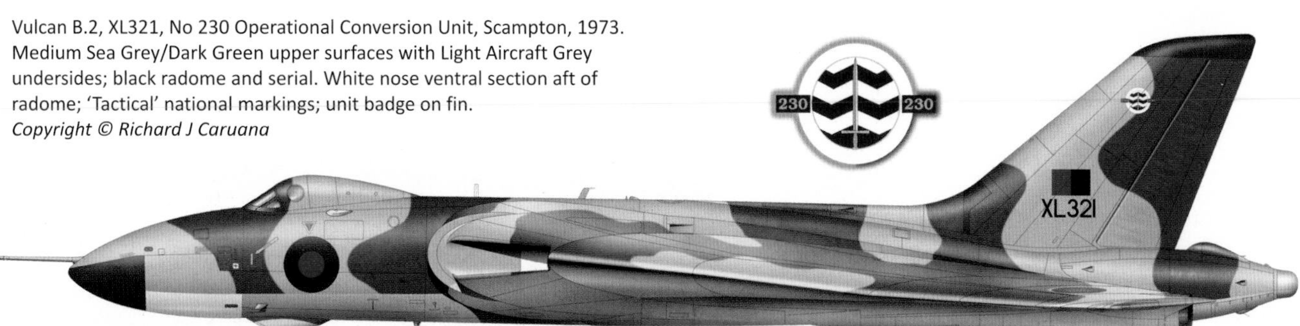

Vulcan B.2, XL321, No 230 Operational Conversion Unit, Scampton, 1973. Medium Sea Grey/Dark Green upper surfaces with Light Aircraft Grey undersides; black radome and serial. White nose ventral section aft of radome; 'Tactical' national markings; unit badge on fin. *Copyright © Richard J Caruana*

32/XL360 Completed 28.2.62. Olympus 201 engines. Blue Steel modifications. 617 Squadron 2.3.62, 230 OCU 13.7.71, Waddington Wing 18.8.75, 230 OCU 21.10.75, 617 Squadron 5.12.77, 35 Squadron 31.5.78, 101 Squadron 5.1.82. Sold to Midland Air Museum 26.1.83, delivered to Baginton 4.2.83.

33/XL361 Completed 14.3.62. Olympus 201 engines. Blue Steel modifications. 617 Squadron 15.3.62, 230 OCU 7.10.70, Scampton Wing 19.11.70, 230 OCU 30.11.70, Scampton Wing 5.4.71, OCU 12.5.71, 27 Squadron/230 OCU 14.1.74, A&AEE 7.8.75, 617 Squadron 3.9.75, 35 Squadron 3.8.77, 9 Squadron 13.4.81. Accident at Goose Bay, Canada, 13. 11.81. Grounded 21.12.81 and placed on display at Happy Valley, Goose Bay, 7.6.82.

34/XL384 Completed 30.3.62. Olympus 201 engines. Blue Steel modifications including 301 engines. 230 OCU 2.4.62, Scampton Wing 5.8.64, Waddington Wing 23.3.70, 230 OCU 27.11.70. Heavy landing 12.8.71. Allocated as escape trainer (8505M) 30.9.76. Later transferred to crash rescue training as 8670M 29.1.81, struck off charge 23.5.85.

35/XL385 Completed 17.4.62. Olympus 201 engines. Blue Steel modifications. 9 Squadron 18.4.62, Scampton Wing 9.10.64. Ground fire at Scampton 6.4.67, struck off charge 7.4.67.

36/XL386 Completed 11.5.62. Olympus 201 engines. Blue Steel modifications including 301 engines. 9 Squadron 14.5.62, Scampton Wing 16.8.65, 230 OCU 1.4.70, 44 Squadron 30.9.77, 101 Squadron 5.81, 50 Squadron 10.81. Delivered to Central Training Establishment, Manston, 26.8.82 as 8760M, scrapped 9.92.

37/XL387 Completed 31.5.62. Olympus 201 engines. Blue Steel modifications including 301 engines. 230 OCU 4.6.62, Scampton Wing 5.2.65, 230 OCU 5.7.72, 101 Squadron 10.1.73, 50 Squadron 8.75. To St Athan for crash rescue training 28.1.82. Sold to T. Bradbury as scrap 2.6.83.

38/XL388 Completed 13.6.62. Olympus 201 engines. Blue Steel modifications including 301 engines. 9 Squadron 14.6.62 (Coningsby Wing), 230 OCU 19.4.71, 617 Squadron 15.9.72, 230 OCU 29.9.72, 617 Squadron 2.2.73, 230 OCU 14.5.73, 617 Squadron 1.11.73, 230 OCU 5.3.74, 617 Squadron 6.3.74, 44 Squadron 1.4.74. To Honington fire dump 2.4.82 (8750M). Sold as scrap 13.6.85. Nose with South Yorkshire Aviation Museum, Doncaster, since 7.4.03.0.

39/XL389 Completed 11.7.62. Olympus 201 engines. Blue Steel modifications including 301 engines. 230 OCU 13.7.62, Scampton Wing 20.5.65, 230 OCU 11.11.70, 617 Squadron 12.70, 230 OCU 7.4.72, 617 Squadron 30.6.72, 230 OCU 12.1.73, 617 Squadron 16.1.73, 9 Squadron 26.6.74, 44 Squadron 7.79, 101 Squadron 6.80. To St Athan 6.4.81. Sold to W. Harold as scrap 31.8.81.

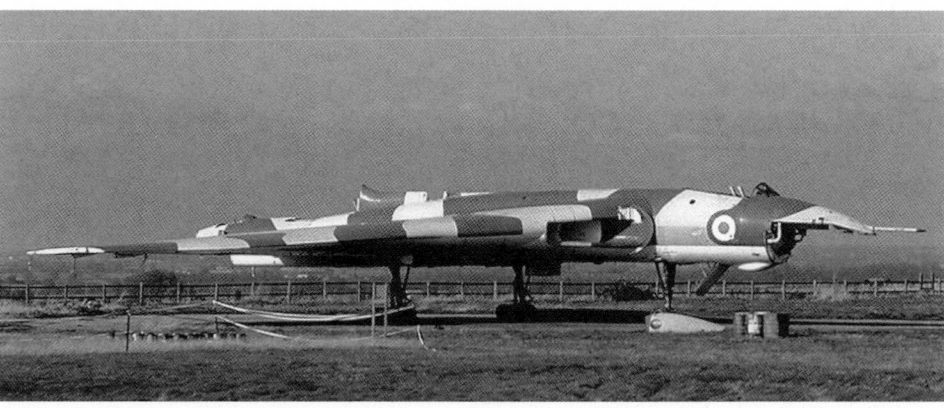
XL384 became a familiar sight for aircraft spotters at Scampton, having languished on the western end of the airfield for many years as a crash rescue airframe. *Terry Senior*

617 Sqn B.2 XL392 at Ramstein, in the then West Germany, in August 1979. *Fred Martin*

40/XL390 Completed 19.7.62. Olympus 201 engines. Blue Steel modifications including 301 engines. First production aircraft with Skybolt hard points. 9 Squadron 20.7.62, Scampton Wing 27.5.65, 230 OCU 30.4.71, 617 Squadron 3.6.71, 230 OCU 31.5.74, 617 Squadron 4.6.74. Crashed during air display, Glenview NAS, Illinois USA, 12.8.78.

41/XL391 Completed 22.5.63. Olympus 301 engines. Blue Steel modifications. A&AEE 22.5.63, BCDU 15.6.65, A&AEE 6.1.66, Cottesmore Wing 31.7.68, Akrotiri Wing 5.2.69, 9 Squadron 17.1.75, 101 Squadron 6.80. Selected for 'Black Buck' modifications, but not used operationally. 44 Squadron 2.6.82. Sold to Manchester Vulcan Bomber Society 11.2.83. Delivered to Blackpool 16.2.83. Scrapped 12.1.06.

42/XL392 Completed 31.7.62. Olympus 201 engines. Blue Steel modifications. 83 Squadron 2.8.62, 230 OCU 11.12.70, Scampton Wing 21.12.70, 230 OCU 12.1.73, 617 Squadron 15.1.73, 35 Squadron 4.1.82. Delivered to Valley for crash rescue training as 8745M 24.3.82, scrapped 8.83.

43/XL425 Completed 30.8.62. Olympus 201 engines. Blue Steel modifications. 83 Squadron 31.8.62, Scampton Wing, 617 Squadron 30.11.72, 27 Squadron 1.11.73, 617 Squadron 1.4.74. Grounded 4.1.82. Sold to Bird Group as scrap 4.82.

44/XL426 Completed 7.9.62. Olympus 201 engines. Blue Steel modifications. 83 Squadron 13.9.62, Scampton Wing, 230 OCU 29.3.72, 617 Squadron 7.4.72, 230 OCU 28.6.72, 617 Squadron 4.7.72, 230 OCU 11.7.72, 617 Squadron 1.8.72, 230 OCU 13.4.73, 617 Squadron 16.4.73, 27 Squadron 6.2.74, 617 Squadron 21.2.74, 50 Squadron 5.1.82, 55 Squadron 1.4.84. Sold to Roy Jacobsen and delivered to Southend 19.12.86. Registered as G-VJET 7.7.87. Maintained in taxi condition by the Vulcan Restoration Trust.

45/XL427 Completed 29.9.62. Olympus 201 engines. Blue Steel modifications. 83 Squadron 2.10.62, Scampton Wing, 617 Squadron 9.69, 27 Squadron 3.71, 230 OCU 29.6.72, 617 Squadron 5.7.72, 230 OCU 25.9.72, 27 Squadron 4.1.74, 230 OCU 11.8.76, 27 Squadron 27.8.76, 9 Squadron 2.5.77, 50 Squadron 4.81, 9 Squadron 10.81, 44 Squadron 6.82. Delivered to Machrihanish for crash rescue training as 8756M 13.8.82. Scrapped 6.95.

XL427 rolls out after touchdown at Waddington in June 1981. *Fred Martin*

XL445 receiving a freshly packed brake parachute within 230 OCU's dispersal area at Scampton.
Joe L'Estrange

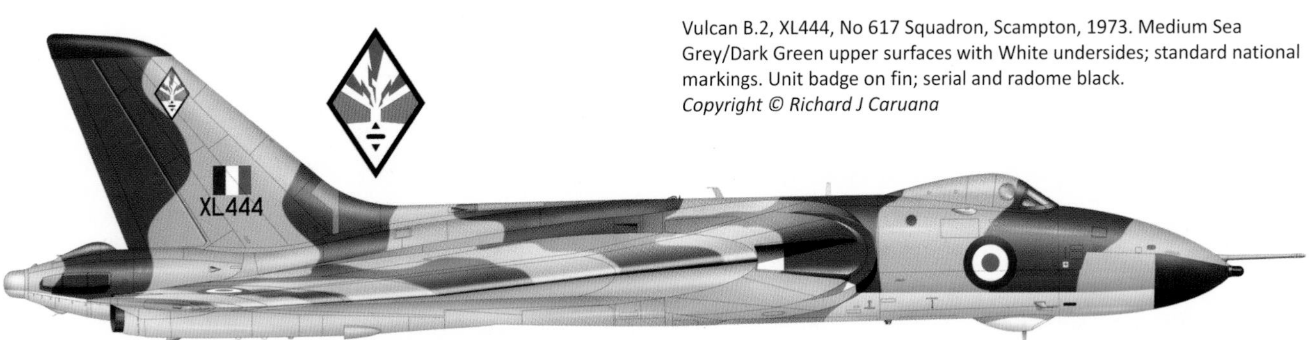

Vulcan B.2, XL444, No 617 Squadron, Scampton, 1973. Medium Sea Grey/Dark Green upper surfaces with White undersides; standard national markings. Unit badge on fin; serial and radome black.
Copyright © Richard J Caruana

46/XL443 Completed 4.10.62. Olympus 201 engines. Blue Steel modifications. 83 Squadron 8.10.62, Scampton Wing, Akrotiri Wing 12.4.72, 35 Squadron 24.1.75. Allocated to RAF Museum 4.1.82. Later sold to Bird Group as scrap 4.82.

47/XL444 Completed 29.10.82. Olympus 201 engines. Blue Steel modifications. 27 Squadron 1.11.62, Scampton Wing, 230 OCU 18.6.66, Scampton Wing 19.6.67, 230 OCU 5.4.71 , 27 Squadron 11 .5.71, 617 Squadron 9.71, 230 OCU/617 Squadron 19.7.72, 617 Squadron 18.12.73, 35 Squadron 31.5.78, 9 Squadron 6.4.81. Grounded 10.9.82. Sold to Bird Group as scrap 8.12.82.

48/XL445 Completed 19.11.62. Olympus 201 engines. Blue Steel modifications. 27 Squadron 26.11.62, Scampton Wing, Waddington Wing 30.9.66, Cottesmore Wing 18.4.68, Akrotiri Wing 15.1.69, 35 Squadron 16.1.75, Waddington Wing 16.6.77, 35 Squadron 1.10.77, 230 OCU 16.10.78, 35 Squadron 1.7.81, 44 Squadron 18.11.81. Conversion to K.2 25.5.82. 50 Squadron 22.7.82. Allocated 8811M for crash rescue training 22.3.84. Delivered Lyneham 1.4.84. Scrapped 1990. Nose at the Norfolk and Suffolk Aviation Museum, Flixton, since 29.7.05.

49/XL446 Completed 19.11.62. Olympus 201 engines. Blue Steel modifications. 27 Squadron 30.11.62, Scampton Wing, Waddington Wing 16.9.66, 230 OCU 28.12.67, Scampton Wing 18.4.72, Akrotiri Wing 31.7.72, 35 Squadron 16.1.75, Waddington Wing 24.5.78, 617 Squadron 31.10.78, 35 Squadron 4.82. Grounded 1.3.82. Sold to Bird Group as scrap 11.82.

Vulcan B Mk 2

Contract KD/B/01/CB.6(a) for forty aircraft, 22.1.58.

50/XM569 Completed 4.1.63. Olympus 201 engines. Blue Steel modifications. 27 Squadron 1.2.63, Scampton Wing, Waddington Wing 17.11.66, Cottesmore Wing 19.1.68, Akrotiri Wing 26.2.69, 27 Squadron 4.7.74, 9 Squadron 23.11.76, 50 Squadron 6.79, 101 Squadron 9.81, 44 Squadron 8.82. Sold to Wales Aircraft Museum 21.1.83, delivered to Cardiff 2.2.83. Scrapped 2.96. Nose to the Jet Age Museum, Staverton 6.2.97.

51/XM570 Completed 26.2.63. Olympus 201 engines. Blue Steel modifications. 27 Squadron 27.2.63, Scampton Wing, Waddington Wing 2.1.67, Cottesmore Wing 26.1.68, Akrotiri Wing 26.2.69, 27 Squadron 8.3.74, 35 Squadron 8.12.76, 230 OCU 28.2.77, 35 Squadron 2.3.77, 617 Squadron 4.9.78, 35 Squadron 31.10.78. Delivered to St Athan 11.3.81. Sold to Harold John & Co as scrap 29.1.82.

52/XM571 Completed 20.2.63. Olympus 201 engines. Blue Steel modifications. 83 Squadron 22.2.63, Scampton Wing, Cottesmore Wing 20.1.67, Waddington Wing 3.7.67, Cottesmore Wing 13.9.67, Waddington Wing 15.12.67, Akrotiri Wing 19.3.69, 27 Squadron 3.1.75, 35 Squadron 9.4.75, Waddington Wing 16.6.75, 35 Squadron 3.11.75, 50 Squadron 15.6.76, 35 Squadron 15.11.76, St Athan 9.1.79, 617 Squadron 27.3.79, Waddington Wing 20.8.79, 617 Squadron 4.12.79, 101 Squadron 8.1.82. Conversion to K.2 11.5.82. 50 Squadron 25.8.82, Waddington Station Flight 4.84. Allocated 8812M 22.3.84. Delivered to Gibraltar for preservation 9.5.84. Scrapped on site 1989.

53/XM572 Completed 28.2.63. Olympus 201 engines. Blue Steel modifications. 83 Squadron 28.2.63, Scampton Wing, Cottesmore Wing 5.4.68, Akrotiri Wing 19.3.69, 35 Squadron 24.1.75, 9 Squadron 2.9.81. Grounded 10.9.82. Sold to Bird Group as scrap 30.11.82.

XM570 enjoyed a successful if uneventful service career, eventually being delivered to St Athan for storage pending disposal in 1981. *Ray Deacon*

Below: An atmospheric shot of XM571 at Scampton prior to conversion to tanker configuration. The aircraft was delivered to Gibraltar on retirement for public display, but by the late 1980s its condition had deteriorated and it was scrapped on site. *Shaun Connor*

B.2 XM572 of 9 Sqn taxiing at Waddington in April 1982. *Fred Martin*

XM572 on display at Finningley's Battle of Britain 'At Home' Day, possibly in 1978. The Vulcan's 'sister' aircraft (the Victor) is just visible on the adjacent V-Bomber dispersal. *Shaun Connor*

A magnificent view of XM573 climbing out on take-off, illustrating the Vulcan's sturdy undercarriage units, the huge main gear doors, and the ECM plates fitted between the jet pipes. *Shaun Connor*

XM573 taxiing at Waddington. The upper surface camouflage scheme was extended down the wing leading edge to the lower lip of the air intake. *Shaun Connor*

54/XM573 Completed 26.3.63. Olympus 201 engines. Blue Steel modifications. 83 Squadron 28.3.63, Scampton Wing, Waddington Wing 25.4.67, 230 OCU 15.2.68, Akrotiri Wing 26.6.70, 27 Squadron 17.4.74, 44 Squadron 9.3.77, 230 OCU 18.12.78, 9 Squadron 7.4.81, Scampton 22.5.82. Delivered to Offutt AFB, Nebraska, USA, 7.6.82, presented to USAF 12.6.82. Displayed at the East Midlands Airport Aeropark.

55/XM574 Completed 12.6.63. Olympus 301 engines. Blue Steel modifications. 27 Squadron 21.6.63, Scampton Wing, 230 OCU 3.5.71, 27 Squadron 12.5.71, 101 Squadron 3.11.71, Akrotiri Wing 24.8.73, 35 Squadron 24.1.75, 617 Squadron 14.8.75. To St Athan 31.8.81. Sold to Harold John & Co as scrap 29.1.82.

XM574 sharing a rather wet hardstanding at Mildenhall with a much older example of classic bomber design. *Robin A. Walker*

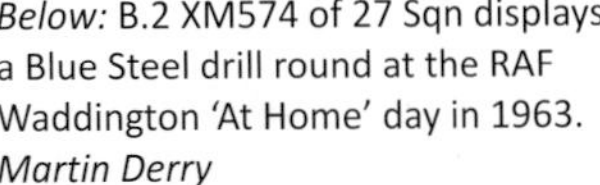

Below: B.2 XM574 of 27 Sqn displays a Blue Steel drill round at the RAF Waddington 'At Home' day in 1963. *Martin Derry*

B.2 XM574 of the Waddington Wing seen at Finningley in September 1972. *Fred Martin*

B.2 XM575 of 44 Sqn took part in the last four-ship scramble of Vulcans at the Finningley air show in September 1981; the aircraft is seen here at Waddington three months earlier. *Fred Martin*

56/XM575 Completed 21.5.63. Olympus 301 engines. Blue Steel modifications. 617 Squadron 22.5.63, Scampton Wing, Waddington Wing 28.7.70, Scampton Wing 27.11.70, 230 OCU 3.5.71, 617 Squadron 7.5.71, 101 Squadron 15.3.74, 50 Squadron 6.78, 44 Squadron 8.79. Sold 21.1.83, delivered to Castle Donington 28.1.83. Displayed at the East Midlands Airport Aeropark.

57/XM576 Completed 14.6.63. Olympus 301 engines. Blue Steel modifications. Scampton Wing 21.6.63. Crash-landed Scampton 25.5.65. Struck off charge 12.65.

58/XM594 Completed 9.7.63. Olympus 301 engines. Blue Steel modifications. 27 Squadron 19.7.63, Scampton Wing, Waddington Wing 24.8.72, 101 Squadron 6.75, 44 Squadron 5.77. Sold to Newark Air Museum 19.1.83, delivered to Winthorpe 7.2.83.

59/XM595 Completed 21.8.63. Olympus 301 engines. Blue Steel modifications. 617 Squadron 21.8.63, Scampton Wing, 27 Squadron 16.8.74, 617 Squadron 9.75, 35 Squadron 11.76, 617 Squadron 2.78, 35 Squadron 4.1.82. Grounded 1.3.82. Sold to Bird Group as scrap 11.82.

60/XM596 Not completed. Aircraft used for static fatigue tests at Woodford, in connection with low-level operations. Scrapped 1972.

B.2 XM594 tucks up the gear at the 1981 Abingdon air show. She is now preserved at Newark Air Museum. *Fred Martin*

Above: Immaculate B.2 XM594 of 44 Sqn posed with drill bombs at Waddington in June 1978. *Fred Martin*

XM595 has a Blue Steel round attached under the bomb bay. This photograph must have been taken shortly after the aircraft was repainted in low-level camouflage as the new paint scheme still extends over the 301-series jet pipes. *Ken Elliott*

Vulcan B.2, XM595, No 27 Squadron, alcombury, 14 August 1971. Medium Sea Grey/Dark Green uppersurfaces with White undersides; black serial and radome. Unit badge on fin on white disc.
Copyright © Richard J Caruana

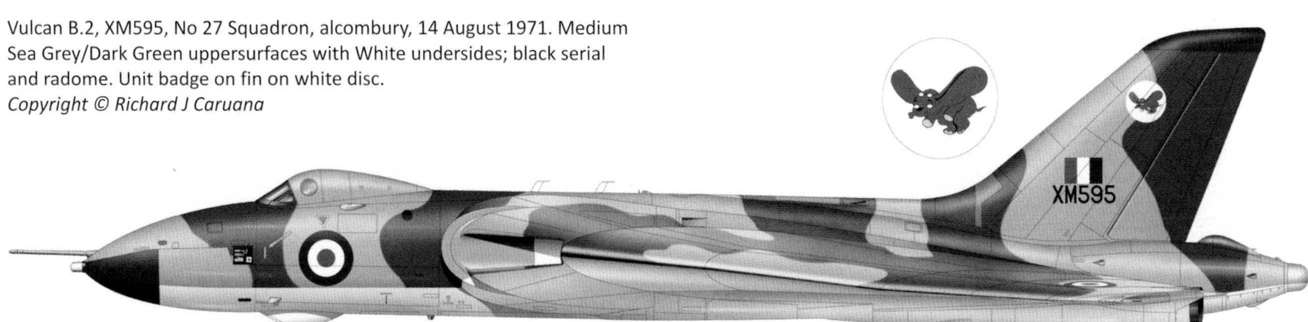

61/XM597 Completed 26.8.63. Olympus 301 engines. Blue Steel modifications. 12 Squadron 27.8.63, Coningsby Wing, Cottesmore Wing 18.12.64, Waddington Wing 18.4.68, A&AEE 29.11.71, 101 Squadron 8.73, 44 Squadron 9.75, 50 Squadron 4.76, 9 Squadron 5.79, 44 Squadron 10.81, 101 Squadron 7.82. Modified for 'Black Buck' operations. 44 Squadron 1.7.82, 50 Squadron 24.12.82. Delivered to East Fortune 12.4.84, displayed at the National Museum of Flight Scotland.

62/XM598 Completed 30.8.63. Olympus 301 engines. 12 Squadron 4.9.63, Coningsby Wing, Cottesmore Wing 11.64, Waddington Wing 9.4.68, 101 Squadron 5.72, 44 Squadron 8.75, 50 Squadron 4.78, 9 Squadron 10.79, 50 Squadron 10.81, 44 Squadron 6.82. Modified for 'Black Buck' operations. Allocated 8778M 4.1.83, delivered to RAF Museum, Cosford 20.1.83. Displayed at the National Cold War Exhibition.

63/XM559 Completed 30.9.63. Olympus 301 engines. 35 Squadron 1.10.63, Coningsby Wing, Waddington Wing 9.12.68, 101 Squadron 5.72, 50 Squadron 3.77, 44 Squadron 6.79. Delivered to St Athan 27.5.81. Sold to H. John & Co as scrap 2.9.82.

64/XM600 Completed 30.9.63. Olympus 301 engines. 35 Squadron 3.10.63, Coningsby Wing, Waddington Wing 3.5.68, 101 Squadron 8.73. Crashed near Spilsby following engine bay fire 18.1.77.

65/XM601 Completed 31.10.63. Olympus 301 engines. 9 Squadron 5.11.63, Coningsby Wing. Crashed on approach to Coningsby 7.10.64.

66/XM602 Completed 11.11.63. Olympus 301 engines. 12 Squadron 13.11.63, Coningsby Wing, Cottesmore Wing, Waddington Wing 24.4.68, 9 Squadron 12.75, 230 OCU 19.10.76, 35 Squadron 29.10.76, Waddington Wing 1.11.76, 101 Squadron 5.80. To St Athan 7.1.82. Transferred to St Athan Historic Aircraft Museum 16.3.83 (8771M). Scrapped 10.92. Nose displayed at Avro Heritage Museum, Woodford, since 20.4.13.

A close-up look at the 35 Squadron markings applied to XM600, an aircraft that was ultimately destroyed after being abandoned near Spilsby following an engine fire on 18 January 1977. *Joe L'Estrange*

Below: XM603 was returned to Woodford upon retirement and the Avro Heritage team (all former Avro employees) worked hard to it, resulting in a beautiful white-finished aircraft. It is now displayed at the Avro Heritage Museum. *Paul Tomlin*

101 Sqn B.2 XM605 touching down at Waddington in June 1978. She was flown to Castle AFB Museum, California in 1981, but later had her refuelling probe reclaimed for Falklands operations. *Fred Martin*

67/XM603 Completed 29.11.63. Olympus 301 engines. 9 Squadron 4.12.63, Coningsby Wing, Cottesmore Wing, Waddington Wing 18.1.68, 50 Squadron 8.75, 101 Squadron 12.80, 44 Squadron 7.81. Sold to British Aerospace for preservation, delivered to Woodford 12.3.82. Mock-up aircraft for K.2 conversions. Handed over to the Avro Heritage Museum, Woodford, 17.2.13.

68/XM604 Completed 29.11.63. Olympus 301 engines. 35 Squadron 4.12.63, Coningsby Wing, Cottesmore Wing. Crashed near Cottesmore following loss of control during overshoot (engine compressor failure) 30.1.68.

69/XM605 Completed 17.12.63. Olympus 301 engines. 9 Squadron 30.12.63, Coningsby Wing, Cottesmore Wing, Waddington Wing 16.12.68, 101 Squadron 8.73, 50 Squadron 5.79. Delivered to Castle California after AFB, USA, 2.9.81, presented to USAF 8.9.81. Displayed at the Castle Air Museum.

70/XM606 Completed 18.12.63. Olympus 301 engines. 12 Squadron 30.12.63, Coningsby Wing, Cottesmore Wing 18.2.65, MoA 14.6.65, Cottesmore Wing 5.4.68, Waddington Wing 13.5.68, 101 Squadron 12.75, 9 Squadron 6.79. Delivered to Barksdale AFB, Louisiana USA, 7.6.82, presented to USAF 14.6.83. Displayed at the Eighth Air Force Museum, Barksdale.

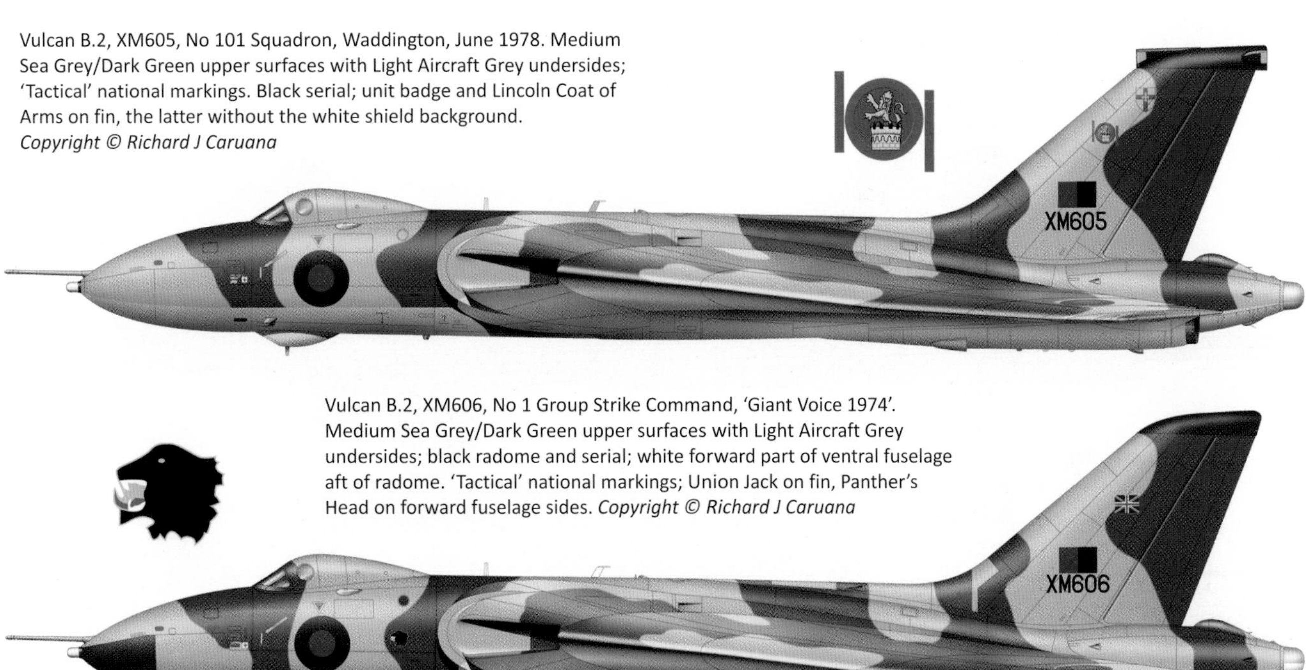

Vulcan B.2, XM605, No 101 Squadron, Waddington, June 1978. Medium Sea Grey/Dark Green upper surfaces with Light Aircraft Grey undersides; 'Tactical' national markings. Black serial; unit badge and Lincoln Coat of Arms on fin, the latter without the white shield background. *Copyright © Richard J Caruana*

Vulcan B.2, XM606, No 1 Group Strike Command, 'Giant Voice 1974'. Medium Sea Grey/Dark Green upper surfaces with Light Aircraft Grey undersides; black radome and serial; white forward part of ventral fuselage aft of radome. 'Tactical' national markings; Union Jack on fin, Panther's Head on forward fuselage sides. *Copyright © Richard J Caruana*

Seen about to touch down at Waddington in March 1982, XM606 is now displayed at Barksdale AFB. *Fred Martin*

An impressive view of XM606 of 9 Sqn as she turns in to her Waddington dispersal in March 1982. *Fred Martin*

Above: B.2 XM607 of the Waddington Wing waits on dispersal at Finningley. *Fred Martin*

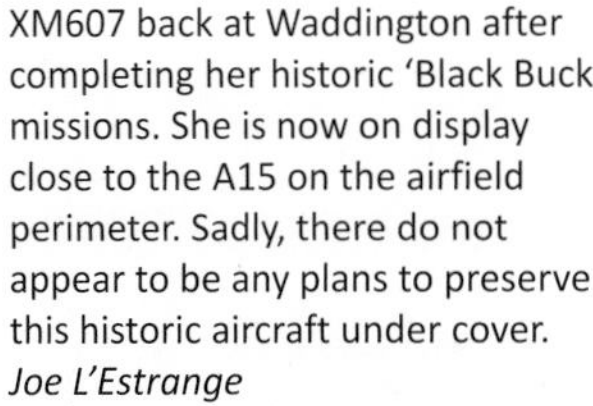

XM607 back at Waddington after completing her historic 'Black Buck' missions. She is now on display close to the A15 on the airfield perimeter. Sadly, there do not appear to be any plans to preserve this historic aircraft under cover. *Joe L'Estrange*

71/XM607 Completed 30.12.63. Olympus 301 engines. 35 Squadron 1.1.64, Coningsby Wing, Cottesmore Wing, Waddington Wing 24.5.68, 44 Squadron 4.76, 9 Squadron 5.79, 101 Squadron 3.81, 44 Squadron 7.81. Modified for 'Black Buck' operations. 44 Squadron 14.6.82. Withdrawn from use 17.12.82, allocated 8779M 4.1.83. To static display at Waddington 19.1.83, displayed on airfield.

72/XM608 Completed 28.1.64. Olympus 301 engines. Coningsby Wing, HSL mods 16.1.65, Cottesmore Wing 9.4.65, Waddington Wing 23.2.68. To St Athan 6.4.81. Sold to Bird Group as scrap 2.12.82.

73/XM609 Completed 28.1.64. Olympus 301 engines. 12 Squadron 29.1.64, Coningsby Wing, Cottesmore Wing 3.3.65, 230 OCU 7.8.67, Cottesmore Wing 1.10.67, Waddington Wing 8.3.68, 9 Squadron 9.75, 44 Squadron 4.76. To St Athan 12.3.81. Sold to W. Harold & Co as scrap 31.8.81.

74/XM610 Completed 10.2.64. Olympus 301 engines. 9 Squadron 12.2.64, Coningsby Wing, Cottesmore Wing, Waddington Wing 5.2.68. Crashed Wingate 8.1.71 following engine bay fire.

75/XM611 Completed 12.2.64. Olympus 301 engines. 9 Squadron 14.2.64, Coningsby Wing, Cottesmore Wing, Waddington Wing 28.5.68, 101 Squadron 5.72. To St Athan 27.1.82. Sold to T. Bradbury as scrap 2.6.83.

Above: B.2 XM608 of 50 Sqn taxiing at Waddington in June 1980. *Martin Derry*

Left: B.2 XM608 employing the brake 'chute. *Martin Derry*

Below: B.2 XM609 of 44 Sqn looking immaculate in the static display at the Queen's Silver Jubilee Review of the RAF at Finningley in July 1977. *Fred Martin*

101 Sqn B.2 XM611 taxiing at Waddington in June 1976. *Martin Derry*

76/XM612 Completed 28.2.64. Olympus 301 engines. 9 Squadron 3.3.64, Coningsby Wing, Cottesmore Wing, A&AEE 5.3.68, Waddington Wing 4.4.68, 101 Squadron 5.75, 44 Squadron 7.81. Modified for 'Black Buck' operations. 44 Squadron 23.5.82. Sold to City of Norwich Aviation Museum 19.1.83, delivered 30.1.83.

77/XM645 Completed 10.3.64. Olympus 301 engines. Coningsby Wing 12.3.64, Cottesmore Wing, Waddington Wing 15.12.67, 230 OCU 5.8.68, Waddington Wing 21.4.71, 101 Squadron 8.73, Akrotiri Wing 12.3.74, 9 Squadron 15.1.75. Crashed Zabbar, Malta, following explosion 14.10.75.

78/XM646 Completed 7.4.64. Olympus 301 engines. 12 Squadron 8.4.64, Coningsby Wing, Cottesmore Wing, Akrotiri Wing 5.2.69, 9 Squadron 17.1.75, 101 Squadron 6.81. To St Athan 26.1.82. Sold to T. Bradbury as scrap 29.6.83.

79/XM647 Completed 15.4.64. Olympus 301 engines. 35 Squadron 15.4.64, Coningsby Wing, Cottesmore Wing, Akrotiri Wing 26.2.69, Waddington Wing 15.1.75, 9 Squadron 1.75, 44 Squadron 9.79, 50 Squadron 9.81. Delivered to Laarbruch 17.9.82 for ground instruction (8765M). Sold to Solair UK as scrap 25.2.85, scrapped 1.3.85.

B.2 XM612 of the Waddington Wing at Finningley in September 1975. *Fred Martin*

Above: 101 Sqn B.2 XM612 touching down at Waddington in June 1979. *Fred Martin*

XM645 suffered a catastrophic accident during an approach to RAF Luqa in Malta on 14 October 1975. The aircraft undershot the runway and was destroyed in a rapidly developing fire shortly afterwards, over Zabar.

XM647 at Waddington, wearing the markings of No 50 Squadron, although paint patches reveal that her previous assignment was to No 44 Squadron on the other side of the airfield. *Shaun Connor*

The leading edge paintwork of XM646 appears to have been recently refreshed, as she catches the Finningley sun in April 1977. *Fred Martin*

XM647 on approach to Waddington in March 1982. *Fred Martin*

80/XM648 Completed 5.5.64. Olympus 301 engines. 9 Squadron 6.5.64, Coningsby Wing, Cottesmore Wing, Waddington Wing 25.1.68, 101 Squadron 5.72, 44 Squadron 5.75, 101 Squadron 3.77, 9 Squadron 9.80, 101 Squadron 10.81. Grounded 10.9.82. Sold to Bird Group as scrap 8.12.82.

81/XM649 Completed 12.5.64. Olympus 301 engines. 9 Squadron 14.5.64, Coningsby Wing, Cottesmore Wing, Waddington Wing 18.1.68, 101 Squadron 8.73, 9 Squadron 4.76, 101 Squadron 8.79. To St Athan 2.9.81. Sold to Bird Group as scrap 2.12.82.

82/XM650 Completed 27.5.64. Olympus 301 engines. 12 Squadron 5.6.64, Coningsby Wing, Cottesmore Wing, Waddington Wing 20.12.67, 44 Squadron 5.75, 50 Squadron 1.77. To St Athan 28.1.82. Allocated 8748M 16.3.83. Sold to Bournewood Aviation as scrap 22.3.84.

83/XM651 Completed 19.6.64. Olympus 301 engines. 12 Squadron 22.6.64, Coningsby Wing, Cottesmore Wing, Waddington Wing 24.4.68, 101 Squadron 5.72, 50 Squadron 9.75, 101 Squadron 9.79. Grounded 10.9.82. Sold to Bird Group as scrap 30.11.82.

XM652 is pictured shortly before retirement at Waddington. The aircraft was subsequently dismantled and transported to Sheffield, where the nose was removed for preservation and the rest of the airframe was scrapped. *Shaun Connor*

Below: XM653 being refuelled, with a Beverley in the background. *Martin Derry*

84/XM652 Completed 12.8.64. Olympus 301 engines. 9 Squadron 17.8.64, Coningsby Wing, Cottesmore Wing, Waddington Wing 24.12.67, 44 Squadron 9.75, 9 Squadron 10.81, 50 Squadron 10.82. Sold to Boulding Group 20.2.84. Dismantled and transported to Sheffield 7.5.84. Scrapped 2.85. Nose privately owned at Welshpool, Wales.

85/XM653 Completed 31.8.64. Olympus 301 engines. 9 Squadron, Coningsby Wing, Cottesmore Wing, Waddington Wing 24.1.68, 101 Squadron 5.72, 44 Squadron 5.75, 9 Squadron 9.75, 101 Squadron 10.78, 9 Squadron 5.79, 101 Squadron 7.79. To St Athan 10.9.79. Dumped 18.12.80. Sold as scrap 28.7.81.

86/XM654 Completed 22.10.64. Olympus 301 engines. 12 Squadron 26.10.64, Coningsby Wing, Cottesmore Wing, Waddington Wing 30.4.68, 101 Squadron 8.73, 50 Squadron 9.75, 101 Squadron 9.81, 50 Squadron 10.81. Grounded 29.10.82. Sold to Bird Group as scrap 30.11.82.

87/XM655 Completed 19.11.64. Olympus 301 engines. 9 Squadron 23.11.64, Cottesmore Wing, Waddington Wing 12.1.68, 101 Squadron 5.72, 44 Squadron 7.81, 50 Squadron 8.82. Sold to Roy Jacobsen 11.2.84, delivered to Wellesbourne Mountford 11.2.84. Registered G-VULC 27.2.84. Re-registered N655AV 1985. Sold to Radarmoor Ltd 1992. Maintained in taxi condition. by XM655 Maintenance and Preservation Society.

88/XM656 Completed 11.12.64. Olympus 301 engines. 35 Squadron 15.12.64. Cottesmore Wing, Waddington Wing 2.2.68, 101 Squadron 9.75, 9 Squadron 12.80. To Cottesmore for display 9.8.82. allocated 8757M, assigned to crash/rescue training. Sold to Bird Group as scrap 30.3.83.

89/XM657 Completed 14.1.65. Olympus 301 engines. 35 Squadron 15.1.65, Cottesmore Wing, Waddington Wing 19.3.68, 101 Squadron 5.72, 50 Squadron 8.75, 101 Squadron 3.77, 44 Squadron 4.80. Allocated to Central Training Establishment 5.1.82, delivered to Manston 12.1.82. Allocated 8734M. Later scrapped. Last production Vulcan B. MK.2. Scrapped by 11.92.

Above: XM653 of the Waddington Wing taxies in. *Fred Martin*

Although supporting information is rare, it would appear that XM654 suffered an undercarriage collapse while landing at Waddington some time around the early 1970s.

XM654 thunders away from Waddington on a training mission. Having spent most of her service life at Waddington (apart from occasional periods of service at nearby Scampton), the aircraft was retired and scrapped there late in 1982. *Shaun Connor*

Appendix Two

Vulcan Squadrons, Second-line and Miscellaneous Units

Note: *For readers unfamiliar with the Centralised Servicing Scheme (aircraft pooling) adopted by the Vulcan Wings, and the effect it had on individual squadron identities during its existence (until 1972 at least) – an early glance at Appendix 3 might prove useful.*

9 Squadron

Following the Second World War, 9 Squadron operated the Avro Lincoln B.2 heavy bomber until May 1952 when the type was replaced by Britain's first jet bomber, the English Electric Canberra B.2, and later the B.6 – the latter being retained until July 1961 when the unit disbanded.

Reformed on 1 March 1962, 9 Squadron received its first Vulcans the following month. Thereafter the unit operated from the bases listed until their last Vulcans were relinquished in April 1982. Formally disbanded on 1 May, the unit reformed at Honington on the Panavia Tornado GR.1 in June 1982.

Above: 9 Squadron Vulcan B.2, XL427, seen at Finningley in September 1978 with an RAF Nimrod MR.1 maritime reconnaissance in the background. *Fred Martin*

Left: XL427 pictured while serving with IX Squadron at Waddington. The gloss finish to the aircraft's camouflage is indicative of a previous period of service with No 27 Squadron in the maritime radar reconnaissance role (although the aircraft was not converted to MRR standard). *Shaun Connor*

1.3.62 – 9.11.64	Coningsby	Vulcan B.2 (from 4.62)
10.11.64 – 25.2.69	Cottesmore	Vulcan B.2
26.2.69 – 31.12.74	Akrotiri, Cyprus	Vulcan B.2
1.1.75 – 9.4.82	Waddington	Vulcan B.2

12 Squadron

12 Squadron received Lincoln B.2s in August 1946 and retained them until April 1952, by which time conversion to the Canberra B.2 was underway. Canberra B.6s arrived in 1955 and were retained until July 1961 when the Squadron disbanded at Coningsby.

Reformed on 1 July 1962 at the same location, the unit received its Vulcan B.2s later the same month, the type being retained until disbandment came on the last day of 1967. 12 Squadron reformed on 1 October 1969, at Honington, on the Buccaneer S.2.

1.7.62 – 16.11.64	Coningsby	Vulcan B.2
17.11.64 – 31.12.67	Cottesmore	Vulcan B.2

27 Squadron

27 Squadron first emerged as a post-war bomber unit on 15 June 1953 at Scampton equipped with the Canberra B.2, although its revival proved relatively short lived as it disbanded on 31 December 1957 at Waddington. On 1 April 1961, 27 Squadron re-emerged at Scampton equipped with the Vulcan B.2, the type being retained until March 1972 when the unit disbanded at Scampton.

On 1 November 1973, the unit reformed at Scampton in the MRR (maritime radar reconnaissance) role equipped with the Vulcan B.2 (MRR) – but occasionally recorded as the Vulcan SR.2 (Strategic Recce). Disbanded again on 31 March 1982, 27 Squadron reformed as a Tornado GR.1 unit at Marham on 1 May 1983.

1.4.61 – 29.3.72	Scampton	Vulcan B.2
1.11.73 – 31.3.82	Scampton	Vulcan B.2 (MRR)

A close-up of the broad sweep of Vulcan XH537's wing. While the stencil used to illustrate Dumbo in both images appear to be the same, small differences in presentation do exist – quite apart from Dumbo's colour. *Fred Martin*

35 Squadron

Established as a heavy bomber unit in the Second World War, 35 Squadron operated the Lancaster until 1949, followed by the Lincoln B.2 and later the Boeing Washington B.1, the latter being replaced by Canberra B.2s from April 1954.

Disbanded at Upwood in September 1961, the unit reformed on 1 December 1962 at Coningsby and received its first Vulcan B.2s later that month. The Vulcans were relinquished in February 1982, while 35 Squadron was formally disbanded on 1 March – presumably (as of 2019 at least) for the last time.

1.12.62 – 6.11.64	Coningsby	Vulcan B.2
7.11.64 – 14.1.69	Cottesmore	Vulcan B.2
15.1.69 – 15.1.75	Akrotiri, Cyprus	Vulcan B.2
16.1.75 – 28.2.82	Scampton	Vulcan B.2

A 35 Sqn B.2, XH559, photographed during the Queen's Jubilee Review celebration at Finningley in July 1977, where, infamously, Labour Government minister Fred Mulley fell asleep in the Queen's presence despite the ear-splitting noise around him. It has a later-style 'numeric' motif and Radar Warning Receiver fitted. *Fred Martin*

The end. Not just the extremity of B.2 XH783's fuselage, as seen on a misty day in February 1982, but the end too of 35 Squadron which disbanded for the last time on the 28th, the last day of the month. *Fred Martin*

Waddington Wing B.2, XM573, landing at its home airfield in June 1978 with the braking parachute is deployed. *Fred Martin*

44 Squadron

Following a similar pattern to other Vulcan operators, this unit operated Lancasters until 1947, Lincolns until 1951 and Washingtons from 1951 to January 1953. Following a brief hiatus, Canberra B.2s arrived in April 1953 which were operated until July 1957 when the unit disbanded at Honington.

On 10 August 1960, 44 Squadron reformed at Waddington from a nucleus provided by 83 Squadron. A front line unit for twenty-two years, 44 Squadron disbanded for the last time on 21 December 1982.

10.8.60 – 21.12.82	Waddington	Vulcan B.1 from 8.60 – 8.62
		Vulcan B.1A from 1.61 – 9.67
		Vulcan B.2 from 9.66 – 12.82

A Waddington Wing 44 Sqn B.2, XM575, banks towards the camera on a misty June morning in 1981 to present a near-perfect plan view of its upper wing surfaces. *Fred Martin*

Waddington Wing 50 Sqn B.2, XM603, in June 1975. *Fred Martin*

50 Squadron

Equipped post-war with the Lincoln B.2 until 1951, followed by the Canberra B.2 from August 1952 until 1 October 1959 (when the unit disbanded), 50 Squadron reformed on 1 August 1961 at Waddington from a nucleus provided by 617 Squadron.

Uniquely, from June 1982, while still retaining a number of B.2s, the Squadron also employed the Vulcan B.2(K), six of which were converted to act as single-point air-to-air tankers (the designation B.2(K) later changing to K.2). 50 Squadron, the last operational Vulcan squadron, disbanded on 31 March 1984.

1.8.61 – 31.3.84	Waddington	Vulcan B.1 from 8.61 – 8.62
		Vulcan B.1A from 8.61 – 10.66
		Vulcan B.2 from 12.65 – 3.84
		Vulcan K.2 from 6.82 – 3.84

83 Squadron

Having operated the piston-engined Lincoln B.2 until as late as December 1955, 83 Squadron disbanded at Hemswell on 1 January 1956. On 21 May 1957, the unit reformed at Waddington to become the RAF's first Vulcan squadron albeit relying initially on aircraft borrowed from 230 OCU until, on 11 July 1957, XA905, the first of an initial allocation of six Vulcan B.1s arrived on station.

During August 1960, 83 Squadron prepared itself to become the first Vulcan B.2 operator and moved to Scampton in October (leaving its B.1s for the newly-reformed 44 Squadron) where it received the first of the new Mark in December 1960. 83 Squadron disbanded for the last time on 31 August 1969.

21.5.57 – 9.10.60	Waddington	Vulcan B.1 from 11.7.57 – 8.60
10.10.60 – 31.8.69	Scampton	Vulcan B.2 from 12.60 – 8.69

Scampton-based Vulcan B.2, XM573, seen at Waddington on 15 September 1963. An 83 Squadron aircraft, sadly we never thought to photograph its tail which, in all probability at that moment in time, would have had the unit's rarely photographed six-tined antler motif emblazoned on the fin. *Martin Derry*

A 101 Sqn Vulcan B.2, XM648, seen at Finningley in September 1981. *Fred Martin*

101 Squadron

After operating the Lincoln B.2, 101 became the RAF's first Canberra squadron in June 1951 and retained them until the unit disbanded at Binbrook in February 1957. Officially reformed on 15 October 1957 at Finningley, the unit became the RAF's second Vulcan squadron although its first Vulcan delivery preceded this date by two weeks.

After relocating to Waddington in June 1961, 101 Squadron disbanded there on 4 August 1982, but within two years had reformed at Brize Norton (on 1 May 1984) as a tanker unit equipped with the VC.10 K.2.

15.10.57 – 25.6.61	Finningley	Vulcan B.1 from 10.57 – 6.61
		Vulcan B.1A from 3.61 – 6.61
26.6.61 – 4.8.82	Waddington	Vulcan B.1 from 6.61 – 5.62
		Vulcan B.1A from 6.61 – 12.67
		Vulcan B.2 from 12.67 – 8.82

617 Squadron

Successively equipped post-war with the Lincoln B.2, the Canberra B.2 and later the B.6, 617 Squadron disbanded at Binbrook in December 1955. Reformed at Scampton on 1 May 1958, 617 became the UK's third Vulcan squadron, the unit remaining at the same location until it disbanded there twenty-three years later on 31 December 1981. 617 Squadron reformed on 1 January 1983 at Marham on the Tornado GR.1.

1.5.58 –31.12.81	Scampton	Vulcan B.1 from 5.58 – 7.61
		Vulcan B.1A from 10.60 – 7.61
		Vulcan B.2 from 9.61 – 12.81

617 Squadron B.2, XL446, seen at Abingdon in September 1979 following its earlier return from a 'Giant Voice' exercise in the USA where it was 'zapped' with the elaborate USAF Strategic Air Command shield. *Fred Martin*

An excellent port-quarter view of 230 OCU Vulcan B.1, XA896, sharing a dispersal with a second aircraft and a solitary guard at Finningley on 15 September 1962. *Martin Derry*

230 Operational Conversion Unit

230 OCU reformed at Waddington on 31 May 1956 tasked with introducing the Vulcan B.1 into RAF service. The first Vulcan received, XA897 on 20 July 1956, was little more than symbolic gesture as the aircraft was almost immediately returned to Avro. Despite temporary visits by other Vulcans, for all practical purposes, the OCU counted 18 January 1957 as the date on which it finally received two B.1s on a more permanent basis.

B.1s, used until 1965, were supplanted by the B.2 from 1960. The OCU continued to train aircrews until, with the eventual reduction of the Vulcan force, the requirement for new crews steadily reduced and as a consequence 230 OCU finally disbanded on 31 August 1981.

31.5.56 – 18.6.61	Waddington	Vulcan B.1 from 18.1.57 – 18.6.61
		Vulcan B.2 from 1.7.60 – 18.6.61
18.6.61 – 8.12.69	Finningley	Vulcan B.1 from 18.6.61 – 1964/65
		Vulcan B.2 from 18.6.61 – 8.12.69
8.12.69 – 31.8.81	Scampton	Vulcan B.2 from 8.12.69 – 31.8.81

Note: *Confusion exists as to whether 230 OCU operated Vulcan B.1As as well as B.1s. One possible explanation for this might stem from the fact that while several B.1s were indeed allocated to the unit, of those later converted to B.1A standard none appear to have ever been reallocated to 230 OCU. For instance, while B.1 XA901 returned to the OCU in 1962 for a further two years' service it is known that, following conversion, former OCU airframes such as B.1A XA910, XA911, XH475 and several others, never did. A second might stem from the fact that for three months 101 Squadron briefly operated B.1As at Finningley in 1961 (q.v.), as did the BCDU (see below) which operated a few near-anonymous B.1As there for about five years.*

230 OCU B.2, XL386, at Leconfield in May 1976. *Fred Martin*

Handley Page Hastings T.5, TG553 '553', seen while serving with the BCBS at Lindholme on 8 October 1961. Like the earliest Vulcans, it is painted High Speed Silver overall with yellow training bands around the wings and rear fuselage only the middle band of the three seen on the wing was painted, the other two yellowish-looking bands were the result of exhaust staining. *Martin Derry*

230 OCU's component units (ex-BCBS/SCBS, Hastings Radar Flight and '1066 Squadron')

RAF V-Force navigators were trained at the Bomber Command Bombing School at Lindholme using various types of aircraft including, from the late 1950s, ten Hastings T.5s fitted with H2S Mk9 radar. On 30 April 1968, the BCBS became the Strike Command Bombing School which moved from Lindholme to Scampton on 1 September 1972.

On 1 January 1974, the SCBS was absorbed into 230 OCU to become the Radar Training Flight. Among the latter's complement of aircraft were four surviving Hastings T.5s, TG503, TG505, TG511 and TG517. Together they formed the Hastings Radar Flight, soon dubbed '1066 Squadron', not in reference to the Hasting's antiquity apparently – merely a reference to some ancient squabble that happened to share the same name!

Although primarily associated with training Vulcan navigators, the T.5s were also used to train Victor, Buccaneer and Phantom navigators, as well as conducting maritime patrols over the North Sea as required until all four Hastings were withdrawn from use on 30 June 1977.

During her relatively brief existence as a static display aircraft at Scampton, XH563 carried the markings of 230 OCU and 617 Squadron on the port side of her tail, with markings for Nos 35 and 27 Squadron on the starboard side. *Terry Senior*

230 OCU Hastings T.5, TG503 '503', seen at Coningsby June 1976. The last of their breed in service, the RAF would withdraw '503' and its sister T.5s one year later. *Fred Martin*

Bomber Command Development Unit

Formed at Wittering on 24 August 1954, the BCDU evaluated aircraft systems for RAF jet bombers. Initially borrowing aircraft as required, from 1959 it began to receive its own aircraft – Valiants and Canberras primarily. A move to Finningley occurred in February 1960 where it operated at least three (or more) Vulcans including B.1As, XA895, XA907 and XA911 at various times between June 1961 and November 1966, plus B.2, XL391, for several months from July 1965, and probably others too. Redesignated Strike Command Development Unit in April 1968, it survived until the last day of the year when it disbanded.

A rare view of a BCDU Vulcan B.1A, XA907, Finningley, 1965. *Martin Derry*

A Vulcan Display Flight B.2, XL426. *Fred Martin*

Vulcan Display Flight

Formed at Waddington on 1 April 1984, having previously been known as the Vulcan Display Team, the Flight's purpose was to exhibit the Vulcan at various air displays. Apart from B.2 XH560 which was used only briefly, the Flight operated two others, XL426 and XH558, until the unit disbanded in 1992.

XM645 on a typically rainy September day at Finningley, being joined by 707C WZ744 prior to displaying to the public at the annual Battle of Britain 'At Home' Day. *Joe L'Estrange*

Appendix Three

Wings, Bases and Dispersal Airfields

Note: *In March 1963, a Centralised Servicing Scheme was promulgated whereby each Vulcan Wing would pool and issue aircraft to its constituent squadrons as required on a daily basis. Intentionally or otherwise, as the new system was progressively implemented, it resulted in the almost complete elimination of unit insignia from the Vulcan force by mid-1964.*

By 1971, only two UK-based Vulcan Wings remained: one at Waddington, the other at Scampton. At the latter, central servicing continued although a return was made to individual squadron allocations during the year, which in turn led to the steady revival of squadron insignia from 1972/73 onwards.

At Waddington, aircraft pooling remained in use until the Vulcan force was finally disbanded. However, in the interest of individual unit pride and spirit, unit insignia was reintroduced there also from 1972.

Coningsby Wing (3.62-12.64)

1.3.62 – 30.6.62	9 Squadron
1.7.62 – 30.11.62	9 and 12 Squadrons
1.12.62 – 16.11.64	9, 12 and 35 Squadrons

Aircraft pooling system commenced 3.1964

Cottesmore Wing (11.64-2.69)

7.11.64 – 16.11.64	35 and 9 Squadrons
17.11.64 – 31.12.67	35, 9 and 12 Squadrons
1.1.68 – 14.1.69	35 and 9 Squadrons
15.1.69 – 25.1.69	9 Squadron

Aircraft were pooled throughout the Wing's existence

Above: Coningsby Wing Vulcan B.2, XM646, seen at Andrew's Air Force Base in 1964. *Martin Derry*

Left: 35 Squadron Vulcans on Coningsby's ORP at the eastern end of the main runway. *Joe L'Estrange*

Scampton Wing B.2, XM595, usefully displaying on the hatch door the motifs of the Wing's three constituent units, namely: 27 Squadron (left); 83 Squadron (six-tined antler – centre bottom); 617 Squadron (right), with RAF Scampton's badge at top centre. *Martin Derry*

Finningley Wing (10.57-12.69)

15.10.57 – 25.6.61	101 Squadron
6.61 – 12.69	230 OCU

Scampton Wing (5.58-3.82)

1.5.58 – 9.10.60	617 Squadron
10.10.60 – 31.3.61	617 and 83 Squadrons
1.4.61 – 31.8.69	617, 83 and 27 Squadrons
1.9.69 – 12.69	617 and 27 Squadrons
12.69 – 29.3.72	617, 27 Squadrons and 230 OCU
30.3.72 – 31.10-73	617 Squadron and 230 OCU
1.11.73 – 15.1.75	617, and 27 Squadrons and 230 OCU
16.1.75 – 8.81	617, 27 and 35 Squadrons and 230 OCU
8.81 – 1.1.82	617, 27 and 35 Squadrons
2.1.82 – 28.2.82	27 and 35 Squadrons
1.3.82 – 31.3.82	27 Squadron

The aircraft pooling system was in use from 1.3.1964. OCU Vulcans were generally assigned separately to those used by the Scampton Wing Vulcans

XL318 at Scampton wearing overall tactical camouflage and 617 Squadron markings. The aircraft flew 617 Squadron's last operational Vulcan sortie and was eventually dismantled and transferred to the RAF Museum at Hendon where it is now on public display.

Deploying its drogue chute this Waddington Wing Vulcan B.1A, XA911, is pictured on 14 September 1963 with a thunderstorm approaching. *Martin Derry*

Waddington Wing (7.56-3.84)

20.7.56 – 20.5.57	230 OCU
21.5.57 – 9.8.60	83 Squadron and 230 OCU
10.8.60 – 9.10.60	83 and 44 Squadrons and 230 OCU
10.10.60 – 25.6.61	44 Squadron and 230 OCU
26.6.61 – 31.7.61	44 and 101 Squadrons
1.8.61 – 31.12.74	44, 101 and 50 Squadrons
1.1.75 – 9.4.82	44, 101, 50 and 9 Squadrons
10.4.82 – 4.8.82	44, 101 and 50 Squadrons
5.8.82 – 21.12.82	44 and 50 Squadrons
12.82 – 3.84	50 Squadron

The aircraft pooling system start date is uncertain but was likely around mid-1963

Akrotiri Wing, Cyprus (1.69-1.75)

15.1.69 – 25.2.69	35 Squadron
26.2.69 – 31.12.74	35 and 9 Squadrons
1.1.75 – 15.1.75	35 Squadron

A Waddington Wing B.2, XL391 with 'Giant Voice' markings at Waddington September 1979. *Fred Martin*

V-Bomber designated Dispersal Airfields

Class One Airfields (i.e. V-Bomber bases)
Finningley
Coningsby
Honington
Scampton
Wittering
Cottesmore
Waddington
Gaydon
Wyton
Dispersal Airfields with ORPs for up to four aircraft
Burtonwood
Bedford/Thurleigh
St Mawgan
Ballykelly
Filton
Kinloss
Shawbury
Cranwell
Middleton St George
Boscombe Down
Pershore
Leeming
Lyneham

Dispersal Airfields with ORPs for two aircraft

Leconfield
Leuchars
Lossiemouth
Yeovilton
Llanbedr
Coltishall
Valley
Manston
Brawdy
Wattisham
Stansted
Elvington
Prestwick
Machrihanish
Bruntingthorpe
Note: the number of available dispersal sites was revised occasionally, as was the designated number of aircraft assigned to each airfield.

XH479 was a Waddington Wing aircraft, and is seen at a rather gloomy RAF Brawdy, one of many dispersal sites to which Waddington's Vulcans regularly deployed. *Ray Deacon*

Appendix Four

Vulcan Losses

XA897	1.10.56	Heathrow: crashed during radar approach
VX770	20.9.58	Syerston: structural failure
XA908	24.10.58	Michigan, USA: electrical failure
XA891	24.7.59	Near Hull: electrical failure
XA894	3.12.62	Patchway: ground fire
XH477	12.12.63	Scotland: crashed during low-level training flight
XH535	11.5.64	Near Andover: crashed after entering uncontrollable spin
XA909	16.7.64	Anglesey: engine explosion
XM601	7.10.64	Coningsby: crashed on landing
XM576	25.5.65	Scampton: crash-landed
XM536	11.2.66	Wales: crashed on TFR trials flight
XL385	6.4.67	Scampton: ground fire
XM604	30.1.68	Cottesmore: engine failure leading to loss of control
XM610	8.1.71	Wingate: engine bay fire
XJ781	23.5.73	Shiraz, Iran: crash-landed
XM645	14.10.75	Zabbar: explosion following undershoot landing attempt
XM600	17.1.77	Near Spilsby: engine bay fire
XL390	12.8.78	Glenview, USA: crashed during air display

Part of the wreckage at the site where XH535 crashed on 11 May 1964 after having entered into a spin while operating from nearby Boscombe Down.

MAIN PLANE KEY DIAGRAM

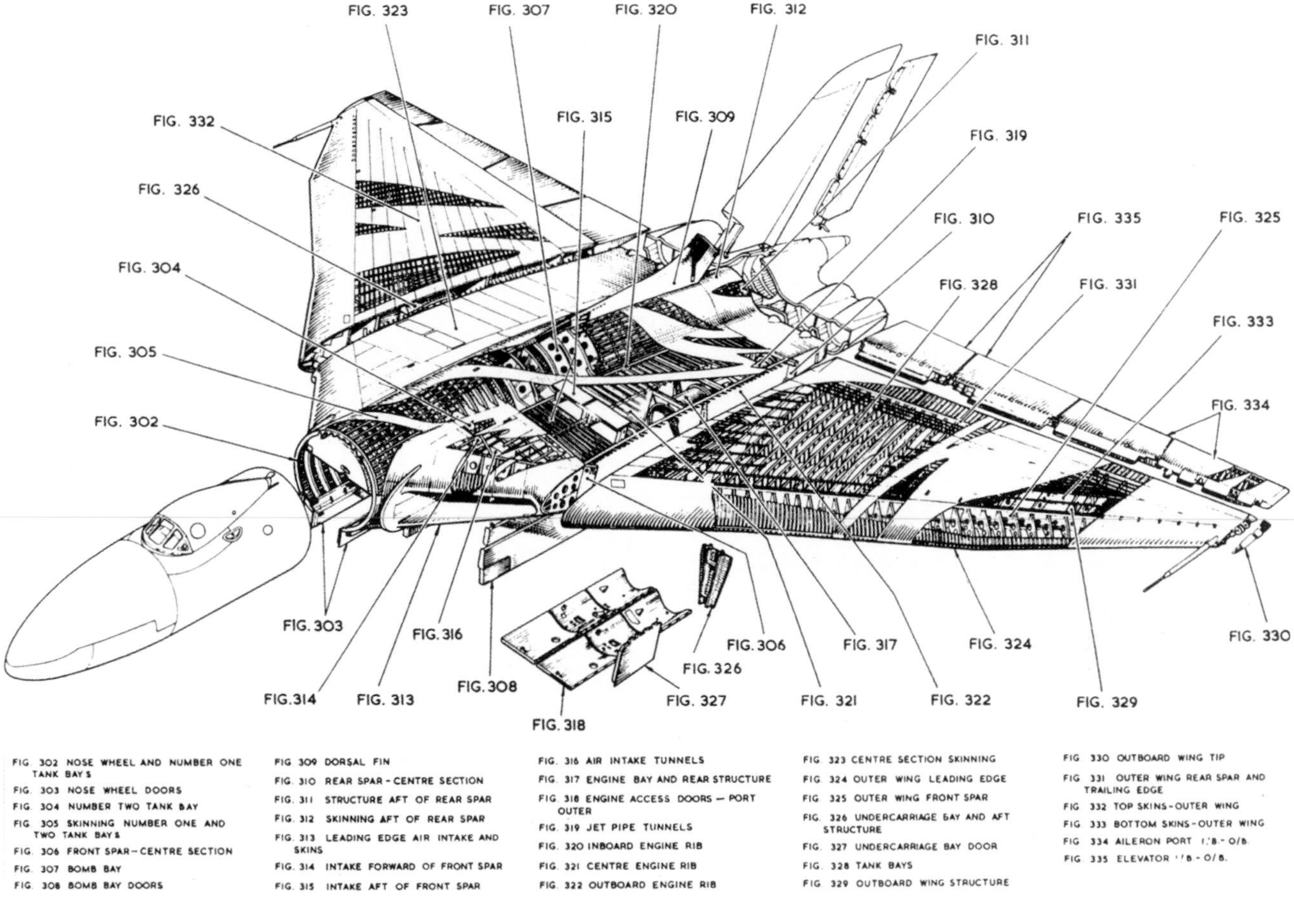

Appendix Five

The Aircrew Manual

The Vulcan B Mk 2 Aircrew Manual was prepared by the Ministry of Defence Procurement Executive, based on information provided by A&AEE test crews at Boscombe Down. Although available space precludes the reproduction of the entire manual, the following (edited) pages are taken directly from the publication and provide some fascinating details of the Vulcan's systems and operating procedures.

The Vulcan B Mk 1.

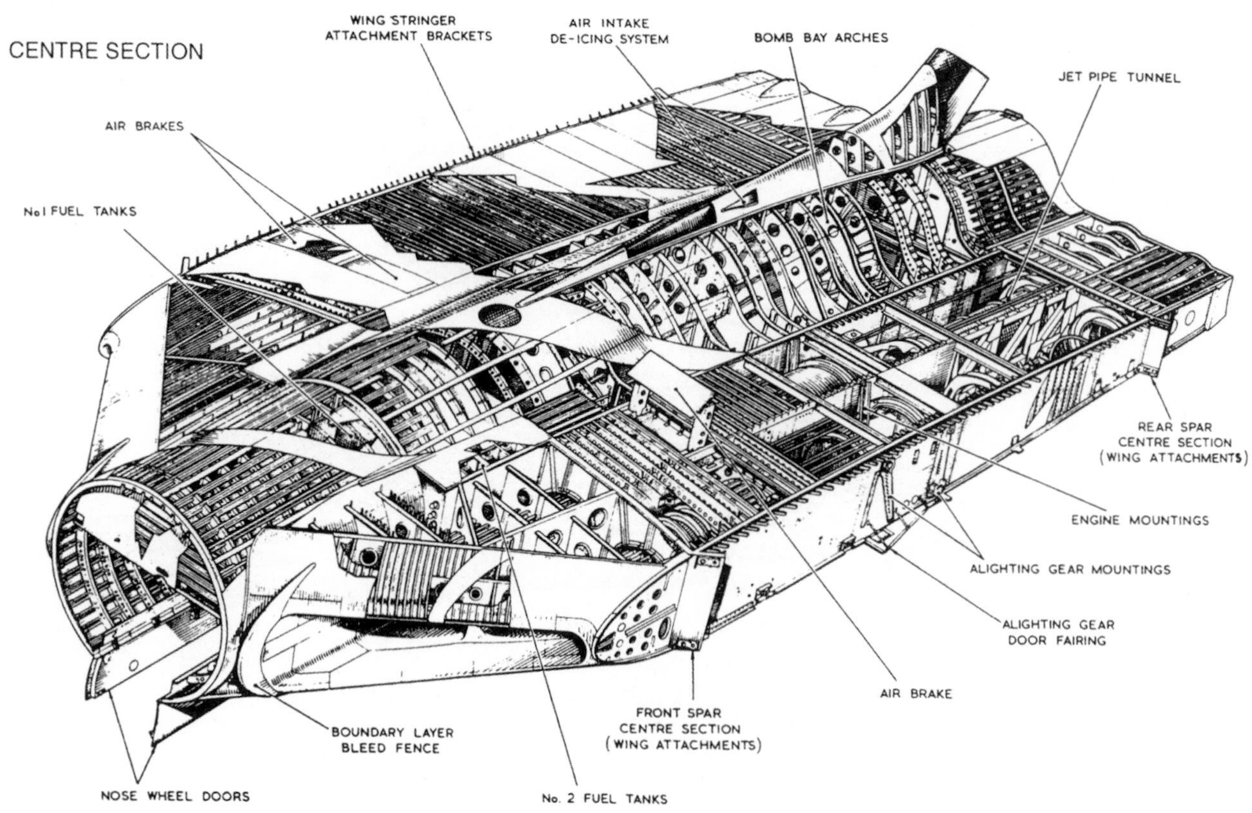

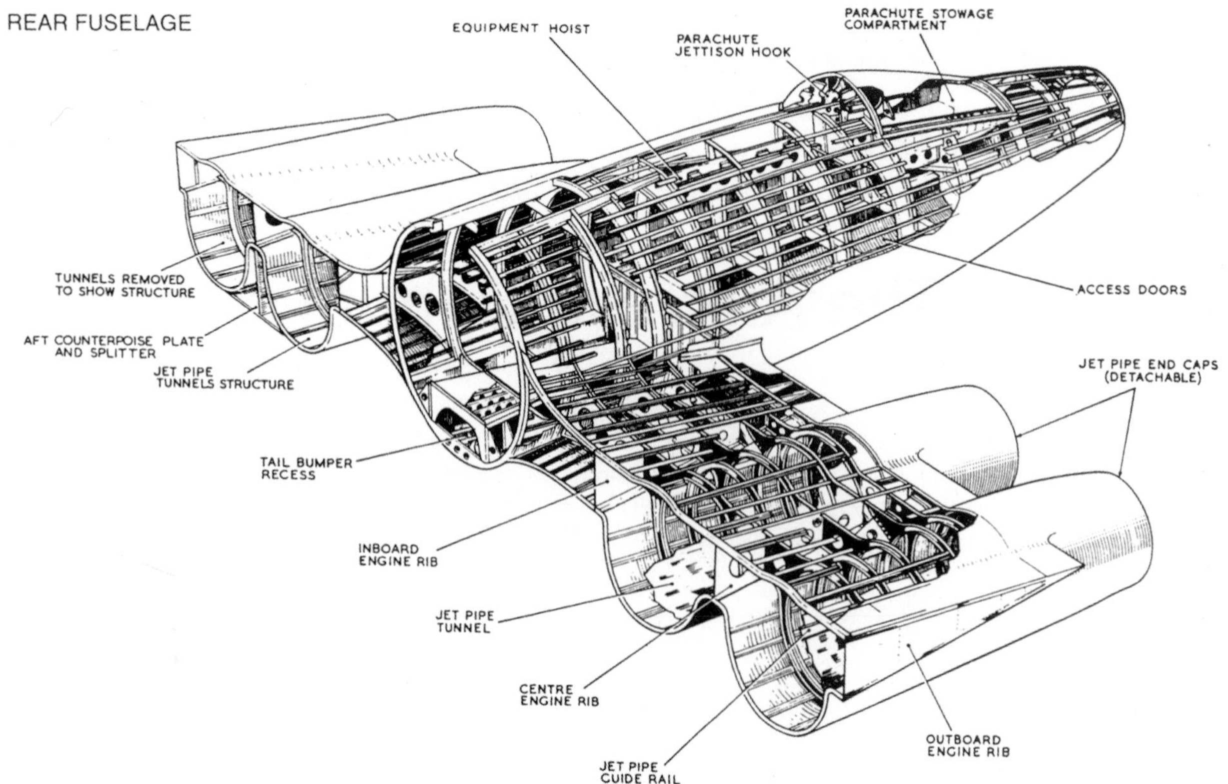

Fuselage and tail structure.

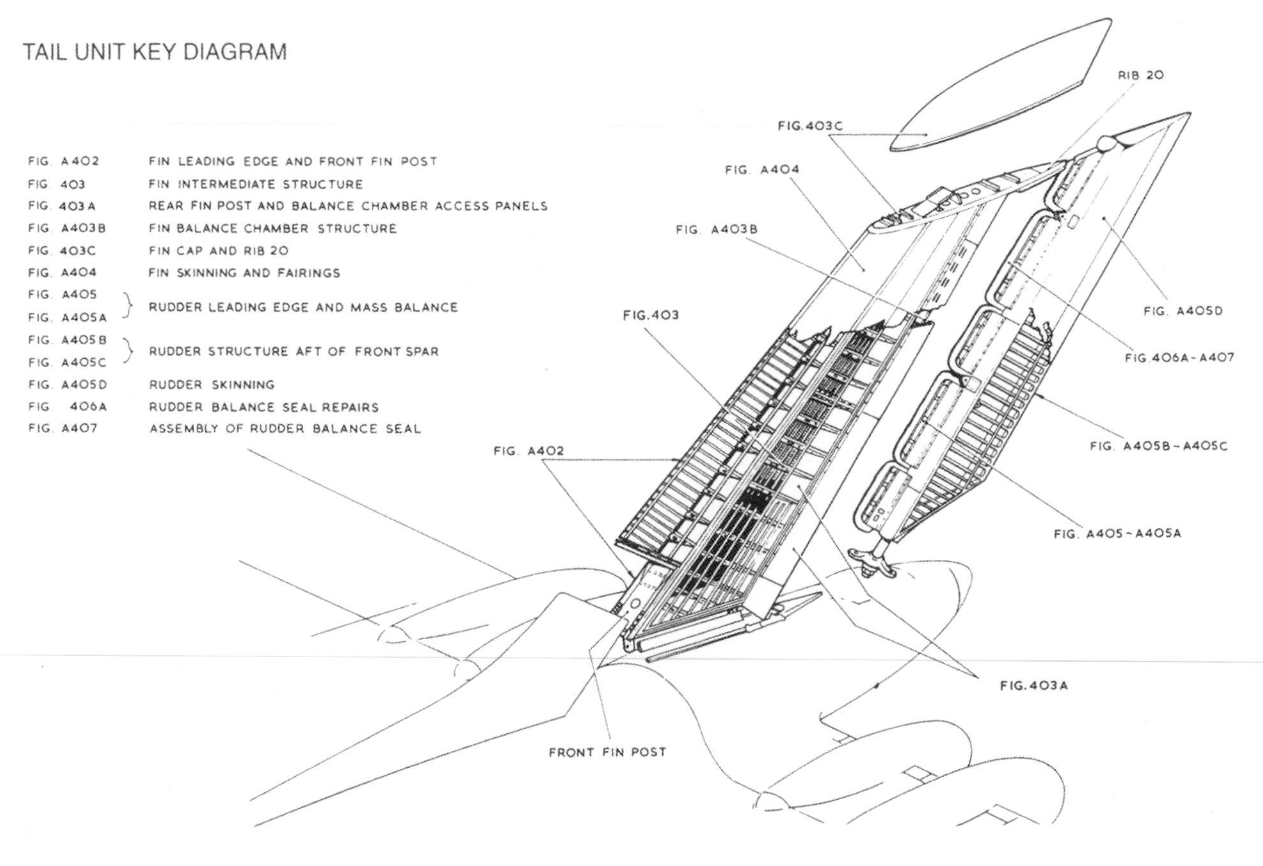

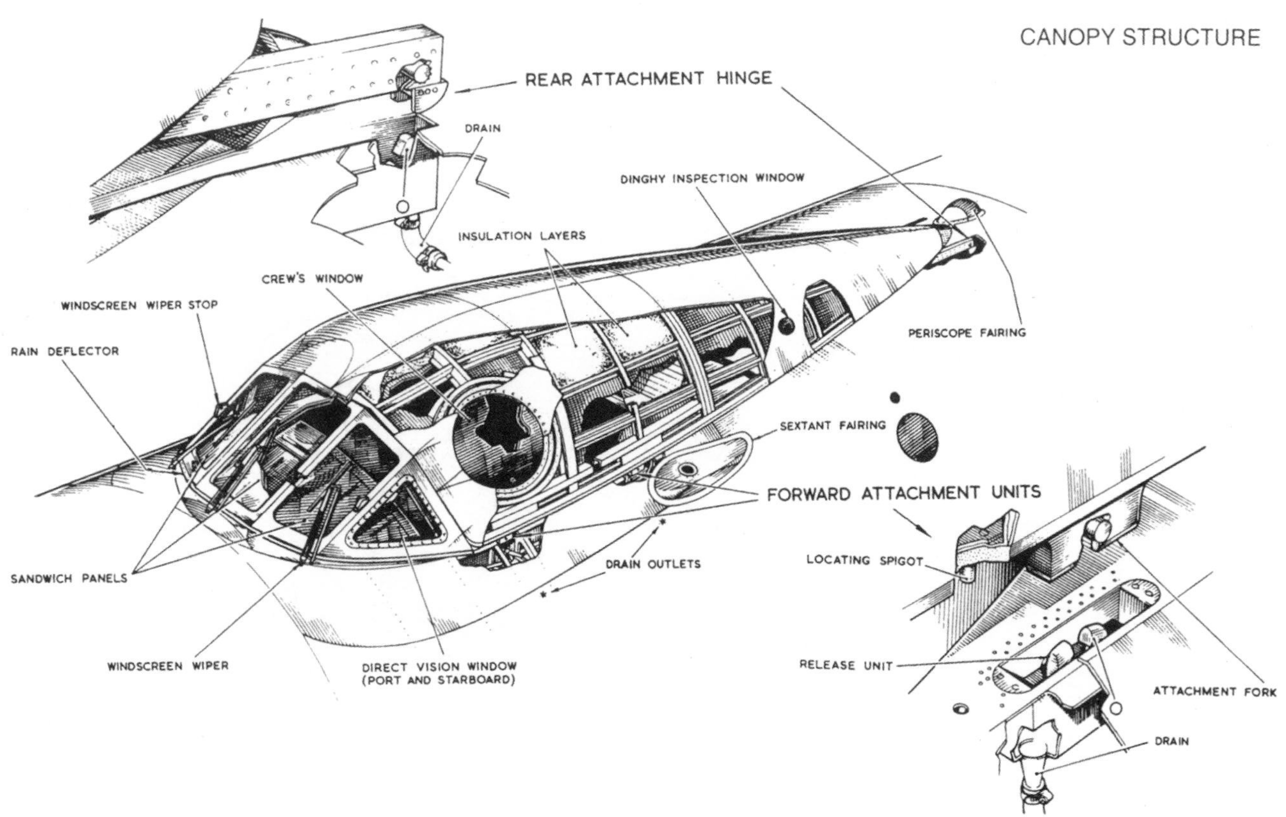

Canopy and tail construction.

FORWARD HOIST ATTACHMENT POINT
PULLEY RELEASE PIN
INCLINOMETER READING TAKEN FROM HERE
FORWARD LIFTING POINT PULLEY
HYDRAULIC INLET PIPE *
GUARD RAIL
HYDRAULIC PRESSURE * DELIVERY PIPE
HYDRAULIC RETURN PIPE *
*HYDRAULIC PUMPS ARE FITTED ONLY ON No. 1, No. 2 AND No. 3 ENGINES
ENGINE TRUNNION BLOCK MOUNTING
FORWARD LIFTING HOIST
SHIM STOWAGE
PORT SIDE No. 1 AND No. 2 ENGINES SHOWN
SHIM CARRIER
ENGINE CONTROL TORQUE SHAFT
AFT LIFTING HOISTS
INFLATABLE SEALS
REFRASIL COVER
FUEL DRAIN PIPE
ENGINE TRUNNION BEARING ASSEMBLY

ENGINE REMOVAL (THREE-POINT HOISTING)

E.C.M. BALLAST AND STOWAGE

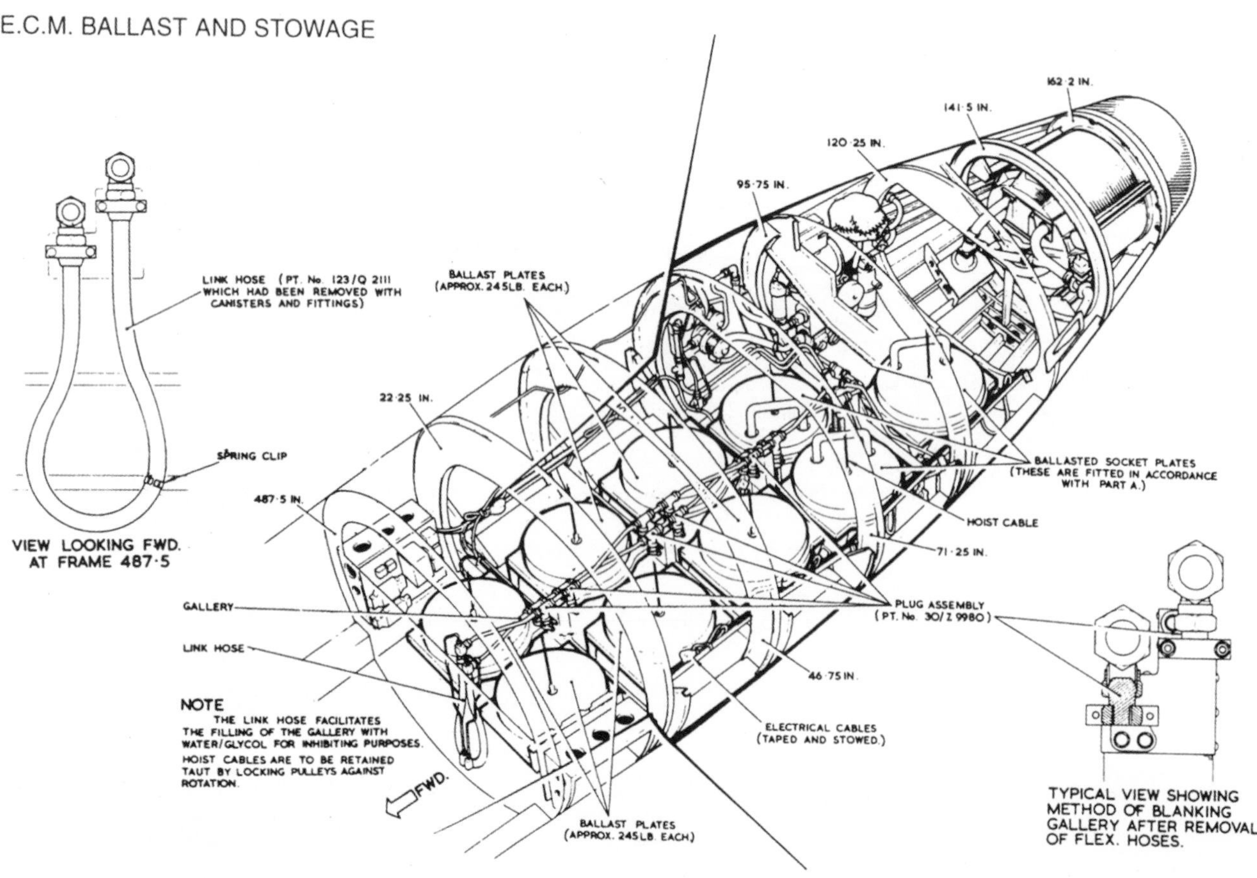

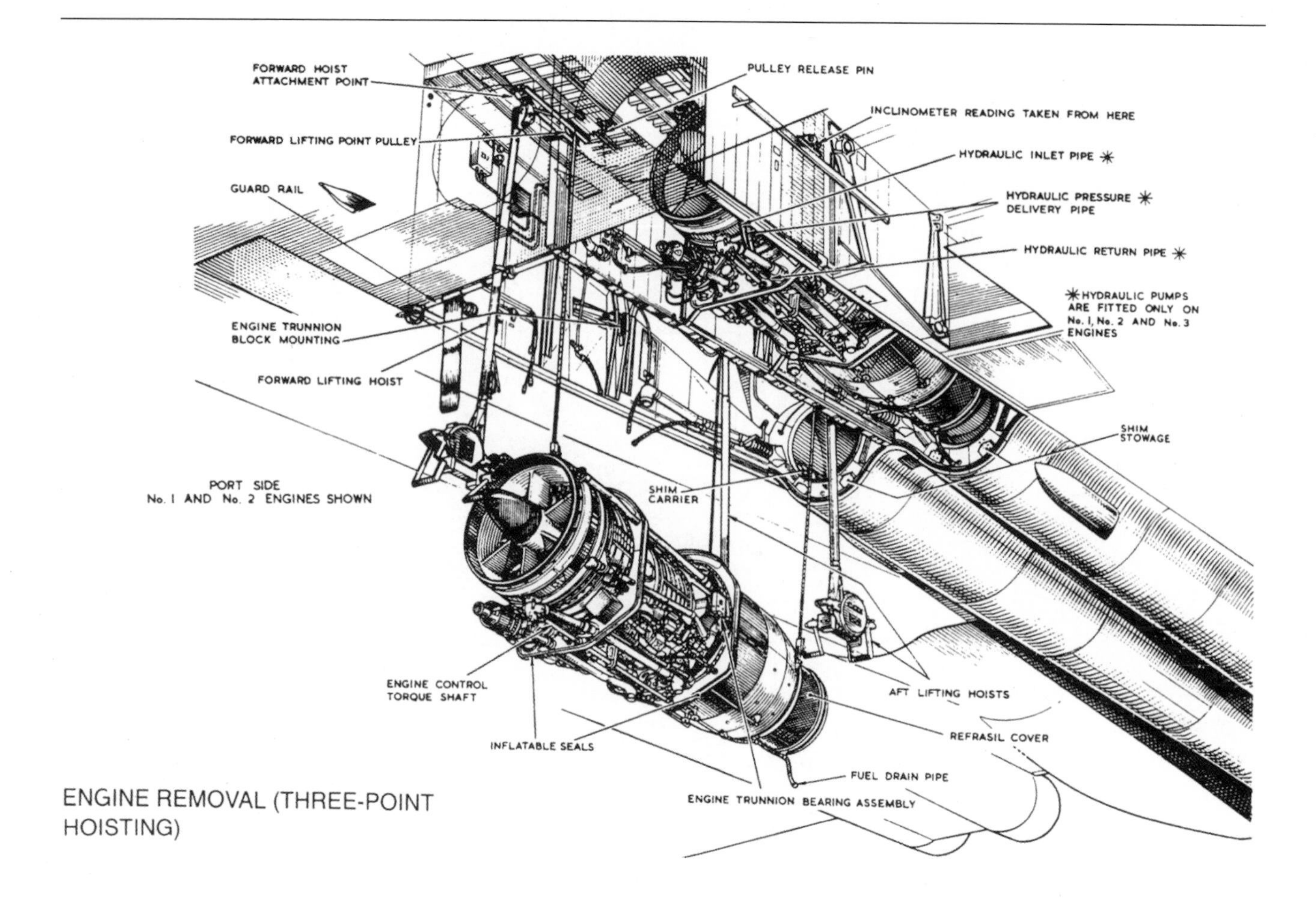

Engine installation and ECM tail cone diagrams.

INTRODUCTION

General

The Vulcan B Mk 2 delta-wing, all-metal aircraft may be powered by four Olympus 200 series engines, each developing 17,000lb static thrust at sea level in ISA conditions, or by Olympus 301 engines, developing 18,000lb static thrust. Provision is made for air-to-air refuelling. A ram air turbine and an airborne auxiliary power plant provide for emergency electrical supplies. Equipment bays outside the pressure cabin contain power distribution and fuse panels in addition to various components of the main services:

a. Nose section.
b. Nosewheel bay.
c. Main undercarriage bays.
d. Bomb bay.
e. Power compartment (aft of bomb bay).

Crew

The aircraft is operated by a crew of five: 1st pilot, co-pilot, navigator/radar, navigator/plotter and air electronics officer (AEO). The pilots sit side-by-side in ejection seats on a raised platform at the front of the cabin normally referred to as the cockpit. Behind the pilots, facing aft, the rear crew members sit on bucket-type seats (the outer two swivelling) facing one long table, behind which is a crate carrying their equipment. A prone bombing position is in a blister below the pilots' floor.

Instrument Layout

The pilots' instruments and controls are on the front panel (divided into four sections), the port and starboard consoles, the centre (retractable) console and the throttle quadrant.

Magnetic indicators and warning lights for the vital systems are grouped across the top of the pilots' front panel and consist of:

a. Two amber MAIN WARNING lights, one at either side, which indicate failure of the PFC units, feel units or autostabilisers (except the yaw dampers).

b. One red alternator failure (ALT FAIL) light which illuminates following a single alternator failure and flashes when two or more alternators fail.

c. Eight magnetic indicators (four each side of the alternator light) for PFC units, artificial feel, autostabilisers, airframes, bomb doors, canopy, entrance door and pressure-head heaters.

On the coaming above the pilots' front panel are four engine fire warning lights which also serve as fire extinguisher pushbuttons.

In addition, two red fire warning lights for the wing fuselage and bomb bay fuel tanks are at the top of the co-pilot's instrument panel. Post-SEM 012, only the bomb bay fire warning light is fitted.

AAP Airborne auxiliary power plant
ADD Airstream direction detector
ADF Automatic direction finding
ADP Azimuth director pointer
AEO Air electronics officer
ARI Air radio installation
AVS Air ventilated suit
BCF Bromochlorodifluoromethane
BRSL Bomb release safety lock
CSDU Constant speed drive unit
DG Directional gyro
DV Direct vision
ECM Electronic countermeasures
EHPP Emergency hydraulic power pack
FRC Flight reference cards
GPI Ground position indicator
HF/SSB High frequency, single side-band
HFRT High frequency radio telephony
HRS Heading reference system
IFF/SSR Identification friend or foe, secondary surveillance radar
ILS Instrument landing system
MFS Military flight system
NBC Navigation & bombing computer
NBS Navigation & bombing system
NHU Navigator's heading unit
NMBS Normal maximum braking speed
NRV Non-return valve
O/H Overheat
PDI Pilot direction indicator
PEC Personal equipment connector
PESJ Pilots emergency stores jettison
PFC Powered flying controls
PFCU Powered flying control unit
PSP Personal survival pack
RAT Ram air turbine
TASU True airspeed unit
TBP Tail brake parachute
TFR Terrain-following radar
TFRU Terrain-following radar unit
TRU Transformer/rectifier unit
VSI Vertical speed indicator

RADAR, RADIO AND ASSOCIATED EQUIPMENT - CREW STATIONS

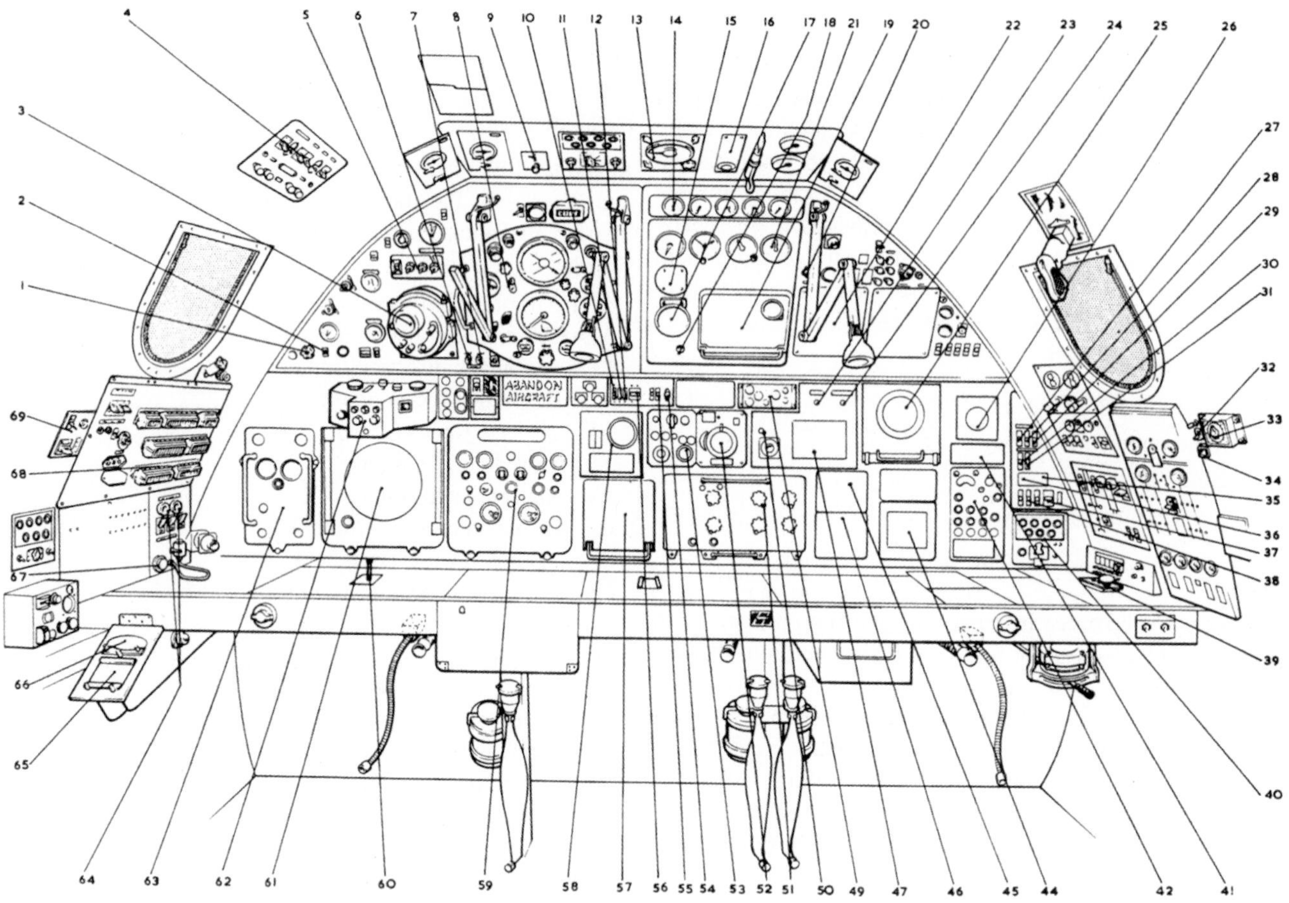

1. SCANNER STABILISATION SWITCH (A.R.I.5928)
2. SCANNER ROTATION CONTROL SWITCH (A.R.I.5928)
3. PULSE ALTIMETER
4. RADIO SWITCH PANEL
5. CONTROL PANEL – TRUE AIR SPEED
6. POWER SUPPLIES INDICATOR – N.B.C.
7. POWER SUPPLIES INDICATOR – H.2.S.
8. NAVIGATION PANEL
9. VARIABLE AIRSPEED UNIT
10. COMPASS ISOLATION SWITCH (PORT)
11. COMPASS ISOLATION SWITCH (STBD.)
12. ON/OFF SWITCH (A.R.I.5972)
13. CONTROL UNIT (A.R.I.23023)
14. D.F. BEARING INDICATOR (A.R.I. 23023)
15. COMPASS – REPEATER – TRUE TRACK MAGNETIC HEADING
16. LOOP CONTROLLER (A.R.I.23023)
17. RADIO ALTIMETER
18. STEERING SIGNAL TEST SOCKET
19. TRACK CONTROLLER UNIT – M.F.S.
20. DIMMER SWITCH – AUTO LAND AND T.C.U. LIGHTS
21. PILOTS DIRECTION INDICATOR
22. WINDOW LAUNCH OR TELE-METERING CONTROL PANEL
23. FAIL INDICATOR (A.R.I.23134/5/4)
24. TEST POINT (G.P.I. Mk.6)
25. INDICATOR, TYPE 6935 (A.R.I.5919)
26. INDICATOR, TYPE 17171 (A.R.I. 18228/1)
27. SUPPLIES SWITCH (A.R.I.5874)
28. OUTPUT SWITCH (A.R.I.5874)
29. SUPPLY SWITCH (A.R.I.23023)
30. PRESS-TO-TRANSMIT SWITCH
31. CONFERENCE INTERCOMM. – MASTER SWITCH
32. ON/OFF SWITCH (A.R.I.18011)
33. CONTROL UNIT, TYPE 705 (A.R.I. 18011)
34. AUTO-THROTTLE ON/OFF SWITCH (Inoperative)
35. DIMMER SWITCH – CONTROL UNIT DIAL LIGHTING
36. DIMMER SWITCH – CONTROL UNIT DIAL LIGHTING (A.R.I.18076)
37. INTERCOMM. MASTER CONTROL
38. EXTERNAL INTERCOMM. SWITCHES
39. MORSE KEY
40. INTERCOMM. CONTROL UNIT (A.E.O.)
41. CONTROL UNIT, TYPE 17172 (A.R.I. 18228/1)
42. CONTROL UNIT, TYPE 9422 (A.R.I. 18076)
43. DELETED (Mod.2501)
44. BLANKING PLATE
45. CONTROL TRANSMITTER/RECEIVER TYPE M53 (A.R.I.23090)
46. ANTENNA CONTROL UNIT, (A.R.I. 23090)
47. CONTROL UNIT, TYPE 16929 (A.R.I. 23134/5/4)
48. NOT ALLOCATED
49. GROUND SPEED SELECTOR UNIT
50. H.R.U. POWER FAILURE INDICATOR
51. G.P.I. MK.6
52. NAV. HEADING UNIT (H.R.S.)
53. H.R.S. CONTROL UNIT
54. EVENT MARKER PUSH SWITCH
55. RECORDER – ON/OFF SWITCH
56. FAILURE WARNING LIGHT (A.R.I. 5959)
57. INDICATOR CONTROL UNIT, TYPE 9869 (A.R.I.5972)
58. CONTROL UNIT, TYPE 7750 (A.R.I. 18107/13)
59. CONTROL UNIT, TYPE 585 (A.R.I. 5928)
60. CONTROL UNIT, TYPE 626 (A.R.I. 5928)
61. INDICATING UNIT, TYPE 301 (A.R.I. 5928)
62. CAMERA, TYPE RX.88 (A.R.I.5928)
63. CONTROL UNIT, TYPE 595 (A.R.I. 5928)
64. N.B.C. ISOLATION INDICATORS
65. CONTROL UNIT, TYPE 12580
66. BOMBING SELECTOR SWITCH
67. N.B.C. SIMULATOR SOCKET
68. BOMBING SELECTOR PANEL
69. SIMULATED BOMBING PANEL

MISCELLANEOUS CONTROLS AND EQUIPMENT - CREW STATIONS (FREE FALL ROLE)

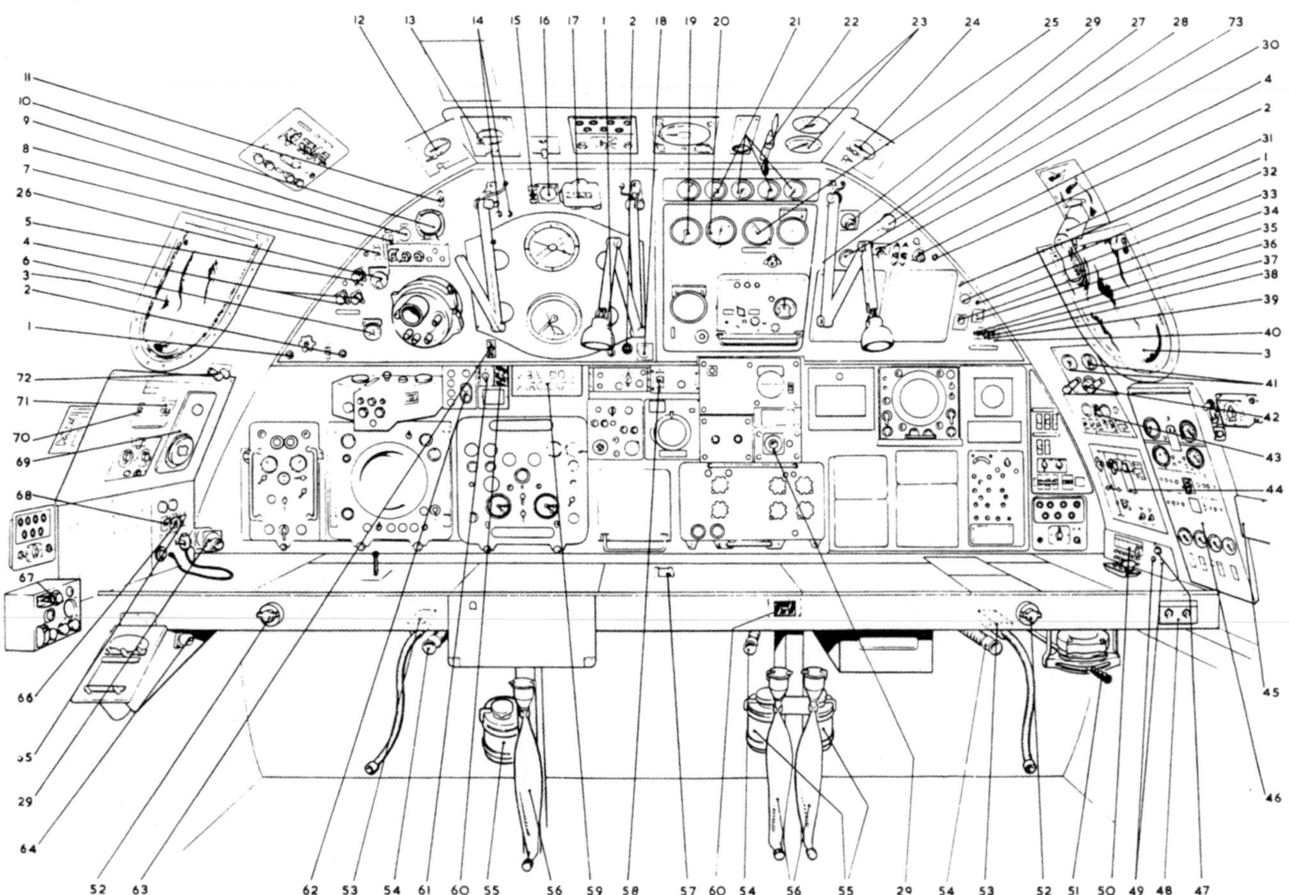

1. OXYGEN PRESSURE INDICATOR
2. LOSS OF PRESSURE WARNING LAMP
3. ANTI FLASH SCREEN
4. DIMMER SWITCHES CHARTBOARD LIGHTS
5. DIMMER SWITCH PANEL LIGHTS
6. BOMB BAY TEMPERATURE INDICATOR
7. BOMB BAY HEATING SWITCH - AUTO-OFF-MANUAL
8. MANUAL HEATING CONTROL SWITCH, INC, -OFF, -DEC.
9. BOMB BAY TEMPERATURE SELECTOR
10. N.B.C. AIR SUPPLY - DIFFERENTIAL PRESSURE GAUGE
11. N.B.C. PRESSURISATION CONTROL - ON/OFF SWITCH
12. OXYGEN REGULATOR - NAV/BOMBER
13. OXYGEN REGULATOR - NAV/RADAR OP.
14. N.B.C./H.2.S. SUPPLIES - BLACK WHITE DOLLS EYES
15. H.2.S. PRESSURISATION CONTROL SWITCH
16. H.2.S. PRESSURE GAUGE
17. AIR NAUT. MILES COUNTER
18. COMPASS DEVIATION CARD HOLDER
19. AIR SPEED INDICATOR
20. ALTIMETER MK.19F
21. FUEL CONTENTS GAUGE No.1, 2, 3 and 4 ENGINE
22. CABIN DECOMPRESSION HANDLE
23. OXYGEN PRESSURE INDICATORS
24. OXYGEN REGULATOR A.E.O.
25. FORWARD THROW INDICATOR
26. OUTSIDE AIR TEMPERATURE INDICATOR
27. DIMMER SWITCH – AUTO LAND INSTRUMENT LIGHTS
28. E.C.M. COOLANT TEMPERATURE GAUGE
29. PUNKAH LOUVRE
30. DIMMER SWITCH, INSTRUMENT PANEL
31. SIGNAL PISTOL
32. ELECTRO MAGNETIC COUNTER WINDOW INSTALLATION
33. WINDOW INSTALLATION SWITCH
34. SWITCH FATIGUE METER
35. R.B.W. CONTROL PANEL, LIGHTS – DIMMER SWITCH
36. STBD. ROOF LIGHT SWITCH
37. PORT ROOF LIGHT SWITCH
38. CABIN LIGHT SWITCH
39. UNDER TABLE LIGHT SWITCH
40. PERISCOPE HEATER SWITCH
41. OIL TEMPERATURE GAUGES C.S.D. UNITS
42. DIMMER SWITCH AND LIGHTS
43. A.A.P.P. CONTROL PANEL 70P (SECT.6)
44. SECONDARY SUPPLIES PANEL 50P (SECT.6)
45. BOMB TIME DELAY EMERGENCY JETTISON PANEL
46. ALTERNATOR CONTROL PANEL 10P
47. EXPLOSION PROTECTION RESET BUTTON (MOD.1744)
48. AUXILIARY A.V.S. SWITCHES
49. EXPLOSION PROTECTION INDICATOR LAMPS PORT AND STBD. (MOD.1744)
50. E.C.M. POWER SUPPLY PANEL
51. PERISCOPE CONTROL HANDLE
52. A.V.S. TEMPERATURE CONTROL
53. A.V.S. FLOW RATE CONTROL
54. AUXILIARY A.V.S. QUICK-RELEASE CONNECTOR
55. RATION HEATERS
56. SANITARY CONTAINERS
57. A.S.I. CORRECTION CARD HOLDER
58. CONTROL SWITCH (A.R.I.5972)
59. ABANDON AIRCRAFT SIGN
60. EMERGENCY DOOR OPENING SWITCHES (2 OFF)
61. ANTI-DAZZLE SWITCH
62. WIND INDICATOR UNIT
63. BOMB DOOR SWITCH
64. DIMMER SWITCH PANEL PILLAR LIGHTS
65. LIGHT SWITCH – AIR BOMBER'S PRONE POSITION
66. BOMB RELEASE SWITCH
67. CAMERA CONTROL UNIT TYPE 551
68. BOMB DOOR CONTROL SWITCH
69. BOMB SPACING UNIT PANEL
70. LIVE JETTISON SWITCH
71. V.T. FUZING SWITCH
72. DIMMER SWITCH AND PANEL LIGHT
73. REMOTE FATIGUE INDICATOR, TYPE M2372

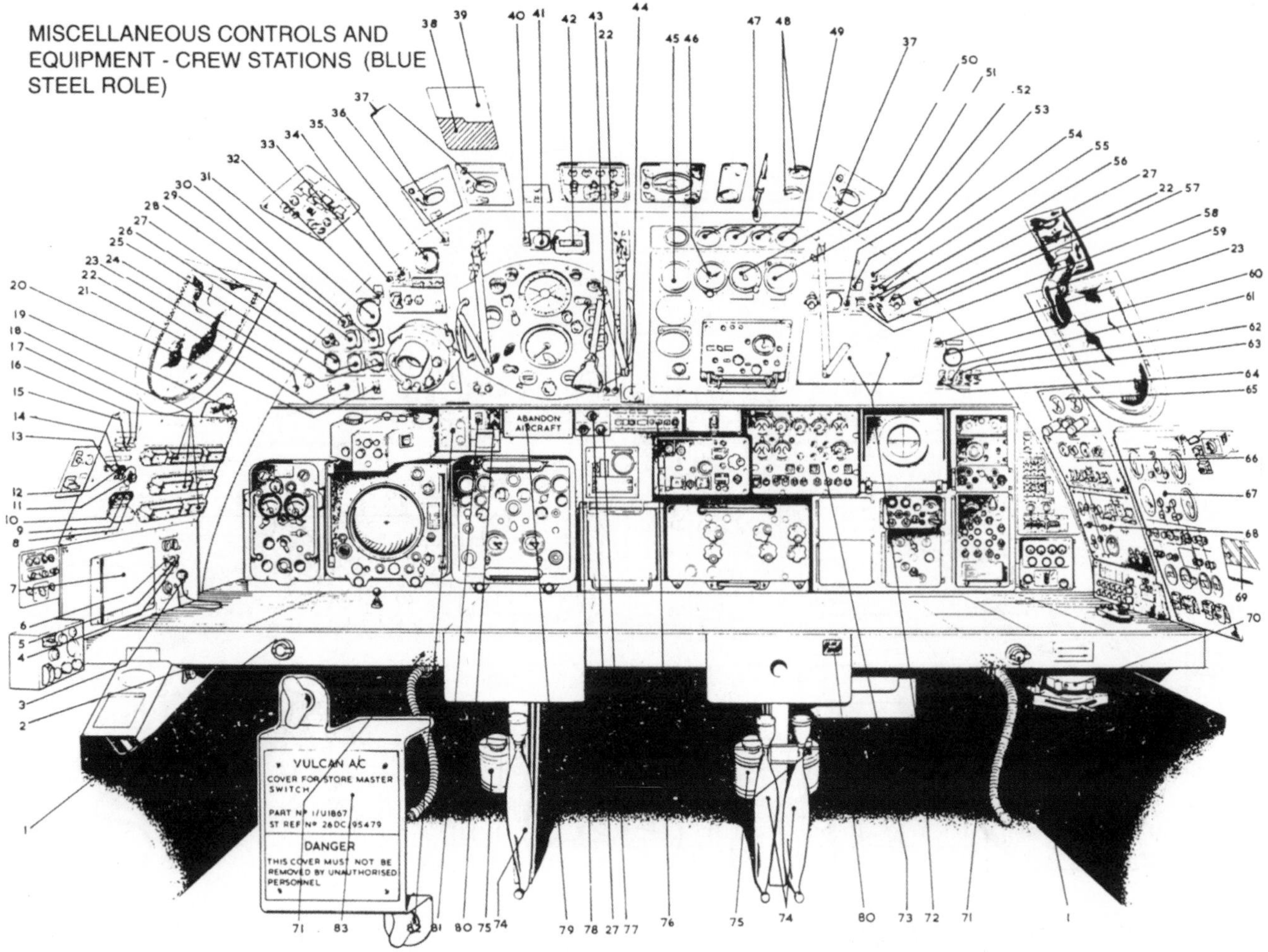

MISCELLANEOUS CONTROLS AND EQUIPMENT - CREW STATIONS (BLUE STEEL ROLE)

1. A.V.S. MANUAL TEMPERATURE CONTROL
2. DIMMER SWITCH - PANEL PILLAR LIGHTS
3. BOMB FIRING SWITCH
4. SWITCH - LIGHT AT AIR BOMBER'S PRONE POSITION
5. BOMB RELEASE SWITCH
6. BOMB DOORS SWITCH
7. MOUNTING PANEL FOR E.Q. CONTROL UNIT
8. PRESS TO TEST INDICATOR - OVERRIDE BREAK UP
9. PRESS TO TEST INDICATOR - COMMAND BREAK UP
10. PRESS TO TEST INDICATOR - SHUTTERED DETONATOR
11. DIMMER SWITCH - E.Q. PANEL LIGHT
12. D.C. SUPPLIES SWITCH - SINGLE POLE, GUARDED
13. RELEASE PRESSURE INDICATOR - ILLUMINATES AMBER - PRESS TO TEST
14. STORE GONE INDICATOR - ILLUMINATES AMBER
15. E.Q. POWER SUPPLIES EMERGENCY OVERRIDE - DOUBLE POLE SWITCH, GUARDED
16. ON/OFF SINGLE POLE SWITCH - RELEASE UNIT HEATER
17. BOMB SELECTOR SWITCH, BALLISTIC - POWER - DOUBLE POLE - GUARDED
18. FUSE BLOCK
19. DIMMER SWITCH AND LAMP
20. STORE HEATING CONTROL
21. ANTI-FLASH SCREEN
22. WARNING INDICATOR - LOSS OF CABIN PRESSURE
23. OXYGEN PRESSURE INDICATOR
24. STORE EXCHANGER OUTLET TEMPERATURE SELECTOR
25. STORE BAY TEMPERATURE SELECTOR
26. STORE ENGINE TEMPERATURE GAUGE
27. DIMMER SWITCH - CHARTBOARD LIGHTS
28. STORE BAY TEMPERATURE GAUGE
29. DIMMER SWITCH - PANEL LIGHTS
30. STORE-EXCHANGER OUTLET TEMPERATURE SELECTOR
31. BOMB BAY TEMPERATURE INDICATOR
32. BOMB BAY HEATING CONTROL SWITCH
33. BOMB BAY HEATING MANUAL CONTROL SWITCH
34. BOMB BAY TEMPERATURE SELECTOR
35. N.B.S. AIR SUPPLY PRESSURE INDICATOR
36. N.B.S. AIR SUPPLY ON/OFF SWITCH
37. OXYGEN REGULATORS
38. ACCESS DOOR - STORE RELEASE HANDLE
39. ACCESS DOOR - SAFETY LOCK REMOVAL HANDLE
40. H.2.S. PRESSURISATION ON/OFF SWITCH
41. H.2.S. PRESSURE GAUGE
42. AIR NAUTICAL MILES COUNTER
43. OXYGEN 'ON' INDICATOR
44. COMPASS DEVIATION CARD HOLDER
45. AIR SPEED INDICATOR
46. ALTIMETER
47. CABIN DECOMPRESSION HANDLE
48. OXYGEN PRESSURE GAUGES
49. FUEL CONTENTS GAUGES - NO.1, 2, 3 & 4 ENGINE
50. FORWARD THROW INDICATOR
51. OUTSIDE AIR TEMPERATURE INDICATOR
52. FIN GAP DOORS SWITCH - NORMAL
53. FIN GAP DOORS - MAGNETIC INDICATOR
54. REFRIGERATION HEATER ON/OFF SWITCH
55. E.P.C.U. START PUSH SWITCH & PRESS TO TEST INDICATOR
56. START/STOP PUSH SWITCH - A/C HYDRAULIC UNIT
57. START/STOP PUSH SWITCHES - A/C REFRIGERATION
58. FIN GAP DOORS EMERGENCY SWITCH-GUARDED
59. SIGNAL PISTOL
60. E.C.M. COOLANT TEMPERATURE INDICATOR
61. CABIN LIGHT SWITCH
62. BELOW TABLE LIGHT SWITCH
63. PERISCOPE HEATER SWITCH
64. PORT ROOF LIGHT SWITCH
65. C.S.D.U. OIL TEMPERATURE GAUGES WITH DIMMER SWITCH AND LAMPS
66. A.A.P.P. CONTROL PANEL - 70P (SECTION 6, CHAP.5)
67. ALTERNATOR CONTROL PANEL - 10P (SECTION 6, CHAP.2)
68. SECONDARY SUPPLIES PANEL - 50P (SECTION 6, CHAP.4)
69. TIME DELAY EMERGENCY JETTISON PANEL
70. PERISCOPE CONTROL HANDLE
71. A.V.S. FLOW CONTROL
72. WINDOW LAUNCH AND/OR TELEMETERING CONTROL PANEL
73. STORE CONTROL PANEL - OPERATING & MONITORING
74. SANITARY CONTAINERS
75. RATION HEATERS
76. STARBOARD ROOF LIGHT SWITCH
77. DIMMER SWITCH - STORE PANEL LIGHT
78. DIMMER SWITCH - INSTRUMENT PANEL LIGHT
79. ABANDON AIRCRAFT SIGN
80. EMERGENCY DOOR OPENING SWITCHES (2 OFF)
81. ANTI-DAZZLE LIGHTS SWITCH
82. WIND INDICATOR UNIT
83. COVER FOR STORE CONTROL SWITCHES

RADAR, RADIO AND ASSOCIATED EQUIPMENT - CREW STATIONS (FREE FALL ROLE)

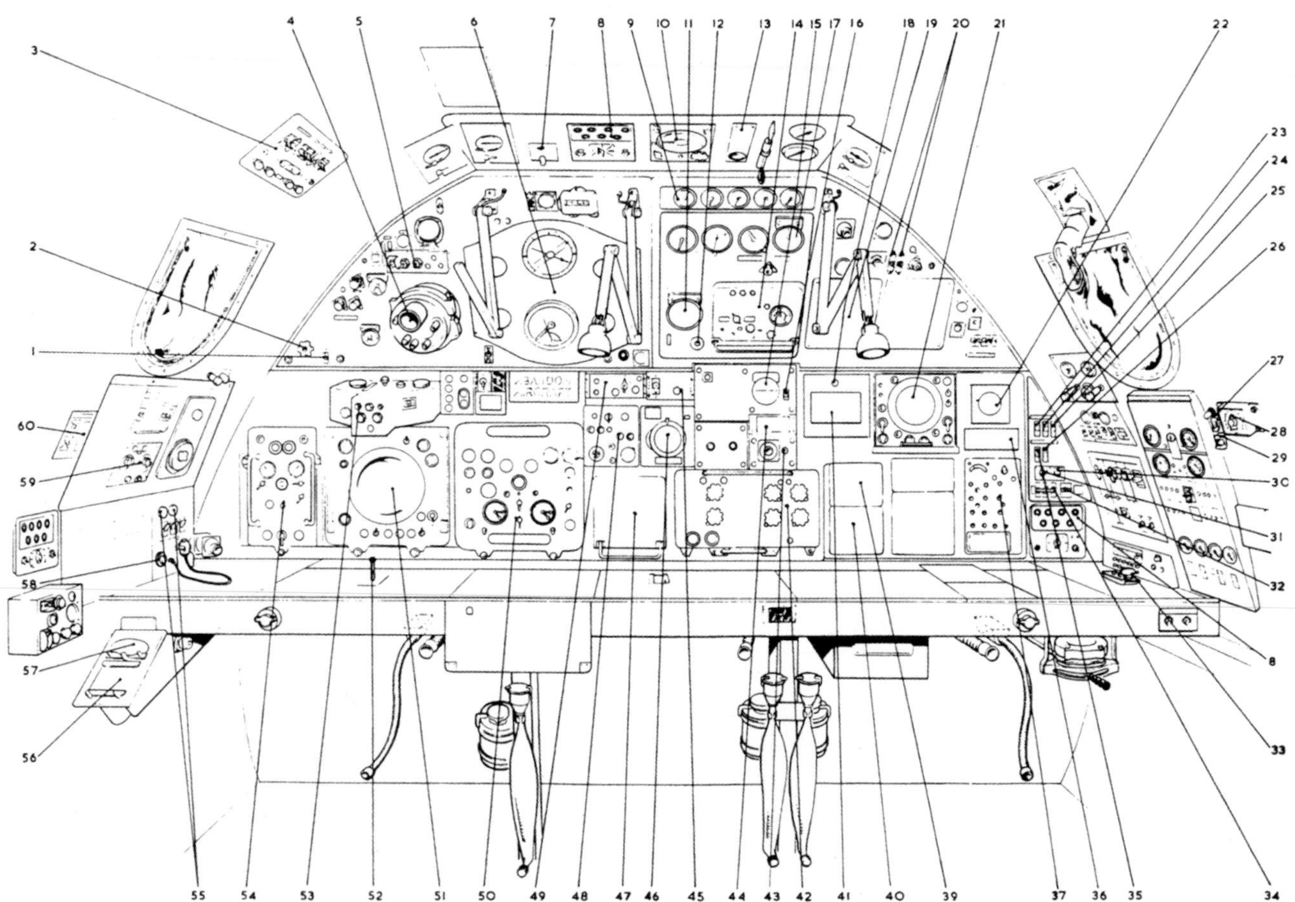

1. SCANNER STABILISATION SWITCH (A.R.I.5928)
2. SCANNER ROTATION CONTROL SWITCH (A.R.I.5928)
3. PULSE ALTIMETER
4. RADIO SWITCH PANEL
5. CONTROL PANEL – TRUE AIR SPEED
6. POWER SUPPLIES INDICATOR – N.B.C.
7. POWER SUPPLIES INDICATOR – H.2.S.
8. NAVIGATION PANEL
9. VARIABLE AIRSPEED UNIT
10. COMPASS ISOLATION SWITCH (PORT)
11. COMPASS ISOLATION SWITCH (STBD.)
12. ON/OFF SWITCH (A.R.I.5972)
13. CONTROL UNIT (A.R.I.23023)
14. D.F. BEARING INDICATOR (A.R.I. 23023)
15. COMPASS – REPEATER – TRUE TRACK MAGNETIC HEADING
16. LOOP CONTROLLER (A.R.I.23023)
17. RADIO ALTIMETER
18. STEERING SIGNAL TEST SOCKET
19. TRACK CONTROLLER UNIT – M.F.S.
20. DIMMER SWITCH – AUTO LAND AND T.C.U. LIGHTS
21. PILOTS DIRECTION INDICATOR
22. WINDOW LAUNCH OR TELEMETERING CONTROL PANEL
23. FAIL INDICATOR (A.R.I.23134/5/4)
24. TEST POINT (G.P.I. Mk.6)
25. INDICATOR, TYPE 6935 (A.R.I.5919)
26. INDICATOR, TYPE 17171 (A.R.I. 18228/1)
27. SUPPLIES SWITCH (A.R.I.5874)
28. OUTPUT SWITCH (A.R.I.5874)
29. SUPPLY SWITCH (A.R.I.23023)
30. PRESS-TO-TRANSMIT SWITCH
31. CONFERENCE INTERCOMM. – MASTER SWITCH
32. ON/OFF SWITCH (A.R.I.18011)
33. CONTROL UNIT, TYPE 705 (A.R.I. 18011)
34. AUTO-THROTTLE ON/OFF SWITCH (Inoperative)
35. DIMMER SWITCH – CONTROL UNIT DIAL LIGHTING
36. DIMMER SWITCH – CONTROL UNIT DIAL LIGHTING (A.R.I.18076)
37. INTERCOMM. MASTER CONTROL
38. EXTERNAL INTERCOMM. SWITCHES
39. MORSE KEY
40. INTERCOMM. CONTROL UNIT (A.E.O.)
41. CONTROL UNIT, TYPE 17172 (A.R.I. 18228/1)
42. CONTROL UNIT, TYPE 9422 (A.R.I. 18076)
43. DELETED (Mod.2501)
44. BLANKING PLATE
45. CONTROL TRANSMITTER/RECEIVER TYPE M53 (A.R.I.23090)
46. ANTENNA CONTROL UNIT, (A.R.I. 23090)
47. CONTROL UNIT, TYPE 16929 (A.R.I. 23134/5/4)
48. NOT ALLOCATED
49. GROUND SPEED SELECTOR UNIT
50. H.R.U. POWER FAILURE INDICATOR
51. G.P.I. MK.6
52. NAV. HEADING UNIT (H.R.S.)
53. H.R.S. CONTROL UNIT
54. EVENT MARKER PUSH SWITCH
55. RECORDER – ON/OFF SWITCH
56. FAILURE WARNING LIGHT (A.R.I. 5959)
57. INDICATOR CONTROL UNIT, TYPE 9869 (A.R.I.5972)
58. CONTROL UNIT, TYPE 7750 (A.R.I. 18107/13)
59. CONTROL UNIT, TYPE 585 (A.R.I. 5928)
60. CONTROL UNIT, TYPE 626 (A.R.I. 5928)
61. INDICATING UNIT, TYPE 301 (A.R.I. 5928)
62. CAMERA, TYPE RX.88 (A.R.I.5928)
63. CONTROL UNIT, TYPE 595 (A.R.I. 5928)
64. N.B.C. ISOLATION INDICATORS
65. CONTROL UNIT, TYPE 12580
66. BOMBING SELECTOR SWITCH
67. N.B.C. SIMULATOR SOCKET
68. BOMBING SELECTOR PANEL
69. SIMULATED BOMBING PANEL

RADAR, RADIO AND ASSOCIATED EQUIPMENT - CREW STATIONS

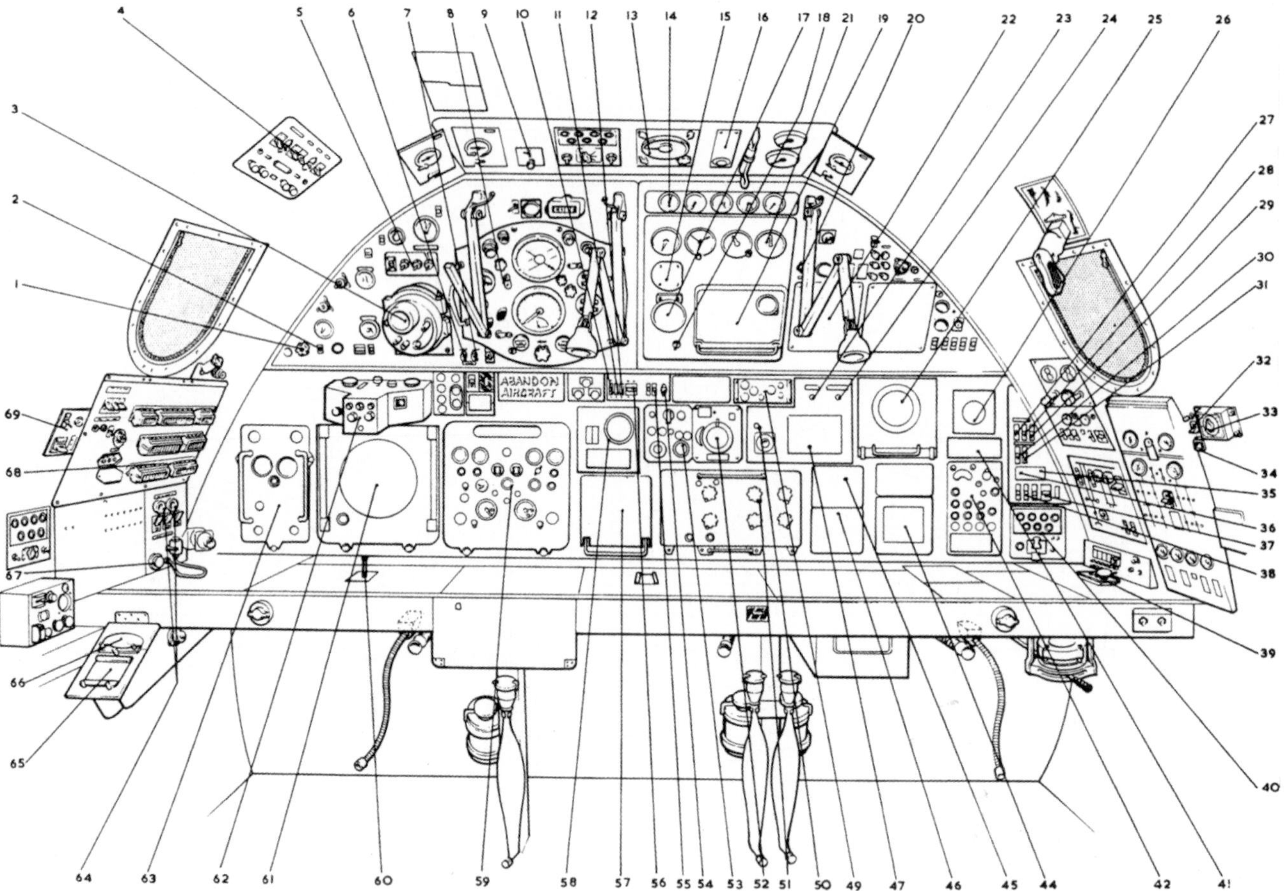

1. SCANNER STABILISATION SWITCH (A.R.I.5928)
2. SCANNER ROTATION CONTROL SWITCH (A.R.I.5928)
3. PULSE ALTIMETER
4. RADIO SWITCH PANEL
5. CONTROL PANEL – TRUE AIR SPEED
6. POWER SUPPLIES INDICATOR – N.B.C.
7. POWER SUPPLIES INDICATOR – H.2.S.
8. NAVIGATION PANEL
9. VARIABLE AIRSPEED UNIT
10. COMPASS ISOLATION SWITCH (PORT)
11. COMPASS ISOLATION SWITCH (STBD.)
12. ON/OFF SWITCH (A.R.I.5972)
13. CONTROL UNIT (A.R.I.23023)
14. D.F. BEARING INDICATOR (A.R.I. 23023)
15. COMPASS – REPEATER – TRUE TRACK MAGNETIC HEADING
16. LOOP CONTROLLER (A.R.I.23023)
17. RADIO ALTIMETER
18. STEERING SIGNAL TEST SOCKET
19. TRACK CONTROLLER UNIT – M.F.S.
20. DIMMER SWITCH – AUTO LAND AND T.C.U. LIGHTS
21. PILOTS DIRECTION INDICATOR
22. WINDOW LAUNCH OR TELE-METERING CONTROL PANEL
23. FAIL INDICATOR (A.R.I.23134/5/4)
24. TEST POINT (G.P.I. Mk.6)
25. INDICATOR, TYPE 6935 (A.R.I.5919)
26. INDICATOR, TYPE 17171 (A.R.I. 18228/1)
27. SUPPLIES SWITCH (A.R.I.5874)
28. OUTPUT SWITCH (A.R.I.5874)
29. SUPPLY SWITCH (A.R.I.23023)
30. PRESS-TO-TRANSMIT SWITCH
31. CONFERENCE INTERCOMM. – MASTER SWITCH
32. ON/OFF SWITCH (A.R.I.18011)
33. CONTROL UNIT, TYPE 705 (A.R.I. 18011)
34. AUTO-THROTTLE ON/OFF SWITCH (Inoperative)
35. DIMMER SWITCH – CONTROL UNIT DIAL LIGHTING
36. DIMMER SWITCH – CONTROL UNIT DIAL LIGHTING (A.R.I.18076)
37. INTERCOMM. MASTER CONTROL
38. EXTERNAL INTERCOMM. SWITCHES
39. MORSE KEY
40. INTERCOMM. CONTROL UNIT (A.E.O.)
41. CONTROL UNIT, TYPE 17172 (A.R.I. 18228/1)
42. CONTROL UNIT, TYPE 9422 (A.R.I. 18076)
43. DELETED (Mod.2501)
44. BLANKING PLATE
45. CONTROL TRANSMITTER/RECEIVER TYPE M53 (A.R.I.23090)
46. ANTENNA CONTROL UNIT, (A.R.I. 23090)
47. CONTROL UNIT, TYPE 16929 (A.R.I. 23134/5/4)
48. NOT ALLOCATED
49. GROUND SPEED SELECTOR UNIT
50. H.R.U. POWER FAILURE INDICATOR
51. G.P.I. MK.6
52. NAV. HEADING UNIT (H.R.S.)
53. H.R.S. CONTROL UNIT
54. EVENT MARKER PUSH SWITCH
55. RECORDER – ON/OFF SWITCH
56. FAILURE WARNING LIGHT (A.R.I. 5959)
57. INDICATOR CONTROL UNIT, TYPE 9869 (A.R.I.5972)
58. CONTROL UNIT. TYPE 7750 (A.R.I. 18107/13)
59. CONTROL UNIT, TYPE 585 (A.R.I. 5928)
60. CONTROL UNIT, TYPE 626 (A.R.I. 5928)
61. INDICATING UNIT, TYPE 301 (A.R.I. 5928)
62. CAMERA, TYPE RX.88 (A.R.I.5928)
63. CONTROL UNIT, TYPE 595 (A.R.I. 5928)
64. N.B.C. ISOLATION INDICATORS
65. CONTROL UNIT, TYPE 12580
66. BOMBING SELECTOR SWITCH
67. N.B.C. SIMULATOR SOCKET
68. BOMBING SELECTOR PANEL
69. SIMULATED BOMBING PANEL

RADAR, RADIO AND ASSOCIATED EQUIPMENT - CREW STATIONS (BLUE STEEL ROLE)

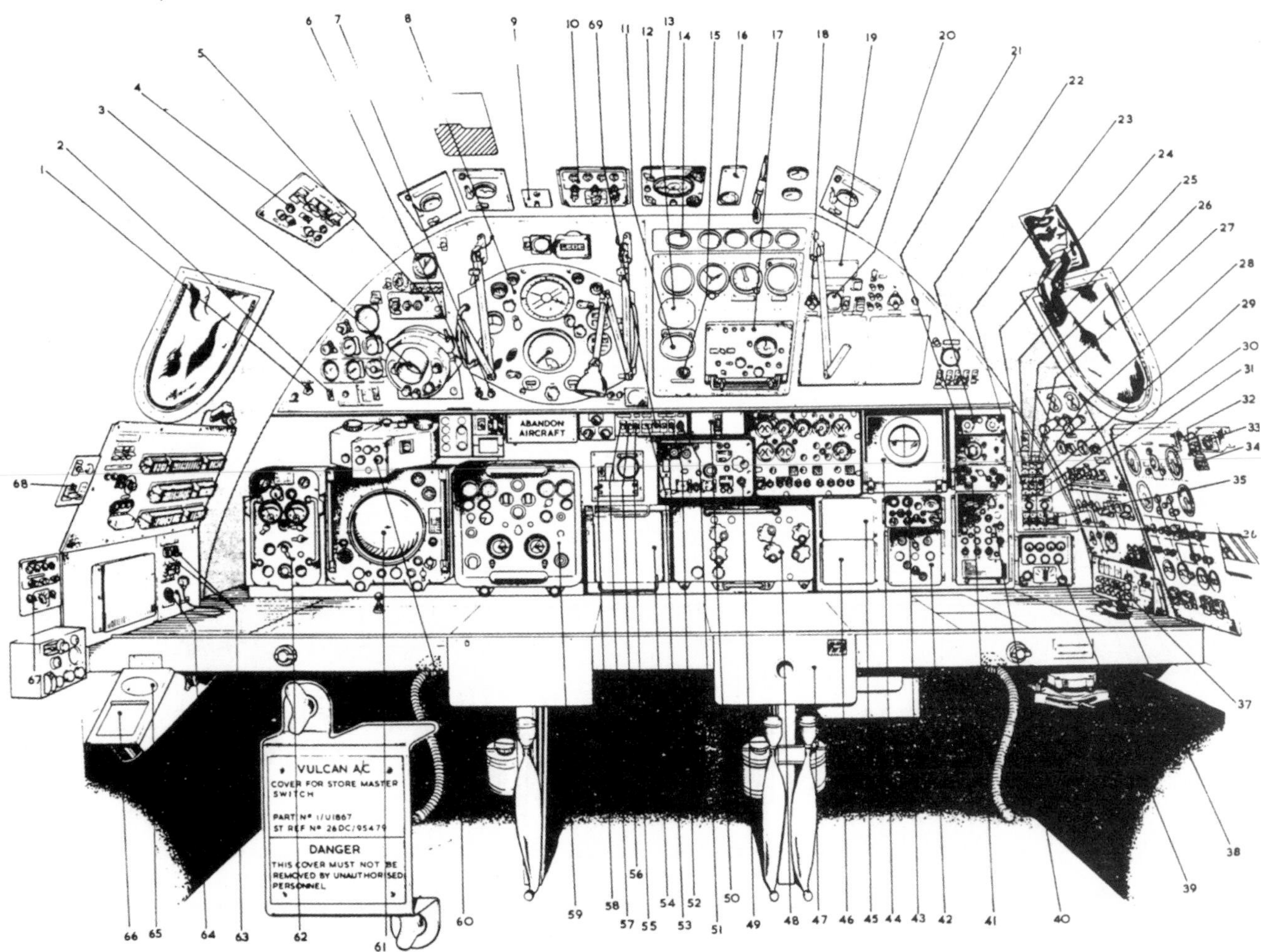

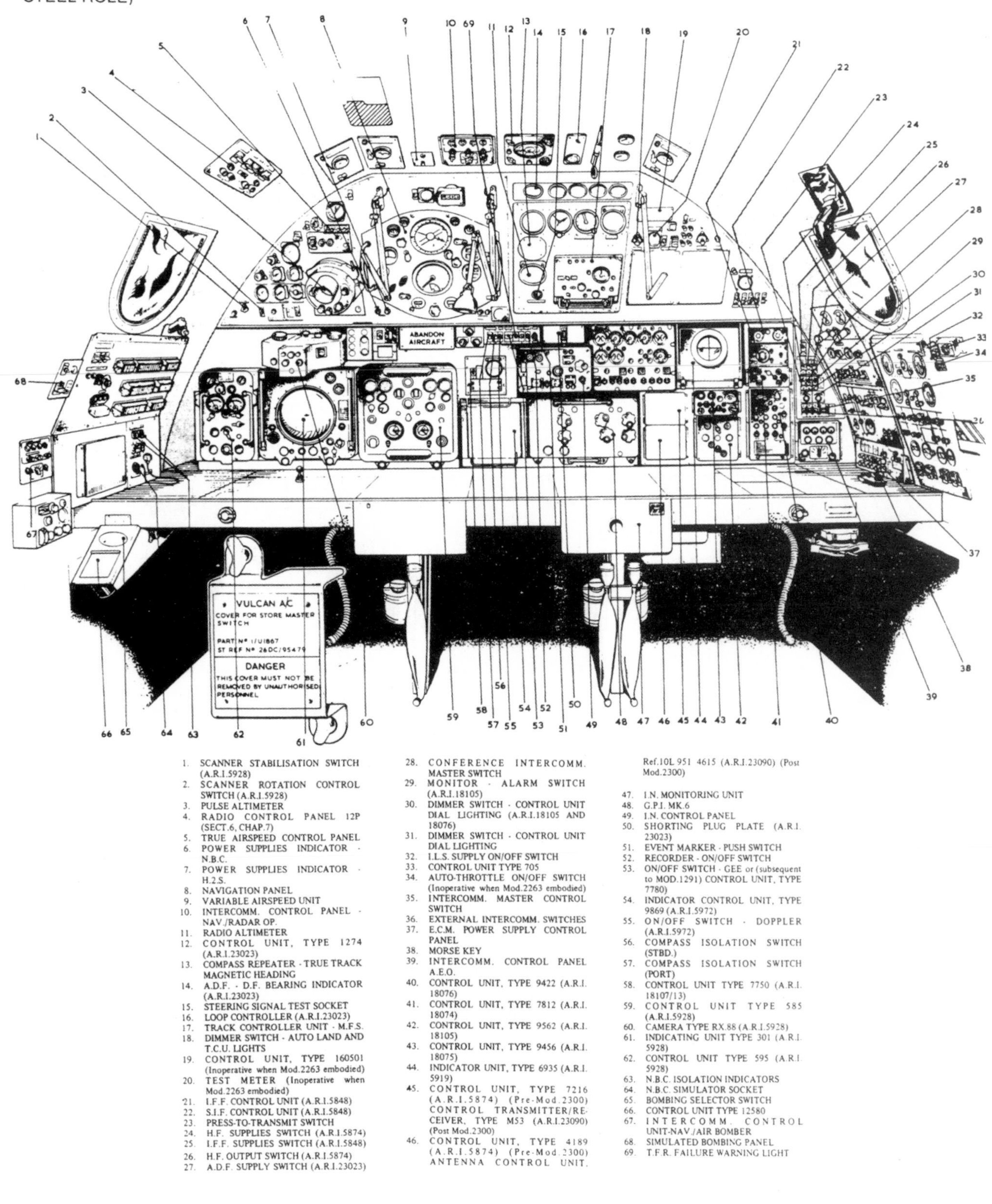

1. SCANNER STABILISATION SWITCH (A.R.I.5928)
2. SCANNER ROTATION CONTROL SWITCH (A.R.I.5928)
3. PULSE ALTIMETER
4. RADIO CONTROL PANEL 12P (SECT.6, CHAP.7)
5. TRUE AIRSPEED CONTROL PANEL
6. POWER SUPPLIES INDICATOR - N.B.C.
7. POWER SUPPLIES INDICATOR - H.2.S.
8. NAVIGATION PANEL
9. VARIABLE AIRSPEED UNIT
10. INTERCOMM. CONTROL PANEL - NAV./RADAR OP.
11. RADIO ALTIMETER
12. CONTROL UNIT, TYPE 1274 (A.R.I.23023)
13. COMPASS REPEATER - TRUE TRACK MAGNETIC HEADING
14. A.D.F. - D.F. BEARING INDICATOR (A.R.I.23023)
15. STEERING SIGNAL TEST SOCKET
16. LOOP CONTROLLER (A.R.I.23023)
17. TRACK CONTROLLER UNIT - M.F.S.
18. DIMMER SWITCH - AUTO LAND AND T.C.U. LIGHTS
19. CONTROL UNIT, TYPE 160501 (Inoperative when Mod.2263 embodied)
20. TEST METER (Inoperative when Mod.2263 embodied)
21. I.F.F. CONTROL UNIT (A.R.I.5848)
22. S.I.F. CONTROL UNIT (A.R.I.5848)
23. PRESS-TO-TRANSMIT SWITCH
24. H.F. SUPPLIES SWITCH (A.R.I.5874)
25. I.F.F. SUPPLIES SWITCH (A.R.I.5848)
26. H.F. OUTPUT SWITCH (A.R.I.5874)
27. A.D.F. SUPPLY SWITCH (A.R.I.23023)
28. CONFERENCE INTERCOMM. MASTER SWITCH
29. MONITOR - ALARM SWITCH (A.R.I.18105)
30. DIMMER SWITCH - CONTROL UNIT DIAL LIGHTING (A.R.I.18105 AND 18076)
31. DIMMER SWITCH - CONTROL UNIT DIAL LIGHTING
32. I.L.S. SUPPLY ON/OFF SWITCH
33. CONTROL UNIT TYPE 705
34. AUTO-THROTTLE ON/OFF SWITCH (Inoperative when Mod.2263 embodied)
35. INTERCOMM. MASTER CONTROL SWITCH
36. EXTERNAL INTERCOMM. SWITCHES
37. E.C.M. POWER SUPPLY CONTROL PANEL
38. MORSE KEY
39. INTERCOMM. CONTROL PANEL A.E.O.
40. CONTROL UNIT, TYPE 9422 (A.R.I. 18076)
41. CONTROL UNIT, TYPE 7812 (A.R.I. 18074)
42. CONTROL UNIT, TYPE 9562 (A.R.I. 18105)
43. CONTROL UNIT, TYPE 9456 (A.R.I. 18075)
44. INDICATOR UNIT, TYPE 6935 (A.R.I. 5919)
45. CONTROL UNIT, TYPE 7216 (A.R.I.5874) (Pre-Mod.2300) CONTROL TRANSMITTER/RECEIVER, TYPE M53 (A.R.I.23090) (Post Mod.2300)
46. CONTROL UNIT, TYPE 4189 (A.R.I.5874) (Pre-Mod.2300) ANTENNA CONTROL UNIT. Ref.10L 951 4615 (A.R.I.23090) (Post Mod.2300)
47. I.N. MONITORING UNIT
48. G.P.I. MK.6
49. I.N. CONTROL PANEL
50. SHORTING PLUG PLATE (A.R.I. 23023)
51. EVENT MARKER - PUSH SWITCH
52. RECORDER - ON/OFF SWITCH
53. ON/OFF SWITCH - GEE or (subsequent to MOD.1291) CONTROL UNIT, TYPE 7780)
54. INDICATOR CONTROL UNIT, TYPE 9869 (A.R.I.5972)
55. ON/OFF SWITCH - DOPPLER (A.R.I.5972)
56. COMPASS ISOLATION SWITCH (STBD.)
57. COMPASS ISOLATION SWITCH (PORT)
58. CONTROL UNIT TYPE 7750 (A.R.I. 18107/13)
59. CONTROL UNIT TYPE 585 (A.R.I.5928)
60. CAMERA TYPE RX.88 (A.R.I.5928)
61. INDICATING UNIT TYPE 301 (A.R.I. 5928)
62. CONTROL UNIT TYPE 595 (A.R.I. 5928)
63. N.B.C. ISOLATION INDICATORS
64. N.B.C. SIMULATOR SOCKET
65. BOMBING SELECTOR SWITCH
66. CONTROL UNIT TYPE 12580
67. INTERCOMM. CONTROL UNIT-NAV./AIR BOMBER
68. SIMULATED BOMBING PANEL
69. T.F.R. FAILURE WARNING LIGHT

MISCELLANEOUS CONTROLS AND EQUIPMENT - CREW STATIONS

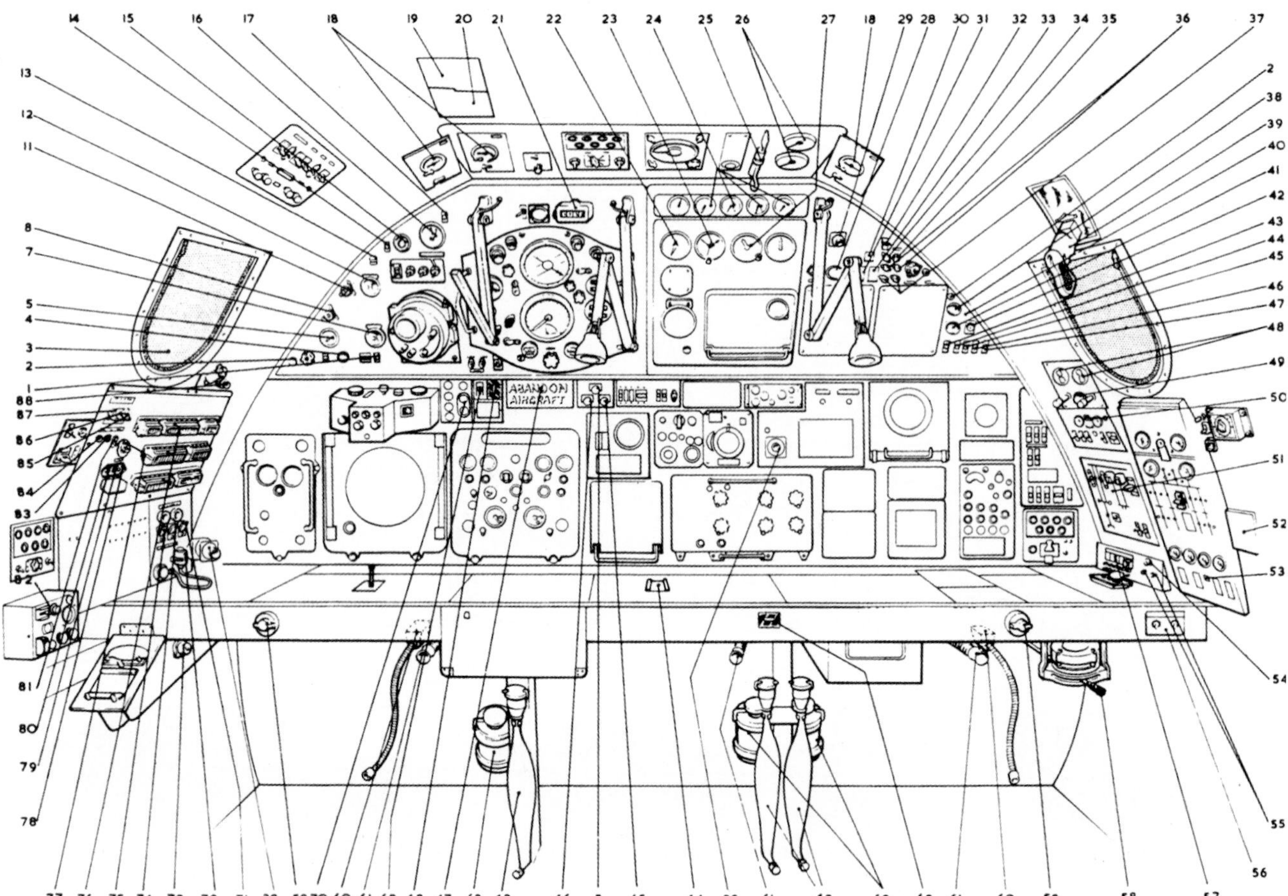

1. STORE HEATING CONTROL
2. OXYGEN PRESSURE INDICATOR
3. ANTI-FLASH SCREEN
4. LOSS OF CABIN PRESSURE WARNING LAMP
5. STORE ENGINE TEMPERATURE GAUGE
6. DELETED
7. OUTSIDE AIR TEMPERATURE INDICATOR
8. DIMMER SWITCH – CHARTBOARD LIGHTS
9. DELETED
10. DELETED
11. DIMER SWITCH PANEL LIGHTS
12. BOMB BAY TEMPERATURE INDICATOR
13. BOMB BAY HEATING CONTROL SWITCH
14. BOMB BAY HEATING MANUAL CONTROL SWITCH
15. BOMB BAY TEMPERATURE SELECTOR
16. N.B.C. AIR SUPPLY PRESSURE INDICATOR
17. N.B.C. AIR SUPPLY ON/OFF SWITCH
18. OXYGEN REGULATORS
19. ACCESS DOOR – STORE RELEASE
20. ACCESS DOOR SAFETY LOCK REMOVAL HANDLE
21. AIR NAUTICAL MILES COUNTER
22. AIR SPEED INDICATOR
23. ALTIMETER
24. FUEL CONTENTS GAUGES No.1, 2, 3 and 4 ENGINE GROUPS
25. CABIN DECOMPRESSION HANDLE
26. OXYGEN PRESSURE GAUGES
27. FORWARD THROW INDICATOR
28. REMOTE FATIGUE INDICATOR TYPE M2372
29. PUNKAH LOUVRE
30. FIN GAP DOORS - MAGNETIC IND
31. FIN GAP DOORS - SWITCH - NORMAL
32. REFRIGERATION HEATERS SWITCH
33. FIN GAP DOORS - EMERGENCY SWITCH - GUARDED
34. E.P.U.C. START PUSH SWITCH
35. PRESS TO TEST INDICATOR
36. START/STOP PUSH SWITCHES - A/C HYDRAULIC UNIT
37. START/STOP PUSH SWITCHES - A/C REFRIGERATION
38. ELECTRO MAGNETIC COUNTER WINDOW INSTALLATION
39. SIGNAL PISTOL
40. WINDOW INSTALLATION SWITCH
41. E.C.M. COOLANT TEMPERATURE GAUGE
42. R.B.W. CONTROL PANEL LIGHTS SWITCH
43. FATIGUE METER - SWITCH
44. PORT ROOF LIGHT - SWITCH
45. CABIN LIGHT - SWITCH
46. BELOW TABLE - LIGHT SWITCH
47. PERISCOPE HEATER - SWITCH
48. C.S.D.U. OIL TEMPERATURE
49. DIMMER SWITCH WITH LIGHTS
50. A.A.P.P. CONTROL PANEL 70P
51. SECONDARY SUPPLIES PANEL
52. TIME DELAY EMERGENCY JETTISON PANEL
53. ALTIMETER CONTROL PANEL
54. EXPLOSION PROTECTION RESET
55. EXPLOSION PROTECTION INDICATOR LAMPS PORT AND STBD. (MOD.1744)
56. AUXILIARY A.V.S. SWITCHES
57. E.C.M. POWER SUPPLY CONTROL PANEL
58. PERISCOPE CONTROL HANDLE
59. A.V.S. TEMPERATURE CONTROL
60. A.V.S. FLOW RATE CONTROL
61. AUXILIARY A.V.S. QUICK-RELEASE CONNECTOR
62. RATION HEATERS
63. SANITARY CONTAINERS
64. A.S.I. CORRECTION CARD HOLDER
65. DIMMER SWITCH - STORE - PANEL LIGHT
66. DIMMER SWITCH - INSTRUMENT - PANEL LIGHT
67. ABANDON AIRCRAFT SIGN
68. EMERGENCY DOOR OPENING SWITCHES (2 OFF)
69. ANTI-DAZZLE LIGHT SWITCH
70. WIND INDICATOR UNIT
71. SWITCH-LIGHT AT AIR BOMBER'S PRONE POSITION
72. BOMB FIRING SWITCH
73. DIMMER SWITCH - PANEL PILLAR LIGHTS
74. BOMB RELEASE - SWITCH
75. BOMB DOORS - SWITCH
76. FUSE BLOCKS
77. DIMMER SWITCH E.Q. PANEL
78. PRESS TO TEST INDICATOR - OVERRIDE BREAK UP
79. PRESS TO TEST INDICATOR - COMMAND BREAK UP
80. PRESS TO TEST INDICATOR - SHUTTERED DETONATOR
81. D.C. SUPPLIES SWITCH - SINGLE POLE GUARDED
82. CAMERA CONTROL UNIT TYPE 551
83. STORE GONE INDICATOR ILLUMINATES AMBER
84. RELEASE PRESSURE INDICATOR - ILLUMINATES AMBER PRESS TO TEST
85. BOMB SELECTOR SWITCH BALLISTIC - POWER - DOUBLE POLE GUARDED
86. ON/OFF SINGLE POLE SWITCH - RELEASE UNIT HEATER
87. E.Q. POWER SUPPLIES EMERGENCY OVERRIDE - DOUBLE POLE SWITCH GUARDED
88. DIMMER SWITCH AND LAMP

FLYING CONTROLS AND INSTRUMENTS

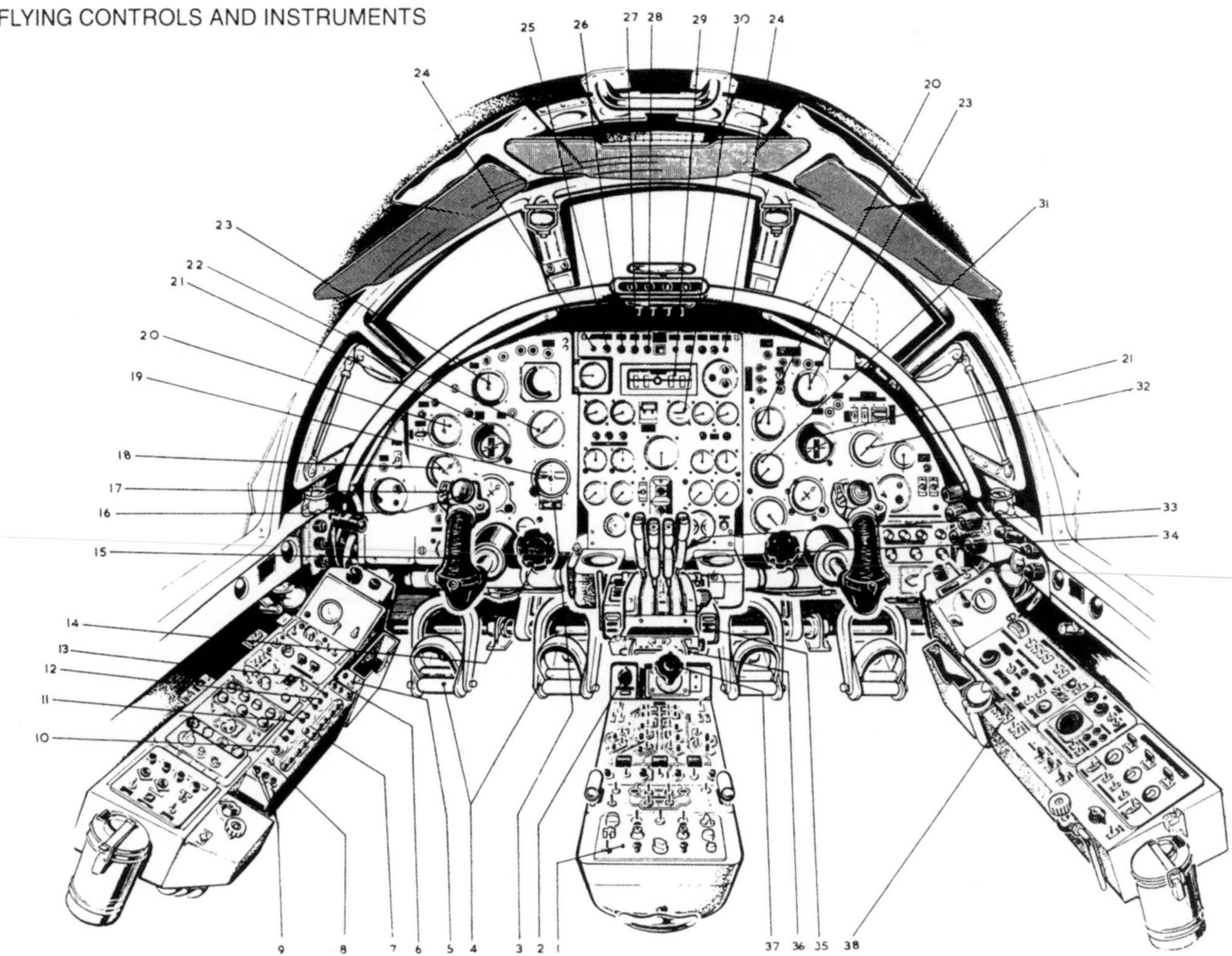

CONTROLS - MANUAL

4. Rudder pedals - incorporating brake master cylinders.
14. Adjuster - rudder pedals.
15. Control handle - elevons.

CONTROLS AND INDICATORS - ELECTRICAL

5. FEEL RELIEF system isolation lock-in switch and indicator. Switch labelled NORMAL-LOCK.
6. ARTIFICIAL FEEL Failure indicator/push switches. Illuminate AMBER when failure occurs. PRESS to isolate failed unit.
7. P.F.C. MOTOR - failure indicator/push switches. Illuminate AMBER for motor failure. PRESS to isolate failed motor.
8. YAW DAMPER - 2 position toggle switch.
   No.1 - rearward - No.1 system engaged.
   No.2 - forward - No.2 system engaged.
9. P.F.C. START - Groups start push switches.
   A. push to start outboard elevon units.
   R. push to start rudder main unit.
   E. push to start inboard elevon units.
10. COMPARATOR RESET - push to reset.
11. PITCH DAMPER - failure indicator push switches. Illuminate AMBER when failure occurs. PRESS to disengage failed system.
12. AUTO-mach - TRIM - individual channel indicators. Illuminate BLUE when failure occurs. PRESS to disengage.
13. AUTO-mach-TRIM - indicator push/pull switch. Illuminates AMBER when failure occurs.
    PUSH to disengage.
    PULL to engage.
16. Elevon trim - 4-way double-pole switch. Operate in natural sense.
17. ARTIFICIAL FEEL RELIEF cut-out switch (guarded). PRESS to disengage system.
24. MAIN WARNING system indicators. Illuminate AMBER for flying control system or unit failure.
25. P.F.C. UNIT failure - magnetic doll's eye indicator.
    BLACK - normal functioning.
    WHITE - displayed when faulty unit is disengaged. Remains white to remind pilot that fault exists after 24 (amber) is extinguished.
26. ARTIFICIAL FEEL failure - magnetic doll's eye indicator.
    BLACK - normal functioning.
    WHITE - displayed when fault occurs in system. Remains white to remind pilots that fault exists after 24 (amber) is extinguished.
27. AUTO-STABILISERS - system failure magnetic doll's eye indicator.
    BLACK - normal functioning.
    WHITE - displayed when fault occurs in system. Remains white to remind the pilots' that fault exists when 24 (amber) is extinguished.
28. AIR BRAKES - magnetic doll's eye indicator.
    BLACK - airbrakes IN.
    WHITE - airbrakes OUT.
33. RUDDER FEEL RELIEF - push switch.
    Press to disengage system.
34. RUDDER FEEL TRIM switch.
    P - lift - trims rudder to port.
    S - right - trims rudder to starboard.
35. AIR BRAKES EMERGENCY switch (guarded)
    UP - NORMAL
    Down - EMERGENCY - energises stand-by motor.
36. AIR BRAKES selector switch (with locking bar).
    UP - IN - air brakes stowed in main plane.
    Centre - MEDIUM DRAG - partial extension of pillars with slats at 35 deg.
    Down - HIGH DRAG - full extension of pillars with slats at 80 deg.
37. EMERGENCY TRIM control
    Press to energise - operate in natural sense.
38. AUTO-PILOT RESET SWITCH (Mod. 2305,2306). Operate to reset d.c. supply.

INSTRUMENTS AND INDICATORS

1. AUTO-PILOT control panel
2. AUTO-PILOT control
3. SIDE SLIP indicator
18. ALTIMETER Mk.19
19. ARTIFICIAL HORIZON (stand-by)
20. AIR SPEED INDICATOR
21. DIRECTOR HORIZON (Beam and G.P.).
22. RATE OF CLIMB INDICATOR
23. MACHMETER
29. CONTROL SURFACE INDICATOR
30. AUTO PILOT TRIM INDICATOR
31. 100,000 ft. ALTIMETER
32. CLIMB AND DESCENT INDICATOR

ENGINE AND FUEL SYSTEM CONTROLS

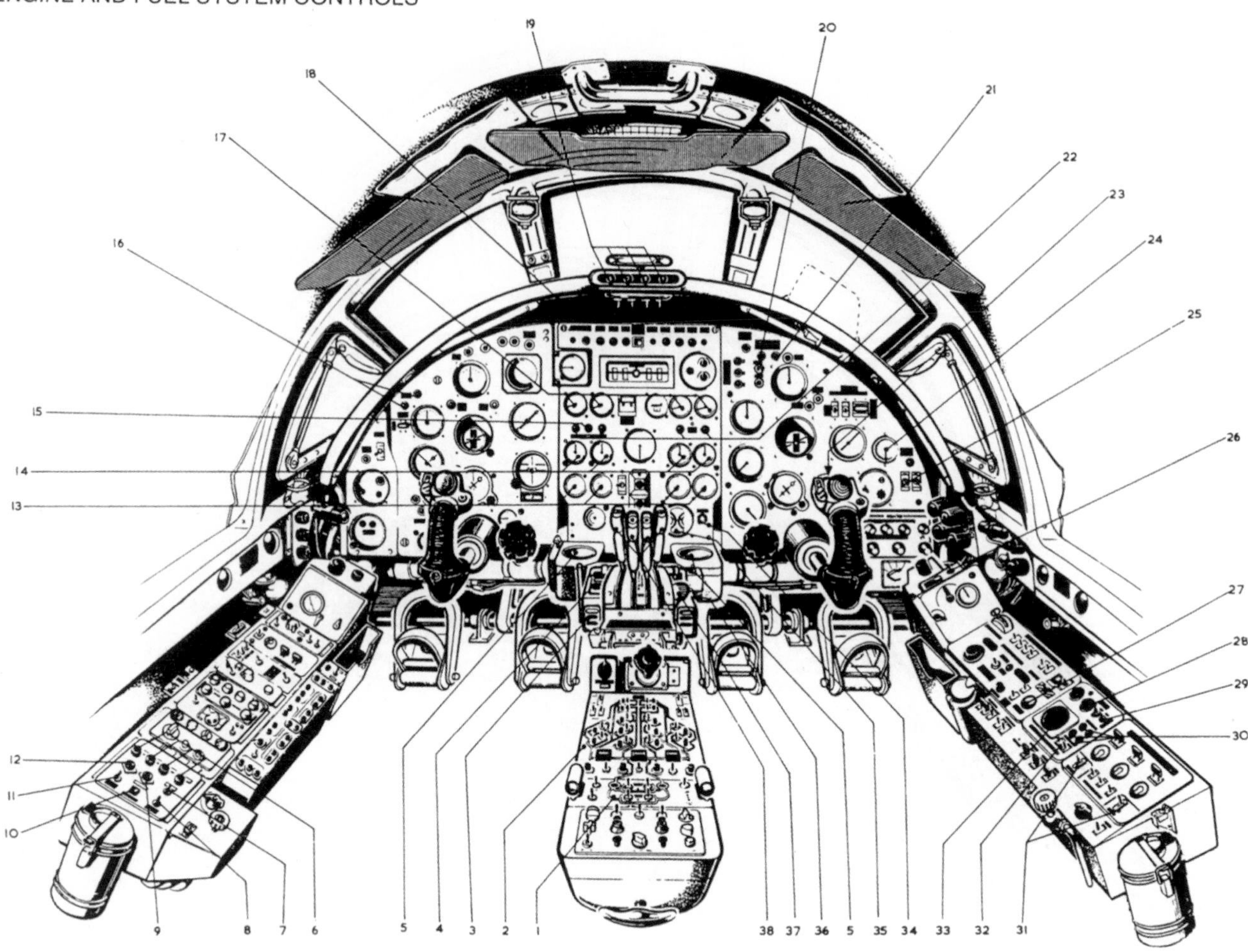

1. BOMB BAY TANKS system control panel.
2. FUEL SYSTEM control panel (Refer to Sect.4, Chap.2).
3. Jet pipe temperature limiter switch (guarded).
4. R.P.M. governor switch (guarded).
5. FUEL TANK CONTENTS panel. Houses four contents gauges (Refer to Sect.4, Chap.2).
6. Individual engine start press switches. PRESS AND RELEASE to start engine.
7. NORMAL/RAPID start selector (lift toggle to select).
8. ON/OFF MASTER switch.
9. Simultaneous RAPID start push button.
10. AIR CROSS-FEED indicator.
11. IGNITION ON/OFF switch.
12. GYRO HOLD-OFF press switch.
13. OIL PRESSURE gauges.
14. Engine R.P.M. indicators.
15. R.P.M. governor indicator - BLACK take-off – WHITE cruise.
16. AUTO-THROTTLE failure indicators. (inoperative).
17. ENGINE TEMPERATURE indicators.
18. L.P. fuel cock switches (guarded). forward - ON, rearward - OFF.
19. Engine FIRE indicator/operating switches. Illuminate RED when engine temperature becomes excessive; press to operate, pull to test filament.
20. FUEL FIRE warning indicator - BOMB BAY TANKS.
21. FUEL FIRE warning indicator - FUSELAGE AND WING TANKS.
22. FUEL LOW PRESSURE WARNING indicators.
23. TOTAL FUEL FLOW indicator.
24. FUEL FLOW indicator. Indicates rate of flow for any engine when relevant switch is selected on fuel control panel.
25. FUEL FLOW indicator switch.
26. TEST switch-flight idling detent, spring-loaded to OFF (Sect.4, Chap.1A).
27. FLIGHT REFUELLING PRESSURE gauge.
28. NITROGEN PURGE - flight refuelling.
29. TANK PRESSURISATION switch.
30. TANK PRESSURE indicators.
31. ENGINE ANTI-ICING system control switches (guarded)
Forward switch - PORT ENGINE
Aft switch - STARBOARD ENGINE.
32. MASTER SWITCHES (2), flight refuelling ON/OFF.
33. TANKS FULL INDICATOR - flight refuelling.
34. FUEL C. of G. indicator switch.
35. FUEL C. of G. indicator.
36. T.F.R. RESET switch PRESS to break.
37. Throttle friction adjuster.
38. Throttle control levers - Re-light press switch in handles.

NAVIGATIONAL, SIGNALLING AND LIGHTING EQUIPMENT

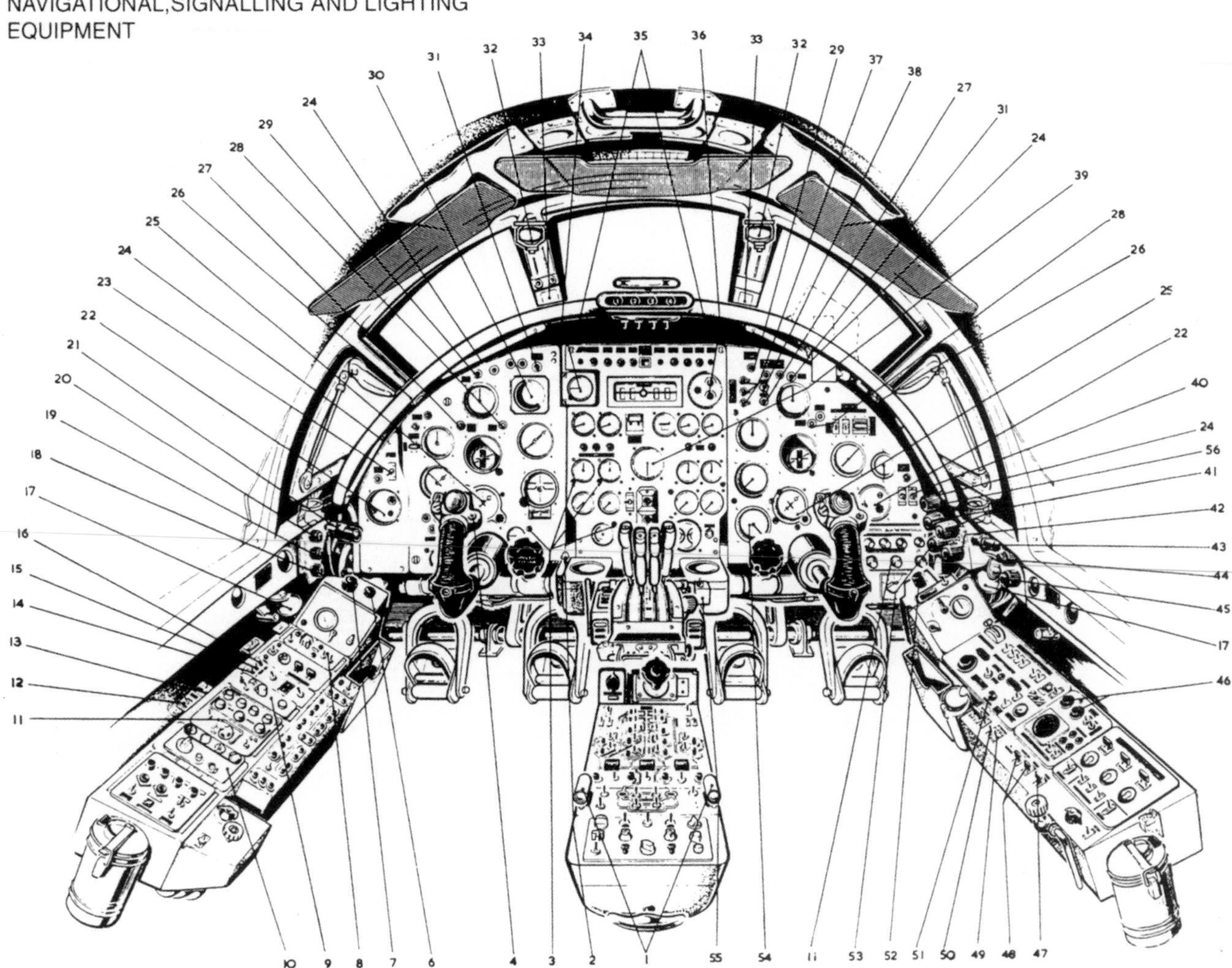

NAVIGATIONAL

5. DELETED
8. CONTROL UNIT TYPE 10695 (A.R.I. 23061).
9. CONTROL UNIT PART NO.T6654 (A.R.I.5959).
14. AUDIO – I.L.S. TACAN A.D.F. CHANGEOVER SWITCH.
22. ANNUNCIATOR UNIT – M.F.S.
25. BEAM COMPASS.
27. T.F.R. VIDEO INDICATOR (A.R.I. 5959).
28. T.F.R. FAIL INDICATOR (A.R.I.5959).
29. T.F.R. WARNING INDICATOR (A.R.I. 5959).
30. ALTITUDE AND AUTO-LAND PHASE INDICATOR.
31. I.L.S. MARKER LAMP (A.R.I.18011).
32. E.2.B. COMPASS.
33. HEAD UP INDICATOR (A.R.I.5959).
34. ACCELEROMETER.
36. M.F.S. SELECTOR UNIT.
37. SELECTED ALTITUDE INDICATOR LAMPS.
    Top – AMBER – above selected altitude.
    Centre – GREEN – selected altitude.
    Bottom – RED – below selected altitude.
38. SELECTOR SWITCH.
    Up – M.F.S.
    Down – T.F.R.
39. TACAN INDICATOR.
54. A.D.F. BEARING COMPASS.
55. READ-OUT HEADING SELECTOR SWITCH.
    Forward – H.R.S.
    Rearward – M.F.S.

SIGNALLING

4. INTERCOMM. PRESS-TO-TRANSMIT SWITCH.
10. CONTROL UNIT 5821-99-942-8543 (A.R.I.18124).
11. INTERCOMM. CONTROL UNIT.
12. R./T.1 TONE SWITCH.
    Forward – ON.
    Rearward – OFF.
13. AUDIO WARNING PRESS-TO-TEST SWITCH.
15. R./T.2 TONE SWITCH.
    Outboard – ON
    Inboard – OFF.
16. AERIAL CHANGEOVER SWITCH.
    Outboard – R./T.1 UPPER.
    Inboard – R./T.2 UPPER.

LIGHTING

1. RED FLOOD LAMPS.
2. ON/OFF SWITCH – HIGH-INTENSITY ANTI-DAZZLE LAMPS.
3. PILLAR LAMPS (TYPICAL LOCATION)
6. DIMMER SWITCH – U./V. LAMPS.
7. DIMMER SWITCH – U./V. LAMPS FOR OXYGEN PANELS.
17. U./V. LAMP.
18. DIMMER SWITCH – FIRST PILOT'S KNEE PAD LAMP.
19. DIMMER SWITCH – E.2.B. COMPASS LIGHT.
20. DIMMER SWITCH – PILOTS' PANEL PILLAR LIGHTS.
21. FIRST PILOT'S KNEE PAD LAMP.
23. ON/OFF SWITCH – WHITE FLOOD LAMPS.
24. U./V. LAMPS.
26. RED FLOOD LAMPS.
35. HIGH INTENSITY ANTI-DAZZLE LAMPS.
40. ON/OFF SWITCH – WHITE FLOOD LAMPS.
41. DIMMER SWITCHES – RED FLOOD ON CENTRE CONSOLE.
42. DIMMER SWITCH – RED FLOOD ON STARBOARD CONSOLE.
43. SECOND PILOT'S KNEE PAD LAMP.
44. DIMMER SWITCH – SECOND PILOT'S KNEE PAD LAMP.
45. DIMMER SWITCH – E.2.B. COMPASS LIGHT.
46. DIMMER SWITCHES – PROBE ILLUMINATION LAMPS.
47. STEADY/FLASH SWITCH – NAVIGATION LIGHTS.
48. LANDING LIGHT SWITCHES – PORT AND STARBOARD – RETRACT, LANDING TAXI.
49. IDENT. LIGHT SWITCH – STEADY, MORSE.
50. EXTERNAL LIGHT SWITCH.
51. DIMMER SWITCH – U./V. LIGHTS.
52. DIMMER SWITCH – U./V. LIGHT-OXYGEN PANEL.
53. DIMMER SWITCH – RED FLOOD ON STARBOARD COAMING.
56. DIMMER SWITCH – S.F.O.M. GUNSIGHT AND PILLAR LAMP.

The co-pilot's position in the Vulcan B.2. This excellent ultra-wide-angle view gives a slightly false impression of spaciousness – the Vulcan's cockpit is, in fact, very cramped. *Paul Tomlin*

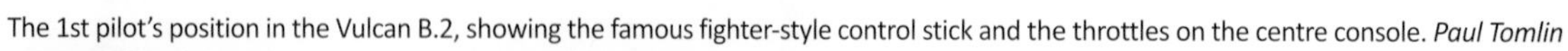

The 1st pilot's position in the Vulcan B.2, showing the famous fighter-style control stick and the throttles on the centre console. *Paul Tomlin*

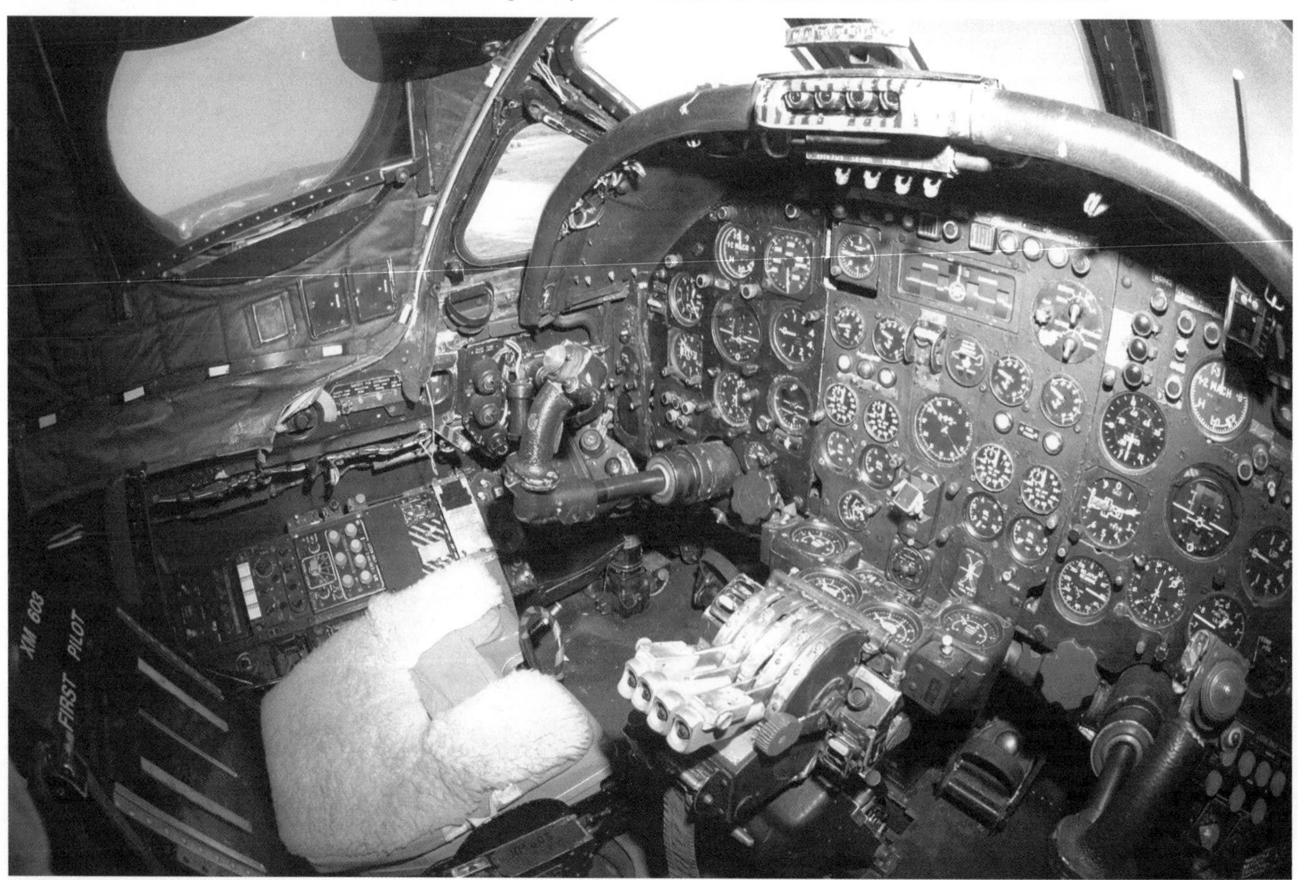

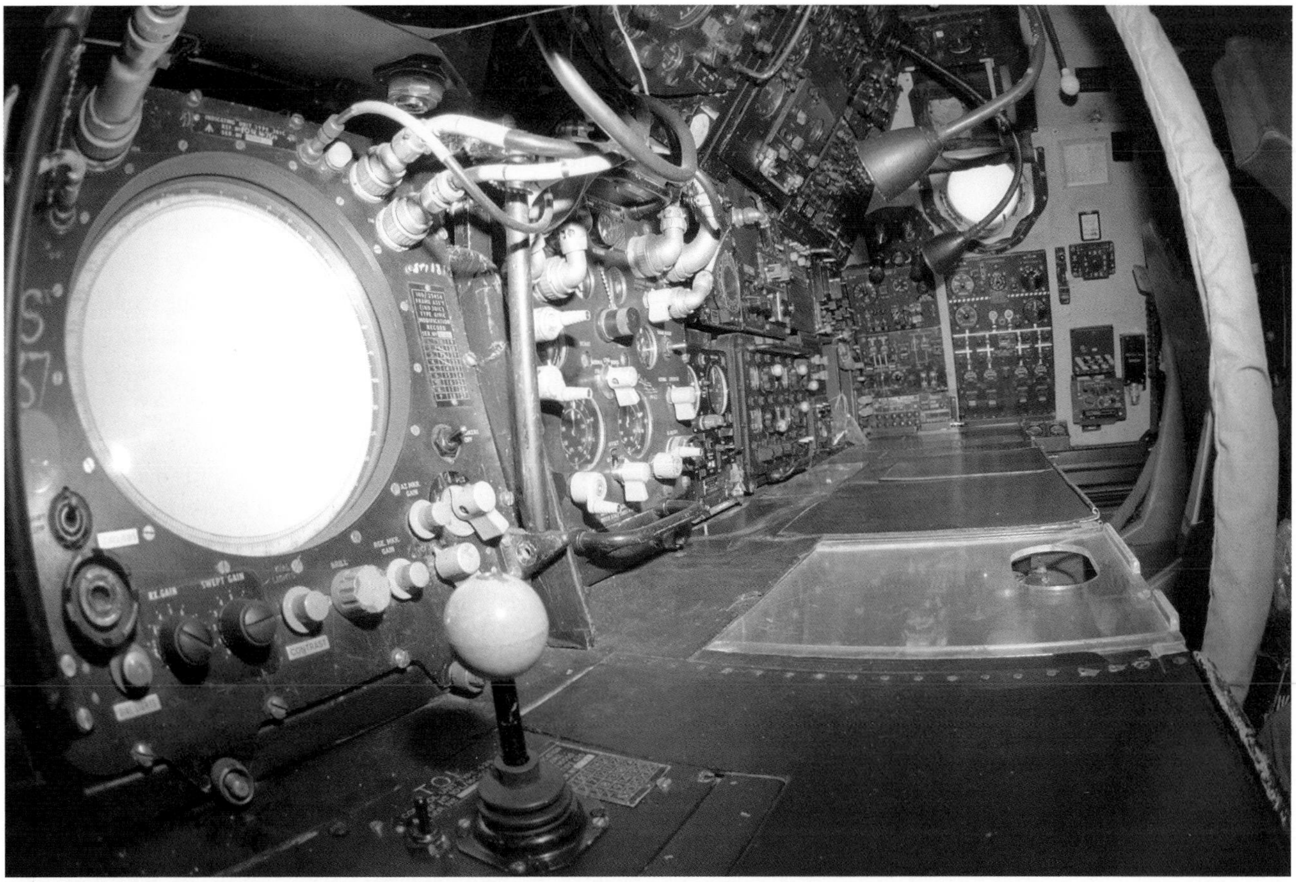

Looking to port and the AEO's position, in the foreground is the Nav Radar's H2S CRT and associated controls. *Paul Tomlin*

Looking across the rear crew's position (often referred to as the 'coal face'), the Nav Radar's seat is visible and much of the equipment available to him and the Nav Plotter, who sat to his right. *Paul Tomlin*

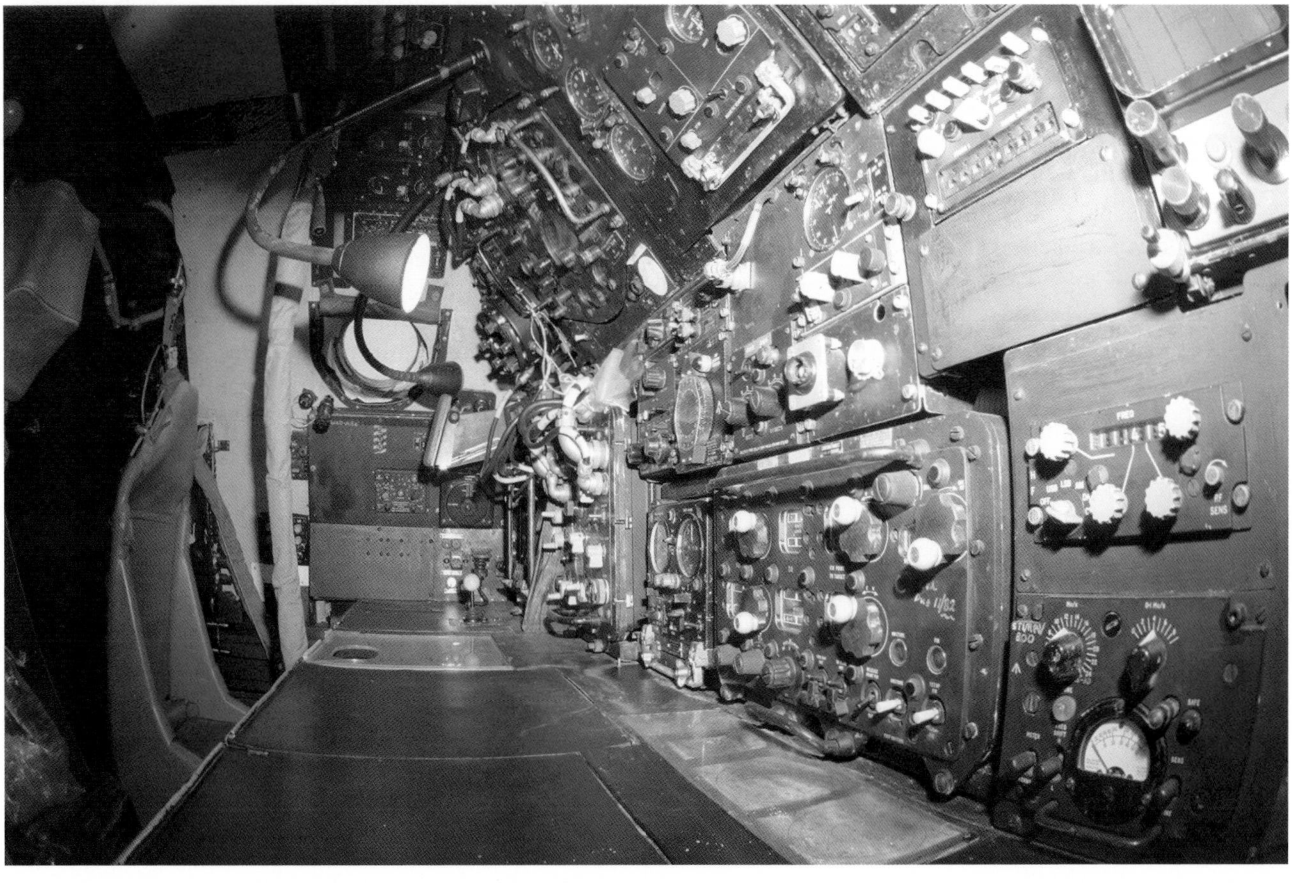

Leading particulars

Principal Dimensions

Overall length 105ft 6in

Wing span 111ft 0in

Height to top of fin 27ft 1in

Wheel track 31ft 1in

Wheel base 30ft 1.5in

PART 1: AIR CONDITIONING

AIR SYSTEM: CABIN AIR

General

The cabin air conditioning and pressurisation system maintains comfortable temperatures and pressures within the crew compartment. Hot pressurised air tapped from the engine compressors, cooled by cold ram air and a cooling turbine, is distributed throughout the cabin via ducting. The temperature of the conditioned air is controlled by varying the proportion of hot air which flows through or bypasses the air cooler or the cooling turbine. The controlled air flow out of the cabin is used to cool equipment in the radome.

Provision is made for conditioning the cabin air both on the ground and in unpressurised flight.

The cabin pressure is determined by the amount of air allowed to flow out of the crew compartment and is maintained by two pressure controllers. Pressurisation can be set for either cruise or combat conditions. Provision is made for emergency depressurisation.

The main controls for cabin heating and pressurisation are grouped together on the starboard console.

Cabin Air Conditioning Unit

The air conditioning unit in the nosewheel bay consists of an air-to-air cooler, a temperature control valve (TCV), a cooling brake turbine unit and a water separator.

The cooler is supplied with cold air from a ram air intake between the cabin and the port engine air intake. The cold air passes through a rearward-facing duct below the unit.

The brake turbine unit is an inward flow turbine coupled to a centrifugal braking compressor. The turbine passes air from the TCV to a water separator and thence to the cabin. The compressor passes filtered air from the nosewheel bay through the exhaust duct. The speed on the turbine is monitored by a pressure ratio switch which, if the turbine overspeeds, automatically selects a warmer setting on the TCV thereby reducing the amount of air passed to the turbine. The COLD AIR UNIT OVERSPEED WARNING MI on the starboard console shows white, reverting to black when the overspeed has ceased.

The temperature control valve is electrically operated, either automatically or manually. When maximum heat is selected, engine air passes directly from the flow valves, through the TCV, to the cabin. As the temperature setting is reduced, air is progressively allowed to pass through the cooler en route to the TCV. At the colder settings, the air passes to the cabin through the cooler, TCV and turbine.

An underheat sensing element opens a bypass valve when the air from the turbine falls below 2 degrees C value, thus allowing warm air to mix with the cold air before it reaches the water separator.

An overheat switch operates to move the control valve towards the cool position when the output temperature rises to 175 degrees C.

Cabin Pressurisation

Cabin pressurisation is achieved by controlling the rate at which the air fed into the cabin is allowed to escape. Each of two pressure controllers, in the nose below the pilots' floor, supplies counter pressure to one of the bellows of the combined valve unit in the front pressure bulkhead. One controller is motorised allowing a CRUISE or COMBAT setting to be selected. The other is unmotorised and acts as a standby in the CRUISE setting. The ground test levers on the controllers must always be fully down in flight.

Pressurisation begins at 8,000 feet and a cabin altitude of 8,000 feet is maintained until the maximum differential pressure is reached. In CRUISE the maximum differential pressure is 9 PSI, attained at approximately 47,000 feet. In COMBAT the maximum differential of 4 PSI is reached at approximately 19,500 feet. Above these altitudes, cabin altitude increases. The change from CRUISE to COMBAT setting takes place at 12 PSI per minute and from COMBAT to CRUISE at 1 PSI per minute.

FLOOD FLOW

Provision is made for flood flow but the system is inoperative.

DECOMPRESSION

To allow decompression of the cabin in an emergency, air release valves (in the lines between the pressure controllers and the bellows of the combined valve unit) can be operated to remove the counter-pressure from the bellows. Operation of the valves is controlled electrically by either pilot and electrically and mechanically by the rear crew.

While the combined effect of the operation of the air release valves, duct relief and decompression flaps ensures a rapid release of cabin pressure, it may take up to 30 seconds for the pressure to fall sufficiently to allow the door to open.

Cabin Ventilation

During unpressurised flight, the cabin can be ventilated via the ram air valve on the port side of the cabin. This allows air from the cabin conditioning ram air intake to enter the cabin but, unless the cabin switches are shut, the effect is negligible. The ram air valve should be closed before pressurising the cabin. The cabin may be ventilated on the ground with air supplies from the cooling air unit, via the normal control valves.

Individual face blowers (punkah louvres) for the rear crew are powered by a 200V AC blower unit under the navigator's table, controlled by a FACE BLOWING ON/OFF switch on the edge of the AEO's table.

AIR-VENTILATED SUITS SYSTEM

Normal Air-ventilated Suits (AVS) System

The air-ventilated suits are supplied from an air conditioning unit, similar to the cabin conditioning unit. The AVS unit uses hot air from the engines or AAPP, and cold air from the cabin system ram air intake. A ground-conditioning connection is provided, so that an external supply may be plugged into the suits.

AVS Air Conditioning Unit

The AVS air conditioning unit, in the nosewheel bay just aft of the cabin conditioning unit, comprises an air-to-air cooler, a turbine and fan, a flow augmenter, a water extractor, a heat exchanger, and a filter. Hot air from the flood flow supply line passes, via an electrically operated on-off cock, to the air-to-air cooler and then to the turbine and the water extractor. A branch line of the hot air bypasses the cooler and turbine and feeds into the water extractor via a temperature control valve, while a further line passes through another temperature control valve to the heat exchanger.

The temperature control valve in the hot-air line to the water extractor is controlled by a sensing element in the line between the water extractor and the heat exchanger. The temperature control valve in the hot-air line to the heat exchanger is controlled by a sensing element, set at 15 degrees C, in the manifold inlet.

A tapping from the air from the cooler passes through a flow augmenter to the forward side of the water extractor, to provide additional pressure at altitude. This air, mixed with the air from the turbine, passes through the filter to the manifold in the cabin via a non-return valve.

AVS Components In Cabin

The temperature and pressure of the air in the manifold are regulated by a sensing element, which operates to regulate the cock supplying hot air to the conditioning unit, and a pressure relief valve and pressure controller. From the manifold, individual lines (each with an electric heater and a manual flow control valve) pass to the crew positions. The AVS hoses for the 6th and 7th seats are fed from the nav/radar's and AEO's AVS lines respectively.

Auxiliary AVS System

The auxiliary AVS system, for cooling, operates by drawing cabin air through the crew's AVS. A 200V AC exhausted unit under the navigator's table is controlled by an AVS ON/OFF switch on the edge of the AEO's table. The pilots select auxiliary AVS by means of individual AVS CHANGEOVER cocks switches which operate changeover cocks under the navigator's table. Rear crew members have a separate line and flow cock. To prevent overheating, the exhausted motor is not to be switched on unless at least one flow cock is open.

BOMB BAY CONDITIONING

Bomb Bay Conditioning System

Twin dorsal intakes provide cold air through ducts via a cold air valve. An outlet louvre in the port bomb door exhausts the air to atmosphere. Controls at the nav/radar's position consist of an OFF (centre) COLD (down) switch which controls the cold air valve, a temperature selector and a manual heat control (both inoperative) and a temperature gauge.

WINDSCREEN THERMAL DEMISTING

Windscreen Demisting Supplies

To demist the inside of the windscreen, hot air is supplied from a sloped duct on each side of the centre panel. Cabin air is supplied to a blower motor, below the pilots' floor on the starboard side.

From the blower motor, the air passes to a 1 kW heater unit and thence to the windscreens.

An overheat switch in the system cuts off electrical supplies to the heater if the temperature of the air in the ducting rises above 70 degrees C and switches supplies on again when the temperature falls to 60 degrees C.

PART 2: AIRCREW EQUIPMENT ASSEMBLIES AND OXYGEN SYSTEM

Warning: The aircraft is safe for parking when safety pins are inserted in both ejection seats and in the canopy as follows:

a. Canopy jettison firing unit sear.

b. Guillotine sear.

c. Canopy jettison and time delay trip lever.

d. Ejection gun sear.

e. Seat pan firing handle.

It is emphasised that pins should not be inserted in the fabric straps above the pilots' heads.

General

The aircrew equipment assemblies comprise the seats, the flying and safety clothing and associated connectors. The pilots are provided with ejection seats while rear crew members have sliding, bucket-type seats, the outer two of which swivel. There are two positions for extra crew members, forward of the nav/radar and AEO seats. There are oxygen and intercom connections at all seven crew stations and static line connections at the five rear crew stations.

EJECTION SEATS

Ejection Seats, General

The ejection seats, type 3KS1 for the 1st pilot and type 3KS2 for the co-pilot, are similar but partially handed.

Each has a horseshoe-shaped parachute pack, a back pad with adjustable kidney pad and a personal survival pack (PSP) type ZC. The seats have a nominal ground-level capability, provided that the speed is at least 90 knots.

Any deviation from straight and level flight at the instant of ejection reduces the seat performance.

The seat pan is adjustable for height by means of a lever on the outboard side of the seat. The trigger in the end of the lever is pressed in to adjust the height and, when released, locks the seat in the selected position.

The adjustable armrests are controlled by either of two levers on each rest, one at the forward end and one at the rear, on the side of the rest.

A lean-forward lever, forward on the pod side of each seat, allows the occupant to lean forward, by unlocking the attachment between the shoulders and the back of the seat. The straps wind in and out, following the pilot's motion, and are locked on application of negative g and/or rapid acceleration.

A negative-g restraint strap is attached to the seat pan and is adjusted by a downward pull.

The seats have pressure-demand emergency oxygen, a Mk 42 parachute and a static-line operated guillotine.

Election Gun and Firing Handles

Each seat is fitted with an 80 ft/sec telescopic ejection gun fired by either the face-screen B-shaped firing handle above the occupant's head or the seat pan firing handle in the front of the seat pan. Either handle must be pulled to its full extent to fire the gun. Safety pins for each firing handle are stowed, when not in use, in a combined stowage for all safely pins, one on each side of the cockpit.

Canopy/seat Connection

An interconnection between the seat-firing mechanism and the cockpit canopy enables the canopy to be jettisoned automatically when any firing handle is pulled. Because of this interconnection, there is always a 1-second delay between pulling an ejection seat handle and the seal ejection gun firing, even if the canopy has already gone.

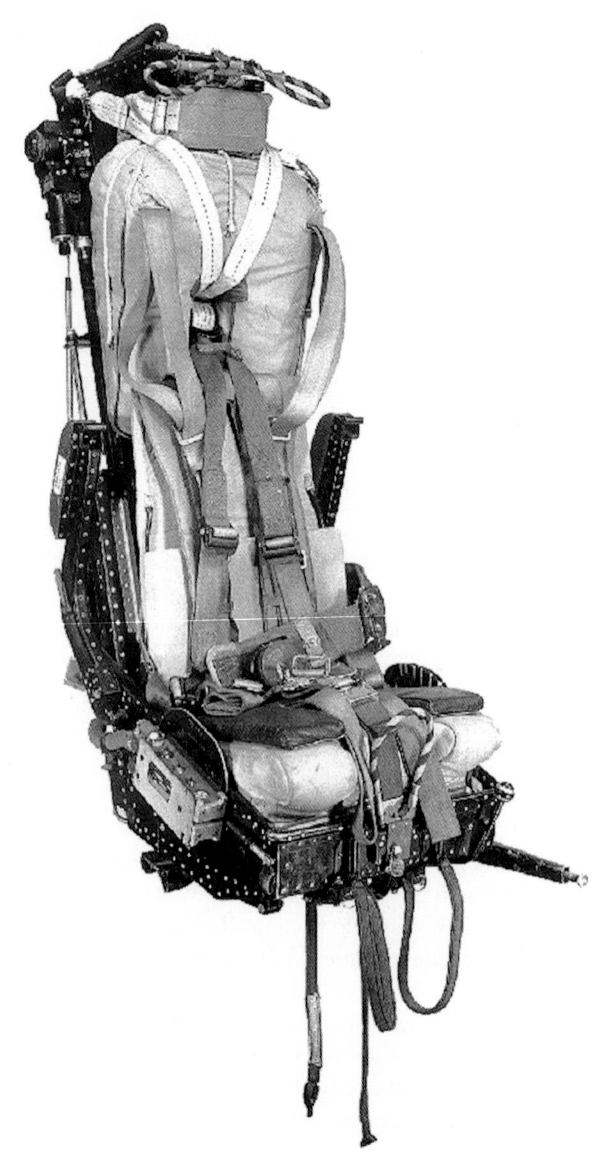

The Vulcan pilot and co-pilot's Martin Baker Mk 4 ejection seat.

THE TYPE 3KS2, Mk 4 SEAT EQUIPPED (STARBOARD VIEW)

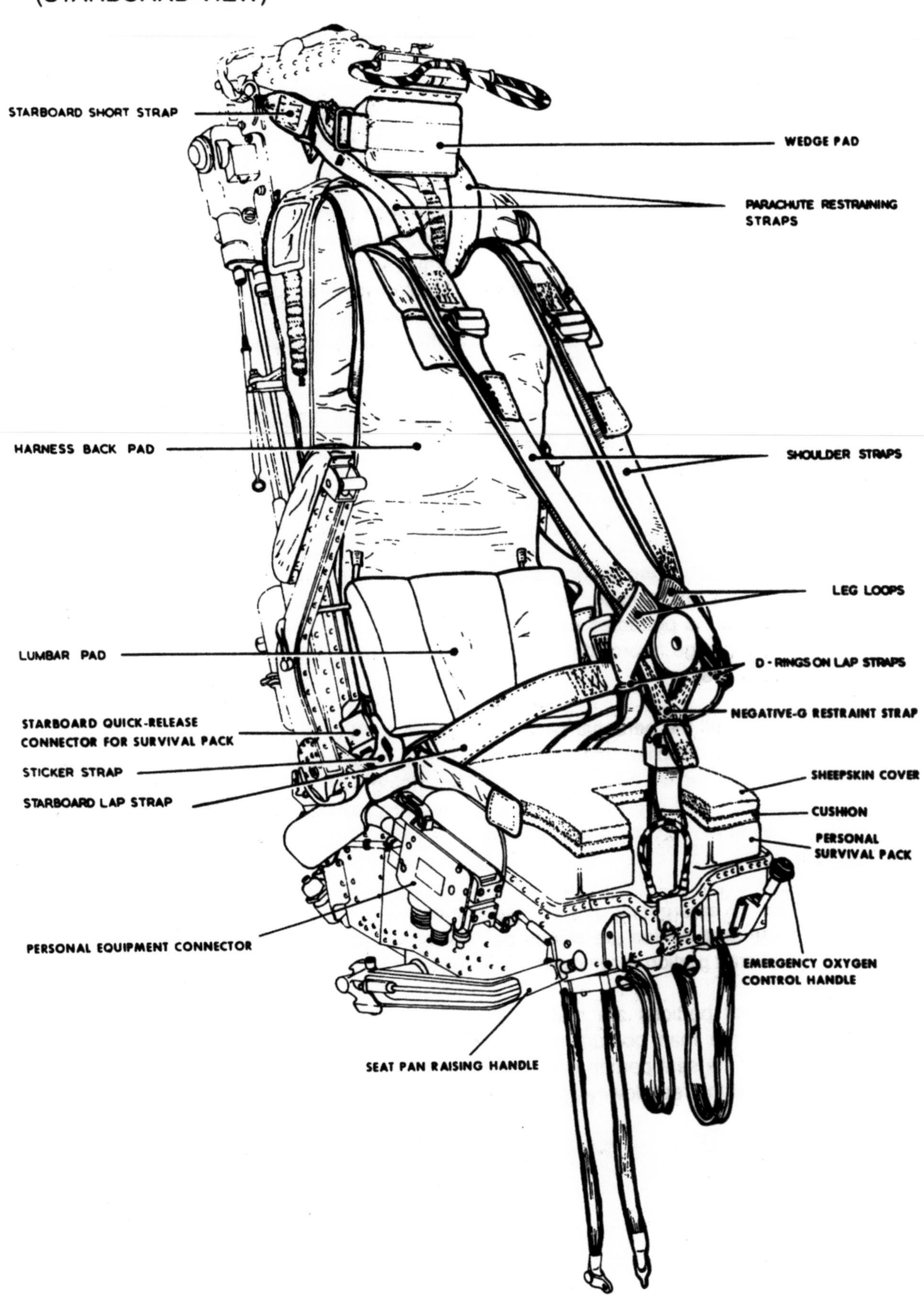

The Martin Baker V-Bomber rear crew ejection seat rig used during trials at Chalgrove. Although the ejection system looks complicated (the outer seats being shunted inwards prior to ejection), the scheme did work, but the cost of the very significant engineering required to refit V-Bombers with the system, when compared to the slim possibility of it ever actually being required, encouraged the Air Staff to abandon the idea. *Both Martin Baker*

The ejection system for the V-Bomber rear crews being successfully tested at Chalgrove, using a Valiant trials aircraft. *Martin Baker*

Leg Restraint

Leg restraint lines ensure that the legs are drawn back and held close to the seat pan during and after ejection. The lines pass through snubbing units below the front of the seat pan and are then fastened to the cockpit floor with rivets which shear at a pull of 400lb. The snubbing units allow the lines to pass freely down through the unit but prevent them passing upwards, except when released by the spring-loaded toggle at the front of each snubbing unit. The leg restraint lines are unfastened when the man portion or cover of the PEC is removed from the seat portion.

Note: Before ejecting, place the feet on the rudder pedals.

Drogue Gun

The drogue gun has a time-delay mechanism and fires half a second after the ejection gun has fired, withdrawing the duplex drogues to stabilise the seat. The time-delay mechanism is operated by a static trip rod, which withdraws a sear from the gun as the seat rises on the rails.

Barostat/g-switch Time Delay

Automatic separation is controlled by a time delay switch, which is inhibited by a barostat and a g-switch.

The time delay runs for 1.25 seconds and is started by a static line as the seat moves up the rails, provided that:

a. The height is not greater than 10,000 feet.

b. The acceleration is not greater than about 4g.

If either condition is exceeded, the barostat and/or g-switch interrupts the running of the time switch until the conditions are satisfactory (seat below 10,000 feet, speed below 300 knots).

When the time switch operates, the harness, leg restraint lines, personal equipment connector, face blind and parachute pack are all released from the seat. Simultaneously, the drogues are detached from the seat but remain attached to the apex of the parachute canopy, withdrawing it upwards. The parachute subsequently develops and pulls the pilot out of the seat.

Note 1: When the seat pan handle is used to initiate ejection, the handle remains attached to the seat and cannot be retained during separation. Conversely, the face-screen handle, if used, remains in the pilot's hands after separation and should be discarded as soon as convenient.

Note 2: For flights over mountainous territory, a 5,000-metre capsule can be substituted for the normal 10,000-foot capsule.

Manual Separation

To allow the occupant to release himself from the seat, if the automatic devices fail to operate, means of manual separation are embodied.

A static-line operated guillotine is provided on the port side of the drogue box. A safety pin is provided for the gun sear and, when not in use, is stowed with the other safety pins on the cockpit coaming.

Separation is achieved by pulling out and up the manual separation lever to the rear of the seat pan on the left side. This releases the harness locks, the parachute retaining straps, the leg restraint lines and the man portion of the PEC and, as the seat occupant falls clear of the seat, a static line attached to the rear of the parachute pack withdraws the sear from the guillotine and severs the drogue link line.

After leaving the seat, the parachute is opened by pulling the D-handle on the waistbelt.

REAR CREW SEATS

Crew Seats, General

a. The navigator/radar and the AEO have seats which are capable of swivelling inwards, to face almost forward (aircraft sense). The navigator plotter's seat does not swivel.

b. Each seat has an assistor cushion to help the occupant rise from the seat in emergency.

The backrests embody spring clips to help retain the top of the parachute pack against the seat back; they also embody clips for stowage of the shoulder straps. A Mk 46 or Mk 49 parachute, including a demand and inflation emergency oxygen set, is used. All three seats can be slid fore-and-aft on rails.

SWIVEL SEATS FOR NAVIGATOR/RADAR AND AEO

a. These seats have a lever to the left of the seat pan to control seat movement. In its spring-loaded central position, the seat is locked. When moved forward, the seat is unlocked for the rake of the seat back and swivelling. The rake is spring-loaded to the forward position to clear obstacles when swivelling; in normal use, the rear position is adopted.

The control lever may be released once either motion has begun and automatically unlocks when the full travel has been reached.

b. When the seat is locked in the fore-and-aft direction to face the table, the same lever may be pulled aft against its spring, to free the seat for sliding. There are several finite positions in which the seat may be locked when the lever is released.

c. For use by persons not occupying the seat, a handle at the top of the back rest, when moved to the right, duplicates the forward motion of the lever for swivelling and raking.

Similarly, a toe-operated lever behind the base of the seat, when moved to the right, releases the seat for fore-and-aft travel (but, unlike the lever, is effective even if the seat is swivelled).

d. In order to clear the table, the seat must be at the forward (aircraft sense) end of its rails before swivelling.

e. Thigh supports at the front of these seats can be adjusted up or down as desired, by a central star wheel beneath them.

f. An S-type Mk 2 personal survival pack in the seat pan is attached to the outside of the parachute harness. It is provided with a lowering line and should be lowered from the harness during the parachute descent.

SLIDING SEAT FOR NAVIGATOR/PLOTTER

a. The lever permitting the seat to slide fore-and-aft is on the right of the seat pan in this seat.

b. An S-type Mk 2 personal survival pack is provided.

c. To prevent trapping the knees under the table, the seat must be at the forward (aircraft sense) end of its rails before operating the assistor cushion.

Assistor Cushions

To assist the occupant to his feet, under conditions of positive g, each seat has an assistor cushion fitted in the seat pan which can be inflated by air at a pressure of 1,200 PSI. A pressure gauge is fitted on the bottle. The air bottle is stowed on the back of the navigator/radar's and AEO's seats and in the back of the seat pan of the navigator/plotter's seat. Pins are provided for the assistor cushion tattles. To make the bottles operative, the navigator/plotter's pin must be removed and the other two must be inserted.

A knob on the right of the seat pan of the navigator/radar's and AEO's seats and a lever on the left of the navigator/plotter's seat pan, when pushed upwards to the full extent of their travel, release the air to inflate the cushion.

The initial inflation, in turn, releases the safety harness lap strap anchorages, freeing the occupant from the seat.

CLOTHING AND PERSONAL EQUIPMENT CONNECTIONS

Personal Equipment Connectors (PEC)

Pilots

a. The pilots' PEC are in three portions, the aircraft portion, the seat portion and the man portion. The aircraft portion is attached to the underside of the seat portion on the right-hand side of each seat. The man portion, an integral part of the clothing, is attached (during strapping-in) to the top of the seat portion. When not in use, the seat portion is protected by a dust cover, for which a stowage is provided on the back of the co-pilot's seat.

b. The PEC connects all the personal services to the man. The aircraft supplies (main system oxygen, ventilation air, mic/tel) are fed into the aircraft portion. As the seat ascends the guide rails on ejection, a static rod causes the seat portion to break away.

e. To prevent loss of emergency oxygen, the lower orifices of the seat portion are closed by valves when the aircraft portion is removed.

d. The man and seat portions are mated by sliding the nose of the man portion into the hooks at the front of the seat portion and then pressing the handle at the rear downwards.

e. To release, press down on the thumb button in the handle and lift the handle. (This also releases the leg restraint lines.)

Rear crew

a. The rear crew and spare seat positions are each provided with an aircraft hose assembly consisting of an oxygen hose, kitten lead and a static line, all taped together in a protective sleeve. A switch and associated electrical wiring, connected to the static line, are also incorporated in each assembly to operate the 'crew gone' lights in front of the 1t pilot.

b. The assemblies are of sufficient length to enable the crew members to move about the cabin and to pass through the door in order to abandon the aircraft.

c. Separate AVS hoses are provided for the crew members, the spare seats and the sextant positions. Quick-release connectors are fitted.

Air Ventilated Suits (AVS)

AVS Mk 2A (nylon) or Mk 2C (cotton) may be worn; the hose from the suit is passed through the clothing to connect with the aircraft supply. The pilots' supply is through their PEC while the rear crew have separate connectors.

Pressure Jerkins and Anti-G Suits (High Altitude)

Pressure jerkins (Mk 4 for pilots and Mk 3 for rear crew) are worn in conjunction with anti-g suits Mk 7B when flying at high altitudes.

These two items form a pressure suit to protect the crew member if cabin pressure fails; they inflate automatically if the cabin altitude reaches 40,000 feet. They are inflated from the oxygen system, both being connected to the pressure jerkin hose assembly. The jerkin connection is permanently made but the hose from the anti-g suit has to be threaded through the outer clothing and then attached. The pressure jerkin embodies a life jacket. The attached hose assemblies terminate in the man portion of the PEC for the pilots and a bayonet connector for the rear crew.

Masks and Helmets

Either a G-type helmet with a Mk 1 protective helmet, or a Mk 2 or Mk 3 protective helmet, may be worn; in all cases either a P2C or Q2C oxygen mask is worn.

These masks are of the chain-toggle, pressure-breathing type with a bayonet hose connection; they are identical apart from size. A toggle on the front of the harness is normally in the up position. The mask should be tested before flight and the knurled screws adjusted so that there is no leakage during operation of the press-to-test facility with the mask toggle in the down position. If cabin pressurisation failure occurs, the toggle is put to the down position, to clamp the mask more tightly on the face for pressure breathing. When pressure clothing is worn, the mask is plugged into the top end of the jerkin hose assembly.

Low-Altitude Clothing Assemblies

For flights below 45,000 feet (cabin pressure), it is not essential to wear pressure clothing with the Mk 21 regulator. In such cases, and when using a Mk 17 regulator, normal flying clothing is worn, with a separate life jacket. To enable the aircraft connections to be made, a special mask hose assembly is required, Mk 2 for pilots, Mk 7 for rear crew (Mk 2 post-Mod 2393).

Rear Crew Safety Equipment

The rear crew members wear back-type Mk 46 or Mk 49 parachutes with type S Mk 2 personal survival packs (PSP). When the static line is used, a barostat control delays deployment of the parachute until below 13,000 feet. This can be overridden by a handle on a strap between the legs.

A demand and inflation emergency oxygen set is provided. The cylinder and opening head are in the top of the parachute pack and the regulator is on the right-hand waist-belt. The operating handle is on the strap between the legs.

OXYGEN SYSTEM

Description of Oxygen System

Oxygen is carried in 12 x 2,250-litre bottles. On early aircraft, all the oxygen bottles are housed in the power compartment, aft of the bomb bay. In later aircraft, four bottles are housed in the power compartment and the remainder in the bomb bay. The bottles are all charged through a connection in the power compartment, the correct charging pressure is 1,800 PSI. Two pressure gauges at the AEO's position show the pressure in each half of the system.

From the oxygen bottles, the high-pressure supply lines pass along the sides of the bomb bay and into the pressure cabin. Master valves, one for each side of the system, are below the crew's flooring; these valves are normally wire-locked to the open position. From the master valves, the supply lines pass along the cabin and are interconnected by transverse lines at four points.

The transverse connections are protected by non-return valves so that, if there is a leak on one side of the system, oxygen is not lost from the other side. From the transverse lines, the supply is fed to pressure-reducing valves, one for each regulator, which reduce the pressure to 400 PSI. The medium pressure lines pass from each pressure-reducing valve to the regulators. From the regulators, oxygen at breathing pressure is fed to the PEC on demand.

Oxygen Regulators, General

a. The oxygen regulators may be Mk 21A, 21B or 17F, one being supplied for each crew member. The 1st pilot's and co-pilot's regulators are at the forward end of the pod and starboard consoles, while the rear crew regulators are at their respective stations.

b. The Mk 17F and 21 series regulators have the same characteristics up to 39,000 feet cabin altitude. Above this altitude, the Mk 21 series automatically delivers a higher pressure than the Mk 17F. The Mk 2 emergency regulators have the same characteristics as the Mk 21 series.

c. In all cases of cabin pressure failure, an immediate descent is to be made until cabin altitude is below 40,000 feet. When flight safety and fuel considerations allow, the descent should be continued, at normal rate, to below 25,000 feet.

Regulator Characteristics

The regulators provide:

a. A mixture of oxygen and air, or 100% oxygen; the flow and volume delivered is in direct relation to the breathing demands of the user.

b. The correct ratio of air and oxygen according to cabin altitude. Above 33,700 feet, 100% oxygen is provided automatically. 100% oxygen may be selected at any height.

c. A safety pressure; the mask cavity pressure is slightly higher than cabin pressure when cabin altitude reaches 11,000 to 14,000 feet and pressure breathing when the cabin altitude exceeds 39,000 feet. Additionally, the Mk 21 series regulators inflate pressure clothing at the same altitude as they give pressure breathing.

d. Positive oxygen pressure by manual selection:

(1) Emergency use (100% and EMERGENCY), ie for toxic fumes.

(2) Mask and regulator testing (Mk 17F).

(3) Mask, regulator and pressure clothing testing (Mk 21 Series).

OXYGEN REGULATORS, CONTROLS AND INDICATORS

Mk 21 Regulator Controls

a. OXYGEN SUPPLY, ON/OFF lever (wire-locked at ON). This lever controls the supply of oxygen to the regulator and must be ON at all times in flight.

b. NORMAL OXYGEN 100% OXYGEN lever. When in the normal position, this lever allows air to mix with the oxygen in suitable proportions, up to an altitude of 33,700 feet. Above this altitude, the air inlet is closed and 100% oxygen is delivered to the mask. With the lever at 100% OXYGEN, the air inlet is closed regardless of the altitude; this position should always be used if toxic fumes are present.

c. JERKIN TEST/MASK TEST/EMERGENCY/ NORMAL lever. When this lever is set to NORMAL oxygen or a mixture of oxygen and air as selected by b. is fed to the mask at the required pressure when the user breathes in. When the lever is set to EMERGENCY, the pressure of the oxygen to the mask is slightly increased above the delivery pressure appropriate to the altitude.

This increase is constant at all altitudes. The EMERGENCY position should be used if toxic fumes are present in the cockpit. To reach the MASK TEST position, the knob in the end of the lever must be pulled out; in the MASK TEST position, the mask can be tested under pressure for leaks. The JERKIN TEST position is similar in operation to the MASK TEST position but gives a higher pressure and is used to test mask, jerkin, and a g-suit simultaneously for leaks.

Mk 21 Series Regulator Indicators

a. A gauge on the regulator shows the pressure of oxygen being delivered to the regulators while two gauges at the AEO's station show the main pressures.

b. A magnetic indicator on each regulator shows white when the user is breathing in. The pilots' and three rear crew members' indicators are duplicated on their respective panels, and the bomb aimer's is also duplicated on a rear central support for the pilots' floor, where it can be monitored by other crew members.

Mk 17F Regulator Controls and Indicators

The Mk 17F regulator carries a pressure gauge, a magnetic flow indicator, a NORMAL/100% selector, an ON/OFF control and an EMERGENCY lever. This last control is pressed in to test the mask for leaks and deflected to either side to obtain oxygen at higher pressure in an emergency.

EMERGENCY OXYGEN

Pilots' Emergency Oxygen

a. Pilots can obtain emergency oxygen automatically on ejection or by pulling up a yellow and black knob inboard of the ejection seat on the forward corner of the seat pan.

b. As the bottle is attached to the seat, emergency oxygen is not available after separation from the seat.

c. A pressure-demand emergency Oxygen Set is fitted to each ejection seat, the cylinder and operating head on the back and the regulator Mk 2 or 3 behind the PEC. The operating head initiates the action and delivers medium-pressure oxygen to the regulator, which then delivers it to the user on demand. Mk 2 and Mk 3 regulators have similar characteristics to the Mk 21 and Mk 17 series regulators respectively, except that safety pressure is delivered from ground level upwards. The cylinder contents are indicated on a pressure gauge. The endurance of the set is approximately 10 minutes provided that the mask is fitted correctly and that there are no leaks in the hose assembly.

Rear Crew Emergency Oxygen

a. The rear crew are provided with Mk 2A and 3A regulators which are identical in performance to the pilots', only the fitting and method of operation being different. An emergency oxygen bottle in the top of each rear crew parachute pack is turned on automatically when the parachute static line is operated. It can also be operated manually by a control on a strap between the user's legs.

b. The pressure demand regulator is on the right-hand waist-belt and a contents gauge is on the bottle in the pack. The bottle is filled to capacity when the gauge needle is on the white line between the sectors marked FULL and REFILL.

USE OF AIRCREW EQUIPMENT ASSEMBLIES

Clothing Assembly

When wearing pressure clothing, the air-ventilated suit and anti-g suit are put on beneath the shirt and trousers and the jerkin is put on top of the flying overalls. The hoses from the anti-g suit and AVS are fed through the outer clothing; the AVS hose is connected to the PEC for the pilots and to an individual connector for the rear crew, the anti-g hose is connected to the main oxygen hose. The oxygen hose from the mask is subsequently attached to the bayonet socket at the top of the main oxygen hose.

When pressure clothing is not worn, the mask hose assembly is used.

PART 3: ARMAMENT AND OPERATIONAL EQUIPMENT

General

There are two methods of bomb aiming in the aircraft, the NBS and the co-pilot's visual sight.

BOMB RELEASE AND JETTISON

Release

Bombs are released automatically by the NBS release pulse or manually by a bomb release button. The 1st pilot has a bomb release button to the left rear of the throttle quadrant. The co-pilot's bomb release button, stowed in a clip below the coaming, operates through a time delay unit at the navigator/plotter's position. The navigator/radar's bomb release button is stowed in a clip adjacent to panel 9P. There is a further bomb release button at the prone bomb aimer's position.

Warning 1: If one or more 1,00lb bombs have fallen on to or against the bomb doors, the doors must be closed (if open) and must not be opened again in flight, except in emergency. If they are opened, a bomb may be trapped against the bomb bay wall, damaging the flying controls and/or the hydraulic lines.

Warning 2: Failure to switch on the time delay unit at the navigator/plotter's station when the co-pilot's release button is to be used results in a failure to release.

Jettison

a. The bombs may be jettisoned live by selection of the live JETTISON switches on the navigator/radar's side panel and on the visual bomb aimer's oxygen panel.

b. An EMERGENCY BOMB JETTISON switch on the 1st pilot's console panel enables the bombs to be jettisoned safely through a wiring system separate from the normal bomb release system.

The switch is a guarded, double-pole switch labelled JETTISON/OVERRIDE, spring-loaded to the off position. When the switch is held momentarily to the Johnson (rear) position, the bomb doors open, the bombs fall and the doors close. Movement to OVERRIDE cancels jettison, provided that the bomb doors have not reached the fully open position. The selection can also be used to close the bomb doors before the time sequence unit does so. The bomb doors close when the time delay unit on the cabin wall at the AEO's position completes the manually preset 24-second cycle.

c. If hydraulic power has failed, the bomb doors must first be opened by the emergency system before attempting to jettison.

d. When nuclear stores and practice bombs are carried, the jettison circuits are rendered inoperative.

Bomb Release Safety Lock.

A bomb release safety lock (BRSL), to prevent inadvertent weapon release, is controlled by a guarded wire-locked switch on the port console marked LOCK/UNLOCK. An amber light comes on if the safety lock is released; a green light shows the lock is engaged. Press-to-test facilities on both lights are used to check a duplicate electrical circuit.

A SFOM bombsight is fitted. The sight is a collimated fixed-angle sighting head, on the coaming in front of the co-pilot. A lighting control is adjacent to the other cockpit lighting controls. The sight line is fixed relative to the aircraft in azimuth but the depression angle can be adjusted to allow for speed, AUW, height and relative air density. There are two scales on the sighting head, one calibrated in degrees and the other in milliradians. In flight, the milliradian setting knob should be used for adjustments to the sighting head. One complete revolution of the milliradian scale adjusts the sighting head by 100. The optical glass is stowed in a container on the starboard console and must be inserted in the sighting head prior to bombing. Care must be taken to avoid scratching the glass while handling.

Window Launchers

The gravity/cartridge window installations are operated by the AEO.

Three gravity window containers (two port, one starboard) and one cartridge discharger are fitted.

Camera Installation

An F95 Mk 9 camera on a frame above the bomb aimer's window is controlled from the navigator/radar's position. The film can be marked to show bomb release automatically by the NBS release pulse or manually by the event marker button.

PART 4: ENGINES AND AAPP

OLYMPUS ENGINES

General

The aircraft may be powered by Olympus Mk 200 series or Mk 301 engines which, due to differences in mechanical details, thrust and limitations, are never installed together on the same aircraft. In the TAKE OFF setting, Mk 201 series and Mk 301 engines produce approximately 17,000lb and 18,000lb of thrust respectively; in the CRUISE setting they produce approximately 16,000lb and 17,000lb of thrust respectively. A rapid starting system is embodied.

The basic engine consists of eight main sub-assemblies. The intermediate casing houses the drives from both compressors for the mechanically driven auxiliaries:

a. LP Compressor Components

(1) LP driven fuel pump

(2) Tacho generator (RPM)

b. HP Compressor Components

(1) HP driven fuel pump

(2) Main oil pump

(3) Main oil scavenge and four auxiliary scavenge pumps

(4) Constant speed drive unit (CSDU)

(5) Hydraulic pump

(6) Tacho generator (not used)

(7) Rotax air starter drive

Engine air is automatically supplied on start-up for the following systems:

(a) Turbine cooling.

(b) Pressurisation of bulkhead seals, oil bearing seals and CSDU oil tank.

(c) Induced cooling of alternators, CSDU oil and Zone 2B.

Air is also available for engine anti-icing when selected.

Engine control is provided by a combined throttle/HP cock. Automatic control is effected by components in the fuel chassis which allow an optimum supply of fuel to the engine, dependent on ambient conditions, airspeed and acceleration stage in the engine. Limitation on engine output is controlled by the position of a Take-off/cruise switch which determines both the RPM governing datums and the corresponding JPT [Jet Pipe Temperature] limiting datums for all four engines. Also, a jet pipe temperature limiter is installed which controls the engine speed below the selected governing speed, if necessary, so that the corresponding jet-pipe temperature limit is not exceeded. When JPT limiting is overridden, manual control of the throttles may be necessary to maintain the engines within JPT limitations.

The Mk 200 series jet pipes are fitted with convergent/divergent nozzles for improved cruise performance; the Mk 301 jet pipes are fitted with convergent nozzles.

To minimise damage to the outer wing and the bomb bay following structural failure of an engine, containment shields are fitted outboard of the outer engines and inboard of the inner engines.

Throttle and HP Cock Controls

The four throttle levers, which also control the HP cocks, are forward of the retractable centre console, in a quadrant marked OPEN/IDLING. The quadrant is gated at the IDLING position and the part of the quadrant below this, which controls the HP cock position, is marked OPEN/HP COCKS/SHUT and has a further gate at the SHUT position.

To move the throttle levers forward from the HP COCKS SHUT gate and to move them aft from the IDLING gate, the sleeves on the levers must be raised.

The throttle friction lever is on the starboard side of the throttle quadrant; forward movement of the lever increases the friction.

Engine Fuel System

The engine fuel system controls the correct amount of fuel the engine requires in the varying conditions of flight. The factors which affect this are:

a. Throttle setting.

b. Engine air intake pressure (P1).

c. Compressor delivery air pressure (P3).

d. JPT limiter isolate switch position.

e. JPT/RPM (datum limits selected by the TAKE-OFF/CRUISE switch).

Fuel from the tank booster pumps flows through the LP cock and fuel filter into the engine fuel system in the following order:

a. LP and HP driven fuel pump. Two engine-driven pumps are inter-connected and driven by the LP and HP compressors respectively.

(1) The LP pump incorporates a double datum hydro-mechanical RPM governor, allowing it to control RPM to take-off or cruise limits with the throttle at OPEN. The two RPM datums are controlled by the TAKE-OFF/CRUISE switch to the left of the throttle levers. In addition to selecting the required RPM datum, operation of the switch also selects the corresponding temperature datum for the engine JPT limiter. Therefore, actual RPM on the engines may be less than the limit selected in order to keep the JPT within limitations. The hydro-mechanical governor maintains the selected RPM datum regardless of changes in temperature and density of fuel; RPM may vary very slightly with increase in altitude.

(2) The HP pump incorporates an overspeed governor set at 3% above maximum HP compressor revolutions should the governor on the LP driven pump fail.

b. Full Range Flow Control (FRFC). Fuel is passed from the fuel pumps to the FRFC, which meters fuel to the engine depending upon:

(1) Throttle-HP cock position.

(2) Engine air intake pressure (P1).

(3) JPT datum limits.

Control is accomplished by the Electro Pressure Control (EPC) which is activated by the JPT amplifier when the JPT approaches the limit selected by the TAKE-OFF/CRUISE switch. In the TAKE-OFF setting a depressed datum is brought into operation which prevents turbine temperature over-swing during rapid throttle opening by slowing down the rate of temperature rise during the last 5 degrees C.

Above about 20,000 feet the TAKE-OFF and CRUISE datums may exceed their respective limits by up to 5 degrees C, therefore in the TAKE-OFF setting at altitude the JPT may have to be kept within limits by use of the throttles. The output from the JPT amplifier may be isolated by means of a JPT limiter ON/OFF switch aft of the TAKE-OFF/CRUISE switch. The switch must always be ON before TAKE-OFF is selected, so that the depressed datum will prevent turbine over-swing during rapid throttle opening. The OFF position of the switch is for use when ground-testing the RPM governors, or for use if a limiter runs away or fails in flight. In these circumstances, the throttles must be handled carefully to prevent JPT rising to the limits. The limiters may be tested on the ground by isolating the governors; two RPM GOVERNOR isolation switches are provided at the top of the port fuse panel behind the AEO's seat. When held on they permit take-off RPM to be selected when the pilot's switch is at CRUISE without altering the cruise setting of the JPT limiter.

Nos 1 and 4 and Nos 2 and 3 engine JPT amplifiers are supplied by the port and starboard main transformers respectively. If automatic load-shedding occurs, the system is de-energised and JPT control must be maintained by use of the throttles.

c. Air Fuel Ratio Control (AFRC). The AFRC is the engine acceleration control. During acceleration it senses compressor delivery pressure (P3) and thus determines how much fuel may be passed to the burners.

(1) In Mk 200 series engines the AFRC over-fuels the engine in the early stages of acceleration. As fuel delivery is proportioned to P3 pressure, it is merely necessary to restrict the P3 pressure fed to the AFRC. This is done by incorporating a P1 P3 switch which measures the

compression ratio of the compressors, and decides how much P3 pressure may be fed to the AFRC. Above 82/83% RPM the overfilling condition ceases and the P1 P3 switch allows full P3 pressure to the AFRC.

(2) In Mk 301 engines overfuelling is not a problem, so a P1 P3 switch is not required. The internal bellows of the P1 P3 switch are removed, thus it becomes an air potentiometer and full P3 pressure is fed to the AFRC. The AFRC is now working at its designed limit, but engine acceleration is still too slow. To enhance acceleration, a Speed Term Switch is fitted. It senses LP fuel pump pressure, and after 65% progressively allows more fuel to bypass the AFRC to the burners, until max RPM is reached.

d. Flow Distribution and Dump Valve. During start up, a fine fuel spray is required and fuel is initially only fed to the primary orifices of the duplex burners. Above about 15% RPM fuel is also allowed to the main orifices of the burners.

The dump valve ensures that when the HP cock is shut, all fuel from the burner pipes is drained into a collector box on the engine doors. The box is drained by the groundcrew.

It will only partially drain in flight.

e. Dipping Valve. This is only used during a rapid start of the engine and ensures that excess fuel returns to the suction side of the pumps. It is automatically de-energised at the end of the start cycle. However, it can be held open by fuel pressure. To ensure that it has closed, the throttle must be briefly returned to the idling gate at the end of the start cycle.

Engine Starting Systems

Each engine embodies its own air starter motor. Air can be supplied from a ground air starter unit, feeding through a connection on the underside of the starboard wing, or from the rapid start system. The ground air supply feeds into the main lines of the engine air system and thence to the starter motors, through electrically actuated valves. Compressed air from a running engine can be used to start the others, singly or simultaneously, provided that the appropriate ENGINE AIR switches on the starboard console are at OPEN.

The rapid starting system is so arranged that the powered flying controls and artificial feel are switched on automatically when the simultaneous RAPID START button is pressed; because of the peak loads involved, electrical power during starting must be supplied by a 60 kVA ground power unit and not from the AAPP.

A gyro hold-off system is embodied, whereby the MFS, JPT limiters, refuelling relays, contents gauges, fire warning, autostabilisers and artificial horizons are de-energised until the engine start master switch is selected ON. The power to the gyros is initially boosted to 200V for approximately 20 seconds, to obtain fast run up.

Controls

a. The starting control panel on the 1st pilot's port console carries the following controls:

Four buttons for individual engine starting, each embodying a light.

A GYRO HOLD OFF button.

A RAPID START button, for starting all engines simultaneously.

A NORMAL/RAPID lock-toggle switch, for selecting Palouste or high-pressure air.

An IGNITION ON/OFF switch.

An AIR CROSSFEED three-position magnetic indicator.

An ON/OFF lock-toggle start master switch.

b. An AIR CROSSFEED indicator shows OPEN whenever the start master switch is ON. When the air selector is at NORMAL, engines can only be started by the individual buttons, using an external air supply or crossbred. When the air selector is at RAPID and the master switch on, all engines can be started simultaneously using the RAPID START button or separately, using the individual buttons. If, for any reason, an engine fails to start on this system, there is sufficient air for one further start on each side; the individual start button must be used for this attempt. With the air selector switch at RAPID and the master switch OFF the gyro hold-off system is effective when the GYRO HOLD OFF button is pressed.

c. Relighting Controls. A relight button in the head of each throttle/HP cock provides a means of relighting the engines in flight. When one of the buttons is pressed, 28v vital busbar power energises the igniter plugs regardless of throttle position or switch selections on the engine start panel. The igniter plugs remain energised until the relight button is released.

Operation (Normal)

With the engine switch ON, master switch ON and air selector switch to NORMAL, pressing a start button energises three circuits:

a. Solenoid to open position on starter air control valve.

b. Engine igniter plugs.

c. Palouste air bleed valve.

The increase in Palouste air opens the air control valve and a pressure switch lights the start button. The air now rotates the starter turbine which drives the HP compressor. Fuel from the HP pump is directed to the primary burners where the igniters initiate combustion. The engine accelerates to self-sustaining speed and disengages from the starter turbine. An overspeed switch on the starter turbine operates the start relay which now de-energises the engine igniter plugs and the Palouste air bleed valve and operates the solenoid on the air control valve to the closed position. When the air control valve closes, the pressure switch breaks, cancelling the light in the starter button.

Operation (Rapid)

The rapid start facility uses a mixture of bottled air and fuel from the booster pumps. When this mixture is ignited, the resulting hot gases turn the starter turbine.

a. Air at 3,300 PSI is stored in four bottles in each of the two engine bays. A charging point is on each wing between the jet pipe tunnels.

Sufficient air is available for three engine starts per side. Air passes from the bottles to the manifold and thence to air bottle solenoid valves on each engine.

b. A solenoid-operated valve opens when the starter button is pressed and allows high-pressure air from the storage cylinder to pass to the reducing valve, which reduces the pressure to 300 PSI for delivery to the combustor. A safety disc, which bursts at 550 PSI, protects the combustor from excessive pressures if the valve fails. If the disc bursts, air exhausts into the same ducting as the normal exhausts from the starter.

c. Air from the bottles, reduced to 300 PSI, flows to the combustor unit and mixes with pressurised fuel in the chamber. The remaining air passes between the two walls of the combustion unit for cooling purposes and also to a non-return valve to prevent air escaping through to the air control valve. The fuel is then atomised, mixed with the air, ignited and ejected on to the starter turbine. A pressure switch then operates the light in the starter button and maintains the electrical circuit to the start components, while a 2-second timer switches off the combustor igniter. The combustor continues to operate until, after a successful start, the cycle is terminated by the starter overspeed switch or the 12-second delay in the time delay unit. The light in the starter button goes out, the air bottle solenoid is de-energised and the time delay units are reset.

d. If the igniter plug fails to ignite the mixture, the pressure switch does not operate. This breaks the electric supply to the time delay units after 2 seconds and the air bottle solenoid and the time delay units are reset.

e. If the combustor ignites but the engine fails to light up, the rapid start sequence is terminated by the 12-second delay and the time delay unit is reset.

Oil Systems

Each engine has its own integral oil system. In 200 series engines, the tank is in the nose bullet, a float-operated contents gauge is on the port side of the air-intake casing, and the tank filling point is on the lower port side of the engine front bulkhead. In 301 engines, the tank is on the port side of the LP compressor casing and has a contents sight glass, the filling point being below the tank. Each engine also has an independent oil system for the alternator constant speed drive unit.

The 14.5-pint tank (3.5 pints air space) is attached to the starboard side of the intermediate casing. A tank sight glass shows tank contents. The tank filling point is on the underside of the constant speed drive unit, at the lower starboard side of the engine. The tank must be replenished within 15 minutes of shutting down the engine. Great care must be taken not to overfill the tank or oil will find its way into the cabin conditioning system.

Magnetic chip detectors (five) are in all oil recovery lines, as follows:

1. No 1 bearing
2. Nos 2 and 3 bearings and auxiliary drives
3. Nos 4 and 5 bearings
4. No 7 bearing
5. No 8 bearing

Engine Instruments

The following engine instruments are grouped together in the pilot's centre panel:

a. Four JPT gauges which are DC operated.

b. Four fuel pressure magnetic indicators, which show white when there is insufficient pressure.

c. Four RPM indicators, of the percentage type. The main scale is calibrated in tens from 0 to 100%, while a small scale gives readings from 0 to 10%.

d. Four oil pressure gauges.

e. One ENGINE CONTROL magnetic indicator (inoperative).

Throttle Detents

a. To reduce the possibility of compressor instability leading to engine flame-out, a detent is provided on the inboard throttle of 301 engines to increase flight idling above 15,000 feet. It gives an increase of about 5% RPM over the outboard engines at 50,000 feet. The detent can be manually overridden in emergency, eg rapid descent or relighting.

b. A detent isolation switch at the forward end of the starboard console permits withdrawal of the detent under air-to-air refuelling conditions, ie at airspeeds not exceeding 250 knots at medium altitude.

c. For use on the ground, a detent test switch is provided on Panel 4P.

AIRBORNE AUXILIARY POWER PLANT

AAPP Description

The AAPP is a gas turbine engine, outboard of No 4 engine. It provides an emergency 200-volt supply below 30,000 feet.

On the ground, the AAPP supplies electrical power and air for the ventilated suits.

The engine has a single-side centrifugal compressor, driven by a single-stage axial turbine on a common shaft. The air intake is on the underside of the engine and air passes through a single, reverse-flow combustion chamber. The jet efflux is directed downwards from the wing. The air intake is held in the closed position by hydraulic pressure, electrically operated. If hydraulic or electrical failure occurs, the intake opens.

No throttle control is provided, as the engine is designed to run at a constant speed.

The air-intake shutter is controlled automatically by operation of the master switch which, with all the other controls, is at the AEO's station. For starting at higher altitudes, oxygen enrichment is provided. The unit carries its own supply in two bottles at the rear of the engine.

AAPP Oil System

Engine lubrication is provided by a gear-type pressure pump, which draws its supply from an oil sump formed by the lower part of the compressor casing. The sump capacity is 4.5 pints. The sump must always be filled to capacity.

The oil pressure varies very rapidly with temperature; the minimum pressure is 4 PSI.

PART 5: FLYING CONTROLS

COCKPIT CONTROLS AND INDICATORS

The flying controls in the cockpit are conventional in operation. Dual interconnected control columns and pendulum-type rudder pedals are provided; these operate powered controls through a series of linkages.

The rudder pedals can be adjusted for reach by a starwheel at the lower inboard edge of each pilot's instrument panel. Toe-buttons are provided on the pedals for brake operation.

The controls for the powered flying controls, artificial feel, artificial feel lock, autostabilisers and mach trimmer are grouped together on a panel on the port console. The elevator and aileron trim and feel relief switches are duplicated on the two control columns, while those for the rudder are on the fuel contents panel.

The emergency trim control is on the forward end of the retractable console.

At the top of the pilots' centre panel is a bank of warning lights and magnetic indicators. The three left-hand magnetic indicators are for the PFC units, the artificial feel and the autostabilisers respectively. The amber MAIN WARNING lights at either end of the group come on if a fault develops in any of the systems (except the yaw damper); this warning is cancelled by pushing in the button of the channel concerned. The appropriate magnetic indicator shows white, as a reminder that a channel is unserviceable. The main warning lights are then available for any subsequent failure.

Below this group is the control surfaces position indicator, representing a view of the aircraft from the rear. There is a separate indicator for each of the control surfaces, with datum lines to show the surface position relative to the take-off position.

All button warning lights can be dimmed by rotating the bezel. With the exception of the mach trim reset buttons, which are blue, all lights are amber.

POWERED FLYING CONTROLS

Elevons

Control of the aircraft about the longitudinal and lateral axes is achieved by eight elevons hinged into the wing trailing edge, four on each side. Each group of four is divided into two outboard and two inboard elevons. For identification purposes they are numbered 1 to 8 from port to starboard. Each surface is operated by a separate electro-hydraulically powered flying control unit (PFCU). Therefore, if a single unit fails, only one of the eight elevons is affected. After failure of a unit, the pilot is unable to move its elevon which will slowly assume the trail position.

The dual interconnected control columns are operated in the conventional sense to control the aircrew in pitch and roll, any movement of the control column causing all eight elevons to move. To simplify subsequent references, the term 'aileron' is used when referring to lateral control and 'elevator' when referring to longitudinal control.

Control column movement is transmitted to a mixer unit which, in turn, transmits the appropriate signals to the elevons. Full elevator and full aileron travel cannot be obtained at the same time. In all cases, the movement of the outboard elevons to the inboard elevons is in the ratio of 5:4.

Rudder

The single rudder is controlled by two powered flying control units, one main and one auxiliary. Normal control is by the main unit with the auxiliary unit idling; changeover occurs automatically if the main unit fails. During ground checks, when the main PFCU is stopped a delay of up to 15 seconds occurs before the auxiliary unit takes over. Since the control inputs to both rudder PFCUs are mechanically interconnected, a restriction of the input to one unit would prevent the pilot's demand being felt at the other. To prevent this, a trip mechanism is incorporated which, when activated, disconnects the faulty unit from the input linkage, gives failure warnings and, if the main unit is at fault, operates the changeover mechanism.

Powered Flying Control Units (PFCU)

Each PFCU consists, basically, of an electric motor driving main and servo hydraulic pumps and a hydraulic jack to move the control surface.

Movement of the cockpit control operates the assembly to supply fluid to the appropriate side of the jack, thus moving the control surface.

When the control surface position coincides with the new position of the cockpit control, jack movement ceases and the control surface remains in the selected position until further control movements are made. A stroke limiter prevents excessively harsh movements of the control surfaces.

Incorporated in the assembly is a surface lock valve. As long as servo pressure is available, the valve is held open to allow fluid to pass to either side of the jack. If, for any reason, this pressure is not available, the valve closes under a spring load and no further fluid can pass to or from either side of the jack. This prevents the surface from flapping in flight and acts as a ground lock. A bleed in each valve allows pressure on both sides of the jack to equalise slowly, thus allowing the surface to trail to the no-load position. When the valve closes it operates a micro-switch and a warning light on the pilot's panel comes on. There is only one control surface lock valve for the rudder assembly and it is housed in the auxiliary unit. An interconnection between the main and auxiliary units holds the lock valve open until no servo pressure is available from either unit.

Indication of an individual rudder PFCU failure is given by a pressure switch in each unit.

Electrical Supplies

a. A 200-volt, 400 Hz AC supply is required to operate the PFC motors. This is supplied from the main busbars, the distribution of loads being as follows:

(1) Elevons: Nos 1 and 8, No 1 busbar

Nos 3 and 6, No 2 busbar

Nos 4 and 5, No 3 busbar

Nos 2 and 7, No 4 busbar

(2) Rudder: Main, No 3 busbar

Auxiliary, No 2 busbar

b. PFC failure warning is operated by 200v DC.

Controls

The 10 push (off) spring-loaded stop buttons for the individual PFC units are arranged along the inboard edge of a panel on the port console, those for the elevons being grouped in pairs of elevons. The inboard button of the rudder pair controls the main unit. Each button incorporates a warning light which comes on if the unit malfunctions. Three PFC START pushbuttons, which also engage the feel systems, are at the rear of the panel and are marked A, R and E: A signifies the outboard elevons and aileron feel, R the rudder and rudder feel, and E the inboard elevons and elevator feel.

The PFC MI shows white if any PFC button is pressed when the servo pressure has fallen below 10 to 15% of normal and a 28-volt supply is available. If the system pressure is normal, the PFC MI shows white only while any PFC button is being pressed.

ARTIFICIAL FEEL

Artificial Feel Units

As the flying control system is irreversible, aerodynamic loads are not transmitted to the pilots' controls. To compensate for this lack of feel, artificial feel units are provided in the elevator, aileron and rudder control runs. Each unit is designed to give a suitable degree of feel for its particular control, the load on the pilots' controls varying with airspeed and/or control surface displacement.

The feel units are positioned alongside the three control runs in the bomb bay and are basically electric actuators which move under the influence of speed change. The starboard pitot-static system provides the feel actuators with the airspeed signal. Movement of the actuator moves a follow-up potentiometer which is compared with a warning potentiometer supplied with port pitot-static information. If the guard system senses a discrepancy equivalent to 30 knots of airspeed, the pilot is warned of a failure in that feel channel.

Rudder and Elevator Feel

The feel for each of these circuits is operated by a combination of spring-loading, related to control surface displacement, and electrical actuation, governed by airspeed, which varies the pilots' mechanical advantage over the feel spring. For the elevator circuit, the airspeed factor varies as the square of the airspeed, starting at 90 knots, with a maximum value of 460 knots. The rudder circuit varies as the cube of the airspeed, starting at 130 knots with a maximum value of 460 knots. In addition, a pre-loaded centring spring is fitted in the elevator and rudder circuits, to overcome the effects of friction in the control runs and the feel spring. The centring spring is on the input side of the feel unit and therefore provides a loading which is constant at all airspeeds but varies with the amount of control movement made by the pilot.

Aileron Feel

Aileron feel is not airspeed-controlled, although the range of control travel is reduced with increased airspeed. Stick loading is produced by resistance from torsion bars and, therefore, increases with increase of control displacement, the load being constant for any given angle. To prevent the pilot applying too great a control angle, variable stops in the feel unit decrease the range of movement as the airspeed increases above 150 knots with a maximum value of 365 knots. This circuit is also pre-loaded to overcome the effects of friction.

Feel Relief

a. Elevator and rudder feel can be reduced to the minimum speed position by use of feel relief buttons. The aileron variable stops can also be withdrawn, thus allowing the full range of control movement.

b. If any part of the artificial feel system malfunctions or if, in an emergency, it is desired to regain low-speed feel conditions, the artificial feel may be relieved using the appropriate button. The button on the control column relieves both the aileron and the elevator systems. The system in which relief is not required may be restarted by pressing the A start button for aileron feel or the E start button for elevator feel. Feel relief on the rudder is achieved by pressing the button on the fuel contents panel.

To regain normal feel, the appropriate start button should be pressed. Relieving the feel also removes power from the system and gives failure warning. When any pad of the artificial feel system is relieved, care in handling the aircraft must be exercised if over-stressing is to be avoided.

c. Initial application of 28v to the aircraft causes feel relief in all three senses.

d. The flap covering the feel relief button on the stick is spring-loaded to closed.

Electrical Supplies

The actuators for the artificial feel are operated by 28v DC.

Controls

The three push (off) pull (on) buttons for the artificial feel warning systems are at the forward end of the panel and are marked FEEL A, R, E. In this case the letters are for aileron, rudder and elevator feel. Each button embodies a warning light; when the button is pushed in, the main warning on the pilots' centre panel is cancelled or inhibited for that channel but feel system operation is not affected.

The artificial feel indicator is either a 3-position indicator which shows black, white or ILS, or a 2-position indicator which shows black or white. It shows black during normal flight conditions and white if any artificial feel channel fails or is relieved.

Feel Locking

To prevent possible feel unit runaway after a feel unit failure a locking facility is provided for all three channels. It is controlled by a single guarded switch. When LOCK is selected, a green light comes on to show that no further movement of normal or relief actuators can take place until NORMAL has been selected, although failure warning is given if the speed is altered by 30 knots from that at which LOCK was selected.

If the speed is changed by more than approximately 30 knots from the locking speed the main warning lights and the lights in the feel indicator buttons come on and the magnetic indicator goes white. If speed has been reduced, out-of-trim and manoeuvring forces are higher than usual. To prevent the main warnings coming on, push in the feel indicator buttons. The lights in the indicator buttons come on and the magnetic indicator shows white.

Before the feel is unlocked, reduce speed to below 250 knots, trim out the control forces and then raise the feel indicator buttons. The main warning lights come on.

After unlocking the feel, ensure that all failure warnings disappear, the green light goes out and the feel forces are at their appropriate level before making large control movements.

TRIMMERS

Description

Control forces felt by the pilot in flight are produced by compression or extension of the feel mechanism, in response to control movement or change of airspeed. Trim adjustment is made by varying the length of the control run between the pilots' controls and the feel unit using an electrically operated actuator which removes the load from the feel spring.

Double-pole wiring and switching is used to prevent runaways.

The trimming systems are duplicated as a precaution against failure and no warning indicator is provided. If the main systems become inoperative, however, the emergency system can be used. Trimming in the elevators sense is not permitted above 0.90M.

Controls

a. Each pilot's control column carries a double-pole 4-way aileron and elevator trim button.

The button-cover covers the two switches controlling the paired trim systems. A catch on the forward edge of the button, when operated, opens the cover and allows each switch to be operated independently, to test the systems. If the 1st pilot and co-pilot attempt to trim in opposite directions, the circuit first selected operates and the other is ineffective.

b. Twin rudder trim switches are on the fuel contents panel, spring-loaded to the centre (off) position. They are marked RUDDER TRIM, PORT/STBD; both switches must be moved for the system to operate.

The emergency trimmer control on the retractable console is moved fore-and-aft for longitudinal trim, sideways for lateral trim and rotated for rudder trim; the button in the top of the control must be depressed during trim selections.

Electrical Supplies

Electrical supplies for the system are 28v DC.

AUTOSTABILISERS

General

Pitch and yaw dampers in the elevon and rudder circuits improve the natural damping of aircraft oscillations. A mach trimmer counteracts the nose-down trim change at high mach numbers.

Controls

At the outboard side of the port console panel are controls for the autostabilisers and automach trimmer.

a. A YAW DAMPERS No 1/OFF/No 2 switch.

b. A RESET COMPARATOR spring-loaded button (for the pitch dampers and automach trim).

c. Four PITCH DAMPERS push (off)/pull (on) buttons, each embodying a warning light (amber).

d. Two AUTOTRIM RESET spring-loaded buttons, each embodying an extension indicator light (blue).

e. An AUTOTRIM ON/OFF pull/push button, embodying a warning light (amber).

The autostabiliser magnetic indicator shows white if any system is switched off or fails.

Electrical Supplies

The gyros and amplifiers in the system are operated by 115V, 3-phase, 400 Hz AC, while the servos, motors and relays are operated by 28v DC.

Yaw Dampers

The yaw damper system is duplicated, the YAW DAMPERS No 1/No 2 switch being selected as required. From each detector rate gyro, signals are passed to the actuator between the feel unit and the rudder PFCU.

Yaw damping is in operation at all heights when either channel is selected but is airspeed monitored; rudder displacement is constant up to 200 knots and then decreases as airspeed increases. The monitoring supplies are from both the port and starboard pitot-static systems.

If a system malfunctions, no warning indication is given. The magnetic indicator, however, shows white when the switch is off or following certain power failures. The switch must be put to the off (centre) position after flight.

Pitch Dampers

a. Pitch dampers improve longitudinal stability at high altitudes and high mach numbers (above 0.90M).

b. There are four channels in the system, each one feeding to one of the inboard elevon PFCs.

A comparator links the four channels and if any one channel differs in operation from the other three, the warning system operates.

c. The system is height-monitored and is inoperative below 20,000 feet. Above this altitude, the amplitude of control movement increases with increase of altitude.

d. The pitch damper servos are electrically heated whenever the undercarriage is retracted, the heating current being connected by a microswitch in the port undercarriage.

The pitch dampers are energised by pulling out the selector buttons on the port console.

The buttons may be pulled out at any stage in flight but the dampers are inoperative until the height switch permits their operation. The buttons must be pushed in after flight.

It may be necessary to use the RESET COMPARATOR button before all or some of the channel lights can be extinguished before take-off.

Automach Trimmer

a. The tandem mach trimmer system operates on the elevator control run, thus controlling all eight elevons. Signals are passed from two transmitting manometers, separately fed by each pitot-static system, through a follow-up system and an amplifier to the servos. The system is brought into operation by a height switch at 20,000 feet.

b. The mach trimmer applies up-elevon as the mach number increases above 0.88 plus or minus 0.01 (200 series engines) or 0.87 plus or minus 0.01 (301 engines). The amount of up-elevon applied is always the sum of the movement of the two actuators. They should be extended by the same amount. Full extension of both actuators represents three-quarters of the total up-elevator movement available but this is only achieved at a mach number of approximately 0.96M, which is outside aircraft limits. A malfunction could result in full travel of the actuators at lower mach numbers.

c. An accelerometer control in each half of the system is set to limits of 1.5 and 0.7g. When g values exceed 1.5 the servos are prevented from extending but are still able to retract; when g values are below 0.7 the servos cannot retract further but are able to extend. The servos return to their normal function when g is between the limits, provided that there is no misalignment.

d. A comparator monitors the system and indicates to the pilot if there is any misalignment of the two servos. Independently wired switches are provided to reset the servos to the minimum position.

Both mach trim channels are energised by pulling out the single ON/OFF button on the port console. The blue lights in the RESET buttons are on whenever the servos are extended.

AIRBRAKES

General

The slat-type airbrakes in the mainplane above and blow the engine air intakes are electrically operated by two motors, using 200V, 3-phase AC; the emergency motor is supplied by No 2 busbar, and the normal motor by No 3 busbar. The supplies to the airbrakes are disconnected if load shedding occurs.

The airbrakes have three extended positions:

a. Medium drag 35 degrees

b. High drag (UC up) 55 degrees

c. High drag (UC down) 80 degrees

The transition from 55 to 80 (77 degrees plus or minus 3 degrees) is automatic when the undercarriage is lowered, but raising the undercarriage does not retract the airframes to 55 degrees

Control and Operation

The airbrakes are controlled by a ganged switch on the rear face of the throttle quadrant. The switch has three positions: IN, MEDIUM DRAG and HIGH DRAG; the button in the centre of the switch must be pressed in before the switch can be moved to the HIGH DRAG position.

a. The airframes are operated by either of two electric motors, connected to torque tubes through differential gearing.

b. The normal motor is isolated when the NORMAL/EMERGENCY switch on the throttle quadrant is selected to EMERGENCY; the emergency motor only is then used for all selections. The emergency motor is isolated when the AIRBRAKE ISOLATE EMERGENCY switch on top of the panel is selected to ISOLATE; the normal motor only is then used for all selections.

c. Without isolation, both motors drive any selection from IN, only the normal motor being used for subsequent selections. However, current operating practice is to use one motor only, usually the normal motor for all selections.

XM571 taxiing back to dispersal at Waddington in 1982. The open airbrakes on the wing upper and lower surfaces are shown to advantage. *Tim McLelland*

d. Pre-flight, the airbrakes are to be checked initially using the emergency motor. After retraction, providing the airbrake slats are flush or slightly proud of the aircraft skin, a further check is to be carried out using the normal motor.

Warning: If, using the emergency motor, the airbrakes retract further than the flush position, remain in EMERGENCY. A NORMAL selection may result in damage to the wing or airframe mechanism.

The reposition magnetic indicator for the airframes at the top of the pilots' centre panel shows black when power is on and the airbrakes are in, or white when no power is available, when airbrakes are selected out or if the airbrakes extend without selection or fail to retract completely. It shows cross-hatched when the airbrakes are in, selected in, and a main connector has welded (a momentary cross-hatched indication is given when the airbrakes are selected from IN to MEDIUM DRAG with the emergency motor isolated; as soon as the airbrakes start to move the indication reverts to white).

The airbrakes must not be operated on the ground when the bomb door access panels and the bomb doors are open.

The airbrakes must not be selected from HIGH to MEDIUM DRAG, or MEDIUM DRAG to IN if iced up, as this may cause damage to the drive mechanism or airframe.

Sustained flight with airbrakes extended against engine power is not recommended.

BRAKE PARACHUTE

A brake parachute in the tail cone aft of the rudder provides additional braking during the landing run. Operation is electrically controlled (28v DC) by a split, 2-pole, JETTISON/STREAM switch on the centre instrument panel. Both halves should normally be used but either half of the switch can stream and jettison the parachute (at a slower rate). If both are used for streaming, both must be used for jettisoning.

A magnetic indicator, beside the external intercom point, is visible when a small access panel on the starboard side of the rear fuselage is raised. The indicator shows black when the parachute door is locked and all switches and relays in the circuit are at their correct setting for streaming.

If unselected streaming occurs, the action of the door opening without electrical selection causes a supply to be fed to the jettison unit and the parachute is jettisoned automatically.

FUEL SYSTEM

General

Fuel is carried in 14 pressurised tanks, five in each wing and four in the fuselage, above and to the rear of the nosewheel bay. The tanks are of the flexible bag type and each tank is enclosed by a metal casing which is part of the aircraft structure. The tanks are not self-sealing but are crash-proof.

The tanks are divided into four groups, each group normally feeding its own engine. A crossfeed system enables the various groups to be interconnected. Automatic fuel proportioning is normally used to maintain the fuel CG position.

Provision is made for carrying two fuel tanks in the bomb bay, either saddle-shaped or cylindrical. Fuel from these tanks passes into the main system through two delivery lines, one each side of the fuselage, to each side of the centre crossfeed cock.

An air-to-air refuelling probe is in the nose and pipes from it join the normal refuelling lines.

A pressure refuelling system is provided for ground use.

The majority of the controls and indicators for the fuel system are grouped in the form of a mimic diagram on the retractable console. The air-to-air refuelling controls are on the starboard console.

Electrical Supplies

All fuel pumps utilise 200V AC but the fuselage tank pumps are controlled by 28v essential DC, while the bomb bay and wing tank pumps are controlled by 28v DC non-essential supply.

FUEL TANKS AND RECUPERATORS

Main Tanks

The tanks on each side of the aircraft are numbered from 1 to 7, Nos 1 and 2 being the fuselage tanks and the remainder the wing tanks. The tank numbers correspond to the CG position of each tank, No 1 having the furthest forward CG and No 7 the furthest aft.

Nos 1, 4, 5 and 7 tanks comprise the outboard tank group (No 1 group port, No 4 group starboard). Similarly, Nos 2, 3 and 6 tanks comprise the inner tank groups (No 2 port and No 3 starboard). Each group normally feeds its associated engine.

The total contents, in gallons, varies between different fuels because the point at which the refuelling valves close is affected by the specific gravity and temperature of the fuel used.

Each wing tank contains a reservoir, with clack valves at its base. These valves are normally open but close when the head of fuel in the reservoir is built up by the auxiliary pumps.

Bomb Bay Tanks

Two fuel tanks may be carried in the bomb bay, one at the forward and one at the rear end. The tanks can be of the saddle or cylindrical type, the former being referred to as the 'A' when fitted forward or the 'E' when fitted aft. If only one bomb bay tank is carried it must be fixed in the forward position.

Saddle tanks: Each saddle tank has four pumps, two on each side; each pair of pumps feeds into the delivery line on that side.

Cylindrical tanks: Each cylindrical tank has three pumps, feeding into a common line, which feeds both delivery lines. The tanks are referred to as the forward and aft tanks.

AAPP Tank

The fuel tank for the airborne auxiliary power plant, in the starboard wing to the rear of the AAPP, has a capacity of 10 gallons.

The tank is filled from the main fuel system via a pipe line from the wing tanks of No 4 group delivery line, whenever the No 4 group wing booster pumps are running. In addition, the tank is supplied from the refuelling line. A float valve in the tank shuts off the supply when the tank is fuel.

Tank Pressurisation and Venting

Each tank group can be pressurised with air from its associated engine. The system maintains a pressure of 1.82 to 2.3 PSI above ambient in the tanks throughout the altitude range of the aircraft, so as to prevent loss by boiling off fuel at high altitude and high fuel temperatures. Pressurisation also prevents negative differential pressure in the tanks, thus preventing risk of tank collapse. The bomb bay tanks are not pressurised.

Pressurisation of the main tanks is controlled by a switch on the air-to-air refuelling panel; below the switch are four magnetic indicators, one for each tank group, which show black when the tanks are pressurised. A switch on the centre console for bomb bay tank pressurisation is inoperative.

Combined inward/outward relief valves are fitted in all tanks, and float valves in the wing tanks. The float valve closes if the fuel level rises, to prevent fuel flowing back into the pressurisation lines. The inward relief valve is set at 0.25 PSI and the outward valve at 0.75 to 1 PSI. Two vent valves for each tank group, one in the bomb bay roof and one under the sailplane, give atmospheric venting when tank pressurisation is not in operation; when pressurisation is switched on, the valve is adjusted by a master control valve which, in conjunction with an air valve, maintains a pressure of 1.82 to 2.3 PSI above ambient. In the event of overpressurisation an outward relief valve in the vent valve relieves at 2.65 to 3.0 PSI.

The AAPP tank is vented to the main system in No 4 tank, No 4 group.

The bomb bay tanks are vented into the fuselage tank lines.

a. When tank pressurisation is selected on initially, the pressure indicators may remain white until RPM reach approximately 80%; they may similarly revert to white after the landing run.

MAIN FUEL SYSTEM

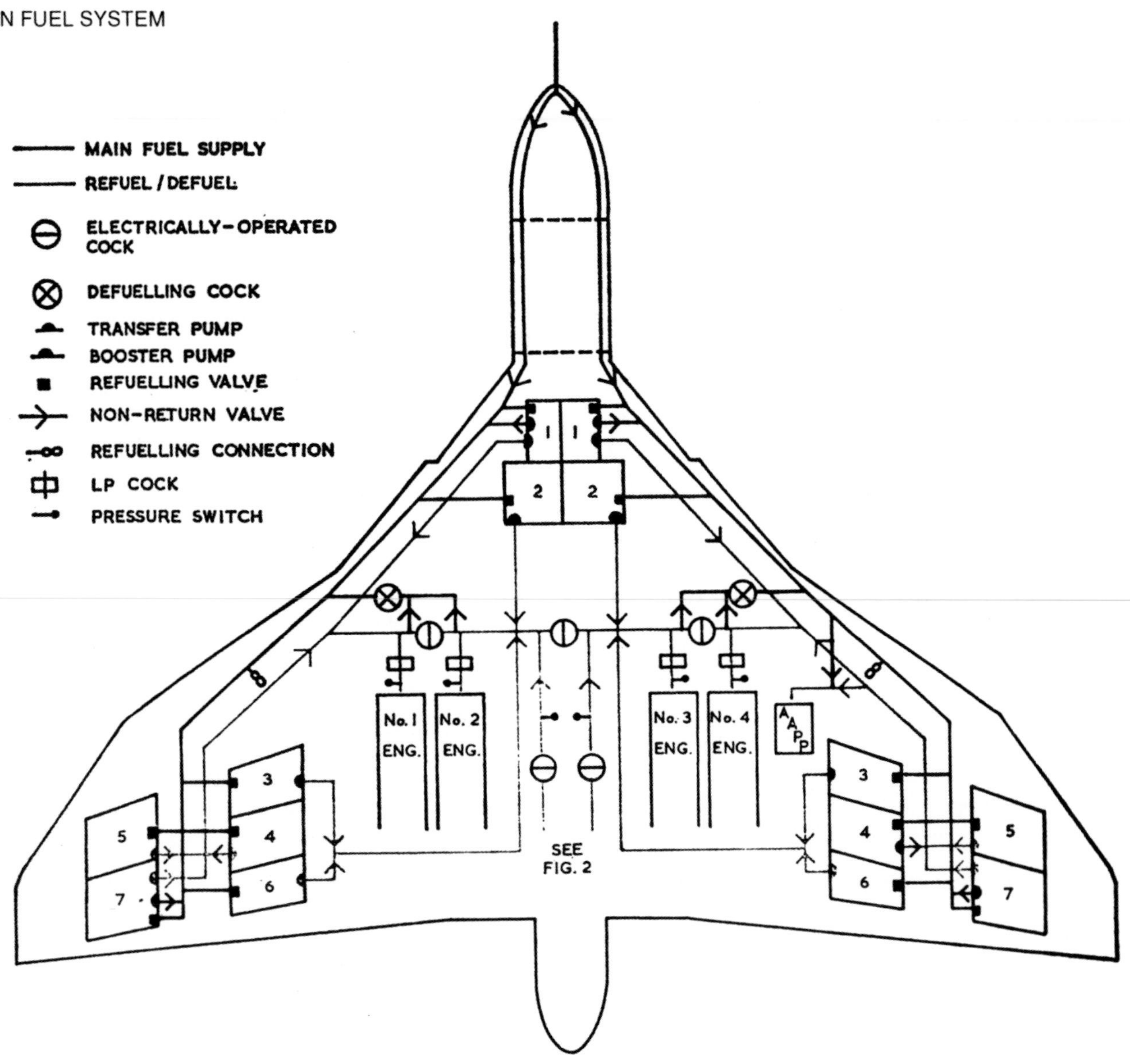

Fuel system diagram.

b. Normally, with power on and pressure off, the pressure indicators show white. If electrical load shedding takes place, the tanks are pressurised regardless of selection and the indicators show black.

c. During descent, with pressurisation on, and after landing when it is selected off, rumbling noises may be heard; this is normal.

Regulators

A 6-gallon recuperator for each tank group supplies fuel to the engines if there is a pressure drop in the fuel supply. Each recuperator is supplied with fuel from its own tank group and with air from its own engine. If the fuel pressure drops, the air pressure (at 6 to 10 PSI) forces the fuel into the engine feed lines. Sufficient fuel is available to supply the engines for approximately 10 seconds at full power at sea level, and up to 2 minutes at altitude at cruising RPM. When negative g is removed, normal fuel flow is resumed and the booster pumps recharge the recuperator.

REFUELLING AND DEFUELLING

Air-to-Air Refuelling

Fuel from the probe flows aft on either side of the cabin, through non-return valves, to join the main refuelling lines. All tanks are refuelled at the same time, the rate of flow being approximately 4,000lb/min.

Probe lighting is provided by two lamps in the nose supplied through separate circuits.

Refuelling on the Ground

Pressure refuelling of the aircraft is via two refuelling points in each main wheelbay. These two points supply a common refuelling line on each side, running to a refuelling valve in the sump plate of each tank. Refuelling instructions are detailed in the Flight Reference Cards.

During refuelling, each tank is filled to the same percentage of its capacity in order to maintain a central CG position. A selector in the port main wheelbay allows selection of quantities from 0 to 100% and operates through the electrical output of the contents gauge amplifiers.

A control panel at each refuelling point carries the switches for the system and indicator lights to show the progress of refuelling.

Only one tank in each group is filled at a time, automatic changeover to the next tank taking place when the first tank is filled to the selected percentage. The order of filling is 1, 4, 5 and 7 in the outboard groups and 2, 3 and 6 in the inboard groups.

Bomb Bay Tanks

a. The bomb bay tanks are filled from the port refuelling point. A double-level float switch is fitted in each tank. The bomb bay refuelling panel is in the port wheelbay and carries a master switch and low and high level indicator lights for each tank. With the master switch ON, the refuelling valves open and the indicator lights come on. When the tank is almost full, the lower float switch partially closes the refuelling valve, reducing the rate of flow and extinguishing the green light. The high level switch completely closes the refuelling valve when the tank is full and extinguishes the red light.

b. The main tanks must be refuelled before the bomb bay tanks.

c. The bomb bay refuelling master switch must be selected off when refuelling is complete.

FUEL SYSTEM CONTROLS AND INDICATORS

Fuel Control Panel: Retractable Console

The fuel panel on the retractable console carries a mimic diagram of the system, including the bomb bay tanks.

Forward of the diagram are three CG control switches, two FWD/AFT transfer pump switches, one for each side of the system, and one PORT/STBD switch for use during air-to-air refuelling. On each side of the diagram are two AUTO/MANUAL switches, one for each group; these switches control sequence timing. In each tank on the diagram is an OFF/ON pump switch (which controls both main and auxiliary pumps) and a CONT pushbutton for contents reading. The crossfeed cocks are represented in the diagram by three magnetic indicators, with OPEN/CLOSE cock switches to the rear of them. Four pushbuttons, marked NO-ENG are provided for flowmeter selection. The bomb bay system diagram has two BOMB BAY/MAIN switches, two ON/OFF pump switches for each tank and a pressurisation switch (inoperative).

Fuel Control Panel: Starboard Console

The air-to-air refuelling controls are grouped on a panel on the starboard console and consist of two probe lighting dimmer switches, a nitrogen switch, a main tanks pressurisation switch, four tank pressurisation magnetic indicators, an ON/OFF split double-pole master switch, a refuelling indicator and a refuelling gallery pressure gauge.

Gauge Panels

The contents gauges for the main tanks, one per group, are on a panel forward of the throttles, while that for the bomb bay tanks is on a panel attached to the inboard guide rail of the 1st pilot's seat, with a pushbutton for individual tank selection.

Fuel Cocks

HP cocks: The four HP cocks are opened by the initial movement of the throttle levers forward from the fully closed position. The sleeves on the levers must be held up to permit movement between the HP COCK SHUT position and the IDLING gate.

LP cocks: The four LP cocks are electrically controlled by four guarded ON/OFF switches on the underside of the coaming above the pilots' centre panel. Each cock is fitted with a bypass through a non-return valve, which acts as a thermal relief for fuel trapped between the engine and the LP cock, when the cock is closed. The LP cocks are supplied with power from the vital busbar.

Crossfeed Cocks and Indicators

a. There are two wing crossfeed cocks, each connecting the tank groups on that side, and a centre crossfeed cock between Nos 2 and 3 groups. The cocks are electrically operated by 28v essential power.

b. Three 3-position magnetic indicators show continuity with the diagram lines when the cocks are open, discontinuity when the cocks are shut and cross-hatch when the cocks are at intermediate position or when no power is available.

Bomb Bay Tank Cocks

Shut-off cocks are provided in the delivery lines from the bomb bay tanks. These cocks are opened together if any bomb bay tank pump switch is selected on.

Ground Servicing Cocks

A manually operated cock in the delivery line from each tank is used during ground servicing. The cock operating levers are so designed that the cover plates cannot be fitted into place unless the cocks are in the open position.

FUEL PUMPS

Booster Pumps and Auxiliary Pumps

a. Each wing tank contains both a booster pump and an auxiliary pump; the fuselage tanks have a single booster pump. The booster pumps are attached to the sump plate at the bottom of each tank and, in the case of the wing tanks, are inside the reservoir.

b. Starvation of the booster pumps with change of aircraft attitude is prevented by the auxiliary pumps at the inboard end of each wing tank.

These pumps run whenever the booster pumps are switched on and supply fuel to the reservoir, thus maintaining a head of fuel for the booster pumps.

Bomb Bay Tank Pumps

a. Each saddle-type (A and E) bomb bay tank has four booster pumps, one pair supplying each feed from the tank. The pumps run in parallel and each pump switch controls one port and one starboard pump in its tank.

b. Each cylindrical tank has three booster pumps, feeding into a common line. The same controls are provided, the right-hand pump switch for each tank controlling the forward pump and the left-hand switch controlling the other two pumps. The tanks are labelled FWD and AFT instead of A and E.

c. When the none BAY/MAIN switches are set to BOMB BAY, the main tank sequence timers are isolated and the main tank pumps run at reduced speed, provided also that the main tanks are selected to AUTO.

Transfer pumps

a. Transfer pumps in Nos 1 and 7 tanks on each side allow the fuel to be transferred in either direction between these tanks, if it is necessary to adjust the fuel CG position. As both tanks are in the same group, transfer does not affect the group contents. Transfer is via the refuelling lines.

b. With a transfer pump switch at FWD, the refuelling valve of No 1 tank opens and No 7 tank pump starts and transfers fuel to No 1 tank; placing a switch to AFT opens the refuelling valve of No 7 tank and starts the No 1 tank pump. The rate of fuel transfer is approximately 100lb/min when transferring from No 1 to No 7 and approximately 50lb/min when transferring from No 7 to No 1; approximately 300lb of fuel must be transferred to alter the slide rule index by 1.2. After transfer, check that the desired amount of fuel has been transferred.

c. If, when transferring fuel, the receiving tank is full when the transfer pump is still running, a float switch closes the refuelling valve.

d. The function of these switches is altered during air-to-air refuelling.

FUEL PRESSURE WARNING INDICATORS

a. Four magnetic indicators on the pilots' centre panel below the IPT gauges show black when the fuel delivery pressure to the engine is satisfactory, white when the pressure downstream of the filter falls below 5 PSI, and black when there are no power supplies.

b. The two bomb bay fuel indicators show black when the fuel pressure is sufficient and the cocks are open. They show white if the fuel pressure falls below 10 PSI.

c. If it is necessary to close the LP cock after shutting down an engine in flight, the magnetic indicators do not always turn white immediately. The time taken for fuel pressure beyond the LP cock to fall below 5 PSI depends on the rate of decay of pressure through the LP cock bypass valve.

Sequence Timers

Because of the configuration of the aircraft, the fuel tanks are disposed forward and aft of the aircraft centre of gravity.

It is therefore essential that fuel should be used at approximately the same rate from all tanks, in order to maintain the fuel CG position.

An electrically operated sequence timer on each side of the aircraft ensures even fuel distribution, by causing the main pumps in each tank to run alternately at full speed and reduced speed (the auxiliary pumps run continuously). The quantity of fuel pumped from any one tank during one cycle of the sequence timer (five minutes) is proportional to the tank capacity; the distribution of fuel is thus maintained throughout the tanks. The sequence timer motors use 200 volt AC.

It will be seen that No 2 tank feeds twice in each cycle, as it is the largest tank.

With all booster pumps on and the AUTO/MANUAL switches at AUTO, sequence timing is in operation; to interrupt the sequence timing in any group, put the appropriate switch to MANUAL and, if it is desired to use fuel from any particular tank in the group, switch OFF all booster pumps which are not required. The AUTO/MANUAL switches should be put to MANUAL after flight, to de-energise the relays.

FUEL CONTENTS GAUGES AND FLOWMETERS

Main Tanks Contents Gauges

a. A capacitor-type system provides indication of fuel contents. The four gauges, one for each tank group, are on a panel forward of the throttle levers. Each gauge is calibrated with two concentric scales, reading in pounds x 1000.

Normally, each gauge reads the contents of its appropriate group, on the inner scale. An individual tank reading is obtained on the outer scale by pressing the pushbutton in the appropriate tank position on the mimic diagram.

b. In no circumstances should two gauge pushbuttons in the same group be pressed simultaneously, as this damages the instrument.

c. Four group contents gauges are at the navigator/plotter's position; there is no means of reading individual tank contents on these gauges.

d. The contents magnetic indicator for the AAPP is at the AEO's station.

e. An accurate 28 volt supply is needed to ensure correct fuel readings.

Bomb Bay Tanks Contents Gauge

The fuel gauge of the bomb bay tanks is on the 1st pilot's seat guide rail, facing towards the co-pilot. The total contents of both tanks are normally shown; individual tank contents can be obtained by pressing the appropriate pushbutton below the gauge.

Flowmeters

a. A Mk 3 flowmeter system is installed. This system is designed to give the following approximate indications:

(1) Fuel consumption by individual engines (lb/min).

(2) Total fuel consumption by all four engines (lb/min).

(3) Total amount of fuel gone (lb).

b. Two indicators, one giving total flow and pounds gone and the other giving instantaneous flow for individual engines, are on the pilot's instrument panel, together with a FUEL FLOW/RESET/NORMAL switch for resetting the total flow indicator. Selection of an individual engine flow is obtained by pressing the appropriate engine pushbutton on the fuel system mimic diagram. The instrument continues to indicate the flow to that engine until another engine is selected.

CG INDICATOR AND SLIDE RULE

A fuel CG position indicator, on the pilots' centre panel, indicates the CG of the fuel system (not of the aircraft). The instrument registers automatically when air-to-air refuelling is in operation but readings can be taken in other flight conditions by pressing the CG CHECK button. This button can also be used, when air-to-air refuelling is in progress, to regain contents gauge readings.

The instrument face has two arcs, one for each side of the fuel system; each arc is divided into three sectors, a central green sector to indicate the safe range and red outer sectors marked NOSE HEAVY and TAIL HEAVY. The needles should be on or near the zero position if the fuel is correctly proportioned. The green sector covers a range of 60,000lb ft, 30,000 lb ft forward of zero and 30,000 lb ft aft. If, for example, both needles were on the forward limit of the green sector, the fuel CG would be 60,000 lb ft forward of the zero position with equally proportioned fuel.

The instrument can be checked before flight by pushing the CG pushbutton and observing any slight movement of the needles. If no movement is observed, it may be because of exact fuel proportioning. This can be checked by transferring fuel from Nos 1 or 7 tanks and checking the indicator for movement while pressing the CHECK button.

The bomb bay tank fuel is not included in the CG indication.

CG Slide Rule

A slide rule for calculating aircraft CG is provided and stowed below the starboard console.

Note: The CG limitations take into account the shift caused by undercarriage retraction.

AIR-TO-AIR REFUELLING CONTROLS

The air-to-air refuelling indicator consists of the outline of the aircraft, with numbered lights in the approximate position of each tank. The lights in the indicator can be adjusted for day or night use by revolving the ring round the indicator. The lights come on when the valves open and go out individually as the tanks fill.

The master switch must be put on before drogue engagement and must not be put off until contact has been broken. With either half of the switch one of the refuelling valves (in all tanks which are not full) are opened, tank pressurisation is switched off, the fuel contents gauges are isolated and read zero, all lights in the indicator come on (except for full tanks) and the CG indicator registers automatically.

As each tank is filled, a double float switch closes the refuelling valve and the appropriate light in the indicator goes out.

100% refuelling may be carried out, provided that the fuel system is depressurised. If pressure remains in the tanks, refuelling may be carried out provided that the tanker main pump is switched off. Contact must be broken immediately if the fuel gallery pressure exceeds 10 PSI.

The master switch must not be set OFF until contact is broken. The nitrogen purge switch is then set ON; this action opens the nitrogen cock and the No 2 tank's refuelling valves and nitrogen pressure at 30 PSI forces fuel from the probe lines into the No 2 tanks.

Contents gauging is regained when the master switch is OFF.

CG control

a. The aircraft CG can be controlled during air-to-air refuelling by three switches on the fuel control panel: the two switches which normally control the transfer pumps, for fore-and-aft control, and the PORT/STARBOARD switch at the top of the panel, marked FR RECEIVER CG CONTROL, for lateral control.

b. When the refuelling master switch is ON, the transfer pump switches are disconnected from the transfer pumps; setting them to FWD closes the refuelling valves in tanks 6 and 7, while setting them to AFT closes the refuelling valves in tanks 1 and 2. In each case, the refuelling valves remain open in all other tanks. If the lateral control switch is moved to PORT or STBD, the refuelling valves in the Nos 6 and 7 tanks on the opposite side are closed.

Air-to-air Refuelling CG Control

The fuel CG indicator must be monitored throughout the refuelling sequence, especially in the initial stages. If a tank refuelling valve fails to open, especially in a tank with a large moment arm, the appropriate needle moves quite rapidly forward or aft; this is the only indication of a valve failing to open. As movement of the needle towards the limit of the green sector becomes apparent, move the appropriate FWD/AFT switch on the fuel panel in the opposite direction from needle movement. Refuelling may be continued for as long as it is possible to keep the needles in the green sector; if it becomes impossible, contact must be broken immediately.

Normally, if all tanks are accepting fuel, the needles move to and fro within the green sector and no action need be taken apart from monitoring. If contact is broken before all tanks are full, the subsequent tank contents check, with the master switch OFF, may show considerable variations in tank percentages. MANUAL use of the fuel system then becomes necessary to balance the tanks.

Even if it is known that the tanks are only partially to be filled, monitoring of the tanks-full indicator is still necessary; in practice, Nos 7, 5 and 1 tanks are the first to fill, in that order, and the other tanks may lag behind by as much as 25%. Any method of adjusting the fuel in individual tanks, in an attempt to make all tanks fill simultaneously, reduces the safety margin provided by this lag and increases the possibility of rupturing a tank if its valve remains open when all the others have closed. Such a failure could be catastrophic.

If the bomb bay tanks refuelling master switch in the port wheelbay is inadvertently left on or an electrical fault produces the same effect, the bomb bay tanks refuelling valves open in flight as the fuel contents decrease.

This results, during air-to-air refuelling, in fuel entering the bomb bay tanks; a careful watch must be kept on their contents gauges in order to keep the CG within limits.

PART 7: GENERAL AND EMERGENCY EQUIPMENT AND CONTROLS

ENTRANCE DOOR, LADDERS AND CANOPY

Entrance to the Aircraft

The aircraft is entered by the door on the underside of the fuselage, below the crew compartment. The door is hinged at the forward end and opens downwards. Door opening can be operated either mechanically or pneumatically; door closing is pneumatically operated.

Door Opening Mechanism

a. The door can be opened from outside by a handle near the rear edge of the door. Operation of the handle deflates the door seal and withdraws the door bolts; the door then opens under gravity.

b. Door opening from the inside is by a lever in a gated quadrant on the port side of the door, at the forward end. Movement of this lever to the gate deflates the door seals and withdraws the locking bolts. If the aircraft is on the ground, the door then opens under gravity. To select the EMERGENCY door opening position it is necessary to move the lever to the fully forward position, in the aircraft sense, via the quadrant gate. The gate is negotiated by moving the lever outwards towards the port side of the aircraft, then continuing forward. In an emergency, the door may also be opened by a switch at the navigator/plotter's position.

An additional door-opening switch is provided, recessed into the edge of the table between the navigator/plotter and the AEO and covered by a flap. The switch has three positions ON/OFF/ON and either on position opens the door through the same circuits as the navigator's switch.

The main door may jam closed if the seal deflation valve plunger is out of alignment. The seal can be deflated by pulling the door inflation pipe from its rubber connector on the starboard side of the seal, or by cutting the rubber connection.

Door Closing Mechanism

a. The door may be closed from the outside by operation of a pushbutton to the rear of the door.

b. The door is closed from inside by a toggle mechanism, the handle of which is below a cover on the crew's floor, between the navigators' seats. The door is closed by pneumatic pressure at 400 PSI and, when the door is closed and locked, the door seal is automatically inflated by nitrogen at a reduced pressure of 25 PSI.

Door Indicators

On the door operating quadrant there are two lights, one red and one green. The green light, marked DOOR SAFE, comes on when the door is locked; the red light, marked DOOR NOT SAFE, comes on whenever the door is unlocked. In addition, there is a magnetic indicator above the pilots' centre instrument panel, which shows white when the door is unlocked. In the nosewheel bay is a green light which comes on when the door is locked.

The crew access door of a Coningsby Wing Vulcan; the Wing comprised Nos IX, 12 and 35 Squadrons, whose badges are displayed with Coningsby's station badge. *Joe L'Estrange*

A view that will be familiar to Vulcan air and ground crews, looking up the ladder to the flight deck past the 6th and 7th seats. *Paul Tomlin*

Ladders

A folding ladder is attached to the entrance door. Before flight it is to be removed and strapped to the stowage on top of the bomb aimer's window cover. On the occasions when this position is covered by extra equipment the ladder should be carried in the pannier.

Warning: Due to the hazard it presents to evacuating the cabin in an emergency, the ladder must not be left in position on the entrance door.

A ladder, which can be stowed by lifting and sliding to port, provides access to the pilots' cockpit.

Canopy, Description

The canopy, of double skin construction, is attached to the fuselage nose section by six attachment points. A seal, inflated by pneumatic pressure, ensures an airtight fit. A schrader valve allows the seal to be inflated by handpump, for weather proofing, when the pneumatic system is being serviced.

Canopy Locking Indicators

a. Two pointers, one on either side of the cockpit, indicate against an arc whether the canopy is locked or unlocked. The locked position is indicated by a small white segment, while the UNLOCKED range is indicated by a larger, red segment.

b. A magnetic indicator, at the top of the pilots' centre panel, shows white when the canopy is unlocked.

Canopy Jettison

Two yellow/black wire-locked canopy jettison levers are provided, one on each side of the cockpit, above the consoles. A pip-pin is provided to lock each lever in the safe position and, when on the ground, the pins are inserted through the hole in the coaming rail to prevent inadvertent operation. The pins must always be removed before flight.

A yellow external release handle is on the pod side of the nose. When this handle or either of the jettison handles is pulled, mechanical linkages open the canopy attachment jaws and operate a torque tube to fire the jettison gun.

The canopy is also jettisoned when any of the seat handles are pulled. In this case, pulling the handle operates the following sequence:

a. A pneumatic valve is opened, which allows air pressure at 1,200 PSI to pass to a jack.

b. The jack operates a torque tube to open the canopy attachments and fire the jettison gun.

c. As the canopy clears the aircraft, it operates a time delay unit which fires the ejection seats 1 second later.

Canopy Jettison Gun Safety-pin

a. A safety-pin, with pip-pin attached, is provided for the jettison gun sear at the rear of the canopy, in the cabin. The pin must be inserted in the sear after flight. Before flight, the pin must be removed from the sear and the pip-pin must be inserted in the adjacent jettison lever mechanism, to link the manual mechanism to the gun.

b. Instructions on the use of these pins is given on two tablets on the perspex cover of the gun, together with a diagrammatic arrangement of the devices. The pins are inserted through sliding panels in the perspex.

Warning: Canopy jettison lever pip-pin must be removed and a check made that the canopy indicator remains black, before the jettison gun is made live.

WINDSCREEN AND ASSOCIATED EQUIPMENT

Windscreen and DV Panels

The laminated windscreen, embodying gold film heating, is divided into three sections.

DV Panels

A triangular DV panel at each end of the windscreen is hinged at the lower edge and opens inwards. The panel is released by pressing a catch in the handle at the top, then depressing the handle, pulling it inwards and sliding it back. No stowages are provided.

When replacing the panel, care must be taken to ensure that the balls at the base are fitted into the guide tube correctly, otherwise a serious cabin pressure leak can occur.

Windscreen Wipers

Windscreen wipers are provided for each pilot's windscreen panel and for the centre panel. The wipers are electro-hydraulically operated and the blades are of the parallel-motion type and are self-parking.

The wipers for the 1st pilot's and centre windscreen share a common hydraulic system and electrical motor, while the co-pilot's has an independent system.

The wipers are controlled by two double-pole, reposition OFF/FAST/SLOW switches, one for the 1st pilot's and centre windscreen panel wipers and one for the co-pilot's panel wiper.

The 1st pilot's switch is on the left of his panel.

The co-pilot's switch is on the right of his panel.

The wipers must never be used on a dry windscreen.

The wipers are operated by 28v DC.

Sun Visors and Anti-Flash Screens

Sun visors are provided for the windscreen and side panels; the side visors are sliding and the front ones are hinged at the top.

They are attached to the lower edge of the canopy.

Anti-flash screens are provided for the windscreen, side screens and crew windows. Those for the windscreens have slide fasteners; when not in use, they are rolled down and stowed. The screens for the side windows are sliding shutters.

INTERNAL LIGHTING

Cockpit Lighting

a. Panel lighting: The pilots' instrument panels and the consoles are lit by a mixture of red flood, white flood (fluorescent) and U/V lighting, while the E2B compasses have red lamps.

Pillar lamps are built into the front instrument panels. The red lamps are controlled by a series of dimmer switches on the cockpit walls, forward of the port and starboard consoles.

One U/V dimmer switch is on each side console. The white flood lighting is controlled by two switches, one on the outboard side of each pilot's instrument panel.

b. Anti-dazzle lighting: Anti-dazzle lighting is provided and is controlled by a BRIGHT/OFF/DIM switch on the port of the fuel contents panel; an additional OFF/BRIGHT switch is provided at the nav/plotter's position.

c. Wander lamp: A wander lamp, with its own integral switch, is attached to the canopy, above the pilots' position.

Crew's Lighting

Both panel lighting and angle-poise lamps are provided for the crew members. This lighting is controlled by a series of ON/OFF switches and dimmer switches at the crew positions.

Servicing Lamps

Servicing lamps are provided in the bomb bay, wheelbase, power compartment and rear fuselage. The master switch for these lamps is in the starboard side of the nosewheel bay. In addition, there are sockets for inspection lamps, one on the front spar bulkhead and one on the rear spar bulkhead, in the bomb bay. The lamps are only operative while an external 28v DC supply is connected.

Electrical Supplies

The U/V and fluorescent lighting uses 115V 3-phase AC from the main transformers; the red flood and anti-dazzle lighting is 28v DC operated.

EXTERNAL LIGHTING

Master Switch

Before any of the external lighting (except the servicing lights) can be used, the EXTERNAL LIGHT master switch must be put ON. This switch is on the inboard side of the starboard console.

Navigation and Anti-collision Lights

Steady navigation lights are provided together with red rotating lights, one on the upper fuselage and two on the underside, below the engine air intakes. The control switch is marked NAVLIGHTS-STDY/FLASH. When FLASH is selected, the navigation lights are steady and the rotating lights operate.

Identification Light

The downward identification light is controlled by a single-pole, STEADY/OFF MORSE switch, on the inboard side of the starboard console. The switch is spring-loaded from MORSE to off.

Landing and Taxiing Lamps

There is a combined landing/taxiing lamp under each wing, the lamp being extended further for the taxiing position than for landing. The lamps are individually controlled by two double-pole, 3-position RETRACT/LANDING/TAXI switches on the inboard side of the starboard console. The landing lamps incorporate a slipping clutch mechanism and blow in if the airspeed exceeds 180 knots. Once the lamps have blown in, the control switches must be reselected to RETRACT and then to LANDING (with the airspeed below 180 knots) before the lamps will re-extend.

Probe Lighting

Two lamps in the nose of the aircraft light the probe for air-to-air refuelling.

They are controlled by individual dimmer switches on the starboard console.

Electrical Supplies

The external lighting uses 28v DC and the navigation and anti-collision lights are essential loads, the remainder of the external lighting is non-essential.

MISCELLANEOUS EQUIPMENT

Sextant

A pressure-tight sextant mounting is provided on each side of the canopy coaming. The sextant is held in either the retracted or the operating position without loss of pressure. A lever in the mounting allows the sealing plate to be opened only when the sextant has been inserted in the carrier tube. The sextant must not be removed until the sealing plate is closed.

Ration Heaters

There are five ration heaters, one at each crew position. The 28v DC non-essential supplies are controlled by two switches at the AEO's position on panel 50P.

One switch controls the pilots' heaters, the other the rear crew's heaters. Sealed food tins are not to be inserted in the heaters; no tin is to be heated continuously for more than two hours.

Periscope

The rearward-facing periscopes are controlled by a handle below the AEO table, raised to select the upper periscope, lowered to select the lower one, and moved sideways to rotate the selected periscope. The periscopes are heated by a 115V 1600 Hz, single-phase supply from No 2 frequency changer, controlled by an ON/OFF switch on the AEO's upper panel.

Rapid Start External Connections

The aircraft is fitted with external pull-off connections, so that the air conditioning hoses, the electrical supply cables, the telescramble line and the static vent plugs are all removed automatically as the aircraft moves forward.

Electrical Supplies

The 28v, 200V and true earth plugs are on a sloping bracket on the pod side of the power compartment, accessible through a spring-loaded access panel in the underside of the port sailplane.

Telescramble

A telescramble and mic/tel socket are on the starboard side of the power compartment, accessible through a spring-loaded access panel. Only the telescramble socket is used for rapid take-off, a further socket being introduced to this line for use of the crew chief. The telescramble system feeds into the intercom station boxes.

Air Conditioning

The cabin conditioning hose connection is on the starboard side of the cockpit and the ventilated suit connection is below the engine intakes on the port side.

Static Vent Plugs

The static vent plugs are pulled out by a cable.

EMERGENCY EXITS AND EQUIPMENT

Entrance Door and Static Lines

The entrance door below the fuselage is the escape exit for the rear crew except in a crash landing. It can be operated by either the EMERGENCY position of the lever on the port side of the door, or by a switch at the navigator/plotter position marked EMERGENCY DOOR OPEN, or by the switch at the edge of the table, or by a combination of these controls. The door can be safely opened in flight at speeds up to 220 knots.

Warning: If the door is opened by either switch, the normal lever may not pass through the emergency gate. If it does not and there is a failure of the 28v supply, the door will close again under the action of the slipstream. The first man at the door must check that the lever is gated in the EMERGENCY position.

After decompression, when the door has been opened, the crew swing themselves out of the opening, using the handle on the back of the nav/plotter's seat.

Static lines for the rear crew and 6th and 7th crew members' parachutes are fitted in the aircraft oxygen hose assemblies. One end of each line is connected to a strong point on the floor beside the oxygen point, while the other end carries the parachute attachment link.

Within each hose assembly, connected to the static line, is a switch and associated electrical wiring. As each crew member abandons the aircraft in an emergency, the pull on the static line operates the switch, illuminating one of the five blue 'crew gone' lights on the bottom of the 1st pilot's instrument panel.

Canopy Jettison

The canopy can be jettisoned by pulling back either of the jettison levers (one beside each pilot). If possible, pull both levers simultaneously. The canopy is automatically jettisoned when any ejection seat firing handle is pulled. It can also be jettisoned on the ground from outside, by pulling the yellow painted handle on the port side of the nose.

First-Aid Kit

A first-aid kit is stowed at the rear of the co-pilot's seat.

Crash Axe and Asbestos Gloves

A crash axe and asbestos gloves are stowed on the cover of the bomb aimer's window.

Life Raft

In addition to the individual life rafts carried by each crew member, a life raft Type MS5, complete with survival equipment, is stowed at the rear of the canopy, below the fairing, outside the pressure cabin. The DINGHY RELEASE handle on the forward face of the stowage container is inaccessible unless the canopy has been jettisoned. Pulling the handle releases and inflates the life raft, which remains attached to the aircraft by a painter.

Signal Pistol and Cartridges

A pressure-tight mounting for the signal pistol on the cabin port walk, above and forward of the AEO seat, has a stowage beside it for 12 cartridges.

External Emergency Equipment

A compartment on the port side of the nose, opened from the outside, carries a first-aid kit, a crash axe, a pair of asbestos gloves and a BCF hand fire extinguisher. Break-in markings, for access to the cabin and to the emergency equipment, are painted in yellow on the outside of the fuselage.

PART 8: HYDRAULIC SYSTEM AND UNDERCARRIAGE EMERGENCY LOWERING SYSTEM

General

The main hydraulic system provides pressure for:

a. Undercarriage raising and lowering and bogie trim.

b. Nosewheel centring and steering.

c. Wheelbrakes.

d. Bomb doors opening and closing.

e. AAPP air scoop closing.

An electrically operated hydraulic power pack (EHPP) may be used for operation of the bomb doors and for recharging the brake accumulators.

A nitrogen system is provided for emergency undercarriage lowering.

Separate self-contained electro-hydraulic systems operate the powered flying control units and the windscreen wipers.

MAIN SYSTEM SUPPLIES

Reservoir

The main system contains 12 gallons of fluid of which 2.25 gallons are contained in a spherical tank set in the roof of the bomb bay, at mid-position on the port side. The combined filling of the main and EHPP reservoirs is through a combined charging point on the inboard wall of the starboard undercarriage bay. For replenishment the undercarriage must be down, bomb doors open, parking brake off, and the accumulators charged with nitrogen and hydraulic fluid. Indications of a full system are given by excess fluid spilling from the overflow adjacent to the charging point and correct level indications on the sight glasses of the main and EHPP reservoirs.

To ensure that a positive head of pressure is maintained at all altitudes, the reservoir is pressurised with air from Nos 1, 2 and 3 engines. A pressure reducing valve reduces the engine air pressure from 105 PSI to 15 to 18 PSI, while a blow-off valve opens at 22 to 27 PSI and closes at 16 PSI. Mod 2321 replaces the engine air pressure with nitrogen pressure from the radio pressurisation system.

Engine-Driven Pumps

Three engine-driven pumps, one on each of Nos 1, 2 and 3 engines, draw fluid from the reservoir via filters. The pumps incorporate an automatic cut-out and, when idling, circulate fluid back to the reservoir through the main return line. The pumps are low-pressure spur gear, high-pressure radial piston pumps.

From the pumps, fluid is delivered via non-return valves to the main galleries at a pressure of 3,600 to 4,250 PSI. In addition to supplying the various services, this pressure is used to charge the wheelbrakes' accumulators.

Test points for the suction, delivery and return lines are in the port main undercarriage wheelbay.

Hydraulic Pressure Gauge

A triple pressure gauge is on the pilots' centre instrument panel.

The left-hand arc shows the pressure in the main gallery, while the two right-hand arcs show the pressure in the two brake accumulators. Each needle is separately wired and fused and has its own pressure transmitter.

UNDERCARRIAGE SYSTEM

General

The undercarriage mainwheel units are four-wheel, eight-tyred bogies; the nosewheel unit is twin-tyred and steerable. When undercarriage retraction is selected, the bogies pivot, to lie parallel to the main oleos.

Hydraulic pressure operates the undercarriage doors, extension mechanism, bogie trimmers and down locks, through electrically controlled selector valves; sequencing of the operation is controlled by microswitches. Each main wheel is fitted with a hydraulically operated down lock; the nosewheel has a mechanically operated down lock.

All down locks are of the over-centre type and will remain in the locked position should hydraulic pressure be subsequently lost.

A close-up view of the massive mainwheel casting and bogie unit, and a glimpse up into the wheel well where the familiar bay roof light is clearly visible. The ground safety locking jack is fitted.

Undercarriage raising and lowering is controlled by an UP and a DOWN button on the pilots' centre panel.

Warning: To ensure that the electrical contacts are made when the landing gear selector is operated, the UP or DOWN button must be pressed fully in.

When the weight of the aircraft is on its wheels, a micro-switch on each bogie is held open and an interrupter pin behind the UP button prevents it from being pushed in. When the weight is off the wheels, the bogies trail and both micro-switches (in series) close. Power is then applied to a solenoid which withdraws the pin, allowing the UP button to be depressed.

This device may be overridden, however, by rotating the flange of the UP button slowly and gently clockwise, through approximately 60 degrees, at the same time exerting positive forward pressure.

Warning: In spite of this safety device, the UP button must always be regarded as operative, as the protective devices may not function.

The undercarriage position indicator is on the pilots' centre panel and indicates as follows:

All wheels up and doors locked closed: No lights

Wheels unlocked: Three red lights

Wheels locked down: Three green lights

A flag indicator, marked U/C, is incorporated in the co-pilot's ASI and shows if the speed is reduced below 160 knots when the undercarriage is not locked down. The absence of the indicator must not be taken as proof that the undercarriage has locked down.

Undercarriage Emergency Nitrogen System

The emergency nitrogen supply for the main and nosewheels is contained in two separate bottles in the nosewheel bay. The bottles are charged to a pressure of 3,000 PSI via charge points adjacent to two gauges on the starboard rear wall of the nosewheel bay. The two controlling valves are mechanically linked and are operated by a single handle on the right of the throttle quadrant. The handle is guarded and wire-locked.

When the handle is pulled to its full extent, nitrogen from the bottles passes to shuttle valves and jettison valves, expelling hydraulic fluid from the lines and allowing nitrogen to pass to the jacks. The undercarriage then lowers, regardless of the position of the normal selector.

Selection of emergency nitrogen also isolates the normal selector solenoids, so that the undercarriage cannot again be retracted once the emergency selector has been operated.

Nosewheel steering, however, is still available.

The alternator CSDU oil system and engine bay (zone 2B) ground cooling system is inoperative when the undercarriage has been lowered by emergency nitrogen. In these circumstances after landing, a maximum individual alternator load of 20 kW may be maintained for a maximum period of 15 minutes subject to CSDU oil temperature, and engine speeds must be kept at idling so far as is practicable.

NOSEWHEEL CENTRING AND STEERING

Nosewheel Steering

a. Nosewheel steering is hydraulically operated and electrically controlled. The nosewheel can move through 47.25 degrees on either side of centre and movement is controlled by a pushbutton on the pilots' control column and by movement of the rudder pedals.

b. With the steering pushbutton depressed, rudder pedal movement causes the nosewheel to move in the appropriate direction; the operation of a drum switch cuts off the electrical supply to the selector valve when the selected angle is achieved.

c. As the nosewheel leaves the ground, a microswitch de-energises the stop valve in the steering circuit. This allows a bypass valve to open, permitting flow from one side of the steering jack to the other. The centring jack is now the only unit exerting any force and the nosewheel is automatically centred.

d. A nosewheel STEERING EMERGENCY OVERRIDE EMERGENCY/NORMAL wire-locked switch is on panel 3P. If steering is not restored after landing, the switch can be put to EMERGENCY to override the microswitch referred to. If the switch is used on the ground, it must be set to NORMAL before the undercarriage is raised after take-off and returned to EMERGENCY when the undercarriage is locked down for landing.

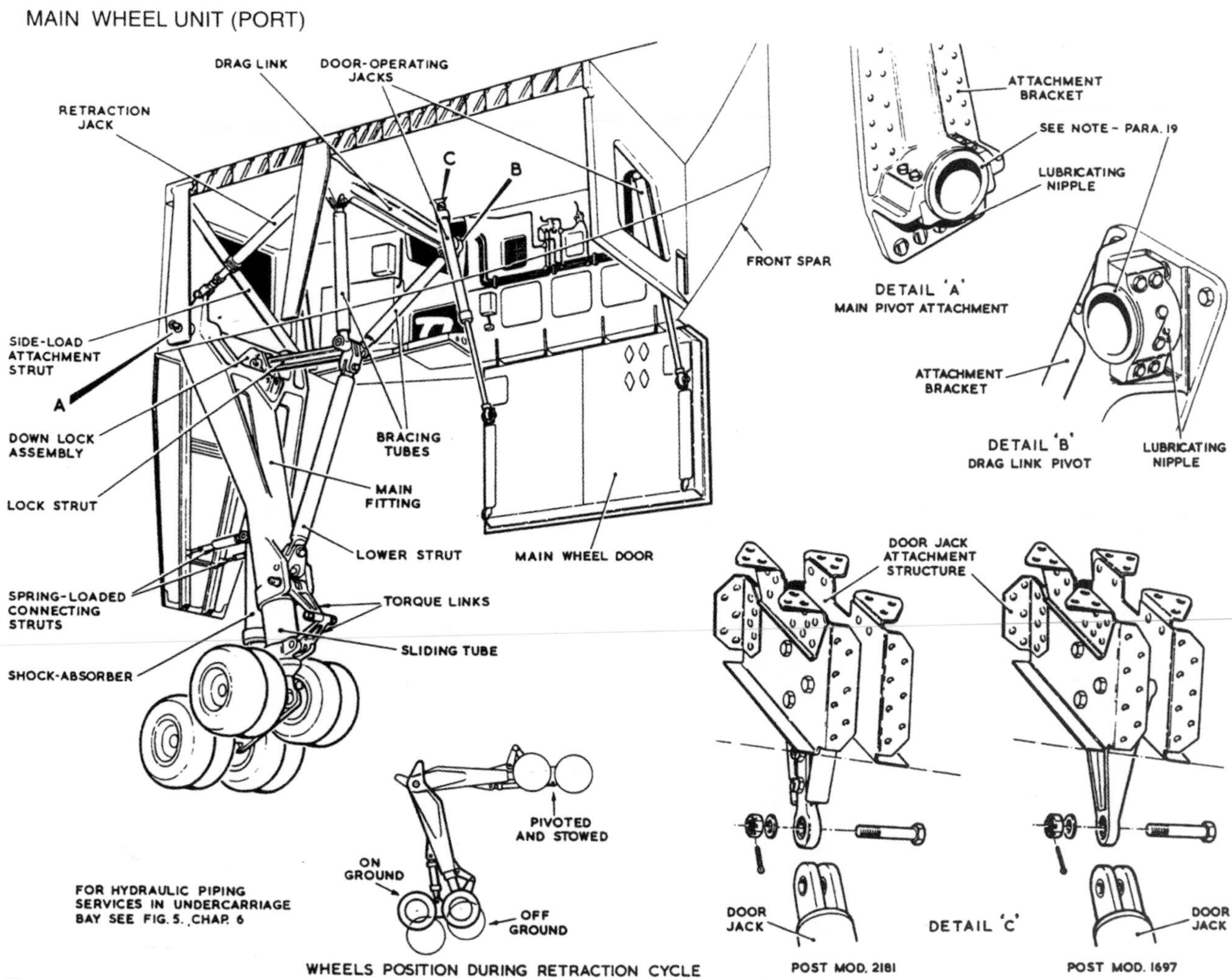

Landing gear diagram.

Nosewheel Centring

Nosewheel centring is hydraulically operated, the main delivery pressure passing through a pressure regulator valve direct to the centring jack. The centring system operates automatically when nosewheel steering is not in use, when the nosewheel is off the ground, or if the main hydraulic system fails.

WHEELBRAKES

General

A maxaret unit in the brake unit of each double-tyred wheel temporarily relieves pressure at the brake if a skid is detected on that wheel. The eight maxaret units operate independently of each other and only operate when the wheels are rotating. The following points must be remembered:

a. If brakes are applied before a wheel touches the ground, the wheel locks and the maxaret unit cannot operate.

b. To stop rotation of the wheels after take-off, it is necessary to apply brake pressure for at least 4 seconds.

c. When landing with a reduced amount of hydraulic fluid (after a line leak) maltreating must be avoided, as fluid under pressure would be bled rapidly to the return lines.

The brake units are hydraulically operated, the main system pressure being reduced to 2,500 PSI at the brakes. Two accumulators, charged to main line pressure, provide instant response and a reserve of pressure for brake operation. These accumulators can be recharged by the hydraulic power pack. A failure of one accumulator does not prevent the other from supplying pressure to both sets of wheels. A drop in nitrogen pressure in one or both accumulators would be disguised as long as the main hydraulic pressure remains normal.

The pressure at the brakes is shown on two dual pressure gauges in the nosewheel bay and the pressure at the hydraulic accumulators is shown on the triple pressure gauge on the pilots' centre panel. The accumulator nitrogen pressure gauges and charging points are also in the nosewheel bay, together with two manually operated pressure release valves, for releasing any residual hydraulic pressure in the accumulators.

A parking brake is provided, which operates through a Bowden cable to open simultaneously all the hydraulic valves in the brakes control valve.

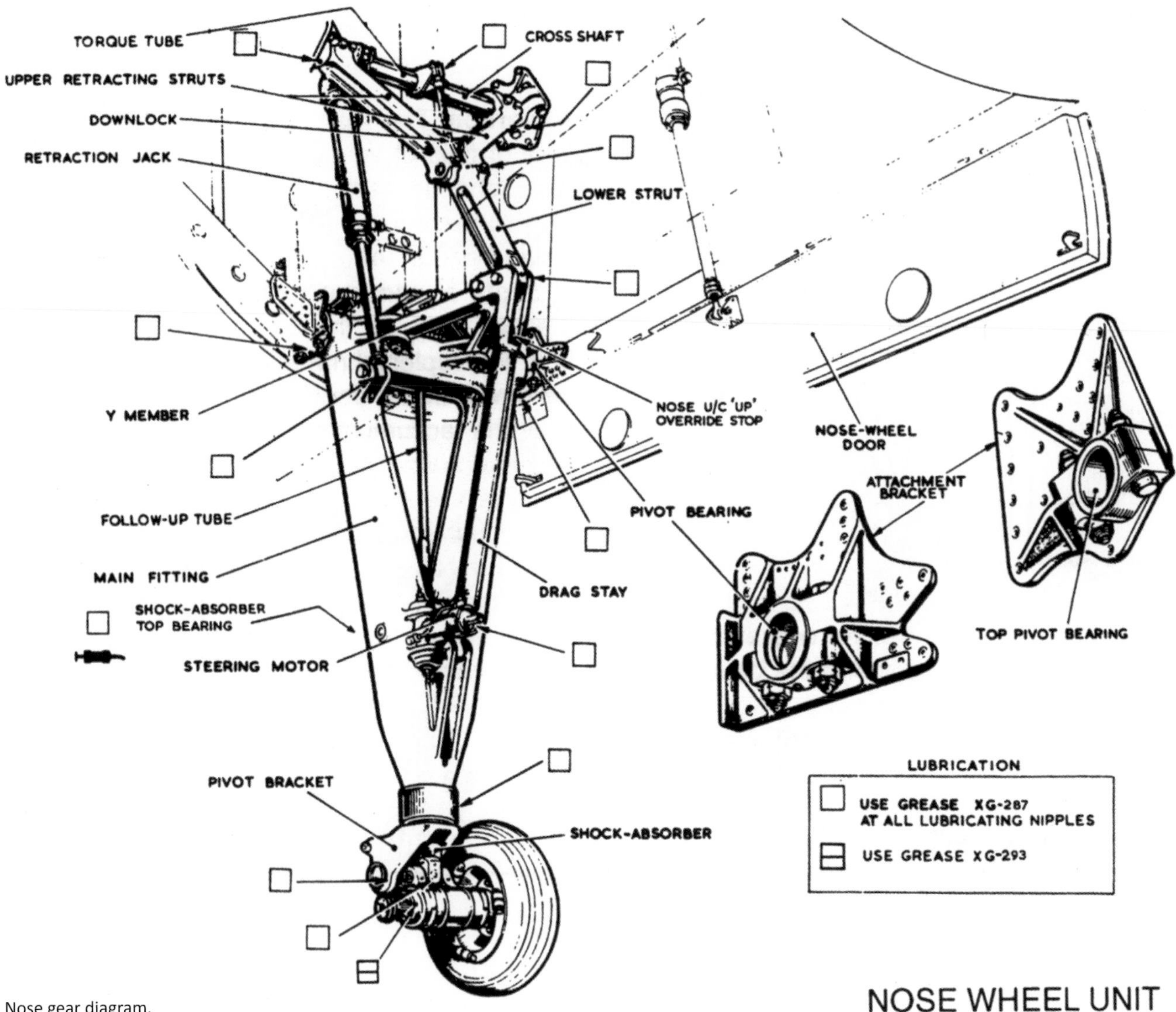

Nose gear diagram.

Operation

Brake selection is controlled by toe-buttons on the rudder pedals. The pressure delivered to the brakes is proportional to the force applied to the toe-buttons; when this pressure is released, the relay in the brakes' control valve closes and the fluid from the brakes is returned to the reservoir.

The parking brake is applied by turning and pulling the lever on the left of the throttle quadrant.

BOMB DOORS

Bomb Door Operation

The bomb doors are hydraulically operated. For normal operation, supplies from the main system are fed through dual selector valves to the four door jacks, via a shuttle valve which forms part of the door locking assembly. If the normal supply fails, the doors can be operated from the EHPP through the emergency selector valve, and the main system is isolated by movement of the shuttle valve in accordance with the power selection.

Controls and Operation

a. The 1st pilot has two switches on the pod console, labelled BOMB DOOR CONTROL NORMAL and EMERGENCY. The NORMAL switch is a rotary type with three positions: OPEN/AUTO/CLOSE.

When OPEN or CLOSE are selected, the bomb doors operate at the time of selection. When AUTO is selected the bomb doors are under the control of the NBS and do not open until they receive the appropriate signal.

b. The EMERGENCY switch is a guarded-double pole three-position switch, labelled OPEN/NORMAL/CLOSE. When this switch is operated, the doors are opened or closed by supplies from the hydraulic power pack and the electrical supplies are cut off from the normal selector.

This switch is inoperative if the power pack is being used to charge the brake accumulator.

c. The nav/radar has a single-pole BOMB DOORS OPEN/CLOSED switch. Operation of either this or the pilot's switch opens the doors but to close them both switches must be operated. The bomb doors can also be operated by the emergency bomb jettison switch.

Indicator

The bomb door indicator is to the right of the alternator warning light. Either a round 3-position or a square disposition magnetic indicator may be fitted. The 3-position magnetic indicator shows black when the bomb doors are closed and white at all other times.

The 3-position magnetic indicator shows black when the bomb doors are closed, white when they are fully open and cross-hatch when they are at an intermediate position or when there is no electrical supply.

AAPP SCOOP

Operation

The AAPP scoop is controlled by the AAPP master switch and operated by a spring-loaded jack. When the master switch is OFF, main line pressure reduced to 1,800 PSI closes the scoop and holds it closed against the spring pressure. When the master switch is selected to ON a solenoid-operated valve is de-energised, opening the hydraulic return line so that the spring pressure opens the scoop. If the 28v DC supply to the solenoid fails, the scoop opens. If the main hydraulic pressure reduces below 1,800 PSI, pressure to the jack is maintained by an NRV. Eventually (approximately 90 minutes), the hydraulic pressure decays and the scoop opens. The circuits are so arranged that the scoop does not close until the AAPP has run down.

PART 9: PNEUMATIC SYSTEMS

There are five separate pneumatic storage systems in the aircraft as follows:

a. Emergency door opening and canopy jettison.

b. Entrance door closing, door and canopy seal inflation and bomb aimer's window de-icing.

c. H2S scanner and NBS.

d. Undercarriage emergency lowering.

e. Engine starting.

The engines supply compressed air for the following systems:

a. Bomb bay seal inflation.

b. Hydraulic reservoir and power pack pressurisation (pre-mod 2321).

c. Fuel tank and recuperators pressurisation.

d. Equipment in rear fuselage.

e. Anti-icing.

Entrance Door System Supplies

Three storage cylinders, charged to 2,000 PSI from an external supply, are on the port side of the crew compartment. Their charging points and pressure gauges are on the front bulkhead in the nosewheel bay. The forward cylinder supplies pressure for emergency door opening and canopy jettison; the remaining two bottles supply door closing, door and canopy seal inflation, and pressurisation of the bomb aimer's window de-icing tank.

A ground servicing cock on the underside of the crew's floor is normally locked in the open position by a red cover. When turned off, it isolates the services supplied by the rear cylinder.

Emergency Door Opening

Pressure at 1,200 PSI for operating the door jacks is controlled by the EMERGENCY position of the door opening lever, on the forward end of the door frame on the port side. With the undercarriage down, the nitrogen passes through a restrictor, to control the rate of movement.

The rear crew switches allow nitrogen to pass to a jack. The jack rotates a camshaft which withdraws the bolts and moves the lower portion of the door handle to the emergency position. This operates the emergency door opening valve, allowing nitrogen to pass to the jacks. The cabin should be depressurised first.

Loads on the door bolts are such that, using the switches alone, the door cannot open until the differential pressure has dropped to 1.5 PSI (30 seconds at 43,000 feet, 9.5 seconds at 27,000 feet); if the manual control is used at the same time, the door opens at 2.65 PSI (20 seconds at 43,000 feet, 5.5 seconds at 27,000 feet). However, unless escape in the minimum time is essential, the manual control should not be used simultaneously with either of the switches.

When the door has been opened by either switch, the door opening lever is carried to the gate while mechanism below the lever moves further to operate the emergency door-opening valve. A 28 volt fault could result in the door closing again under slipstream pressure. To protect against this, the first man at the door must ensure that the lever is gated in the emergency position. The emergency door opening switch is on the vital busbar.

Note: A worn gate may allow the lever to pass to the emergency position.

If the door locking pins become scored, the door may fail to open when the navigator's switch is used. In this case, use the manual control; closing the cabin air switches may assist by reducing the time required to depressurise the cabin.

Canopy Jettison

The canopy jettison pneumatic valve is operated by pulling any one of the pilots' ejection seat firing handles. When the valve is operated, nitrogen at 1,200 PSI passes to a ram which opens the jaws of the canopy attachment and operates the jettison gun.

Door Closing

The door closing valve is operated either by a toggle handle or, externally, by a pushbutton near the door handle. Either selection must be held until the door is locked closed, otherwise nitrogen pressure is lost from the jacks. The toggle handle is stowed under a hinged panel on the starboard side of and below the centre crew seat.

When the valve is operated, nitrogen at 400 PSI is fed to the up side of the door jacks.

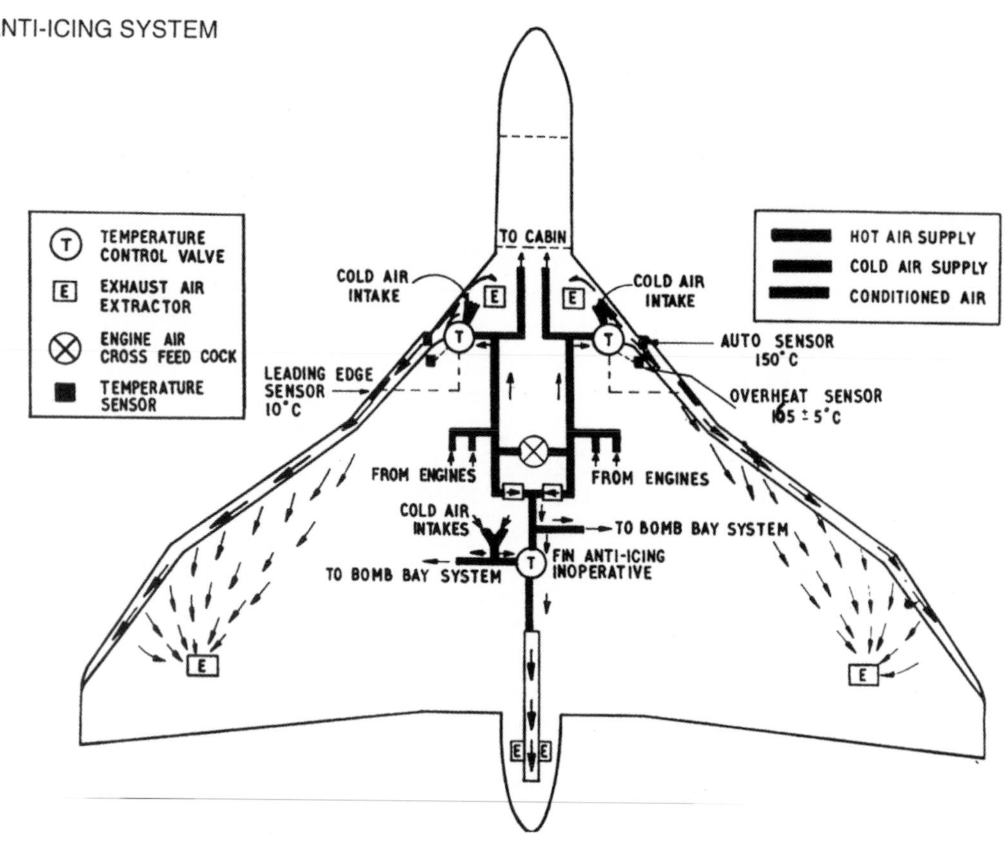

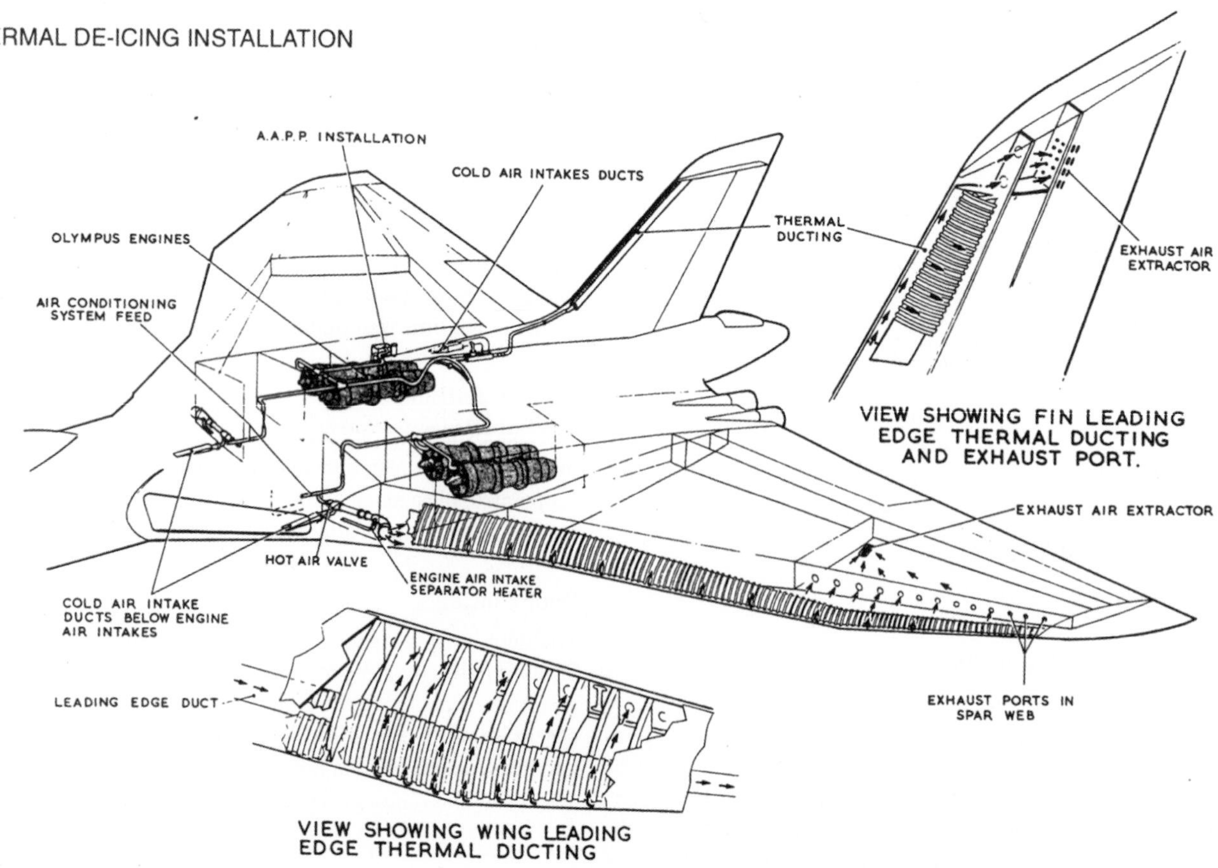

De-icing systems.

Door and Canopy Seals

The door and canopy seals are supplied with nitrogen from the rear cylinder at 25 PSI. The door seal is inflated automatically when the entrance door is closed. The canopy seal is permanently inflated to 25 PSI by a mushroom valve which is held open when the canopy is locked. The seal deflates when the canopy starts to move during jettison.

De-Icing Tank Pressurisation

The de-icing tank is pressurised with nitrogen at 14 to 15 PSI. A pressure-maintaining valve in the line cuts off the supply when the pressure in the main system falls to 150 PSI. The supply is also fed to the probe de-icing tank but this system is inoperative. The ground servicing cock must be closed before filling the tank and the residual pressure of 14 to 15 PSI must be relieved by pressing the pressure release valve on top of the tank.

PART 10: TERRAIN FOLLOWING RADAR

The TFR system is designed to enable the pilot to follow approximately the contours of the terrain at a height selected in flight. The components of the system are:

a. A radar pod in the nose (TFRU), containing a transmitter/receiver, an adjustable antenna, a pitch rate gyro and computing circuits. The pod is cooled by air from the H2S radome.

b. Airstream direction detectors (ADD). Two slotted probes, protruding horizontally from each side of the nose, measure the angle of attack.

c. An airspeed transducer, supplied from the aircraft starboard pitot-static system.

d. A control unit, on the 1st pilot's console.

e. Indicator and warning lights.

The system is used in conjunction with the following items of aircraft equipment:

a. MFS pod vertical gyro.

b. MFS pitch computer.

c. Glide path pointer of the director horizon.

d. Radio altimeter Mk 78.

DESCRIPTION

TFRU

The radar pod measures the slant range forward of the aircraft against a preset datum.

The antenna angle is adjusted by the height selector and the angle of attack measured by the ADD. Dive demands are initiated when range exceeds the datum and climb demands when the range is less than the datum.

Computers in the unit analyse and compare the various signals and pass them to the pilots' instruments as dive or climb signals.

Range Loop

The measured slant range is fed to a range computer which feeds a pitch rate demand signal, proportional to the range error, to the majority circuit.

Height Loop

Over flat terrain or calm water, radar returns may be insufficient to activate the range loop. To prevent this apparent loss of range initiating a dive command, height information from the Rad Alt 7B is modified by aircraft attitude and antenna angle signals to provide a pitch rate command to the majority circuit.

When intending to use the TFR below 500 feet the Rad Alt 7B should not be selected to the 500 feet scale until below 500 feet to avoid spurious climb demands. When the aircraft reaches the set TFR height, Rad Alt 7B can be selected to the 500 feet range to provide improved accuracy when the height loop is in operation.

Programme (Attitude Hold)

As the aircraft climbs towards the summit of a hill, the radar beam clears the crest. To prevent a dive signal being given before the aircraft is over the summit, a delay computer demands a steady attitude computed from the speed, pitch angle, angle of attack and last measured range. After this delay, other sub-systems take command and, in practice, the most positive pitch rate command is a minus 0.5g (applied) dive rate limit known as the pushover. A programme is not initiated if the signals increase progressively through the maximum range as opposed to disappearing instantly. This prevents inadvertent programme being caused by aircraft pitch-up.

Pitch Rate Control

Signals from the range loop, the programme, the dive rate limiter and the fail-safe system are fed into a majority circuit computer, which selects the highest climb demand of these signals. The majority circuit output goes to a comparator as a pitch rate demand for comparison with the output from the pitch rate gyro; any difference is the final pitch rate command fed to the MFS director horizons via the MFS pitch computer when the MFS/TFR switch is selected to TFR. Demands are shown on the pitch director (the GP pointer during non-TFR operations), and the pitch scale is in continuous fast-chase mode.

Climb High Function

If for any reason the aircraft pitch angle is greater than 40 nose-up, the antenna is off-set downwards, thus tending to shorten the range returns and causing the aircraft to fly higher, the effect increasing in proportion to the gradient. This prevents ballooning on the far side of a summit and gives a greater safety margin over steeply rising ground.

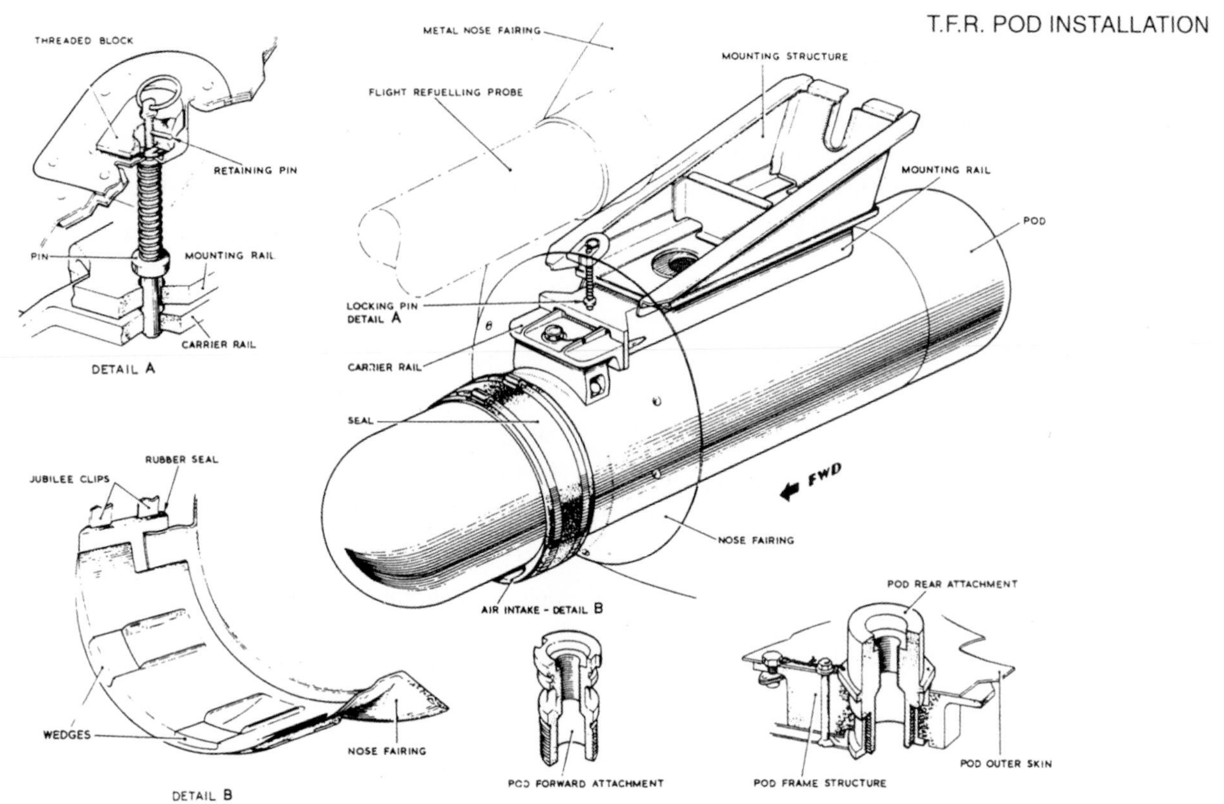

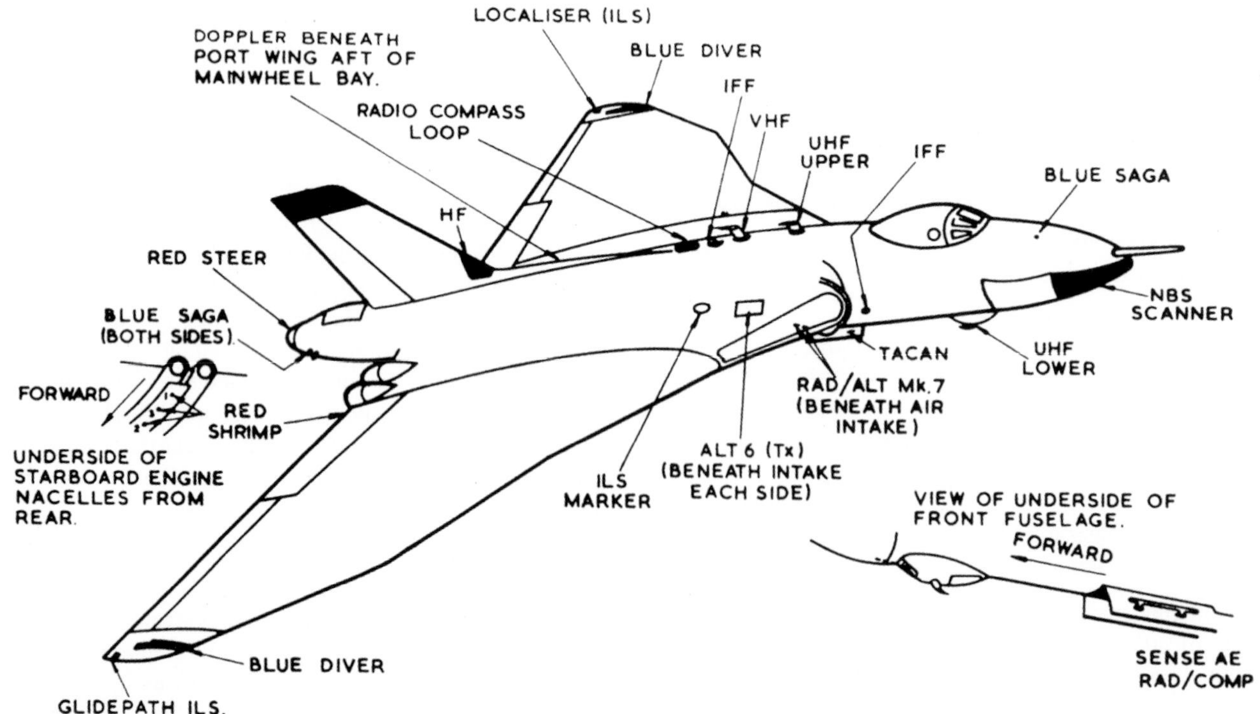

TFR and aerial placement diagrams.

Acceleration Limiters

To limit acceleration to between plus 0.75g and minus 0.5g (applied), airspeed signals are fed into two acceleration limiters. Signals from the limiters limit the pitch rate demands from the height loop to plus 0.75g (applied) and from the dive rate limiter to minus 0.5g (applied).

Safety Devices

Fourteen internal monitors continuously check range loop, antenna servo loop, pitch and pitch rate circuitry, as well as the monitoring system itself. If any monitor detects a failure the green light goes out, the red FAIL light comes on and a pitch up of 1.89 degrees/second is demanded. Those circuits which cannot be monitored are duplicated, both operating simultaneously and either one capable of operating the system. Failure of the radio altimeter activates the fail-safe circuit.

ADD

The ADD provide angle of attack information by rotating two slotted probes to maintain equal pressure either side of internally mounted vanes. This rotation is measured by a potentiometer and passed to the TFRU, where antenna angle is correspondingly adjusted. To compensate for airflow distortion the ADD outputs are paralleled and a mean signal taken.

The ADD probes are electrically heated via the pressure head heater switches.

A discrepancy between the two ADD brings on the ADD MONITOR amber light at the AEO's position and initiates TFR failure indications. If the fault has cleared the warnings may be cancelled; the AEO's by pressing the ADD monitor light and the pilots' by selecting RESET. A TEST/NORMAL switch on the ADD trim panel below the pilots' floor must be at NORMAL before flight. It should be noted that any yaw or sideslip may give an ADD warning and that the ADD warning is operative whether TFR is in the standby condition or on.

Electrical Supplies

The TFRU uses 200V, 3-phase, 400 Hz AC from No 4 busbar via panel 75P. The system also uses 28v DC from panel 4P. The ADD probes are heated by 28v DC, via the pressure-head heater switches. The warning lights use 28v DC from the TFRU while their press-to-test facility is supplied from the main aircraft system. The radio altimeter uses 115V AC from the port main transformer.

If No 1 or No 4 alternator fails or is switched off, the TFR may fail but can be reset immediately.

PART 11: AIRFRAME LIMITATIONS

The Vulcan B Mk 2 is designed for manoeuvres appropriate to the role of a medium bomber, in worldwide conditions. Aerobatics, stalling and spinning are prohibited. Speed must not be reduced below that for the onset of pre-stall buffet and in any case not below the threshold speed for the weight less 5 knots. In manoeuvres at altitude, acceleration should not be increased beyond that for the onset of buffet where this occurs before maximum g is attained.

Height Limitations

There is no height restriction on the aircraft because of airframe limitations. However, the maximum operating altitude is limited by the oxygen equipment.

Note: In the worst case, eg following loss of the canopy, or if the entrance door opens, aerodynamic suction can cause the cabin altitude to exceed the aircraft altitude by up to 5,000 feet.

Arresting Gear

The aircraft has unrestricted clearance to trample the following runway arresting gears, provided that the configuration has not been altered by modifications later than mod 2240:

a. SPRAG.

b. CHAG.

c. RHAG Mk 1

d. PUAG Mk 21

e. Bliss BAK 9, BAK 12, 500S.

Speed and mach Number Limitations

a. With all PFC working and all autostabilisers operative, maximum speed above 15,000 feet – 330 knots or 0.93M (0.92 with Mk 301 engines), whichever is less. (Elevator forces are not to be trimmed out above 0.90M.)

b. With one or more PFC inoperative – 0.90M.

c. With one or both servos of the mach trimmer inoperative – 0.90M unless specifically authorised. If specifically authorised, 0.93M (0.92M Mk 301 engines).

d. With one pitch damper inoperative – 0.93M (0.92, Mk 301 engines).

e. With two or more pitch dampers inoperative – 0.90M.

f. With feel relieved or failed – 250 knots or 0.90M, whichever is less. Extreme care is necessary to avoid overstressing the aircraft.

Maximum Speeds for Operation of the Services

The speed for operating a service also applies to flight with the service extended.

a. Airbrakes: No restriction.

b. Bomb doors: Up to the normal limiting speed of the aircraft.

c. Undercarriage: 270 knots (0.90M above 40,000 feet).

d. RAT: 330 knots or 0.93M (0.92, Mk 301 engines).

e. Tail parachute streaming: 145 knots (max). Any parachute streamed above 135 knots is to be examined before re-use.

f. The tail parachute must be jettisoned at speeds between 50 and 60 knots. In an emergency the parachute may be retained until the aircraft has stopped.

For Low Level Limitations, see Release to Service.

Crosswind Limitations

Maximum crosswind component for take-off, landing or streaming brake parachute: 20 knots.

Weight Limitations

Maximum for take-off and emergency landing: 204,000lb

Normal landing: 140,000lb

If, in emergency, the aircraft is landed at 195,000lb or more, the rate of descent at touchdown must be kept to a minimum and the angle of bank on the approach must not exceed 15 degrees.

Simulated asymmetric flying is not permitted at weights above 195,000lb.

Approach Limitations

The aircraft approach limitations are:

a. PRECISION RADAR 250 feet

ILS, Auto or Manual (in-line localiser) 250 feet

ILS, Auto or Manual (offset localiser) 270 feet

It is advisable that all ILS approaches should be radar monitored.

b. VISUAL COMMITTAL HEIGHTS (VCH)

One engine inoperative 150 feet

Two engines inoperative 200 feet

c. ENGINE OUT ALLOWANCE (EOA)

One engine out 0 feet

Two engines out (up to 185,000lb) 50 feet

Two engines out (above 185,000lb) 100 feet

Autopilot Limitations

The autopilot limitations are:

a. Speeds

Maximum airspeed 350 knots

Maximum mach number: mach trimmer operative 0.90M; mach trimmer inoperative 0.87M.

Maximum airspeed with TRACK and LOC & GP selected (ie feel partially relieved) 180 knots.

b. Minimum altitude (except during ILS approach) 1,500 feet AGL.

c. The artificial feel must be functioning correctly.

With the autopilot engaged, the following conditions must be observed:

a. Longitudinal trim is to be maintained so that the autopilot trim indicator is within the safe range.

b. One pilot is to be strapped into his seat at all times.

c. The autopilot may be used with the elevator channel disengaged (if operationally essential); in the ILS mode the elevator channel must be engaged. Neither the aileron, nor the rudder channel may be disengaged separately.

Air-to-Air Refuelling Limitations

Although the aircraft is cleared for air-to-air refuelling to the following standard, there is no operational requirement at present and therefore it is not to be practised.

The aircraft is cleared for air-to-air refuelling by day and by night using Victor tankers and by day only using Boeing KC-135 tankers, with Mk 8 equipment, subject to the following conditions:

a. Speed

The speed of the tanker at and during contact should be:

(1) KC-135: 260 to 275 knots (at heights up to 30,000 feet)

(2) Victor: maximum speed 300 knots up to 31,000 feet, 0.80M between 31,000 feet and 35,000 feet. Minimum speed 220 knots. Maximum altitude 35,000 feet

b. Airbrakes

Medium drag airframes are recommended for contact.

c. Fuel trim

The CG control switches may be used to maintain the fuel CG. Contact must be broken if either needle of the CG indicator goes into the red sector. The A and E tank gauges are monitored during refuelling to ensure that a spurious EMF has not opened the refuelling valves.

d. Night contacts (Victor)

Before flights involving night contacts, the probe lighting must be correctly focused on the forward third of the probe.

Only one transfer pump in the KC-135 is to be used.

All the necessary flight refuelling system modifications must be embodied in the aircraft, including the Mk 8 probe and probe lighting, and the modification which ensures the ground refuelling master switch is off.

Airframe Anti-icing Limitations

Subject to the embodiment of all necessary modifications, the wing anti-icing system may be used at all levels but the fin system must remain inoperative.

Bomb Bay Tanks

The bomb bay tanks may be used in the following configurations:

a. Two cylindrical tanks.

b. One cylindrical tank in forward position.

c. One A tank in forward position and one cylindrical tank in rear position.

d. One A tank in the forward position of the two forward locations.

e. One A tank forward and one E tank aft.

Bomb Bay Panniers

750lb and 4,000lb panniers may be carried with bomb bay fuel tanks

Opening of Bomb Doors in Flight

The bomb doors may be opened in flight:

a. When the tanks have been emptied of all usable fuel.

b. During an operational emergency with fuel in the tanks.

c. Provided no panniers are fitted in the bomb bay.

d. Provided that time with the bomb doors open is kept to a minimum.

The weapons simulator must not be used on the ground if there are any signs of fuel leaks and may not be switched on in the air until the tanks are empty of all usable fuel.

PART 13: MISCELLANEOUS LIMITATIONS

NAVIGATIONAL, OPERATIONAL AND RADIO LIMITATIONS

NBS Mk1A

The NBS Mk 1A is cleared for use as a navigational and bombing system.

Radio Compass

The radio compass is cleared for unrestricted use.

HRS

The HRS is cleared for use, subject to the following conditions:

a. During start-up, with fast erection, HRS is to be switched on at least 2.5 minutes before taxiing.

b. If airborne starting, the MRG is not to be switched on until initial take-off acceleration is over. The aircraft is to be kept on a steady heading for at least 2.5 minutes while the MRG runs up.

c. Unless it fails, the MRG is not to be switched off until the aircraft is stationary.

Decca Doppler

Decca doppler is cleared for unrestricted use. With the undercarriage down it will operate only in the memory mode. It is compatible with the autopilot when operating normally in the track steering mode. However, prior to doppler unlock spurious drift signals may cause the aircraft to take up a false heading.

ECM

The ECM installation is cleared for use up to 55,000 feet in temperate climates.

MFS

The MFS is cleared for Service use.

IFF

IFF/SSR is installed and is cleared for unrestricted use.

Tacan

Tacan is cleared for use.

Radio altimeter

The following limitations apply to the radio altimeter:

a. Mk 78: High range 100 to 5,000 feet, Low range 0 to 500 feet.

b. Mk 6A: Unrestricted use above 5,000 feet.

Radio communication

a. The PTR 1 75 is cleared for unrestricted use.

b. The ARC 52 is cleared for unrestricted use.

c. The HF/SSB is cleared for unrestricted use.

Electrical System Limitations

The maximum continuous load per alternator is 32 kW, subject to a maximum continuous CSDU oil temperature of 120 degrees C. This oil temperature must not be exceeded and, if necessary, height or loading must be reduced to keep the oil temperature within limits.

AAPP

a. Operating altitude: ground level to 30,000 feet, undercarriage up; ground level to 5,000 feet, undercarriage down.

b. There is no guarantee of a successful start above 30,000 feet. The alternator must never be put on load above 30,000 feet.

RAT

RAT Operating altitude: 20,000 to 60,000 feet.

Max load 17 kW.

Speed range: Maximum speed 0.93M, Minimum speed 0.85M/250 knots, whichever is greater.

Time limit: 10 minutes on load above 30,000 feet, 10 minutes off load above 50,000 feet.

PART 14: STARTING, TAXIING AND TAKE-OFF

Throughout, it must be remembered that the recommended limitations must be observed, and that the relevant checks in the Flight Reference Cards must be made at the appropriate limes.

The air supply required for starting main engines may be obtained from an external source (Palouste) or by crossfeed from a running engine. Alternatively, engines may be started individually or simultaneously using air from the rapid start compressed air installations. As any one, or a combination of more than one of these sources of starting air may be used to start the main engines, the starting order may be varied as required. However, the airflow patterns into the combined main engine air intakes make it advisable to use an outboard rather than an inboard engine to supply air for starting the remaining engines.

Starting a Main Engine using an External Air Supply

When using an external air supply, the engines may be started in any order. If it is intended to start the remaining engines using air supplied by the running engine, this must be No 1 or No 4 engine. After completing the relevant internal checks in the Flight Reference Cards, confirm with the crew chief that the external air supply is connected and that it is clear lo start engines.

The checks before starting are given in the Flight Reference Cards.

Press the starter button and check that the indicator light in the button comes on, showing that the air control valve has opened. Wait for 10 seconds, checking that the oil pressure and RPM are rising, then move the HP cock lever towards the idling gate until a fuel flow of 8 to 10lb/min is achieved. When the JPT rises pause for one second and then move the HP cock lever slowly to the idling gate. During starting, the IPT normally rises to 300C to 400C and then falls to approx 250C as the engine accelerates. If the rate of rise of the IPT indicates that 600C may be exceeded, meter the fuel supply by moving the throttle lever slightly behind the idling gate until the JPT decreases and then slowly advancing it to the normal idling position. If the JPT continues to rise rapidly above 600C and it appears likely that 700C will be exceeded, close the HP cock and isolate the starter motor by switching OFF the engine master switch. After a normal start, the RPM may stabilise below the correct ground idling speed; in order to achieve the correct idling speed, advance the throttle slightly and then return it to the idling gate. The starting cycle is terminated by the overspeed switch. The light in the starter button goes out when the air control valve has closed.

The checks during starting are given in the Flight Reference Cards.

Starting the Remaining Engines

a. The three remaining engines may be started individually and in the same manner using air supplied by the Palouste.

b. The Palouste may be removed and the three remaining engines may be started individually using air crossfeed from the running engine, provided that its speed is set to 70% RPM.

Starting the Remaining Engines Individually Using Air Crossfeed

The checks before starting are given in the Flight Reference Cards.

Set the RPM of the running engine to 70% and check that its engine air switch is open.

Open the engine air switch of the engine to be started, and then start it in the manner described. The AEO should check the voltage and frequency of each alternator and switch on as required, noting that when three alternators are on line the pilots' red ALT FAIL light is steady and when the fourth alternator is switched on the pilots' ALT FAIL light goes out.

If, for any reason, eg sandy airfield, confined space, it is desirable to restrict RPM, when two engines are running their speed may be set to 60% RPM to start the remaining engines as an alternative to one engine at 70% RPM.

Quick Starting of Engines

After starting No 1 or No 4 engine, the three remaining engines may be staged simultaneously using air from the running engine, provided that its speed is set to 93% RPM. Carry out the checks in the Flight Reference Cards.

Start the three remaining engines simultaneously, using the techniques described. As the fuel flow can only be monitored for one of the engines the HP cock levers for the other two must be moved forward carefully using the lever of the monitored engine as a guide. Each engine should start in the normal manner.

Because of the number of indicators to be watched, it is essential that both pilots monitor the fire warning lights and indications of oil pressure, RPM and JPT. If any engine malfunctions, close its HP cock independently of the other engines but leave the engine master switch ON until the starting cycles of the other engines are complete. When all engines are running satisfactorily, the AEO switches on alternators in the normal manner.

Rapid Starting of Engine

a. 200 volt AC and 28 volt DC electrical supplies are required while starting the engines and are normally obtained from an external source. Electrical supplies may be obtained from the AAPP when starting engines individually but this source must not be used when starting all engines simultaneously.

b. The rapid-start compressed air installation, when fully charged, should provide sufficient air for three individual engine starts on each side.

If the engines are started simultaneously, sufficient air should remain in the system for one individual start per side (minimum pressure 1100 PSI). Not more than two combustor starts on any one engine are to be attempted in any one 30-minute period.

c. Owing to the risk of high JPTs and the difficulty of monitoring and adjusting four engines at the same time, simultaneous rapid starting must not be attempted at temperatures below minus 15C. In such temperatures, No 1 or No 4 engines should be started by the combustor starter and the remaining engines started by using air crossfeed.

Simultaneous Rapid Starting

a. In order to gain full benefit from the rapid start installation, a complete combat readiness check should be carried out before engine starting. On completion of the combat readiness check, leave all systems selected as required for take-off. Carry out the checks in the Flight Reference Cards.

Note: The gyro hold-off button relay is only energised when RAPID is selected on the air selector switch and the master switch is OFF.

b. Before starting the engines, at least one booster pump per group must be on. To start the engines, move all throttles to the 50% RPM position, select the master switch ON and press the master RAPID start button.

Combustor ignition is indicated by each individual starter button light coming on. Engine light-up is indicated by rising jet pipe temperature after approximately five seconds. As each engine accelerates, its starter disengages, indicated by the starter button lights going out; on overspeed trip or on timer control after 12 seconds. During acceleration, check the indications of oil pressure, RPM, JPT and fire warning. The JPT should rise to between 400C and 550C in approximately 18 seconds and then begin to fall. When the IPT on any engine has stopped rising, wait a further two seconds and then close all throttles to the idling position; if the IPT on any engine reaches 600C, all throttles must be closed immediately to the idling position. In either event the throttles, having been closed, can be reset immediately to give the required RPM.

c. During the start, the alternators come on line as engine RPM increase and, in addition, the flying control motors start automatically, the rudder first, followed by the ailerons and elevators together. All flying control motors should be started within 13 to 15 seconds of pressing the engine start button. The MFS gyros also start automatically as soon as the master start switch is selected ON.

Individual Rapid Starting

a. The rapid start installation permits individual starting of engines, provided that the air temperature is above minus 26C, without the automatic starting of PFC, etc. One or more engines may be started as required by using the individual start buttons. It must be remembered that, if the remaining engines are to be started using air crossfeed, an outboard engine should be started first.

b. The checks for a single rapid start are given in the Flight Reference Cards.

c. The remaining engines may be started in a similar manner. Alternatively, they may be started by using air crossfeed from the running engine. Select NORMAL on the air selector switch and start.

Taxiing

Ensure that the parking brake is fully off before taxiing.

As visibility from the cockpit is restricted, it is advisable to inspect the area before entering the aircraft, especially if it is intended to taxi in confined spaces. Take particular note of objects likely to be blown by the jet efflux.

Before taxiing, the scanner must be stabilised or secured and the relevant checks in the Flight Reference Cards carried out.

The thrust required to overcome the inertia of the aircraft and tyre set varies with the AUW and the surface but large amounts of thrust are rarely needed. Once the aircraft is in motion, sufficient thrust for normal taxiing is obtained with all engines idling. At light weights, it is difficult to keep the speed down with all engines running. It is recommended, therefore, that on the completion of a sortie, the outboard engines are shut down to reduce the brake load.

As soon as convenient after moving, the brakes and nosewheel steering should be checked. To operate the nosewheel steering, press the selector button at the base of the control handle and offset the rudder by the required amount. Very little is achieved by using asymmetric thrust in turns. Differential braking, however, can be used to assist in tight turns but care should be taken not to turn too tightly by this method, otherwise the steering and centring jacks may be damaged. It is possible to complete a 180 degrees turn fairly comfortably on a 50 yard runway.

a. In emergency, it is possible to taxi using only differential braking to turn; the pilot should always be ready to steer by this method should the nosewheel steering fail.

b. Do not operate the bomb doors while taxiing, as the nosewheel steering becomes ineffective until their operation has ceased (approximately six seconds).

c. The brakes are very effective but it is possible to use them unevenly and thus to overheat one side by inadvertent differential braking, when taxiing or during the landing run.

Take-off

Complete the checks before take-off before entering the runway. Align the aircraft with the runway and, with the brakes applied, open the throttles to 80% RPM, making sure that the brakes hold. Check for significant discrepancies between individual engine indications. When the engines are stabilised, switch on airframe anti-icing if required (30 seconds maximum before take-off). Ensure that the parking brake is off, release the brakes and then open up the throttles to full thrust. If the brakes are released suddenly, the nose tends to rise but it is unlikely that the nosewheel will leave the runway. If, in emergency, the JPT limiter is set to OFF, the RPM must be restricted during take-off in order to avoid exceeding the JPT limits.

There is no tendency to swing and any small deviations in the early stages can be corrected by nosewheel steering. The rudder starts to become effective above 60 to 80 knots. Care must be taken when using either nosewheel steering or rudder to prevent overcontrolling in the early part of the take-off run. Acceleration is good, even at high weights, and is very rapid if full power is used at lighter weights (below 160,000lb AUW). Above 100 knots, nosewheel steering is almost ineffective unless weight is maintained on the nosewheel, therefore the stick should be held well forward of the central position if nosewheel steering is used continuously throughout the take-off run.

At the rotation speed, ease the control column back so that the aircraft becomes airborne. (As weight is reduced below 170,000lb, correspondingly less backward movement of the control column is required).

XM605, fuelled up, plugged in and ready to launch from Finningley's ORP during a Battle of Britain 'At Home' Day.

XM605, almost twenty years on, bears the late-1980s markings that contrast with the earlier late-1960s scheme shown in the previous picture. *Shaun Connor*

Apply the brakes for 4 seconds and select undercarriage up; allow the aircraft to accelerate to the initial climb speed as the undercarriage is retracting and continue to accelerate to climbing speed.

Rotation and Initial Climb Speeds (knots)

150,000 and below – rotation speed 135, initial climb speed 148

160,000 – 139-148

165,000 – 141-149

170,000 – 143-151

180,000 – 148-156

190,000 – 153-160

195,000 – 155-163

200,000 – 157-165

204,000 – 162-170

After Take-off

Keep slip and skid to a minimum while the undercarriage is travelling, in order to reduce stresses on the undercarriage door brackets. The undercarriage retracts in 9 to 10 seconds and no difficulty is experienced in achieving a clean aircraft before the undercarriage limiting speed of 270 knots is reached.

Whenever possible, the undercarriage should be completely retracted before exceeding 200 knots. There is no appreciable trim change during take-off or undercarriage retraction but, as speed increases, a steadily increasing push-force on the control column is necessary, because of the rapid increase in speed. This push-force can be trimmed out easily in increments as the aircraft is accelerated to its climbing speed. Make a visual inspection of the undercarriage after UP selection, using the periscope.

At a safe height, throttle the engines to 93% and select CRUISE on the TAKE-OFF/CRUISE selector. Carry out the after take-off checks as soon as possible after take-off. Engine RPM creep in the climb, and 93% must be maintained by use of the throttles up to FL 300. Above FL 300 set and maintain 95% until top of climb is reached.

95% is the maximum permitted RPM for day-to-day operation in order to prolong engine life.

Under operational conditions, or when specifically authorised, open the throttles fully and climb at maximum continuous power.

During take-off, yaw damper malfunction does not noticeably affect handling characteristics but, as speed increases to about 170 knots, the effects become apparent. Select the other channel immediately or, if both are defective, switch OFF.

Aborted Take-off Procedure

In all instances where the take-off run has to be aborted the following actions are to be taken:

a. Warn crew aborting.

b. Close the throttles.

c. Select airbrakes to HIGH DRAG.

d. Stream the tail brake parachute if speed is between 75 knots and 145 knots.

e. Apply maximum continuous braking at or below normal maximum braking speed.

f. Carry out engine fire/failure drills as appropriate.

g. Inform ATC 'Aborting'.

h. Clear runway if feasible.

i. Carry out brake fire drills before further taxiing.

After an aborted take-off using NMBS it may be possible to clear the runway. In which case the aircraft should be taxied slowly with minimum application of toe brakes; the parking brake should not be used to hold the aircraft stationary during checks or for parking.

Warning: If maximum continuous braking is used above NMBS, the aircraft should be stopped and evacuated as soon as practicable.

Because of the high risk of tyres bursting the aircraft should not be approached for at least 2 hours unless it is necessary to extinguish any fire.

Climbing

The recommended climb speed is 250 knots to 20,000 feet and then 300 knots up to a height where this speed coincides with 0.86M.

If the JPT limiters have been overridden because of unserviceability, the JPT must be watched very carefully if the limits are not to be exceeded during the climb.

Between 10,000 feet and 15,000 feet, the pressurisation failure warning horn may blow because the aircraft rate of climb is greater than the rate of increase of cabin pressure but, by 15,000 feet, the selected pressure should have been achieved. If cabin pressure surging occurs, check that the duct relief flap is in the closed position; if surging persists, close one of the in-use engine air switches and leave closed until the top of climb.

PART 15: HANDLING IN FLIGHT

Engine Handling

Engine life is affected by the frequency and extent of temperature changes caused by increasing and decreasing thrust; therefore, to prolong engine life and maintain performance, throttle movement should be smooth and changes in thrust as few as possible. Slam decelerations to IDLING may be made at all altitudes. With 301 engines, care should be taken to ensure that the inboard throttles are not inadvertently closed beyond the detent position when above 15,000 feet. Slam accelerations should be avoided but, in an emergency, can be made from any throttle setting at all altitudes. It is recommended that throttles should not be opened from IDLING to OPEN in less than 3 seconds.

Engine RPM

Thrust should not be reduced from TAKE-OFF by means of the TAKE-OFF/CRUISE selector switch alone. Before moving the switch to CRUISE, reduce the engine speed to 93% RPM. Then, after setting the switch to CRUISE maintain 93% up to FL300 then set 95% and maintain until top of climb is reached.

If maximum cruise thrust is required, open the throttles fully. Similarly, when increasing thrust from the CRUISE setting to take-off. Close the throttles until the RPM begin to fall before selecting the CRUISE/TAKE-OFF selector switch to TAKE-OFF. The throttles can then be opened fully.

The selector switch must be set to TAKE-OFF (200 series only) before entering the circuit, to ensure that full power is available for over-shoot.

JPT

The JPT limiters must always be switched ON, unless they are proved to be defective.

Each is capable of keeping the JPT within plus or minus 5C of the selected limitation, when controlling, but regular checks should be made for excessive temperatures, especially when taking off in high ambient temperatures, climbing and changing power at high altitudes. The JPT limiters do not prevent excessive turbine temperature during rapidly changing engine conditions with cowls selected but, at TAKE-OFF, their suppressed datum prevents overswing of turbine temperature during acceleration at all ambient conditions.

Under varying ambient conditions and with the JPT limiters working properly, each engine does not necessarily indicate the same RPM and JPT as the others. An engine may reach its governed RPM before reaching the JPT limit or vice versa. However, to be within the limits in steady conditions with all engines at full throttle, no engine should be more than 30C hotter than the mean of the others, or more then 2% RPM slower than the mean of the others.

Engine Idling Speeds

Idling speed varies with altitude and forward speed. The characteristics of the 200 series engines may be different on the incorporation of Mod 1785. They may result in undershoots of idling RPM after a deceleration, or a downward idling reset following a period of idling conditions, both of the order of 5% LP RPM. Engine response from these conditions will be normal in that movement of the throttle will give immediate engine acceleration, but the time to maximum RPM could be extended by as much as 1 second.

General Flying

The elevator stick force varies as the square of the speed but it is possible to overstress the aircraft during manoeuvres in the pitching plane at high airspeeds. At high mach numbers, the elevator stick forces become heavier. Longitudinal stability is noticeably reduced with the CG at the aft limit. For low-altitude training the CG should be kept forward of the mid-position to reduce the possibility of overstressing the aircraft in pitch.

The rudder forces are very heavy except in the low speed range when the use of rudder is necessary for correctly balanced turns.

The ailerons are very light and effective and some experience may be needed before a tendency to overcontrol at higher speeds can be avoided. The stick force is constant over the speed range but the maximum deflection is limited progressively from 150 knots onwards.

a. Coarse use of aileron produces large amounts of adverse yaw and, with the undercarriage down, the resulting sideslip can cause the load limits of the undercarriage doors to be exceeded.

b. Adverse yaw is most marked at 200 knots and heavy, co-ordinated rudder application is required to counteract it. When changing the direction of roll, large angles of sideslip are produced and co-ordination is particularly difficult.

During manoeuvres, the slip indicator tends to under-read due to the heavy damping of the ball.

Trimming

Care is needed to trim the aircraft accurately. The best results are achieved if small increments of trim are applied and time is given for them to take effect before any further adjustments are made.

Airbrake Characteristics

At high airspeeds at lower altitudes, the airframes are very effective and cause only mild buffet in the HIGH DRAG position, with a marked nose-down change of trim. At higher altitudes, close to the limiting mach number, the HIGH DRAG position produces marked buffet, accompanied by severe airframe vibration, as well as the nose-down change of trim. If the limiting mach number is exceeded, this vibration makes cockpit instruments unreadable. At low airspeeds, the airbrakes are much less effective but do assist during the approach and landing. The airframes take approximately 5 seconds to move from IN to MEDIUM DRAG and a further 2 seconds from MEDIUM DRAG to HIGH DRAG.

Bomb Doors

When bomb doors are opened at high airspeeds and mach numbers, moderate buffet occurs. Buffet is correspondingly less at lower airspeeds and mach numbers. With the bomb doors open, a slight increase in engine RPM is necessary to maintain a given airspeed in level flight.

If the aircraft is flown at altitude for more than 30 seconds with the bomb doors open, temperature effect on the control runs may result in gradual but noticeable changes in trim, particularly in the aileron sense.

High Speed Flight

Warning: When flying at high indicated airspeeds or mach numbers, frequent comparisons must be made between the readings of the 1st pilot's, co-pilot's and navigator's ASI and between the pilots' manometers.

At low level/high airspeeds: It is easy to exceed the speed limitations. When approaching the limiting speed, because of reducing longitudinal stability, pitch control becomes extremely sensitive and out-of-trim forces are very small with any change of speed. Extreme care is required to avoid overstressing the airframe.

At high level/high mach numbers:

a. Mach trimmer inoperative: Unless specifically authorised, the speed should be restricted to a maximum of 0.9M if the mach trimmer is inoperative. The aircraft readily accelerates to maximum mach number in level flight, except at high AUW when a slight descent is necessary. Very little change of trim occurs from 0.85M to 0.88M but, beyond this, a nose-down trim change develops which requires a substantial pull-force at 0.93M. At higher mach numbers, the nose-down trim change increases rapidly and there may be very little control left for recovery above 0.94M. If mach number increases further, the nose drops even with the stick fully back. If control is lost, recovery is effected by closing the throttles (leaving the airbrakes IN) and holding the stick fully back until, with decreasing height and attendant drop in mach number, control is regained. The trimmer can be used, if necessary, to relieve the pull-force but do not relieve the elevator feel; over-trimming must be avoided, otherwise there is a strong tightening of the pull-out as mach number is reduced. If the elevator feel does become inoperative, extreme care must be taken during the recovery phase to prevent serious overstressing of the airframe.

b. Mach trimmer operative: No difference in handling is noted from 0.85M until the mach trimmer comes into operation at about 0.87M. Beyond this point a steadily increasing nose-up change of trim develops (due to the progressive up elevator applied by the mach trimmer), which should not be trimmed out above 0.9M. The maximum permitted mach number is 0.93M (0.92M for 301 engines), at which speed a substantial push-force may be required, dependant on CG position. The mach trimmer still operates above 0.93M (0.92) until full extension of the servos occurs at about 0.95M, at which speed the elevators have reached the almost fully up position. The aircraft now behaves, and recovery from above the limiting mach number is effected in exactly the same manner as described. However, overcontrolling is more likely to occur. This is due to the extension and retraction of the mach trimmer with changing speed on recovery.

c. Engine handling: Disturbances in air intake flow during recovery from high-speed runs may lead to engine surge or flame-out particularly with 301 engines. If surge occurs, partially close the HP cock; if surge symptoms persist, close the HP cock fully and take normal cold relight action. If flame-out occurs, attempt a hot relight; if unsuccessful, close the HP cock and take cold relight action.

Pitch and Yaw Dampers Inoperative

Pitch damper: The loss of one channel of the pitch damper is almost unnoticeable and no restriction is imposed but speed must be restricted to 0.90M if more than one pitch damper channel becomes inoperative.

Yaw dampers: The loss of the yaw damper is most noticeable when accurate course keeping is necessary or during an instrument approach.

No limitation is imposed if the yaw dampers become inoperative.

Approach to the Stall

Stalling is not permitted.

Speed must not be reduced below the pre-stall buffet and in any case not below the threshold speed for the weight less 5 knots.

a. As speed is reduced, with airbrakes in and depending on weight, pre-stall buffet may be experienced at the corresponding threshold speed.

b. At high angles of attack the rudder is masked by the tailplane. This results in a reduced rudder response and larger than normal rudder movement is required to maintain directional control.

c. As speed is further reduced, a rate of sink develops which increases rapidly and may exceed 4,000ft/min.

d. At 115 knots all controls are light and effective but rudder response is greatly reduced. In the landing configuration with approach power, any buffet is masked by the airframes.

At speeds below 115 knots at aft CG and with the stick fully back, directional instability may occur, causing considerable yawing and rolling. If this is allowed to develop, it is impossible to control until the stick is pushed forward.

During recovery, the direction of yaw and roll reverse. Excessive height is lost before recovery is complete.

Stalling In Turns

It is not possible to stall this aircraft in turns unless either the g limitations are exceeded or airspeed is extremely low.

However, at high altitudes the loading in the turn should not be increased after the onset of the initial buffet.

Flight In Turbulent Air

Flight in turbulent air should be avoided but, if this is not possible, the speed should be maintained between 180 and 300 knots, preferably at 220 knots.

Flight with the Entrance Door Open

The entrance door may safely be opened in flight at speeds up to 220 knots and higher in emergency. Above 220 knots there is a danger of the forward bulkhead collapsing. Below 220 knots, the handling qualities are not materially affected. There is a slight nose-up change of trim.

The noise level is very high at all speeds and intercommunication is difficult; therefore it is advisable that the captain's orders and intentions are made clear before the door is opened.

PART 16: CIRCUIT AND LANDING PROCEDURES

Circuit Speeds

AUW (lb)	Pattern Speed (knots)	Approach Speed	Threshold Speed
120,000 & below	155	135	125
130,000	160	140	130
140,000	165	145	135
150,000	169	149	139
160,000	173	158	143
170,000	177	162	147
180,000	181	166	151
190,000	185	170	155
200,000	189	174	159
210,000	193	178	163

Landing

If it is necessary to land at an AUW greater than 140,000lb, a runway of 9,000 feet or more should be used. The tail brake parachute (TBC) may be streamed at 135 knots (145 knots maximum) and should be jettisoned between 50 and 60 knots. In emergency, it may be retained until the aircraft has stopped and then removed by ground servicing personnel. However, if it is essential that the TBC be removed before the aircraft has been finally shut down. The inboard engines may be opened up to approximately 40% and jettison selected, or the TBC jettisoned while the aircraft is taxiing (if appropriate). Jettisoning the TBC with the aircraft stationary and shut down is permissible when absolutely necessary but some minor damage to the rear fuselage may result.

a. Fly the circuit and approach at the speeds recommended for the weight. A safe margin for control of the aircraft is allowed with up to 30 degrees of bank angle at pattern speeds and 20 degree bank angle at approach speeds (150 at approach speed above 195,000lb). During the later stages of the approach (approx 200 feet), but not before decision height on an instrument approach, HIGH DRAG airframe may be selected and speed reduced so as to cross the threshold with power on at the recommended speed.

Maintain the correct approach speed by careful use of the throttles. At 195,000lb and above, the rate of descent at touchdown must be kept to a minimum.

b. When the mean surface wind strength is 25 knots or above, irrespective of direction, the threshold speed should be increased by one-third of the mean wind strength. If the threshold speed then exceeds the approach speed, the latter should be increased to equal the former.

Normal Landing

Aerodynamic braking may be used at all weights. After touchdown, when both bogies are firmly on the ground, raise the nose progressively as speed is reduced, until the control column is fully back. Because of the small ground clearance at the wingtips and the high angles of incidence associated with aerodynamic braking, any mishandling in the lateral sense may result in damage to the wingtips.

Bank angles in excess of 3.5 degrees are significant in this respect. Aerodynamic braking must not be continued below 85 knots if the headwind is greater than 25 knots, since there is a possibility of the tail being scraped. Two red lights, on the coaming in front of the 1st pilot, come on when the tail of the aircraft is too close to the runway. With headwind components of less than 25 knots, aerodynamic braking may be continued down to 80 knots. Lower the nosewheel onto the runway and apply the brakes as required.

Overweight Landings

At weights below 140,000lb, normal approach and landing techniques should be employed. At weights above 140,000lb the speed at touchdown may be higher than the normal maximum braking speed (NMBS) and the maximum permissible speed for streaming the TBC. In such cases hold the nose as high as possible after touchdown and use aerodynamic braking until the speed falls to 145 knots. At 145 knots stream the TBC and, when it deploys, lower the nose.

Start braking when the nosewheel is on the runway and the speed is at NMBS. Provided that the AEO has been warned, the outboard engines may be shut down. Under favourable conditions of runway length, headwind or AUW, the pilot may, at his discretion, delay streaming the TBC until 135 knots. (Any parachute streamed above 135 knots may be scrapped.)

Short Landing

Cross the runway threshold at the lowest safe height and at the calculated threshold speed. Provided that the speed is below 145 knots, stream the TBC as soon as the main wheels are on the runway. Use aerodynamic braking until the TBC has developed and the speed is 5 knots above NMBS, then lower the nose. When the nosewheel is on the runway and the speed is at NMBS apply maximum continuous braking.

Braking

Both front and rear wheels of the mainwheel units must be firmly on the ground before the wheelbrakes are applied, as the maxaret units do not operate until the wheels are rotating. As a safeguard against locking the wheels during a bounce after landing, the maxaret units remain inoperative for a few seconds. Apply brake pressure smoothly and progressively when the speed is below the braking speed for the weight. On dry surfaces, the maxaret units normally prevent the wheels from locking when excessive brake pressure is applied but, unless the shortest possible landing run is required, more gentle use of the brakes is recommended.

On wet surfaces, braking action may be severely reduced according to the degree of wetness of the surface. Use light intermittent brake application initially. As speed is reduced, brake application may be progressively increased and held continuously. If slip or skid is suspected, release the brakes momentarily and then re-apply them gradually.

A drastic reduction in braking action must be expected on flooded or icy surfaces; whenever possible, avoid these conditions. If a landing has to be made on a flooded or icy runway, the TBC is to be used, crosswind conditions permitting. If the use of the brake parachute is not possible, aim for a firm touchdown using aerodynamic braking down to at least 90 knots.

Braking action must not be taken until the nosewheel is firmly on the ground; care should be exercised during braking to avoid wheel locking.

To prevent possible damage, it is necessary to limit the speed at which continuous braking is applied. This speed is referred to as the Normal Maximum Braking Speed (NMBS). It is defined as the maximum airspeed from which maximum continuous braking can be applied and the aircraft brought to rest without damage to the brakes. However, heavy braking will markedly reduce brake effectiveness as the heat absorption limit of the brakes is approached; even moderate braking at light weight and slow speed can have the same effect if prolonged, eg, lengthy taxiing using brakes against power. Following any such cases of excessive braking, sufficient cooling time should be allowed to restore the brakes to full capacity.

Warning: If maximum continuous braking is used from NMBS or speeds approaching NMBS, to avoid subsequent damage to the brakes the aircraft should not normally be brought completely to rest. The aircraft should be taxied slowly to dispersal with the minimum use of brakes and parked without applying the parking brake. If it is necessary to stop before reaching the dispersal, the minimum application of toe brakes should be used; the parking brake should not be used.

The emergency maximum braking speed (EMBS) does not apply to landing. Tests have shown that the use of EMBS does not give a corresponding reduction in the landing run.

If the brake parachute fails to deploy, the appropriate NMBS must be used.

Brake Cooling Times

If an intermediate landing is made during a sortie, the following conditions are to be observed:

a. On the intermediate landing, brakes must not be applied above 80 knots.

b. Great care must be taken during taxiing to minimise any further heating of the brakes.

c. A minimum of 15 minutes is to elapse between the end of the landing roll and the start of the take-off run.

d. On the subsequent take-off, the undercarriage must not be retracted for a minimum period of 5 minutes after unstick.

e. If the subsequent take-off is aborted, the TBC must be streamed but the brakes should not be applied until the speed has fallen to the NMBS without parachute figure. If practicable brake application should be delayed until the speed is below 80 knots.

Landing with CG outside Recommended Range

The recommended CG index range for landing is +2 to -2 landing gear down. At low fuel states it may not be possible to maintain the CG within these limits. At any fuel state, if the non-essential loads are not reset after load shedding, the CG will move rapidly aft if the No 1 and 2 tank booster pumps are feeding. An aft movement of about 5 units of index can be experienced during a normal low-level instrument circuit. The tail proximity warning lights are inoperative with non-essential loads shed.

Landing with Aft CG

Landing with the CG at the aft limit presents no undue difficulty. On the approach to land, sufficient elevator control is available and speed stability is not markedly affected. When the CG index is aft of +4, it is recommended that the threshold speed is increased by 10 knots and that the aircraft is flown gently onto the runway rather than flared.

Aerodynamic braking should not be adopted but the TBC may be used as required.

Landing with Forward CG

When the CG index is forward of -4, it is recommended that the approach to land is slightly shallower than normal, and that the threshold speed is increased by 10 knots to improve elevon effectiveness for the flare. After touchdown it may not be possible to maintain aerodynamic braking attitude down to 80 knots.

Crosswind Landing

A crosswind landing, using the crab technique, presents no special difficulty in crosswind components up to the limitation of 20 knots. When yawing the aircraft into line with the runway prior to touchdown, there is a marked tendency for the into-wind wing to rise; this tendency may be countered by prompt application of aileron. Aerodynamic braking may be used within the limits imposed by the tendency of the into-wind wing to rise.

Lower the nose to allow braking below NMBS.

The braking parachute may be used after the nosewheel has been lowered; any swing which develops due to the use of the braking parachute is best controlled by using nosewheel steering and brakes. If any difficulty is experienced in maintaining control, jettison the parachute. Avoid landing on very wet or icy runways in strong crosswinds; if the crosswind component in these conditions exceeds 7 knots, it is inadvisable to stream the parachute.

Landing without using Airbrakes

When landing without airbrakes, use the normal procedure but a longer approach is advisable. To avoid high sink rates developing if the engines are throttled back to the slow response range, any necessary increase in power must be anticipated.

Overshooting

Overshooting from any height presents no difficulties. Engine response is rapid on the approach but is comparatively slow (approximately 5 seconds to maximum RPM in normal temperatures, up to 13 seconds in tropical temperatures) when going round again from the runway. Open the throttles as necessary and climb away. At a safe height, if leaving the circuit, complete the overshoot checks. Under normal conditions, an overshoot from ground level followed by a visual circuit and landing requires 1,200 to 1,500lb of fuel, and an overshoot from ground level followed by a low-level instrument approach and landing requires 2,000 to 2,500lb of fuel. At low AUW, the aircraft accelerates rapidly if full power is applied on overshoot. To avoid an extremely steep climb-away and to prolong engine life, it is recommended that power is restricted to 80% RPM.

Roller Landings

When making a roller landing, hold the nosewheel close to the runway.

Retract the airbrakes and open the throttles smoothly to a minimum of 80% RPM, being prepared for some difference in response from each engine. Avoid any tendency to overcontrol on the rudder. During acceleration, avoid a high nose-up attitude and any tendency to take off below the rotation speed (135 knots up to 150,000lb).

When making a roller landing after an asymmetric approach, lower the nosewheel onto the runway. Before the throttles are opened for take-off they must all be in the idling position; it is essential that RPM on all engines are equal.

PART 17: ASYMMETRIC FLYING

Engine Failure during Take-off

Single Engine Failure: Single engine failure on take-off presents no handling problems. However, in cases of mechanical failure it may result in double failure and should be treated as such.

Double Engine Failure

a. If the failure occurs before the decision speed is reached, abandon the take-off, using the appropriate FRC drill.

b. If failure occurs above the decision speed and the take-off is continued, maintain directional control, using rudder or a combination of rudder and nosewheel steering, until rotation speed is reached. Rotate the aircraft, aiming to reach the initial climb speed at 50 feet. Apply rudder to keep the aircraft straight and, on becoming airborne, apply up to 5 degrees of bank towards the live engines, accepting a slip indication of half a ball's width towards the live engines; the use of coarse aileron must be avoided, as the adverse yaw induced affects directional control. At rotation, be prepared for the aircraft to move sideways towards the dead engines.

Shutting Down or Failure of an Engine in Flight

If an engine fails or is being shutdown in flight, close the throttle to the HP COCK SHUT position, close the appropriate engine AIR switch, adjust the relevant booster pumps and crossfeed cocks as necessary and inform the AEC. The LP cock should be closed if the situation demands it.

If an engine fails in flight when it is unlikely to have suffered, and is not showing symptoms of, mechanical damage, and is windmilling at a steady and reasonable speed, an attempt may be made to relight it. During relighting, keep a careful watch on the engine indicators and, at any sign of malfunction or difficulty, shut the HP cock immediately; do not relight.

a. The buried wing root installation of the Olympus engines makes the adjacent engine of a pair liable to mechanical damage should one engine suffer structural failure. To minimise damage to the airframe, containment shields are fitted outboard of the outer engines to protect the wing, inboard of the inner engines to protect the bomb bay. They are not fitted between each engine and captains must always be aware of the possibility of damage occurring to the adjacent engine whenever fire or failure indications for one engine are received.

b. The following points should be considered whenever engine fire or failures occur in flight.

(1) FIRE/FAILURE indications for both engines: Both engines should be closed down and the appropriate fire/failure drills carried out.

(2) FIRE/FAILURE indications for one engine: The failed engine should be closed down and the appropriate fire/failure drills carried out.

Consideration should then be given to the following:

(a) If accompanied by obvious signs of engine structural failure (eg marked vibration, explosion or ruptured aircraft skin) and circumstances permit, the adjacent engine of the pair should be shut down as a precaution.

(b) If there are no obvious signs of structural failure, the other engine should be left at cruise power and monitored carefully for symptoms of malfunction. It is emphasised that there is no benefit to be gained in this situation from throttling the adjacent engine to flight idle.

Relighting in Flight

If two engines fail simultaneously, relight the outboard first.

Relighting is progressively more certain with reduction in altitude and should always be possible at heights below 35,000 feet, and at airspeeds of 0.9M or less if practicable (preferably below 200 knots). 'Hot relights' may be achieved at any height, provided that relight action is taken within a few seconds of flame-out; hot relights are not recommended if the cause of flame-out is not known, except in the case of a multi-engine flame-out.

a. If, during a cold relight, a rapid rise in JPT occurs, without an accompanying rise in RPM, follow the recommended procedure.

b. If no relight is obtained after 30 seconds, release the button and SHUT the HP cock. A further attempt can be made at a lower altitude after allowing the engine a minimum of 3 minutes to drain out.

c. Do not make more than three attempts to relight any one engine in the same sortie.

d. With Olympus 301 engines, severe buffeting can occur with an inboard engine windmilling. Relighting under such circumstances can be impossible. Buffeting can be reduced by decreasing airspeed and/or reducing RPM on the adjacent outboard engine.

e. Relighting is facilitated by throttling back to idling RPM the adjacent engine of the pair.

After the engine has been relit, inform the AEO.

Whenever an engine is relit in flight, a JPT/RPM comparison check between the engines must be carried out.

General

In asymmetric flight conditions, above 200 knots, there is little difficulty in controlling the aircraft, even with two engines failed on the same side. At and below 200 knots, care is required to co-ordinate aileron and rudder movements to reduce the effects of adverse yaw caused by aileron drag and to maintain accurate control of the aircraft.

If asymmetric, with bomb bay loads of 21,000lb and above, only gentle manoeuvres are to be carried out; rapid and/or full movement of the controls is not to be made.

When flying with one inboard engine windmilling, some airframe buffet may be experienced at speeds above 0.85M. In aircraft powered by Olympus Mk 200 series engines, this buffet is very pronounced at around 40,000 feet and enough to shake the instrument panels vigorously as speed is increased above 0.91 M. At altitudes around 50,000 feet the buffet is less marked and changes little with change of mach number. Aircraft powered by Olympus Mk 301 engines are affected in a similar manner, although the buffet is even more pronounced and becomes most severe if the adjacent outboard engine is set at high power.

When flying at 0.85M at 30,000 feet, the vibration becomes excessive if the RPM of the outboard engine are increased to 88%, while at 48,000 feet/0.85M severe vibration is felt with this engine set at 100% RPM. As height is increased above 55,000 feet, the vibration becomes much less marked and at all altitudes a setting of 80% in the outboard engines reduces the vibration to negligible proportions.

Asymmetric Landing and Overshoot

a. One engine inoperative: The technique and procedure used for approach using the three engines are the same as for a normal approach, except that the power settings are increased by approximately 3%. The visual committal height (VCH) for a three-engine approach is 150 feet.

b. Two engines on one side inoperative: The minimum approach speed during a two-engine asymmetric approach down to VCH must not be below the calculated approach speed for the weight or 145 knots, whichever is higher. High drag airframe must not be selected until a decision to land has been made at the VCH of 200 feet above touchdown.

If the thrust available is marginal, it is recommended that a landing without airbrakes is carried out.

When an asymmetric overshoot is made it must be initiated at or above the VCH and the approach speed. The following procedure will allow an asymmetric overshoot on two engines to be made at any AUW:

a. Level the wings.

b. Increase the power required, counteracting yaw with rudder.

c. Airbrakes IN.

d. Select landing gear up if required. To avoid undue stress on the landing gear door assemblies, only raise the landing gear if any of the following apply:

(1) Rear crew must escape.

(2) At weights above 140,000lb.

(3) Leaving the circuit.

(4) It is essential for safe control.

(5) The AAPP operation must be considered.

e. Allow the aircraft to accelerate to pattern speed before climbing away. Rudder must be applied to counteract the yaw and up to 5 degrees of bank may be applied towards the live engines; coarse use of aileron must be avoided.

At normal landing weights, the power should be increased initially to 93% RPM (full power can be used if required).

Warning: The throttles must be opened carefully so that increase in power is simultaneous with throttle movement. Too rapid throttle movement which entails a lag in power response must be avoided.

PART 18: AIR-TO-AIR REFUELLING (FROM VICTOR AIRCRAFT)

Initial Approach

Warning: When a red or green light, or no lights are showing on the hose drum unit, contacts must not be attempted. Contacts are only permissible if amber lights are showing. If, when in contact, a red light comes on or all lights go out, contact must be broken immediately.

The recommended tanker speed is 220 to 300 knots up to 31,000 feet and 0.8M above 31,000 feet.

The recommended relative closing speed is two to three knots (maximum of five knots); use this speed approaching from below and dead astern, keeping the signal lights in view at all times.

When waiting to start an approach, position the receiver behind and to the starboard of the tanker in case the hose becomes detached while being trailed or wound in.

Turbulence reduces the chances of successful contact. It is advisable to attempt contact at a different altitude, free from turbulence.

If severe turbulence is encountered while in contact, make an emergency break, as there is a danger of the nozzle breaking through hose whip.

When necessary use airbrakes and maintain power between 80 and 90% RPM, where the best throttle response occurs. Do not, however, operate the airbrakes while in contact.

If, during the initial approach, there is any unserviceability of the tanker drogue or floodlights, night refuelling is at the discretion of the receiver captain.

Prior to contact, the following radio and radar checks are to be made:

a. H2S off (NBS left on).

b. HF to standby or receive.

c. Active ECM to standby.

d. Warning receivers off.

e. IFF to standby.

Final Approach and Contact

Make the final approach from dead astern and below the drogue, so that the pilot is looking along the line of the hose. Set power to maintain the correct closing speed and, from about 40 feet, adjust speed by visual judgment rather than by reference to the ASI. Careful engine handling and accurate flying are required and overcontrolling must be avoided. To this end, it is important that the seat is adjusted to a comfortable position so that the pilot does not have to lean forward to get an adequate view of the probe. When about 5 to 10 feet short of the drogue, a moderate buffet is felt; at this point a small increase of power may be needed to maintain the closing speed. As the probe enters the drogue, mild buffeting is experienced, accompanied by considerable noise.

When buffeting is experienced, considerable fluctuation of the pressure instruments may occur and the main and feel button warning lights may flash.

Once contact has been made and the probe is positively coupled to the drogue, fly the aircraft gradually up the line of the hose until the refuelling position is reached; keep the curve of the hose concave to the receiver. A slight reduction of power is then needed to maintain the refuelling position.

The recommended refuelling position is achieved when the forward end of a 10 foot long yellow band on the hose is just entering the serving carriage of the HDU. Seven feet of the hose must be wound in before the tanker fuel valve opens. When the valve is open, the tanker lights change from amber to green and the fuel gallery pressure gauge in the receiver starts to indicate. Continue the approach until the recommended position is reached.

The line of trail of the hose makes it difficult to see the yellow band unless it has been freshly painted; the correct refuelling position is reached when the hose is wound in until the serving carriage of the HDU is one-third of its travel from the right-hand side.

In Contact

Once in contact, make small control movements to hold the correct station, dead astern of the tanker, with the yellow hose markings showing and the signal lights visible; guard against any tendency to over-correct.

Avoid carrying the hose excessively downwards or sideways or probe damage may occur. It is difficult to achieve a permanent in-trim condition. Coarse throttle movement may be necessary to hold station but, normally, make small movements only. With the throttle friction damper fully off, throttle movement is comfortable; always ensure that the throttles are moved together.

XJ825 on an early post-conversion flight after having been modified for tanker operations. The underwing markings were varied quite considerably until a definitive tanker layout was established, comprising a gloss white background with black alignment markings, outlined in dayglow orange.

Breaking Contact: Normal procedure

a. To break contact, reduce power slightly and allow the aircraft to fall astern gradually. Hose unwinding should be controlled at a slow rate by throttle movement. When the last 7 feet of the hose is coming off the drum, the signal lights change to amber if the tanker valve has not already been closed. Aim to break contact with the drogue in its natural position so that it can be watched as it draws away. If contact is broken in any other position, the drogue oscillates over a wide area about its normal position.

b. To ensure a positive break and to prevent the receiver being struck by an oscillating drogue, when the last few feet of the hose (marked with red and white stripes) are being unwound close the throttles to the idling gate.

This ensures a swift deceleration once contact has been broken but (provided that it is not done too soon) should not cause the hose drum brake unit to operate.

c. If the red and white stripes are difficult to see, it should be noted that the serving carriage of the HDU is at the left-hand end of its travel when the last few feet of the hose are being unwound.

Emergency Procedure

If a red light comes on, or if all lights go out, or if it is necessary to break contact quickly for any other reason, close the throttles fully and select airbrakes out to ensure that the deceleration rate from the hose is sufficient to reach a speed of 5ft/sec, when its brake is automatically applied and contact is broken. This method should only be used in emergency conditions or for training purposes, as it throws a heavy load on the hose drum unit.

Clearing the Tanker

When contact is broken, some fuel splash occurs. This should normally be no more than a momentary mist but may be greater following an emergency break. Clear the tanker downwards and to starboard so that the tanker can be kept in view all the time. The H2S is to be left off until 10 minutes after contact has been broken.

Incorrect Contact

If the probe misses the drogue, close the throttles and withdraw to a safe distance along the approach path as the aircraft decelerates.

If the probe hits the outer rim of the drogue, the hose may wind in. If this occurs, withdraw behind and to starboard of the tanker while the hose is retrailed.

If the probe penetrates the canopy or spokes of the drogue, withdraw along the approach path to break contact with the drogue in the natural position; if necessary, wait for the hose to be retrailed.

If the probe appears to enter the drogue but fuel does not flow, a soft contact may have occurred because the closing speed was too low in the final stages of the approach; the hose may wind in. Withdraw and, if necessary, wait for the hose to be retrained.

If speed is too high at contact, the hose loops upwards and the nozzle weak link breaks. A fast approach must be recognised at an early stage and, unless speed can be reduced before contact, the approach must be discontinued. Contact should not be made with the throttles at idling RPM, as no power control is then available for decelerating on contact.

a. If the nozzle is broken through mishandling, the windscreen becomes completely obscured by fuel from the probe body for a few minutes.

The pilot of the receiver aircraft must clear well to starboard and astern of the tanker immediately. Loss of the probe nozzle does not affect aircraft handling but speeds in excess of normal cruising speeds should be avoided.

b. In some cases the fuel splash may be sufficient to flame-out an engine. If this occurs, the alternator of the affected engine must be switched off, to reduce the fire risk, and the cabin air isolation cock must be closed to prevent fuel contamination of the cabin air.

Exceptionally, the engine or airframe may have been damaged by debris from the broken probe and consideration should be given to this before an attempt is made to relight the engine.

Unless an immediate relight is necessary for safety reasons, allow a 1-minute period to elapse before relighting, so that the windmilling engine can scavenge the excess fuel. After relighting, the alternator must be left off for a further 5 minutes, to ensure that its cooling system is scavenged of fuel.

PART 19: EMERGENCIES

CRASH LANDING

The following considerations are recommended if a crash landing becomes necessary:

a. Before landing.

(1) Reduce weight as much as is practicable.

(2) Have the nav/radar make the ejection seats safe. The pip-pin for the canopy jettison gun must not be removed.

(3) If crash landing on an airfield, request foam on the runway as early as possible.

(4) If, in the opinion of the captain, there may be a danger of the navigators and AEO being trapped in the aircraft after the landing, they should be ordered to abandon the aircraft.

Ensure that the undercarriage is in the up position.

(5) A check should be made for obstructions, bearing in mind the direction of expected swing, ie one mainwheel up condition.

(6) If crash vehicles are available, check their positions.

(7) If possible, check that a long ladder is available to expedite the crew's escape.

(8) Ensure that all unnecessary navigational and electrical equipment is switched off.

(9) Uncover the rear cabin windows prior to crash landing.

b. Approach

(1) Make a normal approach with the undercarriage up or down according to circumstances.

The advantages of reducing impact load with the undercarriage down, however, should be carefully considered.

(2) Jettison the canopy and close the HP cocks just before touchdown.

LANDING WITH UNDERCARRIAGE IN ABNORMAL POSITIONS

General

a. If, after using the emergency system, only one leg is lowered, it is recommended that the aircraft is abandoned. In other cases, if a landing is considered feasible, then the general principle is that all crew stay with the aircraft.

b. Where practicable, make the landing at an airfield equipped with foam-laying apparatus.

When landing with one main unit unlocked, the foam strip should be laid along the side of the runway that the wingtip is expected to strike.

The foam acts as a lubricant and so delays the start of the ground loop, which imposes a heavy strain upon the undercarriage.

c. The possibility of major damage is also reduced if, after touchdown, the unsupported wing or nose is lowered at a controlled rate while the flying controls are still effective, rather than be allowed to drop on to the runway.

Landing Techniques

a. Belly landing: If, after the use of the emergency system, all units of the undercarriage remain retracted, it is recommended that the aircraft be belly-landed, as follows:

(1) Reduce weight as much as practicable and switch off all unnecessary equipment.

(2) Have the nav/radar make the ejection seats safe. The pip-pin for the canopy jettison gun must not be removed.

(3) Ensure that the bomb doors and entrance door are closed.

(4) Ensure all loose objects are stowed, that all crew have their harnesses tight and locked, protective helmets on, with 100% and emergency oxygen selected and flowing correctly.

(5) Make a normal approach.

(6) Jettison the canopy while still on the approach.

(7) Make a normal landing, keeping the wings level and the rate of descent to a minimum.

(8) Close the HP cocks. As soon as possible, operate the fire extinguishers and switch off all electrics.

b. Nosewheel up, both mainwheels down.

(1) Move the CG as far aft as possible, within permitted limits of the available fuel.

(2) Carry out a low overshoot, to check wind conditions.

(3) Ensure all loose objects are stowed, that all crew have their harnesses tight and locked, protective helmets on, with 100% and emergency oxygen selected and flowing correctly.

(4) Carry out a normal circuit. Open the entrance door. Jettison the canopy on completion of the final turn. Switch on the landing lamps at night.

(5) Touch down normally at the correct speed.

(6) When firmly on the main wheels, stream the tail parachute (crosswind permitting) and cut the outboard engines.

(7) AEO to switch off Nos 1 and 4 alternators.

(8) Hold the nose up until speed drops to 80 knots, runway length permitting.

(9) While elevator control is still available, lower the nose on to the ground.

(10) As soon as the nose touches, cut the remaining engines; the co-pilot switches off all LP cocks, the AEO switches OFF all alternators and operates the battery isolating switch.

(11) When the nose is firmly on the ground, apply the brakes gently and evenly.

(12) When the aircraft stops, the co-pilot leaves first, followed by the nav/radar, AEO, nav/plotter and 1st pilot in that order. If 6th and 7th seat members are carried they should leave the aircraft after the co-pilot in the order 6th, then 7th.

c. Nosewheel down, one mainwheel down.

(1) Move the CG as far aft and as far away from the failed mainwheel as possible.

(2) Carry out at least one low overshoot, to check wind conditions.

(3) Ensure all loose objects are stowed, that all crew have their harnesses tight and locked, protective helmets on, with 100% and emergency oxygen selected and flowing correctly.

(4) Carry out a normal circuit and jettison the canopy on completion of the final turn. Switch on the landing lamps at night.

(5) Touch down normally at the correct speed.

(6) On touchdown, cut the outer engines and lower the nosewheel onto the ground.

(7) AEO switches off Nos 1 and 4 alternators.

(8) Hold the wing up, using aileron, rudder and nosewheel steering.

(9) Before control effectiveness is lost, lower the wing and cut the remaining engines; hold the aircraft straight for as long as possible. The pilot switches off the LP cocks; the AEO switches off all the alternators and operates the battery isolating switch.

(10) When the aircraft stops, the co-pilot leaves first followed by the nav/radar, AEO, nav/plotter, and 1st pilot

in that order. If 6th and 7th seat members are carried they should leave the aircraft after the co-pilot in the order 6th, then 7th.

d. Nosewheel down, both mainwheels up: In these circumstances, it is recommended that the aircraft be abandoned, as it is considered that the hazards for the rear crew escaping past the nosewheel are less than the danger to the whole crew of the nose section breaking off and the main fuselage overrunning the cabin.

If, for any reason, a landing is imperative, the following technique is recommended:

(1) Reduce weight to the minimum practicable.

(2) Insert the pins in the ejection seats but not in the canopy jettison gun. Switch off all unnecessary equipment. Ensure all loose objects are stowed, that all crew have their harnesses tight and locked, protective helmets on, with 100% and emergency oxygen selected and flowing correctly.

(3) Make a normal approach. Jettison the canopy at the end of the final turn. Switch on the landing lamps at night.

(4) Make a normal landing, keeping the wings level; avoid a high nose-up attitude and land with minimum drift. Do not stream the brake parachute.

(5) As soon as the nosewheel drops to the ground, cut all engines and switch off all services.

ABANDONING THE AIRCRAFT

General

a. Ejections may be initiated in straight and level flight, at any height from ground level upwards. However, runway ejections should only be made when the speed of the aircraft is above 90 knots. At low altitude the aircraft should be straight and level or climbing; any significant rate of descent or nose-up attitude at the instant of ejection reduces the seat performance.

b. Rear crew members can leave the aircraft down to a minimum height of 250 feet at a maximum speed of 250 knots. Whenever possible, speed should be reduced to 200 knots and the aircraft be in a shallow climb with undercarriage raised, prior to escape.

To minimise the possibility of injury by air blast or by loss of equipment, it is recommended that, if circumstances permit, speed is reduced as much as possible before attempting to escape. Escape is made easier if no personal survival pack is worn.

The following amplifies the drills in the Flight Reference Cards:

a. Escape for Rear Crew Members – General.

(1) Whenever possible, altitude should be reduced to below 40,000 feet. Above 40,000 feet the aircraft should only be abandoned in an extreme emergency. Crew members should initiate the demand emergency oxygen set and then disconnect from the aircraft system. Post-Mod 2393 a combined static line, oxygen hose and mantel release coupling is fitted for rear crew members. Speed should be reduced as much as possible. It is most important that the exit is made by sliding cleanly down the door, in a bunched-up attitude. Ground tests also show that rear crew members can clear an extended nosewheel on escape at speeds up to 180 knots.

(2) Move the abandon aircraft switch rearwards to the EMERGENCY position and confirm on the intercom. Post-SEM 027, operation of the abandon aircraft switch automatically illuminates the cabin light.

(3) Nav/plotter and AEO operate the door opening switches. Above 20,000 feet, the normal door opening lever should only be operated in addition to and simultaneously with the switches when escape in the minimum time is essential. Below 20,000 feet, the switches and door opening lever may both be used.

Whichever method has been used to open the door, the first rear crew member to reach the door should ensure the door opening lever is in the gated EMERGENCY position. Ensure that static lines are connected.

(4) When giving the order to abandon the aircraft, the pilot should normally indicate to the rear crew members that the static lines are to be used. However, below 1,000 feet and 200 knots he should order the manual overrides to be used.

(5) The nav/radar's last action before sliding down the door must be to ensure that his oxygen hose passes behind his PSP.

(6) Navigators and AEO leave the aircraft in the order, nav/radar, AEO and nav/plotter. If an experienced 6th seat crew member is carried he will be first to leave the aircraft. In the case of an inexperienced 6th member he will leave after the nav/radar. When an inexperienced 7th member is carried, the 6th member must be an experienced aircrew member or a crew chief.

The 7th member is not to be given any task other than to leave the aircraft when instructed.

The order of abandoning will be 6th, 7th, nav/radar, AEO and nav/plotter. The co-pilot, if possible, should watch the rear crew members leave the aircraft and inform the 1st pilot when the nav/plotter has left. The 'crew' gone lights indicate to the pilot that the rear crew members have left the aircraft.

b. Escape for Rear Crew Members – Undercarriage Raised.

(1) Sit on the floor at the front end of the door aperture facing aft.

(2) Grasping the handle at the bottom of the centre seat, swing forward onto the door and slide down it. At speeds above 200 knots it is advisable to adopt and hold a bunched-up attitude to minimise the possibility of injury from limb flailing. Below 200 knots an extended attitude with the legs straight out and rigid probably gives an easier exit. An upward pull with the arms is necessary to ensure that the PSP is lifted clear of the door edge.

c. Escape for Rear Crew Members – Undercarriage Lowered.

If the undercarriage cannot be raised, the following technique is recommended:

(1) Grasping the handle at the bottom of the centre seat, swing the legs onto the door facing aft. Slide down the door with the legs apart until the feet can be braced against the door-operating jacks. An upward pull with the arms is necessary to ensure that the PSP is lifted clear of the door edge.

(2) Releasing the grip, lean forward with bent knees and grasp the right-hand (port) jack with both hands, as low as possible, thumbs uppermost, right hand on top.

(3) Withdrawing both feet inwards from the jacks, keeping the knees bent and the body close to the port jack, swing down and round the port jack and over the port side at the bottom of the door. Release the hold on the jack as the body swings completely clear. Try to maintain a compact position with the arms close to the body after letting go. Keeping close to the port jack decreases the risk of the PSP fouling the starboard jack.

d. Escape for rear crew members at low altitudes.

Note: Whenever feasible, convert speed to height. If it is necessary to abandon the air-craft at very low altitude (below 1,000 feet), reduction of the time interval between the moment at which the order to abandon aircraft is given and the moment at which the parachute deploys can be of overriding importance and the following points should be borne in mind:

(1) The time taken to open the door can be reduced to a minimum by operating either rear crew switch immediately then, if necessary, operating the manual door control.

(2) The static line arms a barostat, which then withdraws the parachute pack pins after a delay of 2 seconds. Therefore, whether a static line is connected or not, the parachute release handle should be pulled as soon as possible after clearing the door.

The pilots should escape, using their ejection seats, after the rear crew members have escaped.

If the ejection seat automatic system fails after ejection, proceed as follows:

a. When forward speed is sufficiently low, discard the face screen.

b. Pull the manual separation lever outwards and then up.

c. Fall clear of the seat and pull the rip-cord handle.

Ditching

Model tests indicate that the ditching qualities of the aircraft are good, there being no tendency to nose under after impact.

a. The following considerations and actions amplify the drill given in the FRC:

(1) Assessment of sea state: Whenever possible, fly low over the water and study its surface before ditching. It is important to establish correctly the direction of the swell and of the surface wind.

(2) Direction of approach: The aircraft should always be ditched into wind if the surface of the water is smooth or there is a very long swell. However, ditching into the swell or large waves should be avoided because of the danger of nosing under. In practice a direction of approach which is a compromise between swell, wave and wind direction may be the best choice.

(3) Judgement of height: As judgment of height over water can be difficult, the Alt 7 or Alt 6 should be used if possible. The landing lamps should also be used at night.

(4) Fuel weight: Fuel weight should be reduced as much as practicable prior to ditching. Excess fuel may be used to position the aircraft in a more favourable location, eg closer to ships or land, but it is essential that the ditching is carried out while engine power is still available.

b. Ditching drill.

(1) Ensure all loose objects are stowed, that all crew have their harnesses locked and tight, protective helmets on, 100% and emergency oxygen selected and flowing correctly. Uncover the rear cabin windows.

(2) Have ejection seat pins replaced.

(3) Disconnect PSP and lanyards, leg restraint, emergency oxygen and parachute harnesses as appropriate to crew position.

(4) On the approach, stow the fuel console and jettison the canopy.

c. Touchdown: Touchdown should be made in a tail-down attitude at the lowest practicable speed commensurate with the minimum rate of descent.

Touching down at high speed and low angle of attack should be avoided due to the likelihood of the aircraft bouncing and the probable collapse of the bomb-aimer's blister with subsequent flooding of the cabin. In any event the control column should be held hard back after impact.

Aborted Take-off Procedures

a. If an emergency occurs before decision speed, take-off is to be aborted in accordance with the FRC drill. The following emergencies constitute mandatory reasons for abandoning take-off, unless otherwise authorised.

(1) Engine failure.

(2) Any fire warning light coming on.

(3) Double alternator failure.

(4) PFC failure (main warning not accompanied by a white reminder MI).

b. The Captain is to warn the crew 'Aborting'.

c. The pilot flying the aircraft is to:

(1) Close the throttles.

(2) Select HIGH DRAG airframe.

(3) Apply maximum continuous braking at NMBS or below.

d. The non-flying pilot is to:

(1) Stream the tail braking parachute (75 knots to 145 knots).

(2) Carry out engine failure/fire drills as ordered.

e. The AEO calls 'Aborting, Aborting' on the frequency in use.

f. The nav/plotter calls airspeeds down to 50 knots.

Emergency Evacuation on the Ground

The following considerations for evacuating the aircraft in an emergency on the ground amplify the drills given in the Flight Reference Cards:

a. Undercarriage position: The direction of exit depends on the emergency and whether the undercarriage is raised or has collapsed during the emergency. If the nosewheel has collapsed, it may be possible to leave the aircraft through the door; if, however, all the undercarriage legs are retracted, exit will have to be via the canopy aperture. The route from the canopy to leave the aircraft will depend upon the condition of the aircraft and whether a fire exists.

b. Canopy jettison: It will be necessary to leave via the canopy aperture if exit through the door is not possible. When the canopy is jettisoned, if the aircraft is stationary, the possibility exists that the canopy will fall back on to the cockpit and may injure one or other of the pilots.

c. Ejection seat pins: If exit through the entrance door is feasible, the pilots should replace the seat pan firing handle safety pin prior to leaving their seats. If the exit has to be made via the canopy aperture, and time permits, the main gun sear pins are to be inserted to make the ejection seats safe.

d. Crew ladder: If speed of exit is essential and exit through the door is feasible the crew should slide down the door and clear the vicinity of the aircraft as quickly as possible. Only replace the door ladder when the degree of emergency allows.

e. Battery: If conditions permit the AEO should switch off the aircraft battery prior to evacuating the aircraft. It should be borne in mind that with the battery off all cabin lighting is lost.

However, if the battery was left on, when it is safe and if it is feasible, the AEO should return to the aircraft to switch off the battery in order to make the aircraft electrically safe.

Engine Failure above Decision Speed

a. If engine failure or other serious emergency occurs above decision speed, the take-off is normally to be continued and the recommended drill followed.

b. When the aircraft is safely airborne, the pilot flying the aircraft is to close the HP cock(s) of the affected engine(s), simultaneously ordering the non-flying pilot to select undercarriage up.

The flying pilot is then to order the non-flying pilot to carry out the drill required, eg 'Engine Failure Drill, No 3 engine'.

c. The non-flying pilot is to complete, from memory, the engine fire/failure drill as ordered: the immediate actions listed in the FRC as far as 'Fuel Pumps' are to be completed.

d. The AEO declares an emergency on the frequency in use.

e. Once the aircraft is fully under control (pattern speed attained) the FRC checks for engine failure/fire are to be completed.

f. A turn on to the downwind leg is not to be initiated below 1,000 feet above ground level, and until pattern speed is attained. Bank is to be restricted to a maximum of 250 in the turn. It is recommended that an instrument pattern is flown following double engine failure.

g. When the aircraft is established on the downwind leg, the 'Resetting' checks are to be carried out (if required) followed by the 'Pre-landing' checks. The undercarriage is not to be lowered until approaching the glide path.

XM607

XM607

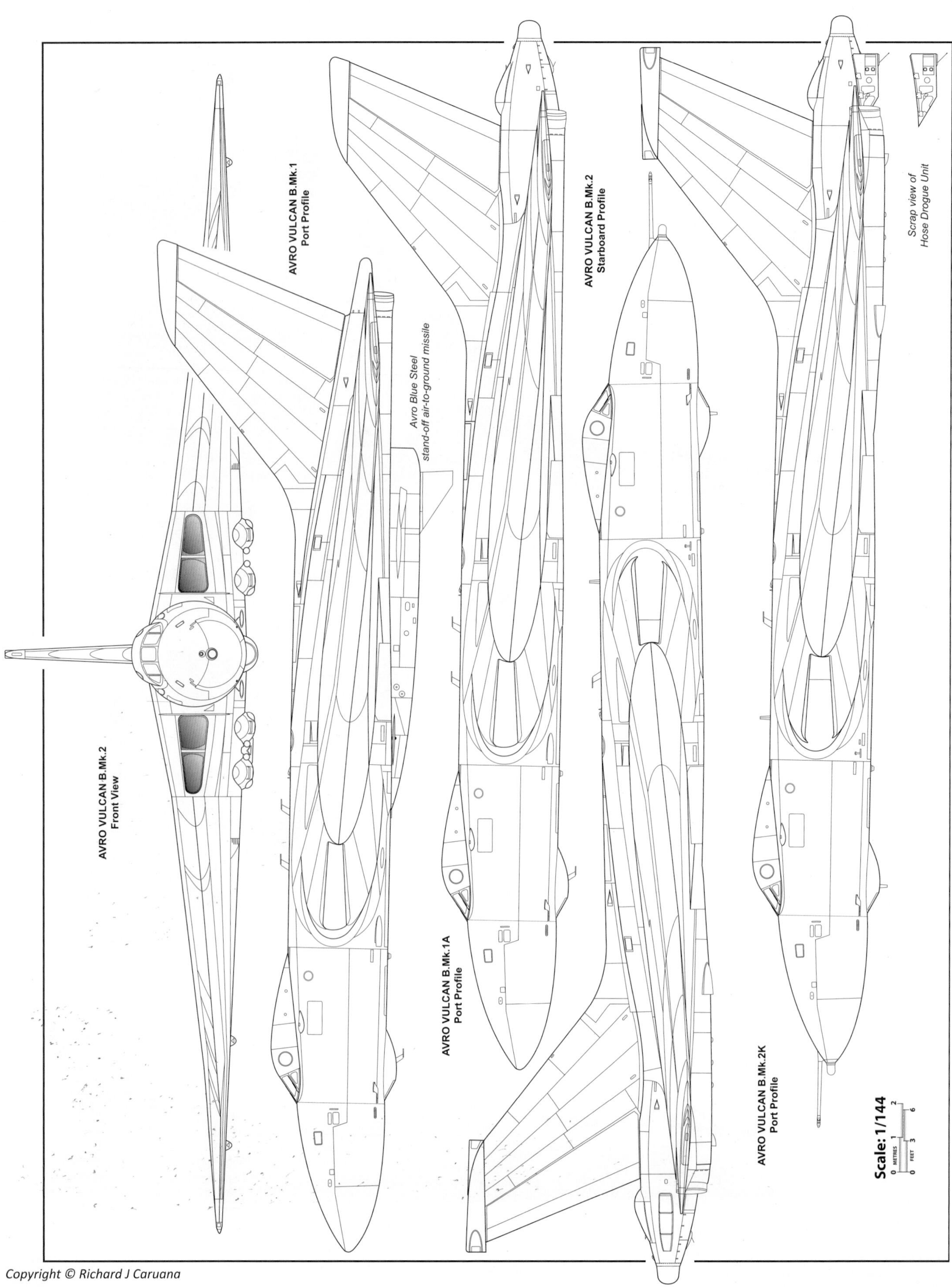
AVRO VULCAN B.Mk.2
Front View
AVRO VULCAN B.Mk.1
Port Profile
Avro Blue Steel
stand-off air-to-ground missile
AVRO VULCAN B.Mk.1A
Port Profile
AVRO VULCAN B.Mk.2
Starboard Profile
AVRO VULCAN B.Mk.2K
Port Profile
Scrap view of
Hose Drogue Unit
Scale: 1/144
0 METRES 1 2
0 FEET 3 6

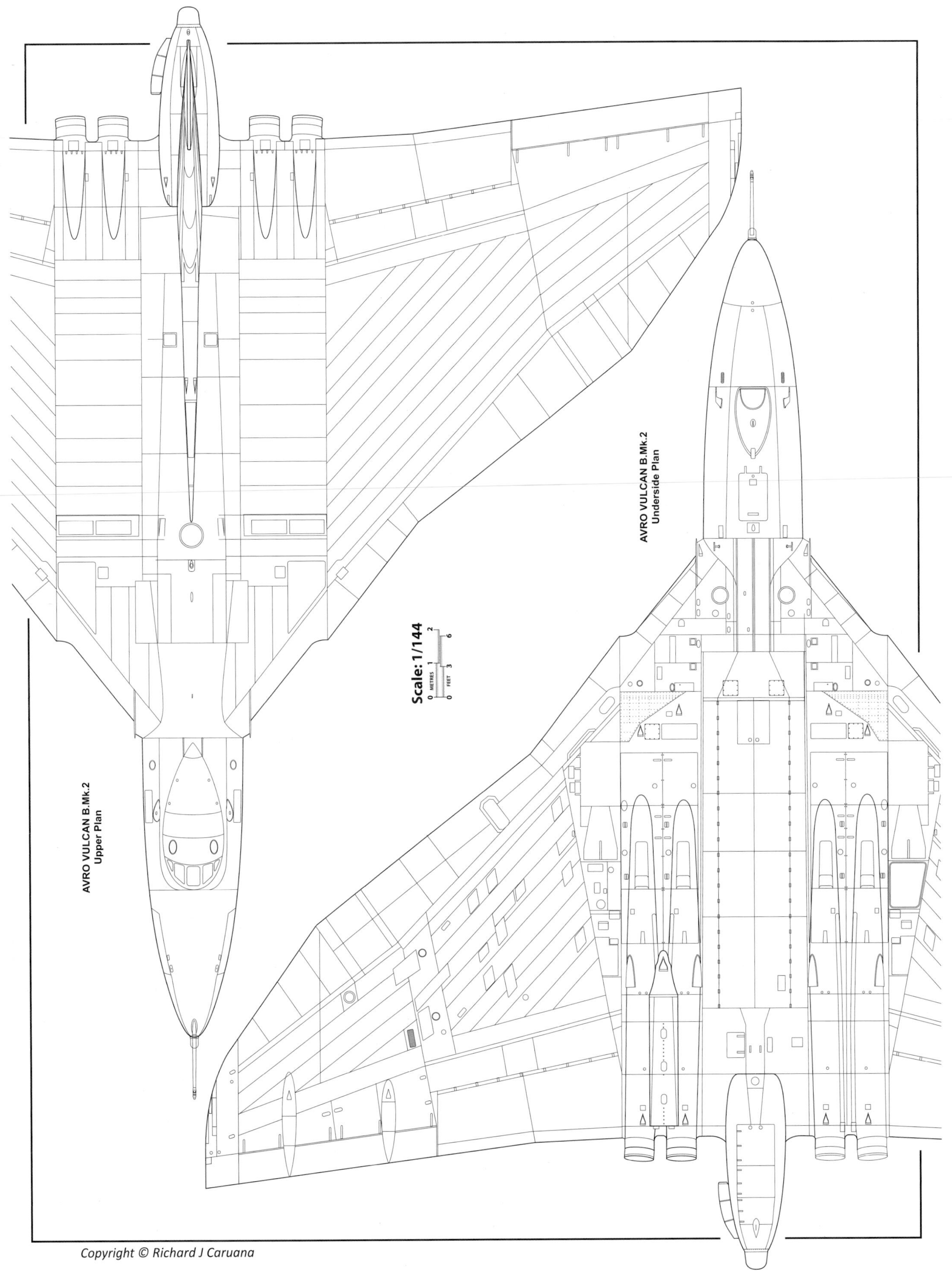
AVRO VULCAN B.Mk.2
Upper Plan
Scale: 1/144
AVRO VULCAN B.Mk.2
Underside Plan

Index

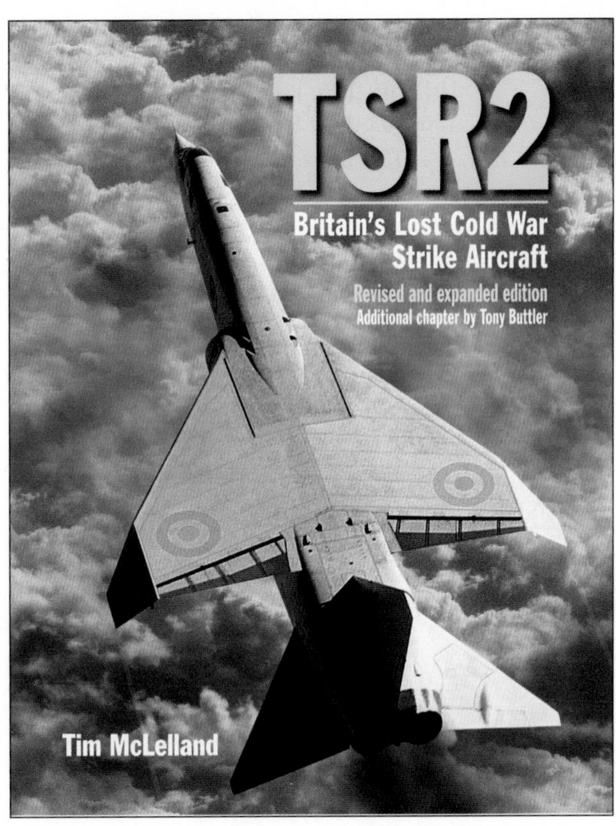

TSR2

Britain's Lost Cold War Strike Aircraft

Tim McLelland

This is a revised and enhanced second edition of a comprehensive, accurate and honest account of the fascinating TSR2 story, tracing the project's development from its origins in the 1950's.

Aimed at aviation historians and those interested in the history of military technology, the book examines the TSR2 project in detail, eliminating the many myths and misconceptions that have surrounded the aircraft for decades. Although much has been written about the TSR2's history, a great deal of misinformation has been published on this subject which this book dismisses presenting the reader with a complete and realistic overview of the entire project.

This book deals with the facts and not the emotion, speculation and fantasy which has plagued the subject for so long. It presents a detailed, factual and very readable account of the development and subsequent demise of TSR2 project.

For this new edition an additional chapter concerning the F-111K, extracts from the TSR2 Crew Manual, and other declassified technical TSR2 documentation has been provided by Tony Buttler, the author of our respected *British Secret Projects* series who has researched this era of British military aviation for many years.

184 pages, hardback

ISBN: 9781910809136, £24.95

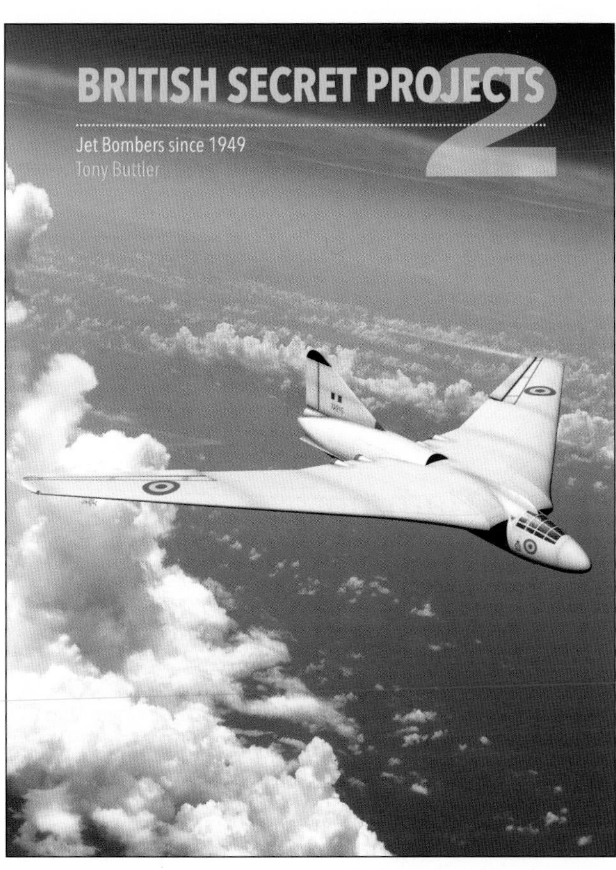

British Secret Projects 2

Jet Bombers since 1949

Tony Buttler

Like fighters, many bomber projects were drawn by British aircraft manufacturing companies in times of potential or actual combat. While names such as Canberra, Vulcan, Victor, TSR2, Harrier and Tornado are known to many as they made it into the skies, the fact that so many other projects from different companies remained on the drawing board provides a rich diversity of 'might-have-been' aircraft designs ripe for coverage.

As with *British Secret Projects 1 – Jet Fighters since 1950*, author Tony Buttler has researched extensively with particular emphasis on the design and development work that took place within various tender design competitions.

Many little-known projects are included that help to illustrate how British bomber development changed against a backdrop of political upheaval, shrinking defence expenditure and technological advancement including supersonic flight, nuclear weapons and VTOL. The story starts with Britain's quest for a jet-powered Mosquito replacement and concludes with reference to the next leap forward, FOAS, an unarmed bomber flown by pilots on the ground.

Accompanied by detailed appendices of all British post-war bomber projects and specifications, colour photographs and artwork, *British Secret Projects 2 – Jet Bombers since 1949* provides a wealth of detailed information on the fascinating world of Cold War secret bomber projects.

352 pages, hardback

ISBN: 9781910809105, £27.50

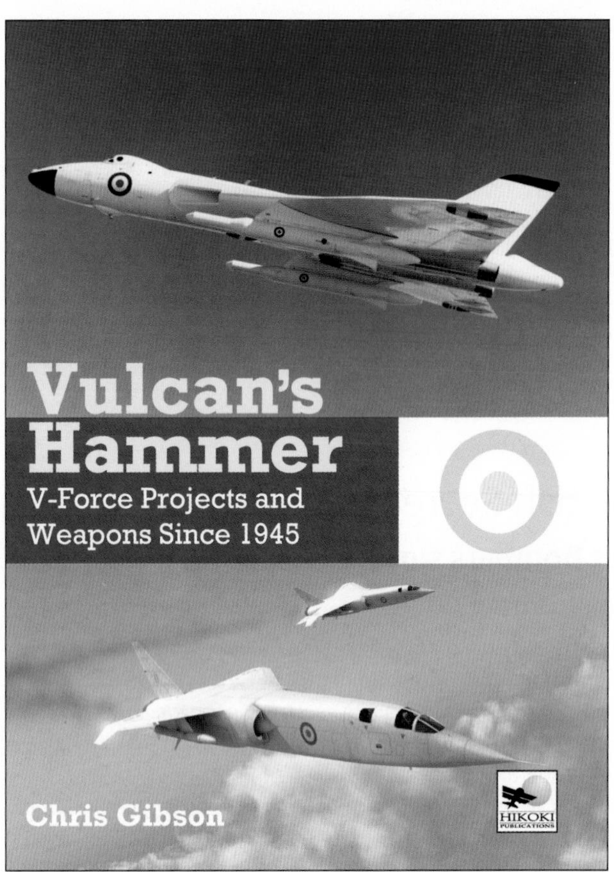

Vulcan's Hammer

V-Force Projects and Weapons Since 1945

Chris Gibson

Following the end of WWII, the United Kingdom embarked on an audacious programme of aircraft and weapons development to maintain its position as a world power. This ultimately led to the V-bombers; Valiant, Victor and Vulcan, that carried the British nuclear deterrent from the mid-1950s until replaced by Polaris in the late 1960s. Prior to the V-bombers, the British aviation industry examined a number of schemes to deliver that deterrent, such as Blue Moon, while their intended replacements, the supersonic Avro 730 and English Electric P.10, could have been the most advanced aircraft in the world in 1960. As political and military circumstances changed, the V-force adopted new concepts, specifically the American Skybolt and the patrol missile carriers: the Pofflers.

Running in parallel with aircraft development were a number of programmes to advance V-bomber weaponry. In addition to free-fall bombs, the UK aviation industry undertook development of missiles and the associated propulsion and guidance systems, and in Blue Steel, created the most complex vehicle ever produced in the UK. As well as arming the Victor and Vulcan, Blue Steel was to form the basis of a range of weapons for TSR.2 and Mirage IV, test vehicles and satellite launchers.

Illustrated with over 200 photographs and drawings plus new colour artwork, *Vulcan's Hammer* presents the story of an alternative V-force and its armament, providing a wealth of fascinating information for historians and modellers alike.

192 pages, hardback

ISBN: 9 781902 109176, £29.95